PANEL PROCEEDINGS SERIES

INTERNATIONAL TOKAMAK REACTOR

Zero Phase

REPORT OF THE
INTERNATIONAL TOKAMAK REACTOR WORKSHOP
ORGANIZED BY THE
INTERNATIONAL ATOMIC ENERGY AGENCY
AND HELD IN FOUR SESSIONS IN VIENNA
5–16 FEBRUARY, 11 JUNE–6 JULY,
1–23 OCTOBER AND 10–19 DECEMBER 1979

INTERNATIONAL ATOMIC ENERGY AGENCY
VIENNA, 1980

EDITORIAL NOTE

The reports received from the various working groups of INTOR have been copy-edited by the staff of the International Atomic Energy Agency to the extent necessary for the reader's assistance, such editing having the particular aim of ensuring that references to certain sections, diagrams or tables can be given unambiguously. The technical content and style of the various chapters are, however, the responsibility of the respective working groups.

Although the individual chapters have all received the approval of the INTOR Steering Committee before publication, the views expressed are those of the respective working groups, and they do not necessarily reflect the views of nominating Member States, nominating international organizations or the International Atomic Energy Agency.

The use in this publication of particular designations of countries or territories does not imply any judgement by the IAEA as to the legal status of such countries or territories, of their authorities and institutions or of the delimitation of their boundaries.

The mention of specific companies or of their products or brand names does not imply any endorsement or recommendation on the part of the IAEA.

INTERNATIONAL TOKAMAK REACTOR: ZERO PHASE
IAEA, VIENNA, 1980
STI/PUB/556
ISBN 92–0–131080–3

Printed by the IAEA in Austria
May 1980

FOREWORD
by the Director General

In placing this report before the public I wish to record my gratitude to the INTOR Workshop participants and the members of the International Fusion Research Council, by whose contribution the initial phase, the so-called 'zero phase', of the International Tokamak Reactor project has now been brought to completion.

PREFACE
by the Chairman of the International Fusion Research Council

The problem of realizing the potential energy of the light nuclear elements for peaceful purposes is one of the most difficult and at the same time one of the most urgent that faces modern physicists and engineers. It is difficult because producing matter in a pure state, thermally insulated and well controlled, at temperatures of 100 million degrees has involved founding a new branch of physics – plasma physics – our knowledge of which still needs further refinement; and the engineering to realize practical applications will require a novel combination of nuclear engineering, electrotechnology and plasma physics. It is urgent because mankind, it appears, is rapidly exhausting cheap fossil fuel resources, so that the question of when controlled nuclear fusion might contribute significantly to mankind's energy needs is one needing early answers for those planning major investment in new energy sources.

All must therefore welcome the step taken by the International Atomic Energy Agency in January 1978, when Dr. Sigvard Eklund, Director General of the IAEA, invited governments sponsoring fusion research to indicate their hoped-for time-scale of fusion development, their interest in increasing international cooperation to speed research, and in particular whether or not they would participate in international studies of the next major step. The response to this initiative was very constructive and as a result the INTOR Workshop was set up to tackle the problem of what would be the most useful major new experimental step, using the so-called tokamak system of creating high temperatures, towards the ultimate objective of practical electricity production by fusion reactors. The present report is the first major outcome of this INTOR Workshop, covering the so-called zero phase.

Dr. S. Mori (Chairman), Dr. W.M. Stacey, Jr. (Vice-Chairman), Academician B.B. Kadomtsev, Dr. G. Grieger and their colleagues in this study have achieved within the required twelve months' time-scale a very remarkable analysis of the possibilities and a synthesis of views. This report and its recommendations are the result of their labours, and in particular of their achievement of a means of working which not merely transcends national barriers but which draws strength from the many and diverse skills within national programmes from all over the world.

The immediate recommendation of the report is that the Workshop should proceed to the next phase, the so-called 'definition phase'. The International Fusion Research Council, which has discussed the report at length, has unanimously endorsed this recommendation, and the Director General has accepted the continuation of the work for this next phase.

The recommendations of the International Fusion Research Council to the Director General contain, inter alia, the following terms of reference for the definition phase. The INTOR Workshop should:

(1) Produce a conceptual design supported by a report;
(2) Keep the parameters under continuous review to ensure that ignition is achieved;
(3) Define a divertor and exhaust system;
(4) Examine the consequences of an RF heating system on the conceptual design;
(5) Examine the possibility of using a lithium-containing neutron shield;
(6) Clarify the boundary between the objectives of INTOR and those of a demonstration reactor;
(7) Specify in detail the engineering test programme, including functional compatibility tests to be carried out in INTOR, and specify the relation of this programme to the conceptual design of INTOR and to the needs of the demonstration reactor;
(8) Detail the R&D support required for the design, construction and operation of INTOR.

Thus, the purpose of the present INTOR report is not merely to place on record a remarkable piece of joint work: it is much more to provide the basis of an actual device, and to provoke discussion to provide the maximum input of ideas and judgements on the optimum objectives of this major, but as yet hypothetical, international fusion reactor.

April 1980 *R.S. Pease, Chairman*

INTOR Group

Euratom	*Japan*	*USSR*	*USA*
G. Grieger	S. Mori	B.B. Kadomtsev	W.M. Stacey, Jr.
F. Engelmann	T. Hiraoka	G.E. Churakov	J.R. Gilleland
R. Hancox	K. Sako	B.N. Kolbasov	G.L. Kulcinski
D. Leger	T. Tazima	V.I. Pistunovich	P.H. Rutherford
P. Reynolds		G.F. Shatalov	

IAEA Scientific Secretariat

R.A. Ellis J.A. Phillips V.S. Vlasenkov

INTOR Terms of Reference
Zero Phase

The following terms of reference were established by the International Fusion Research Council as general guidelines for the zero phase of the INTOR Workshop.

1. To draw on the capability in all countries to prepare a report to be submitted to the IFRC describing the technical objectives and nature of the next large fusion device of the tokamak type that could be constructed internationally.

2. In detail, the Workshop should:
   (a) Review and discuss the results of existing studies of next-step proposals and experimental power reactors;
   (b) Survey the results of experiments, theory and associated technology planned to be available in the early 1980s;
   (c) Make recommendations of the aims, outline technical realization and resource requirements of a possible next step, indicating the alternatives considered;
   (d) Identify the problem areas that need to be tackled before the construction of the next step.

3. The programmatic objectives of the next major international tokamak installation could be:
   (a) Take the maximum reasonable step beyond the present generation of experiments to demonstrate the scientific, technical and engineering feasibility of the generation of electricity by pure D-T fusion;
   (b) Include in a 'primitive sense' all systems and components for practical fusion power plants;
   (c) Provide test facilities for systems, components and materials for practical fusion power.

4. Summary of possible technical objectives:
   (a) D-T burning;
   (b) $1 < Q < \infty$;
   (c) Superconducting magnets;
   (d) Long pulse (steady state, refuelling and exhaust);
   (e) Facilities for testing various experimental blanket modules;
   (f) Electricity generation.

CONTENTS

APPENDICES

4. INTOR SUGGESTED PARAMETERS (cont.)

	SUGGESTED PARAMETERS	NATIONAL SUGGESTED PARAMETERS EC	JAPAN	USA	USSR
Aspect ratio, A	4.0	3.4	3.9	3.8	4.2
Elongation, K	1.6	1.6	1.5	1.6	1.5
Av. ion temperature (keV), $\langle T_i \rangle$	10	10	10	10	10
Av. D-T ion density (10^{20} m^{-3}), $\langle n_i \rangle$	1.3 (1.5)	1.2	1.2	1.4	1.2
Burn energy confinement time (s), τ_E	1.5	1.5–2.0	1.5	1.5	1.5
Effective charge, Z_{eff}(oxygen)[b]	$\leqslant 1.5$	≈ 1.5	1.5($\sim$2.0)	$\leqslant 1.5$	$\leqslant 1.5$
B_T on axis (T)	5.5	5.0	5.2	5.3	5.5
Safety factor, q_I	2.5	2.5	2.5($\sim$2.0)	2.5	2.5
Burn av. (%), $\langle \beta_T \rangle$[c]	5(b)	≈ 6	6	6	5
Plasma current (MA), I_P	6.4	≈ 8	6.0($\sim$7.5)	$\sim$6.5	$\sim$5
Edge toroidal field ripple (%)[d]	$<\pm 0.75$	$<\pm 0.75$	± 0.4	$<\pm 0.75$	$<\pm 0.75$
Centre toroidal field ripple (%)[d]	$<\pm 0.1$	$<\pm 0.1$	± 0.04	$<\pm 0.1$	$<\pm 0.1$
Fast alpha loss (%)	<3	≈ 1	<5	<5	<5

... will be needed in ... which ... to ... not consumed ... injection. Gas cooling is the simplest method of thermal and ... [illegible] ... satisfactory for ... the ... action of the ... coolant ... [illegible] ... uniformly over the ... and is ... For ... [illegible]

[illegible]

[illegible]

[illegible]

[illegible] ... Appendix ...

[illegible] ... Reactor ... TOR ... [illegible]

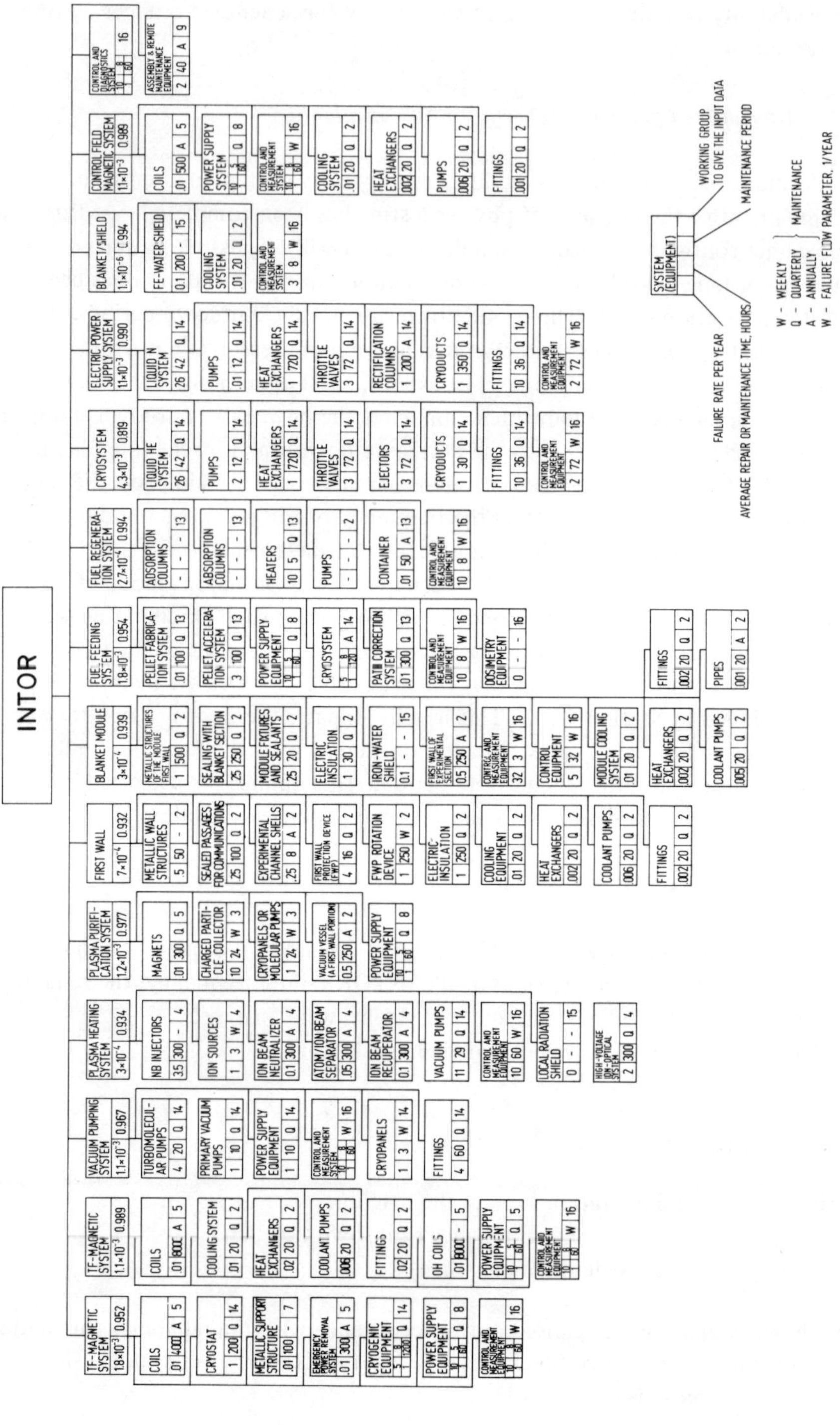

Chapter I
INTRODUCTION

The recent progress in controlled thermonuclear fusion research has been rapid and significant on a world-wide basis, and the prospect of the further development of fusion as a source of power is now considered excellent. Because fusion is one of the few viable energy alternatives for the future, it is of global importance to demonstrate the technical feasibility of fusion power as soon as possible.

Projects are presently under construction in several countries (JT-60 in Japan, T-15 in USSR, TFTR in USA and JET in Europe) which should establish the scientific feasibility of magnetic confinement fusion. These experiments must be followed by one or more experimental reactors which are dedicated to establishing the technological feasibility of fusion, and then by demonstration reactors which will establish the economical feasibility of fusion. The INTOR device is conceived as an experimental reactor which should establish the technological feasibility of magnetic confinement fusion.

Fusion research has long benefited from international cooperation sponsored by the International Atomic Energy Agency (IAEA) and through other, mostly bilateral, arrangements. This cooperation has accelerated the progress in fusion research and established a world-wide working relationship among fusion scientists and engineers.

In view of the excellent progress in fusion research, the Director General of the IAEA in 1978 requested recommendations for an expanded programme that would hasten the development of fusion through international collaboration. In response to this request, the Soviet Union proposed to the IAEA that the next major experiment beyond the experiments presently under construction should be built as a cooperative international enterprise. The International Tokamak Reactor (INTOR) Workshop was organized by the IAEA to identify the programmatic and technical objectives and the nature of such a device, and to determine if the scientific and technical bases for this experiment will be available in the near future. As conceived by the International Fusion Research Council (IFRC), an advisory body to the IAEA, the INTOR device would play a major role in establishing the technological feasibility of fusion and in demonstrating the fusion plasma performance needed for tokamak reactors. The parties to the Workshop are the European Community (EURATOM), Japan, the United States of America and the Union of Soviet Socialist Republics.

The INTOR Workshop was conducted during 1979 as an initial phase of the INTOR project to establish whether or not the INTOR project can be taken into the next phase. The subsequent phases of the INTOR project would be: 1) definition; 2) design; and 3) construction and operation. At the end of each phase, there will be a decision by the governments involved whether to continue to the next phase.

Each party to the INTOR Workshop nominated four participants who met in Vienna four times to define the effort of the Workshop, to review and discuss the contributions of the four parties, and to prepare the report of the Workshop. (When required, these participants were supplemented by experts.) The major part of the work of the INTOR Workshop was performed by experts working in their home institutions. This home effort involved about 100 senior scientists and engineers and about 10–20 man-years of effort for each of the four parties. The personnel involved are listed in Appendices A.1 and A.2.

The basic objective of the INTOR Workshop was defined by the IFRC: "To draw on the capability in all countries to prepare a report to be submitted to the IFRC describing the technical objectives and nature of the next large fusion device of the tokamak type that could be constructed internationally." In particular, the Workshop undertook:

1) to assess the plasma physics and technological bases that are now available for the design of such a device or that are anticipated to be available in the early 1980's;
2) to identify the objectives of such a device;
3) to recommend, as input for the definition phase, reactor characteristics that are consistent with the physics and technological bases and with the objectives, indicating the alternatives considered;
4) to identify the major uncertainties that must be resolved before the construction of such a device can be undertaken, and to identify the required R and D programmes;
5) to identify the resources required and the schedule for such a device; and,
6) to recommend on the technical and scientific feasibility of constructing such a device to operate in the late 1980's or early 1990's.

After discussion of the programmatic and technical objectives of the INTOR project, the principal tasks of the Workshop were assessments of the physics and technological bases for the design of and the identification of the characteristics of INTOR. The data base and R and D needs assessments were carried out in sixteen separate but related categories. The identification of the INTOR characteristics was initially based upon a review of previous conceptual design studies and subsequently evolved during the Workshop to be consistent with the data base assessment and the technical objectives. Appendix A.3 contains a summary of previous conceptual design studies.

Detailed reports on all aspects of the INTOR Workshop activities were developed by experts in the home countries. The content of these reports was then discussed and compared at the Workshop sessions in Vienna and served as the basis for the preparation of this joint report of the INTOR Workshop.

The advantages of this type of international cooperation were quite evident during the INTOR Workshop. Each party to the Workshop provided unique information for the data base assessment, as well as information to confirm the information provided by the other parties. Each party also made unique suggestions for design concepts that affected the reference concept for INTOR. Perhaps even more important, the discussion of the issues and technical factors involved, from the perspectives of the four parties, led to a more broadly based assessment and conceptualization than would have otherwise been possible.

These same advantages of international cooperation will also accrue to subsequent phases of the INTOR project. An experiment on the scale of INTOR, which represents a significant advance from the present state of knowledge and technology, calls for an international mobilization of capabilities. In addition, the pooling of resources in such a collaborative international project will make possible a more rapid development of fusion power without requiring massive infusions of additional funding or impairing the study of non-tokamak concepts in national programmes.

Chapter II

SUMMARY, CONCLUSIONS AND RECOMMENDATIONS

1. ROLE OF INTOR IN THE FUSION PROGRAMME

If successful, the present generation of tokamak experiments, JT-60, TFTR, JET and T-15, on the basis of D-T and non-D-T plasmas, will provide fairly conclusive information on how to achieve ignition plasma parameters. The establishment and sustainment of a D-T plasma burning at least for a certain time seems, therefore, in practical reach. In order to meet the requirements of fusion reactor operation, certain technologies used in previous experiments for heating and maintaining the plasma must either be significantly advanced or replaced by new technologies. In addition, fusion power demonstration requires the development and demonstration of new technology for tritium and electricity production by fusion, the development and testing of materials, the development of reliable components, the demonstration of net electricity production and net tritium breeding and the demonstration of reliable and economic operation of a fusion reactor.

This agenda sets an urgent need for a machine that demonstrates the plasma physics required for fusion power and that provides for the development, testing and, to some extent, demonstration of the technologies required for fusion power. INTOR is intended to fulfil these tasks. In particular, INTOR must develop the technologies needed for a fusion power demonstration reactor (DEMO) which would then demonstrate net electrical power, tritium breeding and economic feasibility. INTOR should be of the tokamak type since the data base and stage of development for this confinement concept are largest and since the gross characteristics of the technologies to be developed for magnetic confinement fusion are relatively independent of the choice of confinement concept. These considerations led to the following programmatic objectives for INTOR:

INTOR should be the maximum reasonable step beyond the next generation of large tokamaks (TFTR, JET, JT-60, T-15) in the world fusion programme.

INTOR should **demonstrate** the plasma physics requirements for DEMO.

INTOR should **demonstrate** those technologies requried for DEMO that must be incorporated as an intrinsic part of the reactor (e.g. superconducting magnets, remote handling technology).

INTOR should be a **test facility** for blanket, tritium production, plasma engineering materials and other technology developments required for DEMO.

INTOR should serve as a test and demonstration facility for magnetic fusion technology in general.

INTOR should demonstrate the reliable operation of a fusion reactor.

It was assumed by the Workshop that the tritium required to operate INTOR would be supplied in large part, if not entirely, from external sources. This is an important issue which must be resolved at an early date.

The technical objectives of INTOR then follow from the programmatic objectives and the physics and technological data base. These objectives are given in Table II-1.

2. INTOR CONCEPT

The concept for a machine that is capable of meeting the technical objectives of INTOR and that is supported by the data base assessment has been developed by the Workshop. The concept recommended as input for the definition phase is characterized by a relatively large (~ 280 m^3) D-shaped plasma heated to ignition by neutral beam injection power of about 75 MW for 5–6 s up to a mean temperature of about 10 keV. Controlled burn should then be achieved for more than 100 s at a plasma-to-magnetic pressure ratio $\langle\beta\rangle = 5\%$, which is marginally stable, theoretically. A divertor is needed to exhaust helium and control impurity influx in order to achieve a burn this long. The plasma current will exceed 6 MA, which is quite sufficient to confine fast alpha-particles. Both the toroidal and external poloidal field coils will be superconducting. INTOR will require about 200 MW_e power. The neutron power load on the first wall will be about 1.3 MW/m^2 (with an option at 1.6 MW/m^2), and the total fusion power will be about 620 MW_t. A stainless-steel, water-cooled shield operating with relatively low (100 – 200°C) maximum structural temperature will be employed on the basic machine. The major suggested parameters for INTOR are given in Table II-2, and more detailed parameters are given in Chapter III.

Since a principal function of INTOR is to demonstrate the feasibility of reactor components and to serve as a test facility for technology development, simplicity and reliability of the basic machine are of the utmost importance, even if this is in conflict with certain aspects of reactor-relevance. The technical feasibility of INTOR is then intimately related not only to the supporting technology development programmes but also to the manner in which each technology is incorporated into INTOR.

TABLE II-1. INTOR TECHNICAL OBJECTIVES

I. Reactor-relevant mode of operation

a) Ignition of a D-T plasma.
b) Controlled $\gtrsim 100$ s burn pulse.
c) Reactor-level particle and heat fluxes ($P_n \gtrsim 1$ MW/m^2).
d) Duty cycle $\gtrsim 70\%$.

II. Reactor-relevant technologies

a) Superconducting toroidal and poloidal coil technology.
b) Plasma composition control (e.g. divertor) technology.
c) Plasma power balance control technology.
d) Plasma heating and fuelling technology.
e) Tritium fuel cycle technology (excluding breeding and extraction).
f) Remote maintenance technology.
g) Vacuum technology.
h) Fusion power cycle technology.

III. Engineering test facility

a) Tritium breeding blanket and extraction technology testing.
b) Advanced structural and breeding materials, coolants, etc.
c) Blanket technology for simultaneous electricity production and tritium breeding.
d) Materials testing.
e) Advanced plasma engineering technology testing.

IV. Demonstration

a) Electricity production by fusion.
b) Tritium production by fusion.
c) Safe and reliable operation of a fusion reactor.
d) Availability $\sim 25-50\%$.

Certain reactor-relevant technologies must be incorporated as intrinsic – not easily replaceable – components (e.g. superconducting magnets, remote handling system) in order to be tested in INTOR. These technologies must be developed to a high level prior to the construction of INTOR, since failure of such intrinsic components would cause INTOR either to fail or to be significantly delayed in achieving any of its objectives. For such intrinsic technologies, INTOR will demonstrate an already-developed technology in a fusion reactor environment.

With certain other reactor-relevant technologies (e.g. electricity and tritium producing blanket), there exists the option of incorporation as intrinsic components or as relatively easily replaceable test components. Inclusion of a new technology as an intrinsic component at the outset places a greater demand upon the development programme supporting that technology and entails an increased risk that failure of that component will cause INTOR to fail or to be significantly delayed

TABLE II-2. INTOR SUGGESTED MAJOR PARAMETERS[a]

Major radius, R (m)	5.2
Plasma radius, a (m)	1.3
Chamber radius, r_W (m)	1.4
Elongation, k	1.6
Burn time (s)	$\geqslant 100$
Duty cycle (%)	$\geqslant 70$
Burn average $\langle\beta\rangle$(%)	$5(6^b)$
Av. D-T density $\langle n_i \rangle (m^{-3})$	$1.3(1.5^b) \times 10^{20}$
Av. ion temperature $\langle T_i \rangle$ (keV)	10
Plasma current, I (MA)	6.4
D-T thermal power, P_{th} (MW)	$620(750^b)$
Neutron wall load, $P_n (MW/m^2)$	$1.3(1.6^b)$
Lifetime (no. of burn pulses)	$5 \times 10^5 - 10^6$
NB heating power, P_B (MW)	75
NB energy, E_B (keV)	175
Fuelling	pellet, gas puffing
Impurity control	divertor
Toroidal field coils	NbTi and/or Nb_3Sn
Toroidal field at centerline, B (T)	5.5
Poloidal field coils	NbTi
Tritium inventory (kg)	3
Maximum availability (%)	25 – 50
Tritium consumption during Stage II (kg/yr)	6

[a] The suggested parameters are tentative and suggested as input for the definition phase, during which the major parameters will be optimized.
[b] For comparison (discussed for Stage III-A).

in achieving any of its objectives. On the other hand, inclusion of a new technology as a test component that could be inserted after INTOR had operated for some time and that could be relatively easily replaced places less of a demand upon the supporting development programme and entails considerably less risk that component failure will significantly delay INTOR in achieving its other objectives. A further advantage of inclusion of a new technology in test components is that INTOR can serve as a test bed for the development of that technology, with successively improved test components being tested in INTOR.

The philosophy of INTOR is to minimize the risk of failure and the demands upon the supporting technology development programmes by including new technologies in test components to the maximum extent that is consistent with achieving the programmatic and technical objectives. This philosophy leads to the concept of staged operation, with the possibility of incorporating some of the technologies that have been tested in test components during an early stage into intrinsic components during the last stage in order to provide a full-scale engineering demonstration.

3. OPERATION SCHEDULE FOR INTOR

The objectives of INTOR could best be achieved in a sequential order by staged operation, as indicated in Table II-3. During Stage I, which might run during the first three years of INTOR operation, the first two sets of technical objectives listed in Table II-1 would be achieved, i.e. a reactor-relevant mode of operation will be established and some intrinsic reactor-relevant technologies will be demonstrated. The availability and the total neutron fluence will be low during this stage, but nevertheless, some preliminary engineering tests will be possible. The production of 5–10 MW_e in non-breeding blanket modules could be achieved towards the end of this first stage.

Stage II, which would run during the following four years, would then be fully devoted to engineering testing. This would include the testing of blanket modules, tritium production and extraction, plasma engineering technologies (e.g. RF launchers), and materials properties. An important criterion for INTOR during this stage is to provide sufficient flexibility to accommodate a variety of experimental programmes at the same time. This would allow the simultaneous testing of various blanket concepts, with different structural and breeding materials, coolants and design concepts, together with the testing of plasma engineering technologies. The reliability of many components will be tested during this stage, and improved components may be installed in some instances. High temperature coolants from tritium-producing blanket modules could also be used to demonstrate simultaneous electricity production and tritium production.

Stage III of INTOR operation will be determined by the choice of alternatives for the utilization of the machine (discussed below) and by the choice of whether or not INTOR should be used to test one or more full blanket sectors. In addition, it is assumed that by the end of Stage II most of the operational problems will have been solved and that a relatively high availability will have been achieved, possibly after the replacement of certain components. Two possible options for Stage III operation of INTOR are shown in Table II-3.

In Option III-A, Stage III is a straightforward continuation of Stage II with accelerated engineering testing in test modules. There is also a strong emphasis

TABLE II-3. INTOR OPERATION SCHEDULE

Stage	Years	Phase	Availability	# D-T shots*	Annual tritium consumption (kg)*	Annual total neutron fluence (n/cm^2)**
I.	1	**Hydrogen plasma, engineering checkout**	10%	–	–	–
	2	**D-T Plasma** (1.3 MW/m^2)	15%	6.6×10^4	4.8	$\lesssim 1 \times 10^{21}$
		1) ignited plasma				
		2) controlled burn $\gtrsim$ 100 s				
		3) $\gtrsim$ 70% duty cycle				
		4) reactor level particle and heat fluxes				
		5) preliminary engineering studies				
		Electricity production				
II.	4	**Engineering testing** (1.3 MW/m^2)	25%	2.2×10^5	6.0	$\sim 2 \times 10^{21}$
		1) blanket modules				
		2) T-production				
		3) combined T and electricity production				
		4) materials properties				
		5) advanced plasma engineering				

III.		**Option III-A**				
	$\geqslant 5$	**Upgraded engineering testing** (1.6 MW/m^2)	50%	$\geqslant 5.5 \times 10^5$	15	$\sim 5 \times 10^{21}$
		6) continue engineering testing				
		7) materials irradiation testing				
		or Option III-B				
	1–2	**Replace blanket**				
	$\geqslant 5$	**Blanket demonstration** (1.3 MW/m^2)	25%	$\geqslant 2.7 \times 10^5$	6.0$^+$	$\sim 2 \times 10^{21}$
		6) full segment or full blanket demonstration				
		7) substantial electricity production				
		8) substantial tritium production				

* Based on 100 s burn pulse, 70% duty cycle and the indicated neutron wall load. The actual number of D-T shots may be quite different during the first stage of operation.

** Primary 14 MeV fluence multiplied by a factor of 6 to account for secondary fluence.

\+ Part of this could be produced by INTOR, if a large-scale breeding blanket is installed.

on achieving the highest possible availability and on operating at a higher neutron wall load (1.6 MW/m^2) to accumulate a significant neutron fluence for materials irradiation testing and to test component reliability. This option would allow for the uninterrupted continuation of engineering testing in INTOR. The experience gained during Stage II would be incorporated in improved components for testing during Stage III. The emphasis upon high availability would require that, to the maximum reasonable extent, the change of experimental modules and components be scheduled to coincide with the scheduled maintenance periods. But it would also allow the provision of the maximum amount of engineering testing and materials irradiation data by INTOR as early as possible. After five years of operation of Stage III (12 years in total) materials would be irradiated to within a factor roughly 2–4, depending on when they were available for insertion, of the probable design limit for DEMO.

Option III-B also assumes that the experience gained during Stage II would lead to improved blanket design concepts, but that the manifold engineering constraints make it necessary to test a full blanket sector before a concept can be applied to DEMO. In addition, this option would continue to provide the possibility of replacing some of the plasma engineering components with improved versions, based upon experience obtained in Stage II, but the information would be delayed by a year or so because of the longer time required for replacing a sector rather than a component. This exchange of a sector would simultaneously provide a test of the remote handling and repair technology. Option III-B also assumes that materials data for DEMO would have to be based almost entirely upon simulation*, being calibrated by INTOR results at neutron fluences lower than in Option III-A. This allows to build INTOR with less redundancy, but also means that materials would only be irradiated to within a factor 10 of the probable DEMO design limit.

In the extreme, Option III-B could be expanded to allow a full blanket replacement (probably excluding the inboard part), which would provide for the demonstration of substantial tritium (kg level) and electricity (100 MW_e level) production in a reactor-relevant way.

Comparing Options III-A and III-B, one has to conclude that Option III-A is based on the viewpoint that INTOR is primarily a test facility for blanket concepts and other technologies that will then be demonstrated on a full scale in one or more demonstration reactors. By emphasizing high availability, Option III-A places added demands upon the development of reliable components. The principal risk with Option III-A is the difficulty of achieving an availability near 50%, in which case it would not be possible to irradiate materials to a level that would provide confidence for construction of DEMO.

---

* Simulation means all methods,using other than the original fusion environment, including damage by particle bombardment and neutron irradiation in fission reactors.

Option III-B, using a segmented blanket, would provide a greater technological basis for the design and construction of a DEMO blanket of that same concept than would be obtained by engineering testing only in modules. In addition, the number of blanket concepts to be tested in the sector approach would probably be limited to not more than two, in which case agreement would be required between the parties when making the choice. The principal risk with Option III-B is that much longer than anticipated may be required to replace a blanket sector, in which case valuable experimental time would be lost at a critical time in the development of fusion. Although INTOR will be designed for remote blanket replacement, it will in fact be the machine on which many of the design concepts for remote maintenance are first tested.

Fortunately, the selection between the Options A and B for Stage III can be made at a later time, provided the design of INTOR offers a high enough modularization and flexibility for keeping the choices open.

4. ALTERNATIVES

As the INTOR concept evolved during the course of the workshop, it became clear that there was a substantial consensus among the participants with regard to the nature of the machine. While there remain certain design options (e.g. the location of the poloidal coils relative to the toroidal coils, the location of the primary vacuum boundary, the type of divertor) to be resolved during the definition phase, it became apparent early in the workshop that the objectives and the results of the data base assessment led directly to specification of the properties of the plasma and the broad physical characteristics of the machine.

The alternatives for INTOR are concerned with its use as a test facility. These alternatives are related to the extent to which INTOR is used for materials testing, tritium production and extraction, and electricity production. Since the technologies associated with all three are essential to the development of fusion power, a balanced programme is intended during Stage II. The alternatives then are related to dedicating the machine to one or more of these technology developments during Stage III, with the consequence of limiting the usefulness of INTOR for further development of the other technologies.

4.1. Materials testing

Many valuable materials studies which require relatively low neutron fluences can be performed on INTOR during Stage III, and to a lesser extent during

Stage II. These studies could be carried out in relatively small test channels or larger test modules. Such studies include investigations of:

1) correlation of the high temperature damage microstructure produced by fusion neutrons with the damage produced by fission reactor neutrons;
2) radiation creep;
3) fatigue and crack growth with thermal cycling and under irradiation;
4) phase stability;
5) plasma-wall interaction and material erosion;
6) radiation effects on yield properties.

Since such studies require incubation doses of roughly 1 MW-y/m^2 of D-T neutron fluences, they can be conducted in INTOR in either Option III-A or III-B.

However, there are certain radiation damage investigations (e.g. embrittlement, swelling) that require threshold damage levels of more than 1 MW-y/m^2. One can only attain such high damage levels if the high-availability Option III-A is adopted for INTOR. In order to appreciate the incentive for high-fluence materials testing in INTOR, it is necessary to consider the requirements and time scale for the development and qualification of a structural material for use in a commercial fusion reactor.

The development of a commercial structural alloy involves first the use of simulation studies to screen promising candidate alloys for extensive testing (tens of thousands of samples) in a fusion environment, followed finally by the development and review of design codes for the chosen alloy. Accelerator-based 14 MeV neutron sources, such as RTNS and FMIT,[+] are unable to provide all the required data for the extensive testing because of their limited test volume. Fission reactors can be used to test stainless steel (or Ni-containing alloys) in which the He/dpa ratio characteristic of 14 MeV neutron damage can be simulated. Other candidate alloys (i.e. Ti, Mo, Al) can be tested in fission reactors, but the interpretation of the results for a fusion environment is less direct because of the different He/dpa ratio. The final testing of commercial alloys must be done in a fusion device.

The incentive for using INTOR to test commercial alloys is related to the time schedule for the DEMO and for the first commercial reactor. A relatively aggressive, but representative, schedule might call for the operation of a DEMO about the turn of the century, followed in about ten years by the startup of the first commercial reactor. If it is assumed that the design basis lifetimes for the structural material in a DEMO and a commercial reactor are 10 MW-y/m^2 and 20 MW-y/m^2 respectively, that the neutron wall loads are 2.0 MW/m^2 and 3.0 MW/m^2 respectively, and that both reactors operate with a 70% plant factor,*

+ RTNS = Rotating target neutron source.
FMIT = Fusion materials irradiation test.

* Plant factor = duty factor x availability.

then the accumulated damage can be estimated as a function of time. (A geometrical neutron peaking factor of 10% should be allowed for both the DEMO and commercial reactors.)

The accumulated damage in INTOR can be calculated for two extreme cases:

1) $P_n = 1.6\ MW/m^2$ and 50% availability in Stage III**, and

2) $P_n = 1.3\ MW/m^2$ and 25% availability in Stage III. Assuming, somewhat optimistically, that construction commences five years prior to initial operation, it can be shown that INTOR can test materials to within a factor ~ 2 by the time of initial DEMO operation, if the 50% availability Option III-A is chosen. Materials designers think they can abide with such an extrapolation for a DEMO. Furthermore, it can be seen that because of roughly a 10-year period separating the DEMO and commercial reactors, the former cannot provide any materials data for the design of the latter, but INTOR could test materials to within a significant factor of the design lifetime of the commercial reactor before the year of construction initiation, assuming that the operation of INTOR can be extended that long. The 25% availability Option III-B for INTOR would provide radiation only to within a factor 10 of the design lifetimes of either the DEMO or commercial reactors. In this case it would be necessary to rely upon simulation to obtain the materials irradiation data needed for the design of DEMO and commercial reactors, with the fluence achieved in INTOR being used for calibration. In either case, Option III-A or III-B, it is not possible to expose materials in INTOR to fluences in excess of the design lifetime limits for DEMO or the commercial reactors.

Dedication of INTOR during Stage III to high availability operation would probably exclude the possibility of replacing blanket segments to demonstrate tritium and electricity production on a large scale.

4.2. Tritium production

As far as tritium production is concerned, the tritium extraction technology is the main object to be studied. Reactor-relevant information will be obtained even if only part of the INTOR surface is used. For testing and demonstrating technologies, there is no incentive to use more than one sector of a given type for tritium production. Already in Stage II the simultaneous production of tritium (1–10 g annually) and heat (20 40 MW) will be studied in test modules. Analysis has shown that the neutronics characteristics of modules larger than $0.7 \times 0.7 \times 0.7\ m^3$ become the same as those of an entire blanket of the same composition. Somewhat larger modules, covering a few m^2, would probably be desirable from tritium extraction and other considerations. If all the engineering

---

** A neutron wall load peaking factor of 1.3 should be used for the INTOR device.

constraints are to be included in the investigation, a full sector of INTOR should be equipped with a blanket. This is the concept of Option B for Stage III.

But in all these cases, practically the full amount of fuel has to be supplied from external sources. The anticipated annual tritium requirements for INTOR are given in Table II-3. The USA will consider the provision of tritium to INTOR as an intrinsic part of its decision on whether to participate in the construction and operation of INTOR. Present tritium cost in the USA is about $8 000 per gram. Analysis also indicates that tritium adequate to meet INTOR's requirements up to the year 2000 will be produced in CANDU heavy water reactors. The provision of tritium by all parties participating in the INTOR project is an important consideration which must be dealt with.

On the other hand, if difficulties are encountered with the supply of tritium, INTOR might be forced to install a full breeding blanket. Analyses indicate that conversion ratios of 0.9 – 1.0 can be achieved if the entire available surface is used for a breeding blanket; and that conversion ratios of 0.6 – 0.7 can be achieved if the relatively inaccessible inboard region is non-breeding and if Li_2O is the breeding material. The use of neutron multipliers (e.g. lithium lead) can enhance the tritium production to achieve breeding ratios above unity. (It is noted that no effort has been made to optimize the design of INTOR for tritium production.)

It must be realized, however, that the requirement for installation of a full breeding blanket adds considerably in complication to INTOR and might reduce its availability. In addition, if the blanket concept has to be selected too early, it might not be the most reactor-relevant.

4.3. Electricity production

An INTOR corresponding to the Suggested Parameters would produce ~ 460 MW_{th} of thermonuclear power averaged over a burn cycle with a 75% duty cycle. The maximum electrical power production capability would then be about 150 MW_e if the total surface area of the plasma chamber could be surrounded with a high-temperature blanket capable of achieving a thermal-to-electrical conversion efficiency of $\sim 30\%$. (The actual capability would be less because of neutron losses through penetrations, but also could be increased by $1.6/1.3 \approx 1.2$ if the $P_n = 1.6$ MW/m^2 option was chosen.)

In order to realize the maximum electrical power production capability, it would first be necessary to test and further develop a high-temperature blanket concept in test modules during Stage II and then to rebuild the entire blanket (inboard as well as outboard) during Stage III. This assumes that the high-temperature blanket concept can be developed early enough in Stage II so that a full blanket can be ready for installation at the beginning of Stage III, and it further assumes that it will then be possible remotely to replace the entire blanket in a reasonable period of time – say 1 to 2 years. Even if these two assumptions are valid, it may be more effective from both cost and programmatic considerations to use the knowledge gained from INTOR to build a DEMO immediately. In fact, total blanket replacement is almost equivalent to building a new machine. With a less technologically advanced, medium-temperature and medium-pressure blanket that could be developed with greater confidence, the likely thermal-to-electrical conversion efficiency is ~ 25%, and the maximum electricity production capability is reduced accordingly.

The lower estimates for the mean power requirements for INTOR are near 200 MW_e . Thus, while INTOR is capable of producing ~ 100 – 150 MW_e of electrical power, that is insufficient to achieve net electrical power. In order for INTOR to produce net electrical power, it would be necessary to increase the size in order to produce more electrical power. This necessary increase in size (and cost) could be minimized by increasing the neutron wall load and by reducing the power requirements. Further reduction of the power requirements for INTOR would force the development of technology in some areas (e.g. neutral beam injection, energy storage and transfer) beyond the level that is needed to meet the primary objectives of INTOR and, in some instances, may force the further development of a technology which will have no reactor relevance. It is the opinion of the Workshop that achievement of net electrical power is **not** an appropriate task for INTOR.

The production of 5 – 10 MW_e should be sufficient to test and demonstrate all the technologies needed for electricity production by fusion, as well as providing a significant demonstration of the latter. This amount of electricity could be produced in a number of ways:

1) two outboard partial segments, each with an area of about 15 m^2 ;
2) eight to ten outboard test modules each with an area of about 3 m^2 ;
3) three inboard partial segments, each with an area of about 17 m^2 ;
4) one complete (inboard plus outboard) segment, with an area of about 40 m^2.

Because of its greater compatability with a flexible blanket technology testing programme and because of its lower risk, the test module option is preferred for the production of about 10 MW_e in INTOR.

5. PHYSICS AND TECHNOLOGY DATA BASES AND R AND D NEEDS

The main part of the present Workshop was devoted to the assessment of the physics and technology bases for the design and construction of INTOR. This assessment was carried out by teams of experts working in their home institutions and then discussed and synthesized by the Workshop. This assessment was based upon a reactor concept as characterized by a set of Guiding Parameters that were developed early in the Workshop and are given in Appendix A.4. This assessment was generally positive with respect to the feasibility of designing and constructing INTOR to operate in about 1990, on the basis of the substantial amount of knowledge and technology that exists today and that will be expanded over the next few years by currently-planned programmes. However, certain crucial information will not be developed by currently-planned programmes; but this information could be developed on the INTOR time scale by the expansion of existing programmes and by the establishment of new programmes. This assessment was based upon information available through October, 1979. The highlights of the assessment and the most critical R and D areas are summarized below.

5.1. Physics

For the parameter range of present-day devices, there is a well-documented experimental data base on energy confinement. Electron energy confinement is found to obey an empirical scaling law, which provides a good basis for predictions of ignition in INTOR. Ion energy transport is found to be within a factor of about 3 of neoclassical predictions and, in any case, constitutes a relatively minor energy loss channel in INTOR. Uncertainties exist in the extrapolation of these scaling laws to the high-temperature regime to be reached in INTOR, but high-power experiments in present facilities, as well as the next generation of tokamaks, will extend the data base toward the INTOR regime in the early-to-mid 1980's. More data will be needed on particle transport (both hydrogenic and impurity ions) in tokamaks in order to understand fully the requirements for D-T fuelling and helium exhaust in INTOR.

INTOR will ignite at $\langle\beta\rangle \cong 4\%$, and must operate at $\langle\beta\rangle \cong 5\%$ to achieve a wall loading of 1.3 MW/m^2. Limitations on $\langle\beta\rangle$, inferred from theory, permit stable INTOR equilibria up to $\langle\beta\rangle \cong 4\%$, which might be increased by further optimization. These equilibria will certainly require careful plasma shape and profile control. There is experimental evidence that tokamaks may be operated somewhat above the theoretical limit, and further high-power injection experiments will determine $\langle\beta\rangle$-limits in INTOR-like non-circular cross-section configurations in the early 1980's.

Neutral beam heating is the primary option for INTOR: 75 MW of 175 keV deuterium neutral beam power applied for 6 s should be sufficient to ignite the plasma, provided that the impurity influx can be controlled. Extension of positive-ion-based neutral beam technology to meet these requirements seems feasible, but requires an accelerated development programme.

RF heating should be retained as an alternative to neutral beam heating because of its potential for technological advantages that will become even more important in a commercial reactor. However, before any RF method can displace neutral-beam injection as the preferred heating technique for bulk heating in INTOR, that method must demonstrate the multi-keV temperatures already obtained with neutral-beam injection, and convincing designs for reasonably efficient and neutron-resistant wave launching structures must be devised and tested. Provision of supplemental electron cyclotron heating at the 10 MW level would be especially useful in the start-up phase and for burn control.

To achieve a burn pulse of at least 100 s in INTOR, an effective magnetic divertor must be provided for exhausting the helium produced by fusion reactions, and for controlling the impurity influx. At present, the poloidal divertor with exterior poloidal field coils seems to satisfy the requirements, and near-term activity for INTOR should be based on this concept. The bundle divertor, in its conventional form, has the drawbacks of a large field perturbation and limited space for shielding. However, suggestions have been made for possible improved bundle divertor configurations, and other local divertor concepts have been proposed. Before any divertor concept can be finally specified for INTOR, detailed design studies and experimental tests are needed.

Gas puffing may prove to be adequate for fuelling INTOR, but pellet injection will provide more flexibility over the deposition profile. Pellet injection at 10^3 m/s is technologically within reach (at least for single pellets) and is adequate for shallow penetration. For deep penetration, which would permit a further increase in the flexibility of the fuelling scheme, about 5 times higher injection velocities will be needed.

Plasma breakdown and current initiation in INTOR requires a one-turn voltage that can be as low as 100 V. Provision of auxiliary RF heating in the early start-up phase may allow the voltage to be lowered somewhat further. For a plasma current of 6.4 MA and a burn pulse up to 200 s, the total flux swing required is about 110 Vs. These requirements, together with those for plasma horizontal position and shape control, can be met with poloidal field designs in which most, if not all, poloidal coils are exterior to the toroidal coils, and such configurations have considerable technological advantages. Control of the vertical position would require passive conducting elements at about the location of the blanket/shield, supplemented by active feedback from the exterior poloidal coils.

Plasma disruptions will probably be unavoidable in INTOR, particularly during initial operation, and all measures should be taken to reduce the frequency of their occurrence, and to control them when they do occur. In any case, INTOR must be designed to withstand the thermal and electro-mechanical consequences of disruptions; the rapid heat deposition may require the installation of some type of sacrificial liner or limiter.

There are several possible methods for active control of the burn temperature in INTOR, such as variation of the field ripple to adjust confinement, and/or operation

TABLE II-4. INTOR PLASMA PHYSICS R AND D NEEDS

R and D needs	Status of world programme	
	Adequate[a]	Inadequate[b]
Energy and particle confinement	x[c]	
β-Limits and effect on transport	x	
Shape and profile control	x[d]	
Disruptions (detection and control)		x
Impurity control and divertor physics and technology		x
Neutral beam heating technology		x
RF Physics and technology	x[e]	
Fuelling	x	
Plasma engineering (start-up, shut-down, burn control)	x[f]	

[a] Programmes are either planned or could be planned on existing or authorized facilities (with some upgrading if required) and using existing trained personnel.
[b] Either the available and authorized manpower and facilities are inadequate or a substantial increase in effort is needed.
[c] Except ripple effects and helium transport.
[d] Except current profile control.
[e] If near-term experiments are successful, an accelerated programme would be needed.
[f] Except burn control.

just below ignition, applying a limited auxiliary heating power throughout the burn pulse. Field ripple effects on energetic ion confinement should be experimentally tested in the next few years.

The most crucial plasma R and D needs in support of establishing an adequate data base for INTOR are given in Table II-4.

5.2. Magnetics and power supply

INTOR will require large-bore toroidal field coils with maximum fields of 10–11 T. Based upon the present state of the technology and upon several existing development programmes throughout the world, the technological basis for the construction of the toroidal field coils incorporating super fluid helium cooled NbTi and/or Nb_3Sn superconductor should be adequate in the mid-1980's.

Pulsed superconducting poloidal field coils external to the toroidal field coils will be used on INTOR. The field strength requirements on these coils can be met

TABLE II-5. INTOR MAGNETICS AND POWER SUPPLY R AND D NEEDS

R and D needs	Status of world programme	
	Adequate[a]	Inadequate[b]
SC Toroidal field coil	x	
SC Poloidal field coil		x
Energy storage and transfer	x	

[a] Programmes are either planned or could be planned on existing or authorized facilities (with some upgrading if required) and using existing trained personnel.

[b] Either the available and authorized manpower and facilities are inadequate or a substantial increase in effort is needed.

by NbTi. It appears possible to develop the technological basis for these pulsed coils on the required time scale.

If normal conducting pulsed coils are necessary inside the toroidal field coils, development of remotely maintainable joints will be required.

Divertors impose severe constraints upon the magnet system. However, preliminary work suggests that design concepts could be developed for which the magnet technology basis would be adequate.

Virtually all of the power supply and energy transfer requirements for INTOR can be met with present technology. Only modest extensions for plasma heating systems are required.

The most crucial magnetics and power supply R and D needs in support of establishing an adequate data base for INTOR are given in Table II-5.

5.3. Blanket and tritium

There are a number of difficult engineering design problems which must be solved before the INTOR first wall and blanket can be designed (e.g. thermal cycling of first wall and/or divertor collector surfaces due to pulsed heat flux, thermal and mechanical consequences of a plasma disruption). Although there is currently little R and D effort directed towards these problems, there is a rather large technological basis from which R and D programmes can be initiated. It should be possible to develop an adequate technological basis to design and build a low temperature,

TABLE II-6. INTOR BLANKET, MATERIALS AND TRITIUM R AND D NEEDS

R and D needs	Status of world programme	
	Adequate[a]	Inadequate[b]
First-wall, limiter and divertor plates		x
Structural materials	x[c]	
Insulators and coatings		x
Integrated tritium fuel handling system	x	
Tritium cleanup, safety and monitoring systems	x	

[a] Programmes are either planned or could be planned on existing or authorized facilities (with some upgrading if required) and using existing trained personnel.

[b] Either the available and authorized manpower and facilities are inadequate or a substantial increase in effort is needed.

[c] For stainless steel only.

non-breeding shield for the basic design and later to design and build hot, tritium-breeding, electricity producing test modules, on the INTOR time scale.

On the basis of the current INTOR concept, the data base either exists now or will exist by the mid-1980's to predict the mechanical and electrical performance of materials in the reactor. There is reasonable optimism that the structural material can be used for the lifetime of the device.

The technology data base for electricity production by fusion should be adequate for INTOR. However, methods of energy storage consistent with the pulsed operation mode must be demonstrated.

The technology data base for tritium-breeding blanket material performance and processing should be adequate to support the engineering testing of tritium extraction and processing in modules in INTOR.

An adequate technology data base for tritium fuel recycle, containment, cleanup, waste disposal, monitoring and analysis is being developed on a time scale compatible with INTOR in several test facilities throughout the world.

Programmes to define the operational environment and performance of insulators and coatings in INTOR are presently inadequate and it is important to increase the R and D efforts in that area as soon as possible.

The most crucial blanket and tritium R and D needs in support of establishing an adequate data base for INTOR are given in Table II-6.

5.4. Engineering

INTOR will be a very complex technological undertaking because of the strong interactions among the major systems and because of the topology. The basic techniques for systems integration and management exist, but must be upgraded considerably.

Certain design choices will have a major impact upon the overall design. The requirement for remote maintenance has the largest impact upon the overall design, and must be an integral part of the design effort for all systems. The inclusion of a divertor and possibly of some poloidal coils interior to the toroidal field coil system, the location of the several vacuum interfaces, the decision on the cryostat configuration for the toroidal field coils, and the determination of the size and number of blanket test modules are other design decisions which will have large impacts on the overall design, on machine reliability and maintainability, and on the required technology development.

The achievement of high availability (25–50%) requires that remote maintainability must be a major design criterion that governs the design of the INTOR systems, that radiation shielding of the penetrations must be adequate to allow the adoption of a partially-remote maintenance plan which is based upon personnel entering the reactor hall for certain ex-reactor maintenance operations, and that component reliability must be a primary objective of the supporting development programme.

INTOR must be designed for remote maintenance to the fullest possible extent. The basic technology needed to remotely maintain INTOR exists today. However, the techniques for designing INTOR for remote maintainability are in a very early stage of development.

The vacuum technology for pumping the neutral beam lines and the plasma chamber between burn pulses should be adequate. However, the development of high speed helium pumps is required. Moreover, the neutron and gamma heat fluxes on the cryopanels could lead to large refrigeration costs if conventional cryocondensation pumps operating at 4.2 K are used.

The technology for plasma diagnostics must be extended for reliable operation in a radiation environment, and diagnostic techniques must be adapted from other fields in order to have an adequate diagnostic basis for INTOR. The control and data processing computer systems that are being developed by industry should be adequate.

The most crucial engineering R and D needs in support of establishing an adequate data base for INTOR are given in Table II-7.

TABLE II-7. INTOR ENGINEERING R AND D NEEDS

R and D needs	Status of world programme	
	Adequate[a]	Inadequate[b]
Remote maintenance and assembly	x	
Penetration shielding for personnel access and component protection	x	
Vacuum pumping of helium		x
Diagnostics, data acquisition and control	x	

[a] Programmes are either planned or could be planned on existing or authorized facilities (with some upgrading if required) and using existing trained personnel.

[b] Either the available and authorized manpower and facilities are inadequate or a substantial increase in effort is needed.

5.5. Safety and environment

Regulations governing the control of radioactivity and toxic materials as well as industrial procedures for dealing with high vacuum technology, cryogenic systems and high voltage systems currently exist. On this basis, specific regulations for fusion need to be developed on a time scale consistent with the design of INTOR.

The data base for conservatively relating tritium release to exposure in humans and for assessing the effects of tritium on the population exists, although some uncertainty may exist concerning the effects of long-term exposure at low levels. This data base can be utilized to develop guidelines on tritium release limits for INTOR.

The existing regulations on tritium accountability and transportation have to be re-examined in light of the multi-kg quantities of tritium which will have to be handled during the operation of INTOR.

It is important to establish criteria for allowable magnetic field and electromagnetic radiation exposure to humans. Such regulations will take a long time to develop and it is urged that programmes be initiated immediately to establish safe working levels for the operating personnel of INTOR.

The most crucial safety and environmental R and D needs in support of establishing an adequate data base for INTOR are given in Table II-8.

TABLE II-8. INTOR SAFETY AND ENVIRONMENTAL R AND D NEEDS

R and D needs	Status of world programme	
	Adequate[a]	Inadequate[b]
Tritium containment, cleanup, conversion and biological effects	x	
Activation product generation, release and transport	x	
Experimental accident data and analytical techniques		x
Magnetic field personnel exposure limits		x

[a] Programmes are either planned or could be planned on existing or authorized facilities (possibly with some upgrading if required) and using existing trained personnel.

[b] Either the available and authorized manpower and facilities are inadequate or a substantial increase in effort is needed.

6. SCHEDULE AND RESOURCES

Preliminary assessments conducted by the Workshop indicate that INTOR can be designed, constructed and put into operation in 9–11 years from initiation of a focused conceptual design. The data base required for properly costing a conceptual design of INTOR exists. The needed management and control techniques have already been developed in previous complex technical projects. Finally, the trained manpower base already in existence today is fully capable of carrying out the INTOR project in a timely and successful manner.

7. CONCLUSIONS

A substantial physics and technology data base for INTOR exists today, and this data base will be expanded over the next few years by currently planned programmes. However, certain crucial information will not be developed by currently planned programmes. Much of this missing information could be developed on the INTOR time scale by the expansion and/or acceleration of existing R and D programmes and by the establishment of new R and D programmes. On this basis, it is concluded that it is scientifically and technologically feasible to undertake the construction of INTOR initially to operate about 1990, provided that the supporting R and D effort is expanded immediately to provide an adequate data base within the next few years in a number of important areas. Furthermore, it is concluded that the construction of an INTOR-like device to operate in the early 1990's is the appropriate next major step in the development of fusion power.

8. RECOMMENDATIONS

The INTOR Workshop participants recommend that the INTOR project be taken into the next phase – definition. The definition phase should have as its objective the development of a single INTOR preconceptual design and such other material as is necessary to allow a decision to be taken to proceed into the design phase.

Chapter III

DATA BASE AND R AND D NEEDS ASSESSMENTS, INTOR SUGGESTED PARAMETERS AND ELECTRICITY PRODUCTION

1. INTRODUCTION

The Data Base and R and D Needs Assessments were done for a set of Guiding Parameters which was developed early in the Workshop and considered to be sufficiently close to the final INTOR parameters. These Guiding Parameters are listed in Appendix A.4.

Concentrated summaries prepared by the sixteen responsible groups for their respective fields are given in Section 2 for the Data Base Assessment and in Section 3 for the R and D Needs Assessment. The cutoff date for new information was October 1979. The full reports of the groups are contained in Chapters IV through XIX.

A more detailed list of the INTOR Suggested Parameters is given in Section 4, together with the proposals of October 1979 made by the four parties.

A more elaborate discussion of the three cases of electricity production by INTOR – 10 MW_e, 100 MW_e and net electricity – is given in Section 5.

2. DATA BASE ASSESSMENT SUMMARIES

2.1. Energy and particle confinement

This report on Energy and Particle Confinement contains an assessment of recent experimental results on energy and particle confinement in tokamaks, in particular those devices with strong auxiliary heating. The report also summarizes the results of many transport computations for INTOR carried out in each of the participating countries, and presents assessments of field ripple effects, synchrotron radiation, and alpha-particle confinement.

The major conclusions of the Energy and Particle Confinement Group are:

1. For the parameter range of present-day devices, there is a well-documented experimental data base on energy confinement. Electron energy confinement is found to obey an empirical scaling law, which provides a good

basis for predictions of ignition in INTOR. Ion energy transport is found to be within a factor of about 3 of neoclassical predictions and, in any case, constitutes a relatively minor energy loss channel in INTOR. Uncertainties exist in the extrapolation of these scaling laws to the higher-temperature regime to be reached in INTOR, and an experimentally-verified transport *theory* could provide a sounder basis for such extrapolations. However, the extrapolations in terms of dimensionless physical parameters are not substantial enough to prevent us from formulating, as a working basis, the requirements for reaching ignition in the INTOR plasma.

2. More data will be needed on the transport of hydrogenic and impurity ions in tokamaks in order to understand fully the requirements for D-T fueling and helium exhaust.

3. Computations for INTOR show that 60–75 MW of 150–200 keV D° neutral beam power applied for 4–6 seconds should be able to ignite the plasma core, provided that it is not excessively cooled by impurity radiation.

4. The volume average β at the time of ignition is in the range of 4–5%, including contributions from beam ions and alpha particles; the value of the volume average β must rise to about 6% in the burn phase to achieve the desired neutron wall loading.

5. However, whether ignition is reached is very sensitive to many details of the physical effects included. A plasma that ignites slowly from the centre outward, while the overall plasma energy density decreases slightly, is the type of ignition that occurs when the device parameters are such that ignition can just be achieved.

6. The D-T fuel mix must be carefully controlled during the heating phase, to ensure an adequate tritium density at the centre of the plasma to balance the substantial injection of deuterium from the beams. The operating scenario must involve supplying pure tritium by gas puffing or pellet injection during the heating phase, and it would be advantageous if some control could be exercised over the recycling rate (probably by a divertor).

7. With respect to impurity effects, the most *desirable* situation is when impurity radiation from the core of the plasma is negligible, while at the same time impurity radiation from the edge region of the plasma is large enough to account for a substantial fraction ($\gtrsim$50%) of the total power outflow. In this way, the edge region is cooled, and the plasma-wall interaction is reduced, so that a steady state would arise in which the influx of impurity atoms is balanced by outward transport of impurity ions. This situation occurs in most present-

day tokamaks, often with carbon or oxygen as the dominant impurity. Computations indicate that this desirable situation may also be achievable in the burn phase of INTOR, the critical issue being to prevent the reacting core of the plasma from being excessively cooled by impurity radiation. In the heating phase, the critical issue will be to prevent excessive cooling of the plasma caused by a sharp increase in impurity radiation, resulting from a lowering of the edge temperature when the beam heating is turned off. In the case of high-Z metallic impurities (Fe, Mo, etc.) the impurity particle density can be quite uniform, although it is always advantageous if the impurity density can be higher in the edge region than in the plasma core. The width of the radiating edge-region is estimated to be 10–20 cm. In the case of low-Z impurities (C, O, etc.) sufficient in number to radiate a substantial fraction of the power, the impurity density must be peaked in the edge region, or else bremsstrahlung from the core will be excessive. Theoretically, the impurity density can become peaked in this way in the edge region by various neoclassical effects, but the lack of experimental verification of these impurity transport models makes quantitative conclusions in this area rather unreliable.

8. The fraction of alpha-particles in INTOR that are born on orbits that intersect the wall/limiter does not exceed 4% for all reasonable plasma profiles.

9. The magnetic field ripple must not exceed ± 0.75% at the plasma edge, ± 0.4% at r/a = 0.7, and ± 0.1% at the plasma centre, or else the ripple contributions to ion thermal transport will prevent ignition. For INTOR, this value of ripple corresponds to a horizontal bore of about 8 m, for a 12-coil design. If the ripple can be adjusted upwards to the range 1.0–1.5% at the plasma edge, then effective burn control is possible.

10. Ripple effects on beam ion confinement may be important. However, if the injection angle is set at 20° to the normal, the ripple losses of beam-ion energy will be diminished, and preliminary estimates indicate that they will certainly not exceed 10% of the beam power. Ripple effects on alpha-particle confinement are not significant.

11. Synchrotron radiation will not be a significant power loss mechanism in INTOR, accounting for no more than 5% of the power.

2.2. Impurity control, fueling and exhaust

The control of the level, species, and distribution of impurities; and the provision for helium ash exhaust, are important and difficult problems in reactor-like tokamaks. We have concluded that provision of some technique for active control of impurities and helium ash is unavoidable in INTOR. Intensive theoretical and experimental studies on INTOR-relevant impurity control and exhaust techniques are, therefore, urgently required. On the basis of our present understanding, our major conclusions are:

1. To ensure that INTOR attains its main objectives in a reactor-relevant manner, a burn pulse of at least 100 s must be provided. Such a pulse length requires an effective technique for removal of the helium ash, and also some means for providing adequate control of the impurity influx.

2. Helium ions must be exhausted from the INTOR plasma at the rate of 2×10^{20} particles/s, i.e., the rate of helium production by fusion reactions. In order to be able to limit the helium concentration in the plasma to the required 10% of the plasma density, the helium particle confinement must not exceed 15 s. If the helium particle confinement time were about 2 s, it would be sufficient if about 10% of the helium transported from the plasma core to the plasma edge could be exhausted. The helium confinement time depends on a variety of transport processes in the plasma and should be investigated experimentall with some urgency.

3. Plasma contamination by impurities adsorbed and absorbed on the wall surfaces can be reduced to acceptable limits by providing proper treatment of the wall surfaces before assembly, and in-situ conditioning after assembly. Further work is needed to identify the optimum method for maintaining a clean surface for very long periods in a high-duty-factor device such as INTOR. The lack of adequate wall cleanliness may also result in the generation of wall-material impurities by unipolar arcs.

4. The prevention of plasma contamination by first wall and limiter/divertor-plate materials will impose severe design constraints on INTOR. The causes will be sputtering of the first wall by charge-exchange neutrals and by unconfined ions, and sputtering of the limiter/divertor-plate by plasma ions, by light impurity ions, and self-sputtering by heavy impurity ions. In fault conditions, impurities can also be produced by evaporation due to high heat loads and/or local heat deposition by runaway electrons.

5. Erosion of the first wall by sputtering may be a serious problem in INTOR. The sputtering erosion rates per power deposited on the wall by hydrogenic particles, for most materials, is maximum at edge temperatures between 200 and 300 eV. E.g., for stainless steel, erosion rates of 0.1 mm/MW-y, or 300 kg/MW-y, may be attained in this case.

6. Some form of magnetic divertor is probably required in INTOR to provide helium exhaust. Depending upon which concept is adopted, a divertor can also provide heat removal from the plasma to the divertor, protection against contamination by impurities generated at the limiter/divertor-plate, and some degree of shielding against impurities sputtered from the first wall. With respect to particle exhaust, the most desirable situation would be to pump the required quantity of helium gas from the divertor chamber, while allowing as much as possible of the hydrogenic gas to recycle to the main plasma chamber. The required pumping speed will depend upon the helium gas density in the divertor chamber. To diminish the exhaust of fuel ions, the hydrogenic gas density in the divertor chamber should be as low as possible. Several divertor concepts are under consideration for INTOR. At present, the poloidal divertor with exterior poloidal-field coils seems to offer a viable solution, and near-term design activity for INTOR should be based on this concept although it is improbable that it could be used in a commercial reactor. Therefore, local divertor concepts have to be developed further. The bundle divertor, in its conventional form, has the drawbacks of a large field perturbation, limited space for shielding, large power consumption, and high stresses. However, suggestions have been made for possible improvement. The magnetic configurations of proposed local poloidal divertors and hybrid poloidal/bundle divertors have been described, but engineering designs of these concepts are not yet available. Before any divertor concept can be finally specified for INTOR, detailed design studies and experimental tests are needed.

7. Non-divertor solutions are attractive for their potential simplicity, but their feasibility has not yet been demonstrated. Three preliminary concepts have been proposed:

(i) ports for ash exhaust adjacent to a toroidal limiter;

(ii) a limiter consisting of a stream of dropped titanium or graphite balls;

(iii) a "wall-lapping" plasma with rotating resonant helical islands to equalize the heat flux to the first wall, and with mechanical valves for ash exhaust. Further work in this field is required.

8. Fueling with a controllable mixture of deuterium and tritium will be needed in the density-rise phase of the startup. Refueling with an equal mixture of deuterium and tritium will be needed in the burn phase, at a much slower rate, to replenish the ions consumed by the fusion reactions. Gas puffing is the simplest method of fueling and refueling. However, there are uncertainties whether it will work satisfactorily for rapid fueling in the density-rise phase in a divertor case, because of the shielding action of the scrape-off and the pumping action of the divertor. Pellet injection will provide much superior flexibility over the deposition profile. Pellet injection at 10^3 m/s is technologically within reach, and is adequate

at least for shallow penetration. For deep penetration, which would permit a further increase in the flexibility of the fueling scheme, about 5 times higher injection velocities will be needed. Assuming that the net pumping of hydrogenic gas from the divertor chamber is very small, so that most hydrogenic ions recycle back to the main plasma chamber, the small amount of refueling needed to replenish the spent fuel can probably be provided by gas puffing.

9. Further research on atomic and molecular data, on surface data, and on plasma-wall interactions would be useful for more reliable predictions of impurity behaviour, and to provide more information on which a selection of the first-wall material might be based.

2.3. Heating

Heating of the INTOR plasma to ignition conditions seems feasible using neutral-beam energies achievable with positive-ion-based technology (150–200 keV per deuteron), but with significant advancements beyond TFTR/JET/JT-60/T-15 beam technology required to obtain long pulses at high duty factor, as well as high currents at the upper end of this energy range. Other development requirements are a high monatomic ion fraction in the ion beam, and possibly an increase in overall efficiency to be obtained by direct energy recovery, and increased reliability and component lifetimes.

Potentially serious disadvantages of the beam heating method are the exposure of the injector internals to the fusion neutron flux, and the expected high capital cost of beam systems. The difficulties incurred by the direct exposure of injector components to neutrons can probably be overcome, especially if the beam injectors can be closed off after start-up.

The following are probable technical parameters for the INTOR neutral beam injection system:

Beam energy	175 keV (the beam development programme should attempt 200 keV)
Beam power	75 MW
Pulse length capability	10 s (the capability should also exist to restart injectors during the burn to achieve an effective 100 s pulse length at 10 MW)
Injection direction	20° to normal
Monatomic ion fraction	About 90%
Bending magnet at ion source	None

Fraction of neutral beam power at full energy	About 80%
Direct energy recovery	Probably required (efficiency ⩾60%)
Power density at port	2–3 kW/cm^2 (depends on detailed design)
Number of injection ports	4–5 (depends on detailed design)
Cross-section of port	0.5–1.0 m^2 (depends on detailed design)
Neutron protection	Plug valve closed after ignition
Lifetime of source	3×10^5 s
Pumping speed per port	5–8 Ml/s
Redundancy	Additional ion sources will be needed to provide required reliability
Negative ion beam	Advantageous for the long pulse capability (10 MW for 100 s); useful energy range 400–500 keV.

The provision of effective heating (if needed) during the start-up phase when the plasma density is low poses some difficulties, due to beam shine through, if the main injector system is to be used. A lower-energy injector may have to be provided, or a larger angle of injection, or perhaps a system capable of variable energy or variable species mix.

There remain some uncertainties about the absolute need for a direct energy recovery system for the INTOR injectors. The advantages of direct recovery (60% efficient) would be an increase in the overall efficiencies of the injectors from an estimated 20% to an estimated 30%, and a reduction in the power to be handled on the beam dumps from about 200 MW to about 85 MW for a 75 MW system. However, to date, there has been very little development effort on direct recovery systems for neutral beam injectors.

Summary of radio-frequency heating

The continued attraction of pursuing the many different RF heating schemes derives in part from the perceived disadvantages of neutral-beam injection mentioned above. To date, only ion cyclotron wave heating (ICRF) has provided a convincing demonstration of substantial plasma heating in tokamaks of significant size. RF heating in the lower-hybrid frequency range has been successful in smaller tokamaks,

TABLE III-1. ADVANTAGES AND DISADVANTAGES OF RF HEATING METHODS

	Advantages	Disadvantages
Very low frequency (TTMP) 100 kHz	Physics understood	No access to coils. Neutron effects on internal coils. Low efficiency of heating
Ion cyclotron 50–70 MHz	Proven effectiveness in large tokamaks. High efficiency	Neutron effects on internal antennae
Lower hybrid 2–3 GHz	Very attractive transmission (waveguides) Efficient sources available (klystrons)	Effectiveness in large tokamaks yet to be proved
Electron cyclotron 120–150 GHz	Very attractive transmission (waveguides) Simple physics.	Substantial tube development needed. High probable cost

but the physical mechanisms of coupling the wave power to the plasma are less well understood. Electron cyclotron heating has demonstrated its heating effectiveness in recent experiments on quite large tokamaks. Transit-time-magnetic-pumping has been found effective on small tokamaks but its extrapolation to INTOR conditions seems technologically too difficult.

The experimental success of ion cyclotron heating, and the potential inexpensiveness of ICRF systems make it an RF technique that presently seems attractive for INTOR. However, lower-hybrid heating has important technological advantages, and should be retained as a viable option. However, before any RF method can displace neutral-beam injection as the preferred heating technique for bulk-heating INTOR, this method must demonstrate the multi-keV temperature increases already obtained with neutral-beam injection, and convincing designs for reasonably efficient and neutron-resistant wave launching structures must be devised and tested.

The technological problems associated with electron cyclotron heating (ECH) lie primarily in gyrotron development, rather than in transmission of the ECH power to the plasma (which is technologically very attractive). Gyrotron at 120–150 GHz, delivering about 500 kW per unit for a 5 s pulse, should be developed for INTOR, for a total ECH power level of at least 10 MW. Such an

ECH system would be an attractive option for providing the needed start-up power at low plasma density.

Table III-1 summarizes the main advantages and disadvantages of the various RF methods.

2.4. Stability control

The problem area "Stability Control" comprises three main topics, namely **β limitations, disruptions,** and **equilibrium and stability control.**

1. In present-day tokamaks the discharge is magneto-hydrodynamically stable. From theoretical considerations it is, however, inferred that going to **higher values of the ratio** β of the kinetic and magnetic pressures, as required for INTOR, at some point **stability limits** will be encountered. From the **theoretical and numerical analysis** performed so far, it can be concluded that in INTOR discharges with a volume average $\langle\beta\rangle$ up to 4% will be stable without a need for wall stabilization. Higher values may be possible, but have not been assessed, yet. They will certainly require careful plasma shape and profile control. **Experimentally**, stable discharges at a value of $\langle\beta\rangle = 2.5\%$ have been obtained in a tokamak with circular cross section (T-11). This value is somewhat higher than theoretically expected for the plasma shape and profile present in this device. Experiments with still larger heating power, already underway, are needed to explore the actual β limitations, in tokamaks.

2. The **requirements** for being able to reach high $\langle\beta\rangle$, following from theory, include:

(1) a D-shaped plasma cross section of strong triangularity;
(2) an appropriate, relatively flat current profile;
(3) operation at low q_a and β_J ($q_J(a) \approx 2.5$, corresponding, for a = 1.2 m, B = 5 T, R/a = 4, and b/a = 1.6, to a plasma current of 5.3 MA, is advisable and, if not excluded for an increased disruption frequency, even lower values of $q_J(a)$ should be considered).

3. During the life time of INTOR, at least 10^3 disruptions are expected to occur. Hence, INTOR must be designed with the necessary protections against the thermal and electromechanical consequences of disruptions. Against fast heat deposition sacrificial elements will have to be installed.

4. On the other hand, all **measures to reduce the occurrence** of disruptions and to **control** them when they appear must be taken. Those presently known include:

(1) very careful position control;
(2) control of the current profile;
(3) use of an appropriate CH transformer system counteracting current decrease;
(4) operation in a regime where disruptions are improbable.

Feedback control of disruptions would require coils close to the plasma and must, hence, be traded-off with the technological complications which it causes.

5. For **horizontal position** and **shape control** mainly a vertical and a quadrupole magnetic field with admixtures of higher multiple components are required. The strength of this control field in the plasma is about 0.5 T. During start-up and shut-down its rate of change is below 2 T/s, except for the first 0.1 s where 5 T/s are reached. The maximum power requirement during start-up is about 400 MWA for a poloidal coil system situated **outside** the toroidal field coils. Poloidal coil configurations of this kind, able to generate the required plasma equilibrium, have been found.

6. For **vertical position control** appropriate **passive conducting elements,** supplemented by a **feedback system** whose **coils** may also be situated **outside** the toroidal field coils, are required. The distance of the conducting elements from the plasma edge has to be smaller than about the minor plasma radius. For INTOR guiding parameters this implies that these elements have to be **part** of, or placed in, the **blanket/shield** structure. This current decay time should be of the order of 100 s, which does not seem difficult to realize at least as long as INTOR is not equipped with a full breeding blanket. The maximum feedback power then required is less than 100 MWA.

7. Although no experiments have been done so far, it is anticipated that current profile control can be satisfactorily performed either by selective electron heating by RF during the ohmic heating phase or, alternatively, by controlled minor radius expansion during current rise. In the former case about 10 MW heating power over about 1 s will be required.

8. The R and D requirements related to the area "Stability Control" are summarized in Section 3.4. From the world tokamak programme, an adequate data base can be expected to be available in time for INTOR check points, when due emphasis is given immediately to investigation of the critical issues identified.

2.5. Start-up, burn and shut-down control

The questions related to scenarios for the main part of start-up and for shut-down in INTOR can be studied, to a large extent, on present-day devices, although the operating regime of INTOR will deviate appreciably from that of tokamaks in operation by now. On the other hand, only limited work is possible on burn scenarios as well as on the last stage of the start-up phase, as reaching, or at least coming close to, ignition is required for this. A possibility of testing the schemes being developed now and in the near future on smaller devices, and extending them to heating a tokamak plasma into a regime were sizeable α-particle heating is present, will be provided by the large devices JET, JT-60, T-15, and TFTR.

Items that require particular attention include:

1. There is now evidence that for **ionization** and **current initiation** electric fields of 1 to 2 V/m may be sufficient, provided that the discharge chamber walls are kept clean from impurities and the stray magnetic fields are small. For INTOR this implies that in this phase a **loop voltage** of 100 V seems satisfactory, if the stray fields are less than 100 G. Providing **pre-ionization** and **auxiliary RF heating** (2 to 5 MW for 10 s) might allow to lower the loop voltage to 50 V.

2. During the **ohmic heating** phase, and in particular in its early stages, the rate of current rise will have to be chosen appropriately to minimize plasma-wall interaction. Impurity, plasma equilibrium and current profile control will have to be provided.

3. During the subsequent phase of **intense heating** and **fuelling** to reach **ignition** large changes of all plasma parameters will be encountered (e.g., the temperature increases by a factor of 10, the density perhaps by a factor of 5). Hence, the heating scheme adopted has to have enough **flexibility** to cope with these variations. Potentially **RF heating is better** suited in this respect than **neutral beam injection**. Therefore, if for INTOR a full RF heating scheme cannot be developed, either a somewhat more complex neutral injection scheme or a mixed RF/neutral injection scenario might have to be considered. Furthermore, in particular in the case of neutral beam heating, for controlling the species mix of deuterium and tritium a flexible **fueling** scheme is required, implying probably the use of both puffing of pure tritium gas and tritium pellet injection. Finally, efficient impurity, plasma position and shape control are needed.

4. **A burn time** of 100 to 200 s requires efficient impurity control and helium exhaust (see Chapter V).

5. The total flux swing necessary for INTOR, as specified by the guiding parameters, is about 90 Vs.

6. There are several conceivable methods for **controlling the burn temperature**, such as

(a) variation of the field ripple up to ± 1 or 2%,
(b) self-regulation by enhanced transport at the β limit,
(c) operation just below ignition, applying a limited auxiliary heating power, e.g. 10 MW, during the entire burn pulse,
(d) periodic adiabatic compression and decompression,
(e) variable fueling (if it is not too slow).

The most reactor-relevant, and probably simplest, burn control methods are the first three of the above, or combinations of them. Only the methods (a) and (b) can (and must) be tested experimentally in reactor-grade H and D plasmas.

7. Shut-down has probably to be approximately the time-inverse of start-up. It will have to be initiated by reducing the reactivity below the ignition point, using the same means as applied for burn control, combined with stopping fueling. Then the plasma will cool down to the ohmic discharge regime. Finally, the current will have to be brought to zero smoothly. The time necessary for shut-down is expected to be at least 5 s. The main concern is that a disruption might occur in an early stage.

8. The R and D requirements related to the area "Start-up, burn and shut-down" are summarized in Section 3.5. From the world tokamak programme, an adequate data base can be expected to be available in time for INTOR check points, when due emphasis is given immediately to investigation of the critical issues identified.

2.6. Magnetics

The main conclusions of the Workshop participants are summarized in this Section. The observations do not present a complete picture of magnetics status. Rather, they represent the principal points of interest and discussion at the Workshops in Vienna. Further information can be found in the various national reports referenced at the end of this report, which describe in detail the supporting studies carried out during 1979.

1. The various development programmes now being conducted are thought to be sufficient for the necessary technology bases to feed INTOR. The programmes appear to explore essentially all reasonable conductor and cooling options.

2. INTOR has a central field of approximately 5 T, leading to 10–11 T maximum at the coil. This leaves open both the Nb_3Sn and the NbTi options. The existing programmes and better definition of the INTOR mission will clarify which conductor and cooling method will work best.

3. The superconducting toroidal field coils are expected to be very reliable over the anticipated lifetime of INTOR. However, our experience is limited and studies with regard to failure modes, repair and replacement must be conducted in order to provide a basis for major decisions on machine design. These studies should attempt to determine the most probable modes of coil failure and the corresponding repair requirements as well as the disadvantages due to a design which allows easy coil removal.

4. Plasma current disruptions change the poloidal field at TF locations resulting in rapid increases in overturning moments and rapid field changes. This will be a major, but solvable, design problem for the SC/TF coils.

5. Shielding is required to protect the insulators and stabilizer from radiation damage and to reduce the radiation-induced heating of the toroidal field coil. The limiting value of radiation to the toroidal field coils is determined by the resistivity increase of the stabilizer. Copper is more resistant to radiation damage than aluminium. However, its recovery when annealed at room temperature is not as good as that of aluminium. Annealing of the toroidal coil takes a great deal of time and therefore is to be avoided if possible. The toroidal field coil can be designed to operate satisfactorily without the need for annealing the stabilizer, provided the fluence is kept below 10^{-4} dpa or 2×10^{17} n/cm^2 (fission neutron spectrum) for copper stabilizer material.

6. The toroidal field coils can be of sufficient size to permit space for proper access to the blanket or the inclusion of a bundle divertor. However, an upper limit on the size is imposed by the requirement of plasma shaping by PF coils placed outside the TF coils and the available power supply for these coils.

7. Poloidal coils outside the TF coils increase the pulse field problems and losses in the TF coils; however, an improved access to the PF coils is more important than the disadvantages of increased refrigeration loads and increased mechanical loads.

8. TF coils that can be split through the use of demountable joints are assessed as impractical for INTOR. This results from considerations of stress and dewar complication, not from considerations of contact resistance.

9. Small scale experiments show that a superconducting induction coil is feasible. The operation of such a coil is less reliable than that of toroidal field coils. The development programmes are limited at present and must be expanded to become a viable option for INTOR. It must be noted that the induction coil probably cannot be used for plasma initiation if the required voltage per turn is greater than 50–100 V. This question requires further conductor and coil development.

10. Designs should favour the placement of poloidal coils outside the TF coil despite increased complexity of structure, higher pulse losses in the TF coils and larger poloidal system power supplies. This results partly from considerations of coil reliability, but mainly from the requirements of access to the blanket.

11. Coils inside the TF coil should be normal conducting coils and must be designed to be replaceable. If coils must be located inside the shield (for fast stabilization), the insulators may have short lifetimes and must be remotely replaceable. It is recommended that design concepts should avoid inside-inboard coils.

12. Present estimates indicate a one-turn voltage for plasma initiation of ~100 V/turn. It may be possible to provide this breakdown voltage with the superconducting OH coil. The use of a special coil for providing this voltage, located near the plasma, is undesirable from the viewpoint of engineering. Another option, suggested by the USSR, is to use a “hybrid” OH coil consisting of a superconducting coil for the slower ramp and a room temperature coil for the faster ramp.

13. For INTOR it is almost impossible to design poloidal divertor concepts with internal coils such as described previously; the design of bundle divertors also entails extreme difficulties. Several new concepts, including poloidal divertors with coils exterior to the TF coils, were discussed at the Workshop, which may allow reasonable engineering practice and still permit acceptable divertor action. Further review of these new ideas is recommended.

14. Divertor designs have received very little attention compared with other magnetic systems. If it is considered necessary to place a divertor coil within the TF coil, then a major new development programme will be required to bring divertors anywhere near the stage of reliability of the rest of the magnetics.

15. A poloidal divertor with external coils and bundle divertors are clearly more accessible and replaceable than an internal poloidal divertor. An overriding consideration in divertor choice must be the ability to provide remote maintenance.

TABLE III-2. POWER REQUIREMENTS FOR INTOR

Assuming: Burn time = 200 s
Dwell + start-up + shut-down time = 40 s

	Peak	Mean (MW)
(i) Poloidal magnetic field (PF coils outside TF coils)		
(a) all superconducting coils	440 MVA	**26**
(b) all copper water cooled coils	440 MVA	200
(ii) Feedback system for plasma control, of the order of	100 MVA	**10**
(iii) Divertor		
(a) poloidal using the superconducting coils ouside TF coils of (1)a and sweeping	250 MVA	**9**
(b) Water cooled copper poloidal inside vacuum vessel		40
(c) Bundle with water cooled coils		100
(iv) Toroidal magnetic field coil		
(a) charging		5
(b) operation		**1**
(v) Neutral beam (10 s pulse, 20% efficiency, no direct recovery)	400 MW	**25**
(vi) Pumped cooling of NB system		**10**
(vii) RF heating. Efficiency not sufficiently known		
(viii) Cryogenic systems*		**30**
(ix) Pumping power for water cooling of shield		**5**
(x) Pumping power for helium cooling of blanket		**20**
(xi) Vacuum pumps		**5**
(xii) Miscellaneous** systems (control systems, diagnostics, buildings)		**25**
TOTAL		
(a) All SC coils, swept external poloidal divertor (items underlined)		**170**
(b) SC coils, water-cooled poloidal divertor		200
(c) SC coils, water-cooled bundle divertor		260

* Lowest estimate: highest estimate is 85 MW
** Lowest estimate: highest estimate is 60 MW

2.7. Power supply and transfer

In the summary of power requirements given in Table III-2, the largest peak powers are for the poloidal magnetic field system and for the neutral beam injectors, and it is proposed that these be supplied by flywheel motor generator sets.

The value of power given for the superconducting poloidal magnetic field system in (i)(a) is for a specific design of poloidal field coil placed outside the toroidal magnetic field coils. By suitable adjustment of the current in these coils, poloidal divertor action can be obtained without using any coils within the toroidal magnetic field coils, and by oscillating the current in a pair of the poloidal coils, the diverted plasma can be swept across the target plate to distribute the heat load.

Total mean power requirements are given for the all superconducting case above and for the cases when a water-cooled internal poloidal or water-cooled bundle divertor are used. These power requirements are probably minimum values and they each might have to be increased by as much as 90 MW if the highest estimates of power for cryogenic systems and miscellaneous systems are used.

In addition to the power supply requirements for the poloidal magnetic field system, the reports from the four countries include discussions of safeguards for the toroidal magnetic field system, power supplies for the neutral injection beam system and protective circuits for the injectors, radio-frequency heating power supplies, alternatives to flywheel generator systems, the system for generating some electric power from INTOR and the capacity of an electric power network to provide the power required for INTOR. In general, the four reports from the four countries showed substantial agreement in the amount and method of supplying the power required for INTOR. A satisfying feature of all the reports was that it considered that the power supply needs of INTOR could be met by relatively small developments of systems already being designed or built for the large tokamaks now under construction. For this reason, the R and D requirements listed in Section 3.7 are in the adequate category.

2.8. First wall, blanket and shield

The main conclusions from the data base assessment were:

1. In general, a wide range of design options has been explored in past fusion reactor studies and wide experience relative to this area is available from fission reactor design and construction. An adequate data base exists for many components of the Reference Design and the only area which could seriously affect the decision to construct INTOR in the mid-1980's is the design of

components which interact directly with the plasma, such as the limiter, first wall or lines, or divertor plate.

2. The data base with regard to structural, thermal hydraulic and vacuum vessel design is considered adequate at this time. Codes exist and experiments verify that first walls can be designed and built to withstand the normal operation heat flux in INTOR. However, the situation is not so promising for limiters or divertor collector plates, and further development work is required. It is not clear that adequate design solutions exist to handle the heat in all cases.

3. The data base required to design the first wall is acceptable, but the exact form of the first wall (i.e. water- or helium-cooled) must await a more detailed design. The data base for coatings to reduce tritium permeation needs to be expanded in this regard. It appears that it will be necessary to protect the first wall from fast plasma disruptions, but, if liners are used, the data base on which to choose liner materials is rather sparse. Perhaps the most critical feature is the lack of information on how many disruptions will occur during the lifetime of INTOR or the magnitude of the power density deposited during a disruption.

4. Adequate codes exist for analyzing the expected life of the first wall and blanket, but irradiated materials data is inadequate, particularly for materials other than steels containing nickel. The results of the analysis will be very design-dependent.

5. The composition and necessary thickness of the shield can be decided on the basis of known data. The form of the structure is not yet clear, but no serious design problems are foreseen.

6. An area of uncertainty in the data base for the first wall, blanket and shield design is the implication of high voltages at start-up and during a disruption. The introduction of conductors in these regions for equilibrium or stability control would create design problems. A better appreciation of this problem will probably only come from further detailed design studies.

2.9. Tritium

Eight major areas of concern have been identified with respect to the use of tritium in INTOR. These areas are summarized as follows:

- Fuel purification and recycle
- Tritium breeding in modular experiments
- Fail-safe storage of tritium
- A minimized tritium inventory outside of storage
- External supply of tritium for a non-breeding INTOR
- Containment of tritium
- Transportation of tritium
- Tritium spill cleanup and waste handling

The status of understanding and of the technical data-base in each area is discussed in detail in the full report of the international team on INTOR tritium systems. A brief outline of the key features of this study is given below; but it is generally concluded that none of the above topics will prohibit the design of INTOR in 1982–3 and construction in the mid-1980's.

The level of knowledge and capability required to design the fuel purification recycle system is expected to be reasonably well in hand by the mid-1980's. Experimental facilities such as the Tritium Systems Test Assembly (TSTA) in the USA, the Tritium Engineering Test Laboratory (TETL) in Japan, and the Vacuum/Tritium Stand (UTS) in the USSR should have provided the necessary technology by that time.

The large-scale production of tritium from lithium-containing materials has been demonstrated on a batch-wise basis in a number of countries, but the ability to continuously extract tritium from a breeder blanket module in a fusion reactor at the rate it is being produced and to simultaneously maintain a low in-blanket tritium inventory remains to be demonstrated. Extraction methods and behaviour of breeding compounds under irradiation could be developed in fission reactor facilities by the mid 1980's.

The required stored tritium inventory is based on that amount corresponding to 30 full-power days of burnup. This amount (2–3 kg) will represent a major fraction of the total INTOR tritium inventory. Although the stored inventory is generally regarded as being of very low vulnerability with respect to catastrophic release of tritium, an optimized, well-contained, failure-proof storage system must be developed. Such development appears to be straight-forward and does not pose any significant problems for INTOR.

The inventory of tritium outside the storage system (e.g. in the fuel fabrication, evacuation, and processing systems) is appreciably more vulnerable to release and, therefore, every effort should be made to minimize this inventory in future INTOR design activities. Presence estimates of this inventory range from several hundred to several thousand grams. A more detailed study of all aspects of this question (including vacuum system approach, divertor, fuel cycle hold-up, and inventory of dissolved tritium in containment structure) is needed.

The supply of tritium for a non-breeding or partially-breeding INTOR appears not to present immediate difficulty in that the USA and Canada have affirmed their capability in this regard. The principal question is one of cost per Curie with the lowest estimation of 0.4$/Curie. International agreements on the shipment of tritium from one country to another may be needed and the format for these agreements should be established at an early stage.

The containment of tritium within the fuel processing system, the blanket and associated processing systems, and the clean-up and waste treatment systems is an area of some concern. The data base concerning hydrogen/metal interactions is reasonably well in hand for INTOR. However, considerable work remains to be done on the development and proof-testing of a variety of tritium permeation barriers, particularly for the heat transfer interfaces and for lining the wall of the primary containment to reduce the level of tritium dissolved in them. A definitive and successful demonstration of the adequacy of INTOR tritium containment practices is absolutely necessary. With appropriate programmatic modifications such containment could be demonstrated in TSTA, TETL, or VTS.

If current tritium-shipping criteria were applied to INTOR (except in the USSR), hundreds of shipments per year would be necessary. A clear need exists to establish and implement criteria that will permit shipments of hundred-gramme quantities within and across international borders. There appears to be no serious problem in this area insofar as all participating countries are aware except that a great deal of time will be required to design and license larger shipping casks. The question of in-plant accountability for tritium (i.e., to perhaps 0.01 of a gramme in the USA) may become a problem area. This is another area of policy which must be examined in more detail.

Finally, the consolidation, handling, and storage of tritiated-wastes present no serious problems for INTOR, in that a reasonably well established technology already exists. It would, however, be of great benefit for INTOR to adopt the philosophy of minimum tritium recycle (from waste steams) for economic reasons and to minimize tritium-active waste volumes.

2.10. Materials

In general, it is concluded that the data base either exists now, or that it will exist by the mid-1980's to be able to make a reasonable judgement on the mechanical and electrical performance of materials in INTOR. Furthermore, there is reasonable optimism that the structural materials for the first wall and shield can be used for the lifetime of the device. More specific conclusions are given below:

Of the four main classes of potential structural materials considered for INTOR (austenitic steels, ferritic steels, titanium alloys and aluminium alloys), the austenitic steels are the most favoured choice. Ferritic steels have better thermal stress properties and more resistance to irradiation-induced swelling, but the post weld heat treatments and the shift in the ductile to brittle temperature makes operations below the 300–400°C range questionable. Titanium alloys present some advantage from the thermal stress and fatigue lifetime standpoint but the general lack of data on irradiated materials and formation of hydrides below 300°C makes them unattractive for use in INTOR at this time. Aluminium alloys have the potential advantage for lower thermal stresses and higher permissable wall thickness, lower long lived radioactivity, and lower tritium permeability. On the other hand, a lack of irradiation data above 100°C, decreased ductility due to post irradiation heating (baking) above 100°C, a lack of irradiation datas on materials with appropriate helium content, and doubts about commercial reactor relevance for electricity production make the Al alloys unattractive for the primary structural alloy in the shield. It may be possible to consider using Al alloys in the first wall operating at about 100°C.

Although none of the alloys considered for INTOR have been irradiated at the appropriate damage rate, or with the appropriate fusion neutron spectrum for INTOR as of 1979, by 1985 it is expected that such information can be obtained on a few properties such as swelling and post irradiation ductility. On the other hand, 316 SS has been irradiated with thermal fission reactor neutrons to approximately the correct helium to dpa ratio. It appears that irradiation of this alloy in the cold worked state at 300–400°C will produce less than 3% swelling and leave total elongation values of about 2% at the end of life. At lower temperature the swelling would be less.

Considerable information on the physical, mechanical and chemical properties of Li_2O has been obtained. This data indicates that tritium can be produced and 90% of it removed from Li_2O pellets at a temperature of about 500°C. Similar removal rates in $LiAlO_2$ require temperatures of 800°C. More in-reactor tritium extraction tests are needed.

The critical component for superconducting magnet lifetimes in INTOR appears to be the stabilizer. Electrical insulators available now indicate useful lifetimes in the 10^9 to 10^{10} rad range. One estimate of the limiting fluence for Al stabilizers is 2×10^{-5} dpa while that for Cu is about 10^{-4} dpa. In complete recovery of the resistivity increase after warming to room temperature (or even 100°C) because of interaction with dislocation cells may present severe problems with respect to the frequency of annealing.

There will be limited data on the electrical and mechanical properties of ceramics ($MgAlO_4$, Al_2O_3, Si_3N_4 and SiC) exposed to the INTOR front wall environmental by 1985. Fast reactor data will suffice for ceramics under displacement damage conditions of interest in INTOR. However, gas production levels will be 1 to 3 orders of magnitude lower than at the first wall of INTOR.

Below 1250°C there are adequate data on fission reactor graphites for the design of INTOR. Present data indicate that operation below 550°C or above 1500°C would be required to achieve a full INTOR lifetime. Unfortunately, the grades studied are not the most thermal shock resistant, and more data on such materials is required.

There are presently sufficient data on the sputtering of candidate wall materials with hydrogen isotopes and helium to predict wall erosion rates in INTOR (within a factor of 2) once the particle spectra and angle of incidence is specified. Wall erosion by this mechanism does not appear to be serious in INTOR if the energetic particle flux is reduced to below a few times 10^{15} cm^{-2} s^{-1} and if the edge temperature of the plasma is kept to ~100 eV. However, if the particle flux is greater than 10^{16} cm^{-2} s^{-1} and the energy is greater than 100 eV serious problems could arise. At the present time, the fluxes and energies of the particles to the first wall are too uncertain at this time to make a more quantitative statement about erosion rates.

A great deal of progress has been made in the area of low-Z coatings for first walls, limiters and particle collection plates. However, as of now there is no clearly superior candidate that has been demonstrated to be compatible with both the plasma and cyclic operation of the structural components. TiB_2 and TiC coatings on graphite, copper or refractory metals hold great promise, but more tests are necessary.

2.11. Systems integration and support structure

1. The system integration goals for the other groups were defined. Requirements concerning evaluation of specific features of main components, design philosophy, goals and procedures were addressed to other groups.

2. Machine topology in connection with maintenance requirements, reliability and experimental component flexibility was considered. The main conclusions are:

(a) TF-coils should be designed and manufactured with such a degree of reliability which should avoid failures through the whole lifetime of the machine.

(b) OH and EC coils should be situated outside of the TF coils. The advantage of access and maintenance in this case essentially outweigh any disadvantages connected with the high overturning loads of TF coils and increased PF power supply requirements.

(c) Inboard reactor components (vacuum chamber and blanket) should be segmented. The dimensions of experimental modules should be sufficient for initial tritium and electricity production tests and for material studies. For express material experiments, small test channels are preferred. Test segments may be required for the third stage of operation to provide adequate reliability.

(d) Manufacturing technology requirements strongly affect the structure design. The major part of the work should be done in fabrication shops and only assembly should be performed on the site. Such an approach is the best method to encourage the use of industrial capacities and technology.

3. The experts discussed the recommendations concerning the location of the cryostat boundary and decided that the vacuum volumes of cryostat and plasma chamber region should be separated to provide the access to the blanket-shield components without warming up the TF magnets.

4. A preliminary assessment of the interactions between various subsystems was carried out. The various interactions were weighted according to the designers' judgement and experience. The most important critical systems in terms of their interactions with other systems were remote maintenance, the TF coils, the vacuum system and the support structures.

5. The experts agreed to use common design codes and regulations.

6. The participants reviewed different variants of proposed INTOR topologies and concluded that they contain significant amounts of identical engineering approaches:

- the number of TF coils is 12;
- OH coils are outside TF coils;
- EF coils are outside TF coils or a hybrid version is possible with the major part of the EF coils outside the TF coils and only small EF coils inside;
- cryostat vacuum volume is separated from the discharge chamber volume;
- discharge chamber and shield are of modular type; modular mounting joints are outside the shield;
- there are identical approaches to the structural requirements from the viewpoint of machine remote maintenance.

However, there are some unsolved problems connected with INTOR topology. Such problems should be solved before the implementation of the detailed analysis of the system integration. Among them are:

- choice of divertor type;
- choice of the types, dimensions and allocation of test modules;
- choice of magnetic coil cryostat (general for all the coils, or separate for each coil);
- engineering decision for EF coil situated inside the TF coils.

All affect the basic nature of the systems interactions and, thus, machine topology.

7. The group members agreed that the design development for INTOR could be made at this time and that no special R and D tasks are necessary in the field of INTOR subsystems interaction.

Investigations should be performed in accordance with technical requirements developed by other group specialists. Important layout decisions should be made as soon as possible to proceed with the design.

2.12. Assembly and remote maintenance

The on-line availability of the INTOR reactor system is based on minimized down-time. Down-time is a function of the ease with which the system can be maintained and returned to service. The complexity of fusion devices poses a significant challenge to reactor designers as well as remote-handling and maintenance disciplines. It is essential that remote-handling considerations be incorporated into the reactor design from the start of the concept design stage. Maintainability is completely dependent upon system design. Incorporation of remote features in the design phase is essential for producing a reactor system capable of performing at a high level of availability.

The methods and technology to maintain a device such as INTOR exist today, provided that the basic design of the machine is based on disassembly/assembly by remote means. The experience being developed during the design and construction of TFTR and JET will be useful during the conceptual design of INTOR. Nevertheless, these techniques and design methods will have to be greatly extended to meet INTOR's objectives.

The requirement for 50% availability presents a significant problem for INTOR. Achievement of 50% availability impacts not only remote handling, but it seriously impacts all systems. Development of machine subsystems that are sufficiently reliable to permit operation under the proposed scenario must be accomplished. Additionally, considerable attention must be given to the remote systems to assure that they are sufficient in number and capable to perform the required procedures in a realistic time frame. Preliminary failure frequency and availability assessments by the USA and USSR indicate that availabilities in the range of 25% to 50% should be achievable.

In summary, the key to 50% availability on INTOR is a design compatible with remote handling methods and development of subsystems sufficiently reliable to perform within the operating scenario. In general, it is believed that a proper data base can be established in the next few years to meet these objectives.

2.13. Shielding and personnel access

The following general agreements were made.

(A) Importance and feasibility of personnel access

(i) Personnel access for maintenance will be necessary for INTOR since not all possible remote maintenance operations can be foreseen. A preliminary estimate presented by the USA indicated that personnel access for some maintenance operations will increase the overall availability of INTOR, and more than compensates for the required additional cost of penetration shielding.

(ii) Personnel access may be feasible for most of the reactor components outside the primary shield, but personnel access for those components at the end of penetrations will be limited.

(B) Assessment of primary shield

(i) The inboard shield should be designed to protect the superconducting toroidal magnetic system from radiation damage. The shield thickness is determined by radiation damage of the stabilizer of SC magnet. Frequent annealing of the stabilizer is difficult and there is the trade-off between frequency of the annealing and additional shield thickness.

For the shield design without stabilizer annealing, a value of the neutron fluence (0.1 MeV) of 1×10^{17} n/cm^2 (5×10^{-5} dpa) for copper is tentatively recommended. However, because this limiting value is affected by a specific design, further investigations are necessary.

(ii) Composition and thickness of inboard shield

To minimize the inboard shield thickness is indispensable for realizing a reasonably-sized reactor. A number of studies have been carried out by the INTOR study teams. In general, these studies indicate the solid shield thickness of 80 ~90 cm and 1.1 ~1.2 m total distance from the first wall to toroidal coil conductor to allow gaps for structural arrangement and machine access for repair.

The shield must be manufacturable as well as have sufficient radiation shielding capability. At present, the most likely shield materials are stainless steel, water, borated water, boron carbide, concrete and lead. It is not clear whether the use of heavy metals (e.g. tungsten) provides sufficient advantage for the inboard shield to outweigh their additional cost.

(iii) Composition and thickness of outboard shield

The outboard shield should be designed to reduce radiation levels to sufficiently low levels after a reasonable time after shutdown to permit maintenance operations. It should be possible to obtain ~2.5 mrem/h within a day or so after shutdown which would provide free access. The required outer bulk shield thickness is estimated to be about 1.5 m, based on calculations without penetrations.

iv) Radiation streaming and additional shield

The presence of many major penetrations (e.g. neutral beams, vacuum ducts, divertor, diagnostic holes, etc.) dramatically reduces the effectiveness of the bulk shield as they cause substantial radiation streaming. Radiation streaming is highly dependent on the size of penetration. In general, additional shield of 30 to 70 cm will be required around all the penetrations to provide permissible dose rates in the main part of the reactor building. However, the dose rate in a straight line at the end of but inside the NBI system one day after shutdown is estimated to be 10^3 to 10^4 mrem/h.

(C) Radiation limits and regulation

At the present state of the INTOR device, it seems difficult to give precise radiation dose-rate limits taking into account the possibility of hands-on maintenance of the machine.

A reasonable guiding criterion for the design could be the choice of the man rem/y exposure. Further study is required to establish a reasonable value.

(D) Other implications of personnel access

It is possible that contamination of the reactor hall could negatively affect personnel access. Further investigations are necessary.

(E) Assessment of biological shield

The biological shield of INTOR must provide a normal environment outside the reactor building. In the main part, the biological shield coincides with the walls, ceiling and floor of the reactor building. It may be made of ordinary concrete with a shield thickness of about 2 metres. No special heat removal from the biological shield is necessary. The design of the shield is conventional for fission reactors and at present no special difficulties are expected.

(F) Nuclear data and codes for shielding design

A modest R and D programme for the nuclear shield design is necessary in the areas of calculational methods, nuclear data, and integral experiments. There is a need for improving Monte Carlo codes to facilitate treatment of specialized complex tokamak geometries. There is also a need for improvement in the nuclear data base which includes specific differential data measurements and evaluation. For induced activation data, theoretical model calculations are needed. Integral and mockup experiments to validate methods and data represent a crucial part of the nuclear design effort for INTOR.

2.14. Vacuum

With the exception of one area, the data base for vacuum systems in INTOR is now available, or it could exist in 1982 to design and construct INTOR. The one area that could present a problem is that of removing helium from the divertor and/or reaction chamber during the burn. The solution to this problem must necessarily wait for the final impurity control method. It is felt that once such a method is determined, an intensive programme (with substantial financial and manpower support) can be developed to design and test such high-speed helium pumps by 1982. Without such an effort, the removal of helium ash during the burn could represent a major obstacle for INTOR.

Proposed particle pumping ranges can be reached by present mechanical equipment only if the total pressure at the pumped chamber (torus or divertor) is greater than 10^{-4} torr. Helium cannot be removed unless the helium pressures are greater than 10^{-5} torr. Unless higher pressures can be tolerated, an excessive number of pumps will be required which occupy an area equivalent to significant fractions of the outboard torus area. Novel pumping techniques such as ion implantation have been proposed, but **these are untested ideas.**

The vacuum system for INTOR should be capable of removing D-T, helium, and other impurities from the scrape-off region surrounding the plasma or from a divertor chamber (poloidal or bundle) at a rate equivalent to the particle loss rate and maintain acceptable pressures in the torus and divertor chambers. Pumps must also be capable of pumping down the torus to base pressures in a short time. All pumped gases will have to be recovered by the fuel recycle system with acceptably low release of tritium to the environment and with the introduction of no impurities to the fuel stream, which would complicate purification and fuel recycle. The pumping system must maintain acceptably low inventories of tritium and regeneration of cyclic pumps must be rapid enough to not seriously reduce the duty cycle or availability of the INTOR experiment.

Much design work, taking into account the requirements of all the INTOR groups, will have to be done before a choice of vacuum boundary location can be made. However, the choice of vacuum boundary location is essential before detailed design can proceed and, therefore, a high priority on this topic is warranted.

It is concluded that positioning the welded joints in the vacuum boundary outside the shield and inside the toroidal magnetic field is preferred.

There is no problem in pumping the torus before and after the working pulse, but difficult problems do arise in the pumping of gases from the NBI systems.

All vacuum seals, measuring instruments, and pumps will have to be capable of remote maintenance (or remote replacement). Seals and valves must be bakeable all-metal units. Surfaces will have to be baked in vacuum for several hours at temperatures of at least 250°C (SS first wall), and/or cleaned by a low-temperature discharge in hydrogen. The possibility of using aluminium alloys inside the vacuum boundary will have to be studied, particularly from the standpoint of baking and cleaning.

No organic elastomers should be used in the vacuum boundaries because of their sensitivity to bakeout temperature, cleaning, radiation and tritium.

Heat generation rates on vacuum pumps in NBI are likely to be high enough that refrigeration costs will be unacceptable if conventional cryocondensation pumps operating at temperatures up to 30 K and getter pumps (Zr–Al) are used. The latter pumps may, however, require relatively long regeneration times which could affect the INTOR duty cycle.

2.15. Diagnostics, data acquisition and control

Although an analysis of the diagnostics, data acquisition and control seems to be necessarily speculative at this early stage of the INTOR development, it is necessary to carry out a preliminary assessment, because this field will take overall system responsibility during the operation phase of INTOR. In addition, new features of INTOR such as reactor-relevant physics and engineering should be considered. Since there is no essential difference in the operation scheme between INTOR and future tokamak reactors, this field will be truly reactor-relevant.

The impact of the general features of INTOR are discussed and the control and data processing system are qualitatively reviewed. Furthermore, new features provided by long D-T burning, accessibility to the tokamak torus, and the reactor environment of diagnostic instruments are briefly discussed. In addition, safety system, reliability, and necessary redundancy are discussed. Probable issues from INTOR's international features are briefly reviewed.

No essential differences were found in Workshop participants' observations in this field of INTOR. However, this area is particularly dependent upon information and input from all other groups, which are required to make further assessments more quantitative and definite.

The number of diagnostics should be minimized during the latter stages of INTOR to be consistent with proto-typical reactor operation. The control of reactor engineering components, such as the blanket, will be similar to industrial plant and fission reactor systems.

Time sequential procedures of plasma control and other separate systems were briefly analyzed. The diagnostics applicable at each operation step is qualitatively reviewed.

No essential technological problem exists regarding the control and data base hardware because of its current flexibility and the expected rapid advances in this field, although there might be some additional technical requirements such as faster data transition time. More development of software and algorithms for feed-back analysis is required for INTOR.

The philosophy of the total control system should be a compromise between hierarchial centralization and redundant de-centralization to prevent trivial faults from reducing the total system availability.

Investigation of effects of irradiation and the tokamak environments on instrumentation including detectors, sensors, cables, optical system, etc., are strongly recommended. Information from fission reactors, accelerators, and the open defense literature is very helpful for irradiation effects on instrumentation.

From past experiences, one can expect conflicts between diagnostic requirements and the limitations imposed by the tokamak structure. Joint efforts by physicists, control engineers, and machine builders are required from an early stage in order to provide adequate access for the instrumentation.

Internationally agreed upon criteria for quality control and assurance for this field of INTOR are strongly recommended to keep its availability high. Furthermore, this field requires international interchangeability, both for hardware and software. It is recommended that INTOR shall impose an international standard for specifications for this field after review by an international group.

Generally speaking, the field of diagnostics, data acquisition and control is feasible for the initiation of INTOR construction in about 1984–85, although there is some remaining ambiguity about irradiation and tokamak environment effects on instrumentation which must be resolved by further research and development efforts.

2.16. Safety and environment

Regulations pertaining to the control of

(1) radioactivity,
(2) toxic materials, and
(3) industrial procedures associated with:
 a. high vacuum technology
 b. large cryogenic systems
 c. high voltage systems

currently exist in the member nations and they are felt to be adequate for starting the design of INTOR. However, there are few, if any, regulations that are specific to fusion devices, and a concentrated effort to establish these guidelines for INTOR must start as soon as possible. It is likely to take 5–10 years to formulate such regulations in compliance with public review and legislative requirements.

The most critical example of where specific attention must be paid to fusion devices is in the area of tritium containment. The analytical models relating release of tritium to exposure to humans currently exist. In addition, the effects of tritium on the population for a given exposure are also fairly well understood at this time. What is required is for a well-established, internationally respected group to put the two together to establish release limits specifically for D-T fusion devices. These guidelines are absolutely necessary before INTOR progresses into the detailed design phase.

There are also two regulations currently in force in some countries which could place undue constraints on INTOR. The first is the 20 g limit on the amount of tritium in approved shipping containers. To supply initial inventories of several kilogrammes could require an unduly large number of shipments. There is no insurmountable problem here, only the time delays associated with the design of larger containers and regulatory licensing of such units. The second problem is with the current accountability practices. It is currently required (at least in the US) that the entire tritium inventory of a facility be established to the nearest 0.01 g twice a year. Considering that inventories of about 3 kg could exist in INTOR and that 5–10 kg/year might be burned up, such a strict accounting may be impossible. The regulations should be examined to see if such a limit can be raised for INTOR or if such a limit is even appropriate for such a test device.

At the present time, the only material in INTOR which might be subjected to safeguards requirements is tritium. It is currently designated as a sensitive, rather than a strategic, material because of its high value (about 4000 to 8000 \$/g). It is anticipated that this status will not change for INTOR but periodic monitoring of IAEA safeguards regulations should be conducted.

Specific regulations for the continued exposure of individuals to high magnetic fields are also urgently needed. At the present, the practice in the field is to use ad hoc values from the accelerator field which may or may not stand up to the close scrutiny of safety review panels.

Guidelines currently exist for the use of non-ionizing microwave radiation, but few, if any, regulations exist for RF heating sources that might be used in INTOR. Such regulations should be established in the next decade.

Considerable investigation has gone into identification of accident sequences for the tokamak fusion industry. Based on conceptual designs of commercial plants, event trees have been constructed to identify accident scenarios as an aid in improving the safety aspects of the next generation of conceptual designs and quantitative assessment of accident consequences. These event trees are not usable at this time to determine defendable values of risk (probability of an accident times the consequences) because of a lack of specific reactor design details and lack of an adequate data base for component failure rates. Such a data base must be developed; however, it will not likely be sufficiently developed in time for quantitative application to INTOR. The state of development of mechanistic computer models for fusion reactor accidents is at an early state. Further development is needed if defendable safety analyses are to be done for INTOR other than "worst case" analyses. The codes will also need to be verified by comparison with experimental data from well-characterized and instrumented safety-related experiments, which are just starting to be performed.

In summary, there are several things that can be done at an early stage to ensure the safety of INTOR. However, neglect of these areas could delay the timely operation of INTOR.

3. R AND D NEEDS ASSESSMENT

3.1. R and D Needs Assessment – Energy and particle confinement
(TABLE IV-6)

	Status*	
	Adequate	Inadequate
Energy confinement scaling		
at high T_e	x[1]	
in noncircular cross-section	x	
Particle confinement and transport		
hydrogenic ions	x[2]	
helium ions		x
impurity ions	x[2]	
Particle and energy transport in divertor tokamaks	x[2]	
MHD effects on transport		
when $\beta \gtrsim \beta$ crit	x[2]	
due to low m resistive modes	x	
Ripple effects on transport		
of thermal ions		x
of beam ions		x
of alpha-particles		x[3]
Demonstration of alpha-particle confinement and heating	x[3]	
Transport and diffusion of		
current density (including bootstrap current)	x[2]	
angular momentum (totation)	x[2]	

1 Assumes achievement of high temperatures ($\gtrsim$5 keV) in present-generation devices (e.g., ASDEX, DIII, PDX) by 1980/2, followed by a confirmation of scaling laws at reactor-like temperatures (~10 keV) in next-generation devices (JET, JFT-60, T-15, TFTR) in 1982/4.

2 Can be carried out on existing experimental facilities but the programme in this area needs increased emphasis.

3 Must rely on simulation studies of alpha-particle behaviour, using energetic beam ions, until initial D-T results are available from JET/TFTR in about 1984.

Adequate: Programmes are either planned or could be planned on existing or authorized facilities (with some upgrading if required) and using existing trained personnel.

Inadequate: Either the available and authorized manpower and facilities are inadequate or a substantial increase in effort is needed.

*The ratings are made on the world-wide fusion effort and not separately on the individual national programmes.

3.2. R and D Needs Assessment – Impurity control, fueling and exhaust
(TABLE V-2)

	Status *	
	Adequate	Inadequate
Experimental data on plasma-wall interations and edge plasma control in discharges with and without a divertor, in the presence of strong auxiliary heating	x[1]	
Experimental and theoretical studies of physical processes in:		
(a) poloidal divertors	x	
(b) local, e.g. bundle, divertors		x
Design studies of divertors suitable for INTOR		x
Experimental study of non-divertor techniques for ash exhaust and impurity removal		x
Experimental studies of pellet injection both in ohmically heated plasmas and in the presence of strong auxiliary heating	x	
Technology of D and T pellet injection at the required high repetition rate, at a minimum speed of 10^3 m/sec, with a design goal of 5×10^3 m/s		x
Surface physics studies of impurity generation and particle recycling processes	x[1]	

[1] Programmes and facilities exist in these areas but there should be more emphasis on INTOR-specific problems.

Adequate: Programmes are either planned or could be planned on existing or authorized facilities (with some upgrading if required) and using existing trained personnel.

Inadequate: Either the available and authorized manpower and facilities are inadequate or a substantial increase in effort is needed.

*The ratings are made on the world-wide fusion effort and not separately on the individual national programmes.

3.3(a) R and D Needs Assessment – Heating – neutral beam

(TABLE VI-8)

	Status*	
	Adequate	Inadequate
Ion beam. Development of high-current (~100 A) ion beam at 175 keV (preferably 200 keV if practicable)	x[1]	
Pulse length. Development of a 10 s pulse capability for the ion source and the beam dump, which must handle ~0.5 kW/cm^2 with direct recovery and ~2 kW/cm^2 without direct recovery. In addition, certain of the units should have a 100 s capability	x[1]	
Monatomic fraction. A 90% D^+ fraction in the ion beam should be a design goal, with 80% a minimum requirement	x[1]	
Direct recovery. Development of a direct recovery system for the unneutralized beam ions with at least 60% efficiency is almost certainly required		x
Efficiency. Simultaneous achievement of high neutralization efficiency, high transmission efficiency, long ducts and low gas flow to the plasma		x
Radiation hardening. Adequate lifetime of all injector components against the expected neutron bombardment and other radiation must be insured. A plug valve shield should be developed for insertion in the beam ducts at the end of the heating pulse		x
Reliability and remote maintenance		x
Negative ion beams. At present, no absolute need is foreseen for this capability, but for purposes of control of a slightly subignited INTOR plasma, it would be desirable to have a portion of the beam power (~10 MW) at 400–500 keV, and with 100 s pulse capability		
Power supplies.	x	

[1] An expansion of the present neutral-beam development programme, probably including upgrading of a present beam test stand or construction of a new test stand, is required to achieve these objectives.

Adequate: Programmes are either planned or could be planned on existing or authorized facilities (with some upgrading if required) and using existing trained personnel.

Inadequate: Either the available and authorized manpower and facilities are inadequate or a substantial increase in effort is needed.

* The ratings are made on the world-wide fusion effort and not separately on the individual national programmes.

3.3.(b) R and D Needs Assessment – Heating – RF

(TABLE VI-19)

	Status*	
	Adequate	Inadequate
MW-level experiments to prove heating effectiveness, followed by demonstrations of heating larger plasmas to reactor temperatures (ICRF. LHH)	x	
Experimental verification of absorption and heating by ECH at the 200–400 kW level, followed by MW level experiments	x	
ICRF and LHH launching structures resistant to thermal loading and neutron radiation		x
Vacuum windows resistant to thermal loading, neutron radiation and high RF power		x
Gyrotron development at 120–150 GHz, 500 kW, 5 s pulse	x	
ICRF, LHH power supplies	x	

Adequate: Programmes are either planned or could be planned on existing or authorized facilities (with some upgrading if required) and using existing trained personnel.

Inadequate: Either the available and authorized manpower and facilities are inadequate or a substantial increase in effort is needed.

*The ratings are made on the world-wide fusion effort and not separately on the individual national programmes.

3.4. R and D Needs Assessment – Stability control

(TABLE VII-1)

	Status*	
	Adequate	Inadequate
β limits and transport at high β with the aim of demonstrating $\langle\beta\rangle \approx 6\%$ at $R/a \approx 4$	x	
Occurrence, early detection and avoidance of disruptions, including low-q operation, discharges with intense heating, and shut-down	x[1,4]	
Space and time dependence of energy deposition as well as transport during disruption	x[1]	
Position and shape control in large noncircular tokamaks in reactor-relevant conditions	x[1,2]	
Feedback control of kink modes and disruptions[3]		x
Current profile control		x

1 More specific emphasis required.
2 Extrapolability to INTOR to be analysed.
3 Such feedback control systems might not be practicable in INTOR.
4 Early detection of disruptions, on a time scale on which operating a mechanical protective system is possible, is essential.

Adequate: Programmes are either planned or could be planned on existing or authorized facilities (with some upgrading if required) and using existing trained personnel.

Inadequate: Either the available and authorized manpower and facilities are inadequate or a substantial increase in effort is needed.

*The ratings are made on the world-wide fusion effort and not separately on the individual national programmes.

3.5. R and D Needs Assessment – Start-up, burn and shut-down control

(TABLE VIII-4)

	Status*	
	Adequate	Inadequate
Ionization and current rise at low loop voltage	x	
Optimum current-rise scenario (experiments in large, in particular noncircular tokamaks)	x[1,2]	
Maximum density in clean ohmic discharges	x	
Density rise and profile control during heating to ignition	x	
Compatibility of heating schemes (neutral beam, RF or mixed) with parameter variation during heating to ignition	x[1]	
Burn temperature control, including diagnostics		x
Smooth shut-down, in particular after intense heating	x[1]	

[1] More specific emphasis required.

[2] JET, JT-60, T-15 and TFTR will provide full-size information from 1982 on.

Adequate: Programmes are either planned or could be planned on existing or authorized facilities (with some upgrading if required) and using existing trained personnel.

Inadequate: Either the available and authorized manpower and facilities are inadequate or a substantial increase in effort is needed.

*The ratings are made on the world-wide fusion effort and not separately on the individual national programmes.

3.6. R and D Needs Assessment – Magnetics[1]

(TABLE IX-1)

	Status*	
	Adequate	Inadequate
Superconducting toroidal coil development and 12 tesla (Nb_3Sn) conductor development	x[2]	
Superconducting induction coil development		x
Divertor development		x
Internal poloidal field coil development, if required		x
Radiation effects on superconducting TF coils components	x	
Superconducting magnetic systems integration experience	x	
Development of large, reliable cryogenic systems	x	

[1] It is assumed that a prototype INTOR TF coil will be built as part of the project.
[2] Additional R and D for high field coil demonstration, similar to large coil task at ORNL (USA), is required by Japan.

Adequate: Programmes are either planned or could be planned on existing or authorized facilities (with some upgrading if required) and using existing trained personnel.
Inadequate: Either the available and authorized manpower and facilities are inadequate or a substantial increase in effort is needed.

*The ratings are made on the world-wide fusion effort and not separately on the individual national programmes.

3.7. R and D Needs Assessment – Power supply and transfer

(TABLE X-1)

	Status*	
	Adequate	Inadequate
Energy storage		
flywheel generators	x	
superconducting storage systems	x	
Switches		
HV switch tubes	x	
vacuum circuit breakers	x	
SCR switches	x	
superconducting switches	x	
Neutral beam injectors		
transmission lines	x	
ion source protection	x	
Radio-frequency system		
tubes	x[1]	
replaceable antennae	x	

[1] Only the ECRH tube needs development.

Adequate: Programmes are either planned or could be planned on existing or authorized facilities (with some upgrading if required) and using existing trained personnel.

Inadequate: Either the available and authorized manpower and facilities are inadequate or a substantial increase in effort is needed.

* The ratings are made on the world-wide fusion effort and not separately on the individual national programmes.

3.8. R and D Needs Assessment – First wall, blanket and shield

(TABLE XI-2)

	Status*	
	Adequate	Inadequate
Development of methods to protect the first wall against plasma disruption (e.g. liners or mechanical devices)		x
Determination of heat flux limits for limiters, divertor dumps and targets		x
Determination of heat flux limits under normal operation for first wall and liner materials, theoretical and experimental	x	
Clarification of the needs and operating conditions of electrically insulating parts in the first wall and shield region		x
Out of reactor tests of standard blanket, shield and first wall components and mockups (thermal hydraulics, pressure and thermal load response)	x	
Study of heat storage devices for smoothing out the power output	x	

Adequate: Programmes are either planned or could be planned on existing or authorized facilities (with some upgrading if required) and using existing trained personnel.

Inadequate: Either the available and authorized manpower and facilities are inadequate or a substantial increase in effort is needed.

*The ratings are made on the world-wide fusion effort and not separately on the individual national programmes.

3.9. R and D Needs Assessment – Tritium

(TABLE XII-7)

	Status*	
	Adequate	Inadequate
Demonstrate performance, reliability and maintainability of integrated fuel handling system. Demonstrate performance of auxiliary safety (clean-up and back-up) systems and tritium monitoring instrumentation (TSTA, TETL, UTS)	x	
Study dynamics of enclosure clean-up operations following tritium spills	x	
Determine concentration and chemical composition of impurities exhausted from the plasma		x
Characterize tritium interaction with materials under realistic conditions		x
Develop effective barriers to tritium permeation		x
Develop and demonstrate removal of tritium from the coolant	x	
Resolve tritium shipping, accountability, and safeguard issues		x
Develop and demonstrate extraction of tritium from breeder materials	x	
Study tritium aspects of fuel fabrication and injection		x
Study the influence of a divertor on the tritium system		x

Adequate: Programmes are either planned or could be planned on existing or authorized facilities (with some upgrading if required) and using existing trained personnel.

Inadequate: Either the available and authorized manpower and facilities are inadequate or a substantial increase in effort is needed.

* The ratings are made on the world-wide fusion effort and not separately on the individual national programmes.

3.10. R and D Needs Assessment – Materials

(TABLE XIII-17)

	Status*	
	Adequate	Inadequate
Generation of irradiation effects data (e.g. mechanical properties) on nickel containing alloys for design purposes	x	
Environmental effects on fatigue of stainless steel at elevated temperatures	x	
Determine synergistic effects of stress/strain and irradiation on key properties of stabilizer materials	x	
Generation of irradiation effects data on super-conductor and insulator materials as required for engineering design	x	
Development of fabrication techniques and data base on bulk irradiation effects for non-stainless steel structural materials and non-structural first-wall or liner materials (ceramics, graphite, insulators)		x
Determine chemical erosion characteristics of materials and coatings exposed to plasma at high temperature and high fluence conditions	x	
Develop retention/release characteristics of T and He on materials exposed to plasma (first-wall, liners, limiters, divertor collectors) under design conditions		x
Development of low-Z coatings for first-wall or liner components (fabrication, irradiation effects and thermal shock resistance)		x
Determine chemical and radiation stability of breeding materials under irradiation at elevated temperatures and chemical reactivity with structure and/or coolant	x	
Study the effect of gaseous transmutants on mechanical properties of structural materials which do not contain Ni		x

Adequate: Programmes are either planned or could be planned on existing or authorized facilities (with some upgrading if required) and using existing trained personnel.

Inadequate: Either the available and authorized manpower and facilities are inadequate or a substantial increase in effort is needed.

*The ratings are made on the world-wide fusion effort and not separately on the individual national programmes.

3.11. R and D Needs Assessment – Systems integration and support structure

No R and D needs were assigned specifically for this area, since this activity is concerned with the integration of the other areas.

3.12. R and D Needs Assessment – Assembly and remote maintenance

(TABLE XV-1)

	Status*	
	Adequate	Inadequate
General remote handling apparatus	x	
Develop reactor mockup	x	
Develop remote welding and cutting systems	x[1]	
Develop viewing and inspection systems for remote welding		x
Methods for reliability analysis and verification of components		x
PF coil joints if required for internal coils		x

[1] Further development for specialized conditions in fusion devices is required.

Adequate: Programmes are either planned or could be planned on existing or authorized facilities (with some upgrading if required) and using existing trained personnel.

Inadequate: Either the available and authorized manpower and facilities are inadequate or a substantial increase in effort is needed.

*The ratings are made on the world-wide fusion effort and not separately on the individual national programmes.

3.13. R and D Needs Assessment – Shielding and personnel access

(TABLE XVI-2)

	Status*	
	Adequate	Inadequate
Integral experiments for bulk shield	x	
Integral experiments for penetrations		x
General shield design methods	x	
Multi-dimensional transport techniques for streaming		x
Nuclear data for shielding	x[1]	

[1] Additional data required on induced activation levels.

Adequate: Programmes are either planned or could be planned on existing or authorized facilities (with some upgrading if required) and using existing trained personnel.

Inadequate: Either the available and authorized manpower and facilities are inadequate or a substantial increase in effort is needed.

*The ratings are made on the world-wide fusion effort and not separately on the individual national programmes.

3.14. R and D Needs Assessment – Vacuum

(TABLE XVII-8)

	Status*	
	Adequate	Inadequate
Development of high speed helium pumps		x
Development and testing of large dry turbomolecular pumps		x
Test of the performance of cryopanels in the presence of high neutron and gamma fluxes		x
Development of baking, degassing, and discharge cleaning methods for stainless steel surfaces inside the vacuum vessel	x	
Development of baking, degassing and discharge cleaning methods for Al alloys surfaces inside the vacuum vessel		x
Evaluation of radiation heating in cryopumps for NBI and development of alternate pumping techniques	x	
Remote leak detection		x
Testing of vacuum seals suitable for remote disassembly and maintenance	x	
Design and fabrication of large cryosorption pumps and confirmation of their performance		x
Development of vacuum components with large cross section ($\gtrsim 1\ m^2$)		x

Adequate: Programmes are either planned or could be planned on existing or authorized facilities (with some upgrading if required) and using existing trained personnel.

Inadequate: Either the available and authorized manpower and facilities are inadequate or a substantial increase in effort is needed.

*The ratings are made on the world-wide fusion effort and not separately on the individual national programmes.

3.15. R and D Needs Assessment – Diagnostics, data acquisition and control

(TABLE XVIII-9)

	Status*	
	Adequate	Inadequate
High-β control, density and α-particle plasma diagnostics	x	
Transient noise and damage measurements of detectors and transducers		x[1]
Radiation damage and thermal studies of optics system, vacuum seals and isolator materials		x
Blanket nuclear measurement technique	x	
First wall and blanket instrumentation techniques in tokamak environment		x
Feedback control computer algorithms	x	
Reliability evaluation of control systems	x	

[1] The plasma and engineering monitors must be assessed in detail to determine precisely what the detailed R and D work will be.

Adequate: Programmes are either planned or could be planned on existing or authorized facilities (with some upgrading if required) and using existing trained personnel.

Inadequate: Either the available and authorized manpower and facilities are inadequate or a substantial increase in effort is needed.

*The ratings are made on the world-wide fusion effort and not separately on the individual national programmes.

3.16. R and D Needs Assessment – Safety and environment

(TABLE XIX-11)

	Status*	
	Adequate	Inadequate
Develop reliable tritium handling, processing, containment and cleanup systems	x	
Obtain activation product generation, release, transport and shielding date	x	
Develop probabilistic and mechanistic accident analysis techniques		x
Determine tritium conversion rates and biological effects	x	
Obtain lithium and lithium compound fire consequences, prevention, extinguishment data.	x	
Obtain magnet accident experimental data		x
Develop definitive magnetic field exposure guidelines for humans		x
Assess the consequences of earthquakes on the safe operation of INTOR		x

Adequate: Programmes are either planned or could be planned on existing or authorized facilities (with some upgrading if required) and using existing trained personnel.

Inadequate: Either the available and authorized manpower and facilities are inadequate or a substantial increase in effort is needed.

*The ratings are made on the world-wide fusion effort and not separately on the individual national programmes.

4. INTOR SUGGESTED PARAMETERS

	SUGGESTED PARAMETERS[a]	NATIONAL SUGGESTED PARAMETERS			
		EC	JAPAN	USA	USSR
OPERATING MODE					
Peak D-T power (MW_{th}), P_{th}	620 (750)	620	690	720	500
Burn time (s), t_{burn}	>100	>100	200	>100	100–(200)
Dwell time (s), t_{dwell}	20	30	35	20	20
Startup and shutdown time (s), t_{ss}	15	20	15	15	15
Max. availability (%)	25–50	25	25	50	40
No. pulses/lifetime	$5 \times 10^5 – 10^6$	3×10^5	3×10^5	10^6	10^6
DIMENSIONS					
Major radius (m), R	5.2	5.5	5.5	5.0	5.0
Plasma chamber radius (m), r_w	1.4	1.7	1.5	1.4	1.3
Plasma chamber volume (m^3)	320	500	400	310	250
Plasma chamber area (m^2)	380	480	415	360	320
Reactor radius (m)	11	–	10	11	12
Reactor height (m)	15	–	16	15	16
PLASMA					
Plasma radius (m), a	1.3	1.6	1.4	1.3	1.2

[a] The parameters suggested as input for the definition phase.
For Footnotes b–r, see page 79.

4. INTOR SUGGESTED PARAMETERS (cont.)

	SUGGESTED PARAMETERS	NATIONAL SUGGESTED PARAMETERS EC	JAPAN	USA	USSR
Aspect ratio, A	4.0	3.4	3.9	3.8	4.2
Elongation, K	1.6	1.6	1.5	1.6	1.5
Av. ion temperature (keV), $\langle T_i \rangle$	10	10	10	10	10
Av. D-T ion density (10^{20} m^{-3}), $\langle n_i \rangle$	1.3 (1.5)	1.2	1.2	1.4	1.2
Burn energy confinement time (s), τ_E	1.5	1.5–2.0	1.5	1.5	1.5
Effective charge, Z_{eff} (oxygen)[b]	⩽1.5	≈1.5	1.5(~2.0)	⩽1.5	⩽1.5
B_T on axis (T)	5.5	5.0	5.2	5.3	5.5
Safety factor, q_I	2.5	2.5	2.5(~2.0)	2.5	2.5
Burn av. (%), $\langle \beta_T \rangle$[c]	5(b)	≈6	6	6	5
Plasma current (MA), I_P	6.4	≈8	6.0(~7.5)	~6.5	~5
Edge toroidal field ripple (%)[d]	<±0.75	<±0.75	±0.4	<±0.75	<±0.75
Centre toroidal field ripple (%)[d]	<±0.1	<±0.1	±0.04	<±0.1	<±0.1
Fast alpha loss (%)	<3	≈1	<5	<5	<5

HEATING[e]							
Mode	NBI (RF backup)	NBI	NBI (neg. ions)	RF (preferred)	NBI, RF backup	NBI, RF backup	NBI, RF backup
Power/area (MW/m^2)	20–30	25	25	25–100	30	25	20
Power (MW), P_B	75 (startup); 10[f] (continuous)	75(+50)	50(+30) (depending on impurities)	50(+30)	75	75 (startup) 10 (continuous)	75 (pulsed) 10 (continuous)
Pulse length needed for ignition (s)	6	5–10	5–10	5–10	5–7	6	5
Pulse length capability (s)	10	–	–	–	–	–	–
Neutral beam energy (keV), E_B	175[g]	200	400	LH or ICRH	200	175	200
IMPURITY AND PARTICLE CONTROL							
Particle flux at the edge of the plasma[h] (s^{-1})	~10^{23}			–	10^{23}	10^{23}	10^{23}
Fueling mode	gas puffing, pellets			gas puffing, pellets	gas puffing, pellets	gas puffing, pellets	gas puffing, pellets
Impurity control mode	improved divertor[i]			cold blanket, divertor	poloidal divertor, wall lapping plasma	improved divertor	divertor
TOROIDAL FIELD COILS							
Number	12			12(12–16)	12	12	12
Conductor	Nb_3Sn and/or NbTi			Nb_3Sn or NbTi	Nb_3Sn	NbTi and/or Nb_3Sn	Nb_3Sn and/or NbTi

4. INTOR SUGGESTED PARAMETERS (cont.)

	SUGGESTED PARAMETERS	NATIONAL SUGGESTED PARAMETERS EC	JAPAN	USA	USSR
Max. field (T), B_{TFC}	10.6	≈10	~11	10.6	10.6
Bore height/width (m)		11/7.5	10/7	10/8	10/8
Max. nuclear heating (kW)	5	≈2	–	5	<5
Max. tolerable radiation (rads) – epoxy	10^{10}	5×10^9	1×10^9	10^{10}	10^9
Stabilizer	Cu	Cu	Cu	Cu	Cu
Max. tolerable radiation to stabilizer before anneal (n/cm^2)	10^{17}	2×10^{17}	2×10^{17}	2×10^{17}	5×10^{17}
POLOIDAL FIELD COILS[j]					
Volt-seconds	110	130	115	90	90
OH current ramp time (s)	5	5 (faster, if possible)	~5	5	5
OHC conductor	NbTi	NbTi	NbTi	NbTi	NbTi
OHC location[k]	outside	outside	outside	outside	outside
OHC max. field at coil (T)	8	6	6	6	5
OHC max. field rise (T/s)	10	24	~6 ($t < 0.1$ s) ~1 ($0.1 < t \leqslant 5$ s)	8	3
EFC conductor	NbTi	NbTi	SC/Cu	NbTi	NbTi (slow) resistive (fast)
EFC location[k]	outside	outside	outside	outside	outside/inside

FIRST WALL					
Material	SS (Al alloy)[l]	0	316 SS	SS	SS or Al alloy
Coolant	H_2O	0	H_2O	H_2O (pressurized)	H_2O, He
Max. structure temperature (°C)	<300(~150 for Al alloy)	0	~150	300	200–250 SS 150–200 Al
Av. neutron wall load (MW/m^2), P_n	1.3 (1.6)	≈1	1.3	1.6	1.3
Av. surface heat load (MW/m^2)[m]	¼ neutron wall load	0.25–0.3	0.3	0.4	0.35
No. normal pulses/lifetime	$5 \times 10^5 – 10^6$	3×10^5	3×10^5	10^6	10^6
No. disruptions/lifetime	10^3	some	–	10^3	10^3
Protective mechanisms[n]	liner or limiter		low-Z protective wall	low-Z (coated) protective wall	protection limiter
TEST FACILITIES					
Type	channels, modules, segment	segments	modules	channels and modules	channels, modules or segments
Location	outboard	outboard	outboard	outboard	outboard
Purpose	electricity generation, tritium breeding, materials test	(n)	tritium breeding and materials	electricity generation, tritium breeding, materials	blanket component and integral tests[q]

4. INTOR SUGGESTED PARAMETERS (cont.)

	SUGGESTED PARAMETERS	NATIONAL SUGGESTED PARAMETERS EC	JAPAN	USA	USSR
Material – structure	SS	SS	SS	SS	SS
Maximum structural temperature (°C)	100–200	500–700	400	600	600
Materials to be tested		Li (enriched/ natural), Li compounds, Be/Pb multiplier	Li_2O, advanced alloys	Li compounds, Li metal, LiPb alloys, advanced alloys	$LiAlO_2$, $LiSiO_4$, Li_2SiO_3, neutron multipliers, advanced alloys
SHIELD					
Inboard – material	SS, H_2O, B, (Pb)	SS, borated H_2O, Al_2O_3, Pb	SS, borated H_2O, heavy concrete	SS, H_2O, B, (Pb)	SS, H_2O, Pb, B
Inboard – thickness (m)[p]	1.1	1.1	1.1–1.2	1.1	1.1
Outboard – material	SS, H_2O, B, (Pb)	SS, borated H_2O	SS, borated H_2O, heavy concrete	SS, H_2O, B, (Pb)	SS, H_2O, Pb, B
Outboard – thickness (m)	1.4	1.3–1.4	1.4	1.4	1.4
Coolant	H_2O	H_2O	H_2O	H_2O	H_2O
Max. structure temperature (°C)	100	⩽80	~100	100–200	100
TRITIUM AND VACUUM					
Tritium flow rate (g/hr)	55 (68)	60	70–360	37	50
Inventory (kg)	2.5 (3)	2	3–7 depending on burnup	3	5
Consumption (kg/yr)[r]	6 (7.5)	6–7	7.4	7.5	6

Tritium cleanup time (days)	<3		<1	1–2	3	2–3
Plasma chamber exhaust comp.			10% He, unburned D, T, impurities	40% D, 40% T, 20% He, light impurities	45% D, 45% T, 10% He, light impurities	45% D, 45% T, 10% He, light impurities
Pre-shot base pressure (torr)	10^{-6}		10^{-7}	$\leqslant 10^{-7}$	10^{-5}	10^{-6}
Initial base pressure (torr)	10^{-7}–10^{-8}		10^{-8}	10^{-8}	10^{-7}–10^{-8}	10^{-7}–10^{-8}
Fractional burnup (%)	5		5	0.3–1.5	5–10	5
External tritium fueling rate (s^{-1})	4.4×10^{21} (5.4×10^{21})		4.4×10^{21}	5×10^{22} maximum	3×10^{21}	3.6×10^{21}

[b] Value that can be tolerated.
[c] Allows 1.5% for impurity and alpha contributions.
[d] Peak/average maximum.
[e] For heating to ignition.
[f] May be needed to control burn pulse. Not an additional requirement if beams used for heating can be refired during burn.
[g] The beam development programme should try to develop 200 keV beams.
[h] Indefinite to within a factor of 10.
[i] Near-term design should concentrate on poloidal divertor with external coils.
[j] OHC = ohmic heating coils, EFC = equil. field coils.
[k] Relative to toroidal field coils.
[l] Stainless steel is preferred, but aluminium alloys are not ruled out.
[m] Without divertor.
[n] The protective mechanism should last at least 5 years at 50% availability of INTOR.
[o] Depending on design and tritium permeation.
[p] Blanket plus bulk shield thickness, including inboard gaps, dewars, insulation and 85 cm of actual shielding.
[q] To be determined in the future.
[r] At 25% availability in the final stage of operation.

5. ELECTRICITY PRODUCTION

Electricity will be produced in INTOR for the purposes of demonstrating the associated technology and demonstrating the production of electricity by fusion. The alternatives in this regard have to do with the amount of electricity to be produced.

An INTOR corresponding to the suggested parameters would produce ~460 MW_{th} of thermonuclear power averaged over a burn cycle with a 75% duty cycle. It is desirable to operate the first wall relatively cool to maximize its lifetime, and it would be technically difficult to convert the energy deposited on the divertor plate⁺ to usable heat, so that only the energy deposited in the blanket will be considered in this Section, except as otherwise noted. The energy deposited in the blanket is increased typically by ~10% in H_2O/SS blankets and by ~20% in tritium breeding blankets by exoergic neutron reactions. This energy would be removed from the blanket and delivered to a steam turbine system via a heat exchanger and energy storage system that smoothed out the pulsed energy source.

With a neutron wall load of 1.3 MW/m^2 and a thermal conversion efficiency of 25%, which is consistent with the medium coolant temperatures and pressures envisioned for early stages of INTOR, the electricity production capability from the blanket alone is about 0.25 MW_e/m^2 of blanket surface. If the heat deposited in the first wall is included, this number may reach about 0.30 MW_e/m^2. For more advanced blankets operating at higher temperatures and pressures, thermal conversion efficiencies of about 30% are reasonable. To obtain these higher efficiencies, coolant outlet temperatures of ~400°C are required, which in turn implies maximum structural temperatures of 450–500°C. For these more advanced blankets, about 0.35 and 0.40 MW_e/m^2 could be obtained, depending upon whether the heat from the first wall is included. The poloidal distribution of neutrons is peaked towards the outboard side of the plasma chamber. In INTOR, the neutron energy deposition averaged over the outboard half and inboard half is 30% greater and 30% less, respectively, than the average neutron energy deposition over the entire blanket.

The maximum electrical power production capability of INTOR would be about 150 MW_e if the total surface area of the plasma chamber could be surrounded with a high-temperature blanket and the heat deposited in the first wall is used.* (If the P_n = 1.6 MW/m^2 option was chosen, the maximum electrical power production capability would be about 180 MW_e.) The production of about 150 MW_e in INTOR is predicated upon two crucial assumptions. First,

---

⁺ Assumed to be about 10% for the purposes of this section.

* The actual capability is less than 150 MW_e because of neutron losses through penetrations, etc.

it is assumed that an advanced, high-temperature blanket concept capable of producing usable heat at about 400°C coolant outlet temperature can be satisfactorily tested at an early time in Stage II and that the basic INTOR configuration is compatible with a full scale blanket based upon that concept. Second, it is assumed that it will be possible remotely to replace the entire INTOR blanket in a reasonable length of time. Although INTOR will be designed so that such a replacement is possible, in principle, INTOR will in fact be the machine on which the design concepts for remote assembly and disassembly are put to their first test. Even if these two assumptions are correct, it may be more effective from both cost and programmatic considerations to use the knowledge gained from INTOR to proceed immediately to build a DEMO. In fact, total replacement of the blanket is almost equivalent to building a new machine.

Three alternatives for the production of electrical power were discussed by the Workshop.

5.1. Production of 10 MW_e

Production of 10 MW_e in a medium temperature H_2O/SS blanket could be accomplished in a number of ways, without having to utilize the heat deposited in the first wall. Test modules or partial segments located between the toroidal field coils on the outboard part of the torus would have to cover about 30 m^2 of the plasma chamber surface area. Two partial segments, each with an area of about 15 m^2, or ten test modules, each with an area of about 3 m^2, would be needed. Three partial segments, each covering about 17 m^2, would be needed on the inboard side. Finally, of course, one complete (inboard and outboard) segment, located between adjacent toroidal field coils and covering a surface area of about 40 m^2, could be used to produce 10 MW_e. (Each of these areas would be reduced by the factor 1.3/1.6 ≅ .8 if the $P_n = 1.6\ MW/m^2$ option was used.)

There are certain advantages associated with each of these options. The test module option has the greatest flexibility in accommodating different types of blanket design concepts, in allowing for relatively fast removal and/or replacement of a module in the event of malfunction, and in allowing for the subsequent replacement of the medium-temperature, electricity-producing modules with other blanket test modules, including high temperature, tritium and electricity-producing modules of an advanced design. Partial outboard segments are somewhat less flexible than test modules because of the longer down time required for their removal and replacement and because of the smaller number of blanket design concepts that can be studied, but they do have the advantage of providing a full scale test of integrated blanket engineering systems. Both the test modules and the partial outboard segment options have the advantage of allowing the relatively inaccessible inboard blanket and shield to be conservatively constructed to minimize the risk of malfunction and the long down time that would be required to replace or repair an inboard portion of the blanket. The use of complete segments is less flexible even than the use of outboard partial segments, but has the advantage of providing a more comprehensive full scale test of integrated engineering systems. Both the inboard partial and complete segment options share the disadvantage of requiring that the inboard blanket be constructed less conservatively, thus increasing the risk of long periods of machine unavailability in the event of a malfunction in that region.

Because of its greater compatability with a flexible blanket technology testing programme, the test module option is preferred for the production of 10 MW_e. It should also be noted that established steam turbine technology can be used to produce 10 MW_e from medium-temperature blanket modules with outlet coolant temperatures of about 300°C or lower.

5.2. Production of 100 MW_e

The production of 100 MW_e in a medium temperature blanket would require that about 90% of the surface area of the plasma chamber be covered with blanket and that the heat deposited in the first wall was also used. With a high temperature blanket, about 75% or 65% of the surface would be covered, depending upon whether the energy deposited in the first wall was used. Thus, it is clear that the production of this level of power requires that a large part, if not all, of the available surface area of the plasma chamber be used for an electricity-producing blanket. (Each of these areas would be reduced by the factor $1.3/1.6 \cong .8$ if the $P_n = 1.6$ MW/m^2 option was used.)

To equip INTOR initially with a medium temperature blanket capable of producing 100 MW_e would increase the risk, relative to the recommended use of a cold blanket, of machine unavailability for engineering testing because of blanket malfunction, without a commensurate gain in information. In addition, established steam turbine technology for 100 MW_e units is only compatible with higher blanket coolant temperatures, so that some added cost would be associated with ordering a one-of-a-kind unit.

Thus, production of 100 MW_e is only considered for INTOR during Stage III, with the original cool (100–200°C structural temperatures) shield being replaced between Stages II and III by a high temperature blanket. This would adversely impact the continued use of INTOR during Stage III for engineering testing, but would provide a full scale demonstration of a reactor-relevant blanket system.

5.3. Production of net electrical power

If the total surface area of INTOR is equipped with a high temperature blanket, during Stage III, and if the energy deposited in the first wall is converted to usable heat, then about 150 MW_e could be produced. The lower estimates of the mean power requirements for INTOR are near 200 MW. This comparison suggests that with a modest, but nevertheless costly, increase in size and/or neutron wall load and some work on reducing the power demands INTOR could produce net electrical power. (About 180 MW_e could be produced at a neutron wall load of $P_n = 1.6\ MW/m^2$.)

An attempt to further reduce the power consumption in INTOR would require additional technology development beyond what is needed to meet the primary objectives of INTOR. While the technology developed for INTOR will be reactor-relevant, it may not be necessary to carry the development so far for INTOR as for later reactors. In addition, in several areas (e.g. power supplies) INTOR can use existing technologies.

The major power sinks in INTOR are neutral beam injectors, the superconducting poloidal field coils, the cryogenic systems, and in Option III-B, the He cooling pump system. Replacement of the positive ion source neutral beam injection system with a negative ion source system or possibly with an RF system could reduce the power demand for plasma heating. Any such replacement possibility would have to be incorporated into the initial INTOR configuration. At the present time, there is neither a reasonable assurance that a negative ion source can be developed nor that any RF method will be demonstrated and the associated technology developed on the INTOR time scale. The power consumption in the poloidal field coil system is large because the coils are located outside the toroidal field coils, but to move them inside would introduce severe engineering design problems. There are many other similar considerations which relate to the difficulty of reducing the mean power demand.

Replacing the original blanket, or a large fraction of it, with a high-temperature blanket would preclude or diminish the use of INTOR for further engineering and materials testing during Stage III.

Finally, the benefit and significance of net power production must be contrasted with the additional cost and risks and with the additional demands upon the supporting development programmes that are entailed. The amount of technical information that would result from producing 200 MW_e is certainly not much different than would be obtained by producing 100 MW_e and probably not significantly more than would be obtained by producing 10 MW_e under similar blanket conditions in test modules.

Chapter IV

ENERGY AND PARTICLE CONFINEMENT

P.H. RUTHERFORD – US, Chairman
F. ENGELMANN – EC
B.B. KADOMTSEV – USSR
H. MOMOTA – Japan, Expert
V.I. PISTUNOVICH – USSR
T. TAZIMA – Japan

1. EXPERIMENTAL STATUS OF CONFINEMENT IN TOKAMAKS

1.1. Confinement in ohmically heated tokamaks

1.1.1. Energy confinement

Examination of the results of energy confinement studies in a large number of ohmically-heated tokamaks reveals that the simple confinement law

$$\tau_E = 5 \times 10^{-21}\,\bar{n}a^2 \qquad \text{s} \tag{1}$$

(where a is the minor radius in metres and $\bar{n}$ the mean density in particles per cubic metre) summarizes the data remarkably well, although results from individual tokamaks may differ from this law by as much as a factor two. Remarkably, this empirical law seems to apply in spite of the fact that the physical regimes encountered in the different devices, e.g. with respect to the importance of radiation losses and the strength of MHD activity, varies considerably. There are, however, indications that the numerical coefficient in Eq.(1) may depend on q_a. Two effects are believed to play a role in determining this dependence on q_a. As q_a is decreased, the discharge MHD stability is dominated by the "sawtooth" oscillations within the region $q < 1$, and energy transport increases markedly in this region. However, the effect can be counterbalanced by improved stability to m = 2 and m = 3 modes, resulting in an overall improvement of τ_E. Thus, although an empirical law $\tau_E = 2 \times 10^{-21}\, q_a^{1/2}\bar{n}a^2$ was once proposed on the basis of Alcator data [2], more recent results from the same device [3] have shown energy confinement times equal to, or greater than, those predicted by Eq.(1) for $q_a \cong 2$. Recent results from T-10 at $q_a \cong 1.6$ also show no decrease in confinement from that predicted by Eq.(1) [4].

A number of ohmically heated tokamaks (DITE, DIVA, ISX-A, JFT-2, Pulsator, TFR) have reported energy confinement times about a factor two larger than would be predicted by Eq.(1). Often, such favourable results have been obtained by careful attention to control of impurities, for example by extensive gettering of the vessel wall accompanied by controlled puffing of hydrogenic gas into the discharge, as in ISX-A [3] and JFT-2 [2], or by reducing plasma-wall interactions by means of a magnetic limiter, as in DIVA [2].

Some alternative empirical scaling laws have been proposed, in particular by the Frascati Group [1], based on results from the FT device, and by Mercier [1], based on results from TFR. When extrapolated to INTOR, these scaling laws predict an energy confinement time somewhat better than that predicted by Eq.(1). Other scaling laws have been obtained by regression analysis of extensive data sets compiled by Hugill and Sheffield [1] and by Pfeiffer and Waltz [1]. These, too, give energy confinement times for INTOR about equal to, or somewhat better than, those predicted by Eq.(1). The same conclusion is reached [1] when the range of empirical models is narrowed to those that obey certain plasma-physical constraints expressed in terms of dimensionless parameters.

Although some of these scaling laws may contain a T_e-dpendence, it is important to remember that the **true** T_e-dpendence of the energy confinement can never be determined from ohmically heated tokamaks alone, due to the intrinsic relation between the ohmic heating power and T_e. Indeed, it is easy to see that any confinement law of the form $\tau_E = na^2 f(T_e^{5/2}/B_\theta^2 Z_{eff})$, where f is any function, is consistent with Eq.(1) for an ohmically heated plasma. The "hidden" dependences on T_e and B_θ can be determined only through additional direct experimental control of the input power, such as is provided by neutral beam injection. Since the largest extrapolation from present-day tokamaks to INTOR is in regard to T_e, the results from strongly beam-heated toakamaks are of paramount importance in assessing the confinement to be excpected in INTOR, and these results are treated separately in Section 1.2.

There is evidence from Alcator [3] and ISX-A [3] that Eq.(1) for the gross energy confinement time holds until τ_E approaches the value determined by ion neoclassical heat transport considerations. In this interpretation, one must consider only that part of the plasma dominated by thermal conduction, and one must exclude the outer region of the plasma which is dominated by atomic processes. When this is done, the scaling and magnitude of the temperatures are in close agreement with a model in which there is a transition from empirical losses through the electron channel to neoclassical losses through the ion channel accompanied by strong electron-ion temperature equilibration. This model is somewhat contradicted by results from Pulsator, where the transition occurs at too low a density to correspond to the transition to neoclassical-dominated losses. On the other hand, initial results from the FT device show no saturation of τ_E as the density is increased up to 4×10^{14} cm^{-3}. Densities in INTOR are, however,

not expected to approach this regime where the gross energy losses are dominated by ion conduction.

Overall, energy confinement results in ohmically heated tokamaks are consistent with a model in which the electron thermal diffusivity χ_e is given by

$$\chi_e = 5 \times 10^{17}/n_e \qquad \text{cm}^2/\text{s} \qquad (2)$$

where n_e is in cm^{-3}. The form given in Eq.(2) is employed in most of the computational studies discussed below.

As experimental data from present-day tokamaks becomes more extensive, there are increasing indications that the numerical factor in Eq.(2) should be an increasing function of radius within the discharge. In particular, recent results from PLT [3], JFT-2 [2], and T-10 [4], as well as earlier results from ORMAK, all indicate that the value of χ_e in the central part of the discharge is smaller than that given in Eq.(1) by a factor of 2–4, in cases without sawtooth oscillations or in the slow temperature – rise phase of the sawtooth. However, during the temperature-drop phase of the sawtooth, the energy transport must be very rapid, so that the effective time-averaged thermal conductivity in sawtooth discharges may not differ from Eq.(2) by such a large factor.

1.1.2. Particle confinement

While there is a wealth of consistent data on energy confinement in tokamaks, the experimental understanding of particle confinement and fluxes is poor. The experimental particle confinement time τ_p is defined by $\tau_p = N/S$ where S is the volume ionization rate and N is the total number of particles. The result of measuring τ_p depends on the plasma opacity to neutrals, i.e. on the relation between the mean-free-path of a neutral λ and the plasma radius a. For opaque plasmas ($\lambda \ll a$, typical of high density operation in most present-day devices) τ_p can be considerably smaller than τ_E. This is due to strong localization of S near the vacuum chamber wall. The neutral density in the center of the plasma is then determined by recombination, and production of plasma by ionization plays no role in the support of the central plasma density.

In general, measurements of this gross particle confinement time reveal very little in regard to the particle flux since, in opaque plasmas, the **source** of plasma is confined to the outermost region of the discharge. Since a nonzero density gradient is maintained near the centre, it is clear that a diffusive flow is not the only component of the particle flux. From the theoretical view, the neoclassical (or Ware) pinch can provide such an additional component, but is quantitatively inadequate to explain the inward particle transport in gas-puffing experiments in large tokamaks.

Measurements of particle confinement time by observations of hydrogen ionization light do not usually include the outermost part of the plasma. Direct measurements of the particle fluxes from the plasma have been made on DIVA and JFT-2, and these indicate that the overall particle confinement time is somewhat less than the energy confinement time [2].

It is tempting to employ a particle diffusion coefficient D related to the empirical electron thermal diffusivity χ_e, for example $D = \chi_e/4$ as proposed in the reference transport model formulated at the INTOR Workshop in February, 1979. However, in some devices (notably DITE, JFT-2 and Pulsator) better agreement with experimental data is obtained by choosing a smaller coefficient $D = \chi_e/12$ together with the neoclassical pinch. The fact that well-rounded density profiles are observed in large tokamaks, such as PLT and T-10, suggests that D must decrease more strongly as $r \to 0$ than would be the case with an n_e^{-1} dependence alone. Much of the data from PLT and T-10 would be reasonably well fitted by the choice

$$D = 4 \times 10^{16} (1 + 2r^2/a^2)/n_e \qquad cm^2/s \qquad (3)$$

Given the importance of the shape of the density profile to ignition, as well as the question of fuelling from the plasma periphery, further study of the particle fluxes in present-day tokamaks is clearly required.

1.1.3. Impurity confinement

While much experimental effort has been devoted to studying impurity particle transport in tokamak plasmas, relatively little is presently known about the details of the mechanisms affecting this transport and a number of key questions remain unanswered. In particular, neoclassical theory predicts that, under certain conditions, high-Z impurities can accumulate in the centre of the plasma, an effect which would be devastating to high density, long pulse operation. The experimental evidence for any such accumulation has not been firmly established and a number of results have contradicted the accumulation scenario.

On ST [3], ORMAK [3], and JFT-2 [2], observations of the soft X-ray continuum have been used to measure $Z_{eff}(r)$, and in each case no profound on-axis peaking was found. On the other hand, on the JIPP T-II device [2], measurements of the continuum in the visible imply a rather strong peaking of Z_{eff} on axis. Similarly, results from ATC [3], where a diagnostic neutral beam was used to determine the current profile, also imply that Z_{eff} is more peaked on axis than one would expect simply from the fact that the impurity ions are more highly stripped in the hotter central region. On TM-3 [4], observations of carbon indicated both accumulation and inward transport throughout the discharge duration. In Pulsator, inward transport of impurities was observed at high collisionality.

The results imply a diffusional velocity of 1–3 cm/ms, a value which appears to apply universally to impurity ion transport in tokamaks. Similar speeds were found on ATC with pulsed argon injection [3], as well as with aluminium [3], and scandium [3] injection; Fe and Mo injection on Alcator [3] also yield penetration velocities of approximately 1 cm/ms. Evidence for the inward transport term proportional to the gradient in the proton density has been found in the ATC aluminium injection experiments.

Conversely, the majority of experiments on larger tokamaks show that Z_{eff} is approximately uniform over the plasma cross-section. In the steady state phase of the discharge, Z_{eff} varies very little in time, despite the introduction of impurities at the plasma periphery. This indicates an intense mechanism of impurity removal. X-ray observations of argon-seeded plasma in T4 [4] and PLT [3] show a rapid circulation of impurities at the plasma periphery, by an anomalously high diffusion. This result was also found in experiments on carbon injection in DIVA [2]. Recent measurements of argon-seeded plasmas in T-10 indicate an outward diffusion velocity of ~0.5 cm/ms for highly stripped argon ions near the centre of the discharge.

According to calculations carried out for T-4, PLT, and TFR, the short lifetime of ions at the periphery may only be explained by an anomalously high diffusion. The time of impurity ion diffusion into the plasma is also shorter than the neoclassical time. This anomalous diffusion may be due to participation by the impurity ions in the same physical processes that result in anomalous diffusion of the hydrogenic ions. The Fe injection on Alcator has shown a finite confinement for the injected impurities, the decay time being about twice the global energy confinement time.

In summary, the following points regarding impurity transport appear to be well established: inward diffusional velocities are of the order 1 cm/ms; inferred Z_{eff} profiles sometimes show a significant peaking on axis, but often do not: both inward (as predicted by neoclassical theory) and spreading (larger than neoclassical) transport processes exist; and impurity particle confinement times appear to be of the same order as that of the working gas.

On the basis of these results, all four INTOR delegations proposed exactly the same model for impurity transport, namely that the radial flux of impurities Γ_Z should be a combination of the neoclassical terms together with an anomalous term of the form $-D\delta n_Z/\delta r$, where D is the same diffusion coefficient as is used for the hydrogenic ions. The predictions of this model are found to be in good agreement with measured oxygen profiles in TFR and DITE [1], and measured carbon profiles in DIVA [2], with the results of aluminium injection experiments in DIVA [2], and with results from T 4.

Very little data is available yet on the shielding of impurities by a divertor scrape-off layer, but results from DIVA (operated with a divertor) show that the accumulation of injected aluminium can be reduced by a factor ~7 compared with

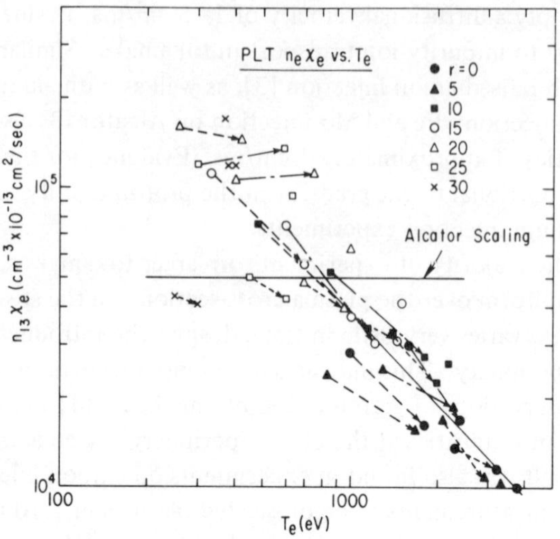

FIG.IV-1. Electron energy balance in PLT. Electron thermal conductivity (product of electron density and thermal diffusivity) versus electron temperature. Lines of a given type connect measurement with and without injection on the same discharges.

the non-divertor case, and that impurity backflow from the divertor is very little. A fuller discussion will be found in Chapter V.

1.2. Confinement in tokamaks with strong auxiliary heating

1.2.1. Neutral beam heating results on PLT

This INTOR data base assessment comes at a time in the fusion programme when our learning curve is at its steepest, particularly as regards tokamaks with strong auxiliary heating. First results from PLT [3] are very encouraging, but they raise many questions. If high temperature tokamak plasmas really prove to have an electron thermal diffusivity given by $\chi_e \propto (n_e T_e)^{-1}$, as a detailed analysis of the electron energy balance in the beam-heated PLT indicates Fig. IV-1, then the heating power in INTOR could be decreased, although the electron temperature profile would tend to be subject to thermal instability. On the other hand, the PLT results also suggest that while $n_e \chi_e$ is below Alcator scaling in the plasma core, it may rise significantly above Alcator scaling in the outer regions of the plasma. If we permit only a radial dependence in the coefficient in Eq.(2), rather than a temperature dependence, then the best fit to the PLT results would be given by

$$\chi_e = 10^{17}/[n_e(1-r^2/a^2)^{3.5}] \qquad \text{cm}^2/\text{s} \qquad (4)$$

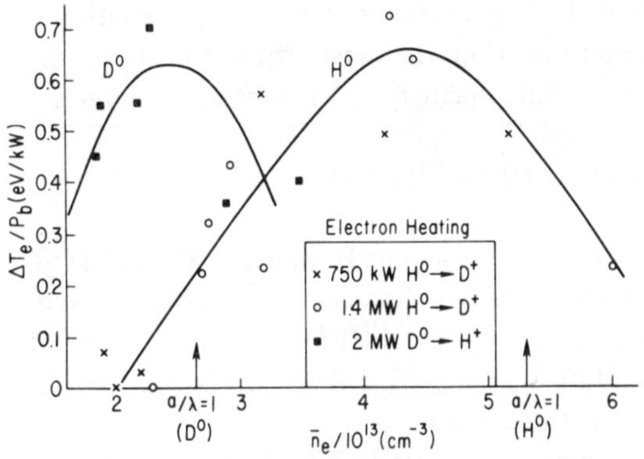

FIG.IV-2. Results from PLT. Neutral beam heating efficiency versus density for hydrogen (H°) and deuterium (D^{0}) injection. The density axis is also scaled in terms of the ratio of the minor radius (a) to the integrated mean free path (λ).

within $r/a \leqslant 0.6$, with χ_e of order, or larger than, Eq.(2) for $r/a \gtrsim 0.6$. In the trade-off between high voltage and high power in neutral beam systems, the optimum from the point of view of heating efficiency is then shifted towards the higher voltages, as compared to the optimum determined by $n_e\chi_e = 5 \times 10^{17}$.

However, since the electron heat conduction is smaller in the central part of the discharge, less alpha-particle heating is required in this region to reach ignition. On balance, at least according to modelling calculations, without impurities, ignition seems to be slightly easier to achieve in INTOR when a properly normalized shape factor, as in Eq.(4), is introduced. The mean-free-path of a beam-injected neutral is given approximately by

$$\lambda_{mfp}(m) \sim E_b(keV)/[180\; n_e(10^{20}\; m^{-3})A_b] \qquad (5)$$

where A_b is the atomic number of the beam neutral. For 40 keV H° and D° parallel injection into PLT, Fig. IV-2 shows that the optimum heating efficiency is obtained when

$$a/\lambda \equiv \int_0^a (\lambda_{mfp})^{-1}\; dr \leqslant 1$$

Taking into account the improvement of ~1.5 in penetration obtained by going to near perpendicular injection, the **most efficient** beam heating in INTOR with $\bar{n}_e \cong 1.4 \times 10^{20}\ m^{-3}$ should occur for D° beams with energies

$$E_b(\mathrm{keV}) \cong 240\ \bar{n}_e(10^{20}\ m^{-3})\ a\ (m) \qquad (6)$$

(i.e. about 400 keV), based on scaling from PLT. With INTOR beams in the energy range $E_b \cong 170-200$ keV, a heating scenario must be developed in which the initial density is lower ($\bar{n}_e \sim 7 \times 10^{19}\ m^{-3}$), with the final density being reached only as ignition is approached and alpha particle heating begins to play a role in the centre of the plasma.

At central ion temperatures up to 7 keV, corresponding to collisionality parameters ν_i as low as 2×10^{-2}, the ion temperature increase remains proportional to beam power, and inversely proportional to line-average density, although the most collisionless regime is characterized by a marked enhancement in the level of density fluctuations measured by microwave scattering. The ion energy balance has been investigated [3], on the basis of a Monte Carlo simulation of the beam-heating process, and numerical calculations of energy loss by neoclassical ion thermal conduction, ion-electron coupling, empirical thermal convection, and charge-exchange. Because of the uncertainties in Z_{eff} (r), q(r), and especially the neutral density $n_0(r)$, and because ion thermal conduction is only a small fraction of the total energy flow, it is found that for the case of injection into a low density plasma an enhancement of thermal conduction over the neoclassical value by as much as a factor 5 cannot be ruled out. In cases of neutral injection into higher-density plasmas, charge-exchange losses are reduced, but ion-electron coupling becomes the dominant term in the ion energy balance. Thermal conduction remains a small term in the total ion power balance and, even in this case, there is a possible enhancement over the neoclassical value by as much as a factor 3–4. At the same time, it is important to note that the results are also consistent with completely neoclassical ion thermal conduction, even in the low-density high-power cases with strong fluctuations. On the other hand, the T_i-dependence of the ion thermal conduction cannot be considered established as yet.

1.2.2. Neutral beam heating results on ISX-B

At the same time that PLT has operated at high plasma temperatures, ISX-B has begun studies of high-β plasmas (see Fig. IV-3). With 1 MW injection (not shown on the figure), $\beta_p = 2.8$ has already been achieved in a circular cross-section plasma with A = 3.5. $\beta_T(0)$, including an approximately 20% contribution from beam ions, was 8%, while $\langle\beta_T\rangle$ was 1.7%. The measured profiles were calculated to be unstable over half the plasma volume to high-n ballooning modes,

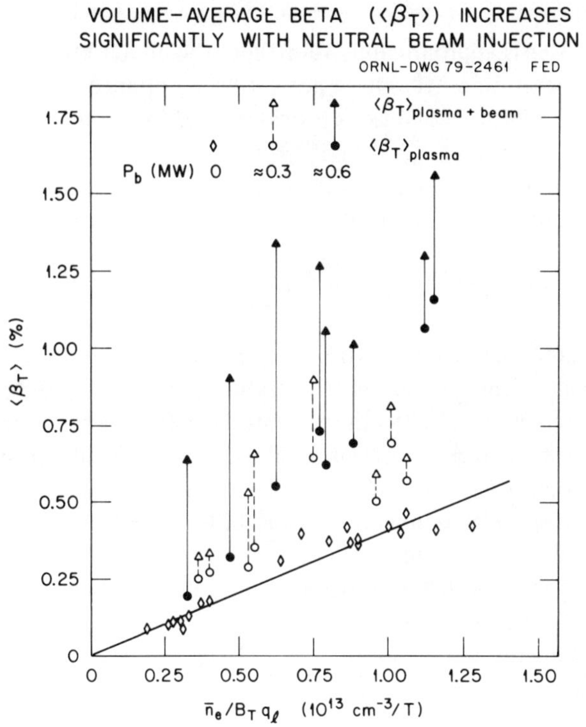

FIG.IV-3. Scaling of volume average β_T *with density and neutral beam power in ISX-B.*

and a change was observed in the characteristics of the MHD signals observed on magnetic loops and by soft X-rays. However, the overall energy confinement time was not diminished. In addition, preliminary ISX-B results indicate some improvement in τ_{Ee} in the plasma core with injection, as in PLT. Ion heating at high n_e, and therefore high β, is consistent with neoclassical theory. At low densities, however, ΔT_i does not increase, as on PLT, suggesting that charge-exchange may be a serious limitation. ISX-B shows ΔT_i independent of n_e while PLT shows $\Delta T_i \propto n_e^{-1}$. In a much larger machine, such as INTOR, the charge-exchange losses should be further reduced.

1.2.3. Neutral beam heating results on T-11

The neutral beam power in the T-11 device has recently been raised to 600–650 KW at about 35 keV [4]. The target plasma was clean ($Z_{eff} \cong 1$), and had a relatively low q-value ($q_a \cong 2.6$). At the end of the injection pulse, the neutral beam power is approximately four times the ohmic heating power. Although density and temperature profiles have not yet been obtained, diamagnetic measurements and the equilibrium conditions provide values for β_p, and hence β. On this basis, values $\langle\beta\rangle \cong 2.5\%$ and $\beta_p = 4\pi\,(\langle P_{\parallel}\rangle + \langle P_{\perp}\rangle)/B_p^2 \cong 2.1$ have been achieved. During the injection pulse, the total radiation and charge exchange increases from about 10% to about 40% of the total power input.

1.2.4. Neutral beam heating results from other tokamaks

Approximately 0.5 MW of neutral beam power at about 35 keV has been applied to TFR [1]. The central ion temperature was raised by 0.8–1.0 keV, with smaller increases (~0.2 keV) in the central electron temperature. In TFR, the ion temperature increase is proportional to beam power and inversely proportional to density, just as in PLT, so that the gross ion energy confinement time does not change with temperature. The TFR group believes that the ion thermal conductivity is definitely enhanced over the neoclassical value (up to 5 for H and 3 for D), and they find the absolute value to be about 5×10^3 cm^2/s.

Injection of about 1 MW at 30 keV into lower density plasmas in DITE [1] raised the ion temperature by about 0.5 keV, and produced a central β-value of about 1.8%.

1.3. Recommended INTOR-relevant confinement experiments

Experiments relating to energy confinement are aimed at evaluating the electron thermal diffusion coefficient $\chi_e(n, T_e, T_i, B_\theta, \ldots)$ under further extension of the plasma parameters. Of particular value to INTOR would be additional data on the dependence of χ_e on T_e and T_i, since a favourable trend increases confidence in the ignition scenarios for INTOR within the framework of the present reference design. Confinement studies in noncircular tokamaks will, during the coming year, provide information on the scaling of τ_E with plasma shape. Information is also needed on the nature of the power flow to the wall/limiter. There is evidence (see Section 1.2) that χ_e increases strongly with radius; is this simply through dependence on the n and T profiles, or is it, for example, a consequence of the q profile? Finally, more information is needed on the **ion** thermal diffusion coefficient χ_i; the evidence that χ_i is close to neoclassical predictions is somewhat indirect; additional measurements of χ_i at higher T_i and as functions of B_θ and B_ϕ would be valuable, especially for the "hot-ion" modes of operation.

As discussed in Section 1.1.2, the present understanding of **particle confinement and diffusion** is inadequate and experiments on current tokamaks should address this issue. Pellet injection experiments would be useful since in these experiments a source of particles can be created well within the discharge. Further understanding of gas-puff experiments is also required. An important question is whether the neoclassical pinch effect is in fact responsible for the observed inward flux of particles. An additional useful experiment would be the study of a mixture of hydrogen isotopes (H into D, D into H) in order to ascertain whether there is a natural segregation (i.e., a preference for H or D to concentrate in the centre). This also has a bearing on the degree to which the fuel mix can be controlled by external gas puffing.

Finally, continued experimentation on the **transport of impurities, especially helium,** is required. In particular, the question of impurity accumulation must be addressed further. Divertor experiments should also be helpful in investigating the transport of impurities, because of their ability to change the rate of influx of impurities at the plasma edge.

The operation and diagnosis of TFTR, JET, T-15, and JT-60 should provide a major extension and/or test of the confinement data base. Specifically, these devices should, by virtue of larger physical size and high power heating, be able to extend into higher temperature regimes, closer to those needed for INTOR, the empirical scaling laws of energy, particle, and impurity confinement. At the $Q \gtrsim 1$ objective of TFTR, the alpha particle heating is small compared to the overall plasma energy losses. Also, the pulse lengths are somewhat short ($\lesssim 1.5$ s) for seeing any alpha-particle effects. JET, on the other hand, if provided with sufficient heating power, is expected to reach a regime in which alpha-particle heating prevails in the plasma core. Thus, while it is unlikely that these devices will be able to explore the physics of alpha particle heating of ions to ignition, they should provide a first look at alpha-particle production and (electron) heating effects in tokamaks, and also first results on alpha-particle confinement.

1.4. Summary – Applicability to INTOR

As the preceding overview indicates, the data base for energy confinement scaling is quite large and yields a reasonable empirical scaling law for energy confinement. Particle and impurity confinement are not as well understood and it has not been possible to develop useful scaling laws for them. Nonetheless, some models for particle and impurity transport have been developed.

It is important to realize that the bulk of our understanding of energy, particle, and impurity confinement is empirical in nature. (Ion heat transport, which appears to be neoclassical, is the notable exception.) That is, the confinement data is not as yet closely coupled to any particular theory or set of theories. Thus, given that the confinement scaling relationships have been derived over a relatively narrow range of parameters (primarily small, circular, not particularly hot plasmas), the extrapolation of these scaling laws to the larger, noncircular, hot plasmas in INTOR is not entirely persuasive. In particular, since there is no unifying theory available at present, it is not known how large an extrapolation is being made in terms of the critical dimensionless parameters. In the absence of such a unifying theory, we can only examine the extrapolations involved in terms of known dimensionless parameters. In this regard, INTOR represents a modest extrapolation in collisionality (ν_* down from $\lesssim 10^{-1}$ in PLT to $\sim 10^{-2}$), β (up from 2.5% now to $\gtrsim$5%) and noncircularity (from unity up to ~1.6); experimental tests are planned in each of these areas.

2. POINT-MODEL STUDIES OF INTOR

2.1. Ignition requirements on energy confinement time

We first consider the idealized case in which ion and electron temperatures are assumed equal, and the energy confinement time τ_E is treated as an independent parameter. We define an energy multiplication factor Q as

$$Q = \frac{5(P_{bf} + P_{tf})}{3\langle n\rangle\langle T\rangle/\tau_E + P_{brem} - (P_{bf} + P_{tf})} \tag{7}$$

Here P_{bf} and P_{tf} are the alpha-particle power released by fusion reactions of the nonthermal beam ions and the thermal plasma ions, respectively. The factor of five accounts for the total energy released per fusion event, which is five times the energy of the alpha particle. Curves of constant values of Q can be plotted in the plane spanned by $\langle n\rangle\tau_E$ and T(0), as shown in Fig. IV-4. To relative average and central values of density and temperature we have assumed profiles of the form $n(r) \sim (1-r^2/a^2)^{\alpha_n}$ and $T(r) \sim (1-r^2/a^2)^{\alpha_T}$, and we define $\langle T\rangle_n$ to mean the density-weighted mean temperature such that $\langle n\rangle = n(0)/(\alpha_n + 1)$ and $\langle T\rangle_n = T(0)\,(\alpha_n+1)/(\alpha_n+\alpha_T+1)$. In Fig. IV-4, $\alpha_n = 0.5$ $\alpha_T = 2$, and we have displaye Q = 1 (breakeven), Q = 5, and Q = ∞ (ignition). For the INTOR reference design case of $\langle n\rangle = 1.4 \times 10^{14}$ and T(0) = 23.3 keV ($\langle T\rangle_n$ = 10 keV), we see from Fig. IV-4 that an energy confinement time of about 1.3 seconds or greater is required for ignition.

2.2. Ignition requirements on plasma beta

If we assume some particular scaling law for the energy confinement time, then curves of constant $\langle\beta\rangle$ can be drawn in as shown on Fig. IV-4. Here we assume (cf. Sec. 1.1.1) $\tau_E = 5 \times 10^{-21}\,\langle n\rangle\,a_{eff}^2$ s where a_{eff} is the effective plasma radius in metres, and $\langle n\rangle$ is in m^{-3}, and $\langle\beta\rangle = 16\pi\langle n\rangle\langle T\rangle_n/B^2$. Since INTOR has a non-circular cross section, we use a_{eff}^2 = ab giving a_{eff} = 1.47 m. We assume the same density and temperature profiles as before to relate the peak and average

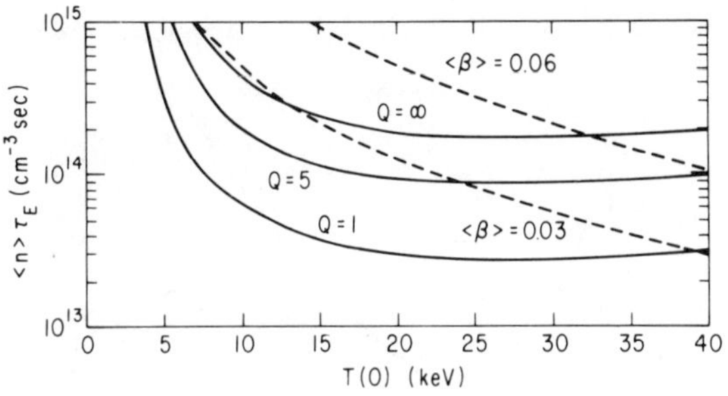

FIG.IV-4. Point-model requirement for breakeven (Q = 1), a driven system (Q > 5), and ignition (Q = ∞). Also shown are the corresponding values of plasma ⟨β⟩ assuming the empirical scaling law for energy confinement.

temperatures. Note that we do not include the pressure of beam ions, α-particles, and impurities (~1%) in this crude estimate. On Fig. IV-4 we show two curves: $\langle\beta\rangle = 0.03$ and 0.06. Note that $\langle\beta\rangle = 0.03$ is very marginal for ignition, but $\langle\beta\rangle = 0.06$ would allow a margin to compensate for uncertainties in the assumed scaling law for the energy confinement time.

Indeed, as we have discussed in Section 1.1.1, there is some evidence from present-day tokamaks that a factor 2 improvement over the scaling law employed here can be achieved. Moreover, the total energy confinement time in a high-temperature plasma with $T_i \cong T_e$ should be almost twice that given by the empirical law, which is thought to govern only electron energy transport, with ion energy confinement being much better (provided that it remains approximately neoclassical). On the other hand, there is no experimental evidence to justify the assumption $\tau_E \sim nab$ rather than $\tau_E \sim na^2$. In addition, the presence of even a small percentage of radiating impurities can significantly raise the ignition requirement on $\langle n\rangle\tau_E$ and $\langle\beta\rangle$ above those shown in Fig. IV-4.

2.3. Neutron wall loading

The neutron wall loading in INTOR can be expressed as a function of β and T, for fixed B = 5 T and fixed device size. This is shown in Fig. IV-5, in which the density and temperature profiles have both been taken to be parabolic. Again assuming some particular scaling law for the energy confinement time, the ignition curve can be added to Fig. IV-5. In fact, we show two ignition curves in Fig. IV-5, corresponding to the scaling laws $\tau_E = (5\text{–}8) \times 10^{-21}\ \bar{n}_e ab$; the upper ignition curve corresponds to the standard scaling law, and the lower curve represents a factor 1.6 improvement in τ_E. Note that impurity radiation losses are disregarded in this discussion.

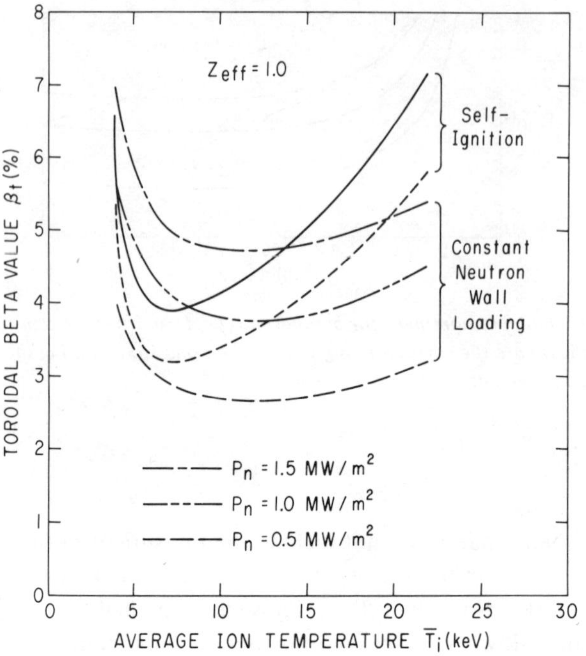

FIG.IV-5. Required toroidal ⟨β⟩ value for self-ignition and various neutron wall loading as a function of ⟨T_i⟩. The two self-ignition curves describe the range $\tau_E = (5-8) \times 10^{-21}\,\bar{n}_e ab$.

If it is desired to achieve a neutron wall loading of 1.2 MW/m^2, implying a thermal wall loading at ignition of 0.3 MW/m^2, Fig. IV-5 shows this to be possible, with the standard scaling law and negligible impurity radiation, by operating just slightly into the ignited regime, with ⟨β⟩ just above 4% (again excluding beam ion and alpha-particle contributions).

Suppose, however, that the confinement is 1.6 times better than that given by the standard scaling law. Then, the lower ignition curve in Fig. IV-5 is the applicable one. In this case, one could operate the lower ⟨β⟩ (closer to 3%), but the neutron wall loading would fall to the uninteresting value of about 0.6 MW/m^2. Alternatively, one could lower the confinement artificially (for example, by increasing the field ripple) to restore a 1.2 MW/m^2 neutron wall loading.

Suppose, next, that the confinement is 1.6 times worse than that given by the standard scaling law. In this case, ignition could be restored by raising the density by a factor 1.3, at the expense of a higher thermal wall loading (increased from 0.3 MW/m^2 to 0.5 MW/m^2) and a higher ⟨β⟩ (increased from 4% to 5% without beams ions, alpha-particles, or impurities). The higher thermal wall loading might be acceptable, especially with a divertor, and the higher ⟨β⟩ could be compensated by an increase in field strength from 5T to 5.7T. Otherwise, the only remaining option would be a somewhat larger machine.

TABLE IV-1. POINT-MODEL IGNITION CALCULATION

E_b keV	P_b MW	t_b s	$n_e(10^{20}\ m^{-3})$ initial	final	$\beta_{ign.}$ %	$\beta_{p,\ ign.}$
200	75	3	1.15	1.3	4.5	4.5

2.4. Heating power needed for ignition

Time-dependent point-model calculations of heating to ignition in INTOR have been presented in the reports by the EC, Japanese and USSR delegations. In all cases, analytic procedures are used to obtain appropriate averages over the radial profiles. The calculations presented in the European report [1] have employed the most complete point model, which includes:

(i) an alpha-particle treatment that accounts for orbit losses, finite slowing-down time, beam-plasma reactions, and the proper distribution of alpha energy to electrons and ions;

(ii) empirical electron thermal losses $\tau_{Ee} = 5 \times 10^{-21}\, \bar{n}_e\, a_{eff}^2$ and neoclassical ion thermal losses;

(iii) hydrogen and impurity radiation, and synchrotron radiation;

(iv) a neutral-beam heating treatment that takes account of penetration effects by multiplying the beam power by the coupling efficiency $C_{eff} = C_{pen} \exp(1-C_{pen})$, where

$$C_{pen} = 180\,(2-0.5\,\delta)\, \bar{n}_e(10^{20}\ m^{-3})\, a(m)/E_b(keV) \tag{8}$$

Optimum penetration is given by $C_{pen} = 1$. The parameter δ describes the profiles, which are taken of the form $n \sim (1-r^2/a^2)^{\delta/2}$ and $T \sim (1-r^2/a^2)^{\delta}$. The criterion $C_{pen} = 1$ for $\delta = 2$ (parabolic density profile) corresponds to $E_b\ (keV) \cong 180\ \bar{n}_e\ (10^{20}\ m^{-3})\ a(m)$, a slightly less severe requirement on the beam energy than would be suggested by the PLT results (see Section 1.2.1).

The results of one calculation of ignition in INTOR [1], taking profiles given by $\delta \cong 1.5$, are summarized in Table IV-1 (smaller values of δ require longer injections pulses).

The increase in density is due to the beam fueling. Other results obtained with this code are that the range of densities for which ignition can be achieved at fixed beam power and fixed beam pulse length depends sensitively on the beam energy. For $P_b = 75$ MW and $t_b = 4$ s, ignition can be achieved at $E_b = 160$ keV only within the range $0.8 < \bar{n}_e\,(10^{20}\ \mathrm{m}^{-3}) < 1.2$, whereas at $E_b = 200$ keV the ignitable range is extended to $0.8 < \bar{n}_e\,(10^{20}\ \mathrm{m}^{-3}) < 1.7$.

3. COMPUTATIONAL STUDIES OF INTOR USING TRANSPORT CODES

The one-dimensional dynamic evolution of INTOR discharges was simulated by the 1–D transport code groups in the EC (Fontenay, Garching, Culham), Japan (JAERI), the USA (General Atomic, Oak Ridge, and Princeton), and the USSR (Kurchatov). The present Section is devoted to a summary of the results obtained by these groups. Emphasis will be placed on heating to ignition, in particular by neutral beam injection. The early history of the discharge (including ionization and current rise phases) will not be discussed here, but it is treated in Chapter VIII.

3.1. Reference transport model

The transport simulation studies have generally used the INTOR reference concept parameters. Since the INTOR plasma is proposed to have a noncircular cross-section, but the codes are only one-dimensional, the noncircular (1.2 × 1.8 m) 4.5 MA INTOR tokamak plasma is modeled with the equivalent 1.5 m circular plasma carrying a current of 3.8 – 4.0 MA.

The reference transport model used is as follows. For energy transport the electron and ion thermal diffusivities are taken to be of the form

$$\chi_e(r) = 5 \times 10^{17}/n_e(r) \qquad (\mathrm{cm}^2/\mathrm{s})$$
$$\chi_i(r) = 3\,\chi_i^{\mathrm{neoclass}}(r) \tag{9}$$

The particle diffusion coefficient is often taken to be $D = \chi_e/4$, but a variety of other models (mostly involving a smaller value of D) have also been employed. Most codes include the neoclassical trapped particle pinch term in the particle transport.

For alpha-particle deposition, different assumptions were made, ranging from instantaneous local deposition to treating slowing–down and non-local effects fairly rigorously. In the reference case, Z_{eff} is assumed to be essentially unity (i.e., little or no impurities). Also, the effects of helium ash accumulation, magnetic field ripple, MHD instability profile flattening effects, and other nonideal effects are neglected in the reference model, but can be included in variations of the model. One interesting effect that has emerged is that, if the small ($\lesssim$10%) fraction of the alpha-particle energy transferred to the ions is neglected, the results are quite different. This is because the ion temperature would then be less than the electron temperature and make ignition more difficult. This observation illustrates the sensitivity of the ignition condition to the fraction of the alpha-particle energy transferred to the ions by classical, or possibly anomalous, processes.

Indeed, the differences in the results obtained by the different groups demonstrate that the conditions for achievement of ignition in INTOR are very sensitive to many details, both of the physical model and of the operating scenario.

3.2. Variations of the reference transport model

Because of its fundamental importance in the determination of the overall confinement and performance of INTOR, numerous models of the electron heat conduction were considered. Simply increasing the coefficient in the scaling of χ_e with density by a factor f decreases the energy confinement time by the same factor. Further, the time to reach ignition and the β at ignition are similarly increased. The addition of a favourable temperature dependence in the electron heat conduction scaling, for example as suggested by the recent PLT data (cf. Section 1.2), $n_e\chi_e \sim T_e^{-1}$, can make ignition easier. The introduction of a further multiplicative radial profile factor $f(r) \sim (1-r^2/a^2)^{-3.5}$ [see Eq.(4)] which is small at the plasma centre and large at the plasma edge, such as has been suggested by PLT data, produces complicated results even when it is normalized to yield the same overall energy confinement time. Specifically, the neutral beam heating is less effective in the outer edge of the plasma, where the heat conduction coefficient is large and the beam must penetrate further to achieve optimum heating, as illustrated in the PLT data shown in Fig. IV-2. However, at the same time, since the electron heat conduction in the centre is smaller, less alpha particle heating is required to achieve ignition. On balance, ignition seems to be slightly easier to achieve when a properly normalized shape factor f(r) is introduced, as long as impurity effects are disregarded.

The ion heat conduction coefficient was varied from 5 down to 1 times neoclassical, and ignition was found to be only slightly easier in the latter case, because the dominant energy loss at ignition tends to be via the more lossy electron component. The effects of charge-exchange neutral energy losses were found to be roughly equivalent to reducing the effective injected power by about 10 MW. When the finite slowing-down time ($\sim$0.3 – 0.5 s) of the alpha particles is taken into account, the time of beam injection required to reach ignition is extended by roughly this slowing-down time.

TABLE IV-2. COMPUTATIONAL RESULTS FOR INTOR USING THE REFERENCE TRANSPORT MODEL, AND NEGLECTING IMPURITIES

Code	Beam Power MW	Beam Pulse s	Beam Energy keV	Full Energy Fraction	$\langle n_e \rangle$ 10^{20} m^{-3} Initial	Final	Separate D and T	Recycle Factor	Ignition Achieved	$\langle \beta \rangle_{ig}$ %	Comments
Garching	72	6	200	0.95	1.1	1.7	Yes	1.0	Yes	4–5	D:T = 40:60
Fontenay	70	4	200	1.0	1.1	~1.5	No	~1.0	No	–	Pulse too short
	90	4	200	1.0	1.1	~1.5	No	~1.0	Yes	7*	
Culham	75	5	200	1.0	1.1	1.6	No	1.0	Yes	7*	
	75	7	150	1.0	1.1	>1.6	No	1.0	Yes	7*	
	50	5	100	1.0	0.27	1.23	No	1.0**	Yes	3.7	Beam switched off when $\langle n_e \rangle$ = = 0.85 × 10^{20} m^{-3}
Kurchatov	50	∞	200	0.5	0.7	–	No	1.0	No	–	Parallel injection, pellet fueling.
	75	4	200	0.5	0.7	1.0	No	1.0	Yes	4	as well as gas puffing.
Princeton	60	4	150	0.8	0.7	1.4	Yes	0.9	Yes	4	Controlled tritium puffing
Oak Ridge	60	4	150	1.0	1.0	1.0	No	<1	Yes	4	Density held constant
JAERI	75	5	200	1.0	0.7	1.3	Yes	0.5	Yes	7*	Pellet fueling ripple, ±0.8%
	90	4.5	200	1.0	0.7	1.2	Yes	1.0	Yes	12.5*	gas puff ripple, ±0.5%

* These are β-values attained when the beam is switched off, or later during the burn, not at the moment of ignition.

** While beam is on; additional gas puffing after beam is off.

3.3. Particle transport

Owing to the lack of sufficient experimental data on the subject, the modeling of the density profile evolution is a major area of uncertainty. The basic simulation problem is to reconcile the edge localisation of the gas puffing particle sources in present machines, which for ordinary diffusive particle fluxes should lead to very flat density profiles, with the rather peaked density profiles that are observed experimentally. The US groups usually use a combination of the Ware pinch effect and a diffusion coefficient $D \sim \chi_e/4$. Most of the models predict very flat density profiles for INTOR plasmas. The profiles are slightly peaked at the plasma centre during the intense neutral beam heating phase; they can become slightly hollowed out during the burn phase because of the burnup of the fusion fuel.

The USSR groups use a diffusion coefficient rather similar to Eq.(3) with an extra factor that increases towards the periphery [4]. The USSR groups also consider fueling both by gas puffing and by pellet injection (in the latter case, assuming no recycling). Pellet fueling yields narrower density profiles [4] that are easier to ignite with 150–200 keV beams.

Several codes (Garching, JAERI, Princeton) have the capability of treating the separate densities of D and T in the background gas. Results from these codes draw attention to the need for carefully controlled tritium fueling in the face of substantial injection of deuterium from the beam. Indeed, in some calculations with the JAERI code, in which a small initial density (about 2×10^{13} cm^{-3}) was assumed and a rather small diffusion coefficient employed ($D = \chi_e/30$), the tritium supplied at the edge by gas puffing is unable to mix sufficiently rapidly with the deuterium injected to the centre by the beams. Ignition fails for this reason. In the JAERI calculations reported in Table IV-2, $D = \chi_e/12$ was employed.

As an example of computations of this kind, in calculations with 60 MW of beam power at 150 keV [3], the particle current represents a flux of 3.2×10^{21} particles/s or a rise rate of the volume-averaged density of 1.3×10^{13} $\bar{cm}^3$/s. For five seconds of injection this will raise the volume-averaged density by 6.7×10^{13}/cm^3. Since this source is all D^+ and is somewhat peaked at the centre (certainly more peaked than the source from gas puffing and plasma recycling), it can seriously dilute the tritium in the main body of the plasma [3]. This effect can be reduced by recycling only 90% of the plasma (D and T) as neutrals (~10% is pumped by a divertor, etc.) and by gas puffing tritium at the plasma edge at a rate designed to keep the volume average density at a programmed level. For the standard "variable density startup" case, the relative fraction of tritium $[n_T/(n_D + n_T)]$ in the centre goes from 0.9 at t = 0 to 0.37 after 5 seconds of beam injection. Injection for longer times continues to reduce the central tritium fraction. After the beams are turned off a 50% D–50% T mixture is puffed in at a rate programmed to keep the volume averaged density constant. Without the partial recycling and tritium gas puffing during beam heating, either the density will rise at a rate $\sim 1.3 \times 10^{13}$/cm^3 s (faster at the centre), or there will be very little tritium at the centre.

3.4. Requirements for heating to ignition

The first problem in discussing heating to ignition is to define ignition. Ignition occurs when, without further auxiliary heating power being applied, the alpha particle heating will, perhaps after an extended "smoldering" phase, cause the plasma energy content to increase. If the neutral beam heating is turned off too soon, the energy content decreases monotonically and the plasma does not ignite. If the beams are kept on to just past the minimum ignition time then the energy content starts to decrease, then the plasma smolders a while at a given energy density level, and finally slowly ignites. This "smoldering" approach to ignition occurs because the centrally peaked alpha particle heating ignites the central region of the plasma before the "burn" expands to encompass the entire plasma.

For ignition of this type, it is important that switching the beams off leads to a decrease in losses, due to a flattening of the temperature gradients near the plasma edge [1]. Thus, this type of ignition may not be possible in the presence of impurities in the plasma edge region whose radiation loss increases as the electron temperature decreases [1].

3.4.1. Beam power, energy, and pulse length needed

The results from the various groups using the reference transport model, and neglecting impurity effects, have been compiled in Table IV-2. In all cases, the density profile is extremely broad (except for the USSR cases with pellet fuelling) so that the calculations may well be underestimating beam penetration. The results displayed in Table IV-2 again draw attention to the importance of control of the density and of the fuel mix. The lowest beam power at which ignitio was achieved (60 MW at 150 keV) was in "variable density startup" calculations at Princeton, [3], in which controlled tritium puffing, and recycling held to 90%, enabled the density to be raised from $\langle n_e \rangle = 0.8 \times 10^{14}$ cm^{-3} initially, to $\langle n_e \rangle = 1.1 \times 10^{14}$ cm^{-3} at the moment of ignition, keeping a favourable DT fuel mix all the time. As in the pointmodel studies discussed in Section 2, the Princeton calculations show that 150 keV beams are adequate to ignite $10^{14} < \langle n_e \rangle$ (cm^{-3}) $< 1.3 \times 10^{14}$ plasmas, while 200 keV beams are needed for $1.3 \times 10^{14} < \langle n_e \rangle$ (cm^{-3}) $< 1.8 \times 10^{14}$ as shown in Fig. IV-6. In these results, the $\langle \beta \rangle_{ig}$ values are those pertaining to the moment of ignition, i.e., when the energy balance is favourable in the central half of the plasma volume. This explains the difference in the $\langle \beta \rangle_{ig}$ values quoted in Table IV-2 between the US, European and Japanese calculations. In fact, the Culham Group [1] also finds that switching the beam off at $\langle \beta \rangle = 4.5\%$ would still yield an ignited plasma.

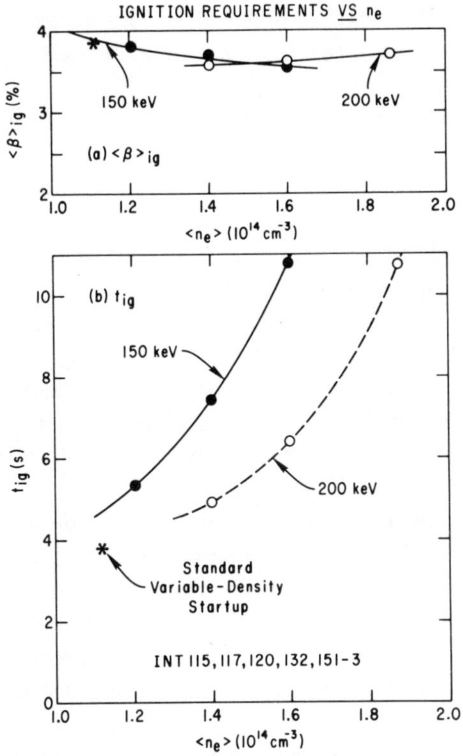

FIG.IV-6. (a) Volume average pressure $\langle\beta\rangle_{ig}$ *and (b) heating pulse* t_{ig} *for ignition at constant density with 60 MW of 150 keV and 200 keV beam injection, and variable density with 150 keV beam (*).*

3.4.2. Beam penetration

The majority of the neutral beam heating calculations for INTOR have been carried out under the assumption that the available beams would be positive-ion based with energies in the 150–200 keV range. Within this range the highest energy, and more importantly the highest atomic species fraction, produce the best penetration and, hence, most efficient heating.

As a counter-example, a case of ignition with beam energy below this range was presented in the US report, taken from the Oak Ridge TNS studies [3]. Here, ignition was possible with 60 MW of 120 keV D° beams because of the reduced distance of the magnetic axis from the plasma edge due to finite β effects in these 1–1/2–D calculations, and the fact that alpha-particle heating takes over in the centre before the density builds up to cut off the beam penetration. Also, in calculations at Culham [1] using 50 MW of 100 keV beams, ignition was achieved by heating at low densities ($\langle n_e \rangle \lesssim 0.85 \times 10^{20}$ m^{-3}), followed by gas puffing. Such a scheme is likely to require tritium injection.

3.4.3. Power and particle outflux to wall and limiter

The total thermal power from an ignited INTOR plasma (excluding neutron power) is in the range 80–100 MW. Of this, charge-exchange and radiation (bremsstrahlung corresponding to $Z_{eff} \cong 1.5$) account for about 10 MW each. The rest is in conduction and convection losses.

The total flux of hydrogenic ions from the plasma interior in INTOR is likely to be of the order of 10^{22} particles/s, corresponding to a central particle confinement time of the order of 2–3 s. The actual flux of particles crossing the plasma boundary will be much larger than this, since it will be dominated by a high rate of particle recycling in the edge region. We might call this flux at the boundary the "recycling flux".

The magnitude of the recycling flux is highly dependent on conditions encountered at the edge of the plasma, in particular the local temperature and density gradients, and on the energy of the incoming neutrals. One way to estimate the recycling flux is to observe that plasma energy must be transported across the separatrix (or limiter scrape-off) mainly in the form of convection, so that a given amount of energy transport can be associated with a certain particle transport. For example, if 50 MW of power is lost from INTOR by electron and ion transport processes (i.e., not radiation on charge exchange), and if the edge temperature is about 100 eV, then the ion recycling flux must be about 3×10^{23} particles/s. For a plasma diffusion coefficient in the edge region of about 10^4 cm^2/s, [see Eq.(3)], this recycling flux would correspond to a density profile dropping from about 10^{14} cm^{-3} to zero over a 10 cm wide edge region. The edge density of 3 eV neutrals would be of the order of 10^{11} cm^{-3}. In the case of pellet fueling, the recycling flux could in principle be reduced, at the expense of a higher edge temperature, although a significant reduction would occur only in the case of pellets that penetrate deeply into the plasma, which requires a pellet speed greater than that considered practical in the near term [5]. In the case where there is a divertor, the situation is quite complicated and requires a careful calculation of the energetics and particle balance in the scrape-off plasma.

3.4.4. Impurity effects on heating to ignition

All of the delegations have presented estimates [5] of the maximum tolerable impurity levels in INTOR, and there is general agreement between the various estimates. Since the INTOR plasma must be heated from $T_e(0) \sim 1$ keV to $T_e(0) \sim 20$ keV by about 60–100 MW of neutral beam power, after which the alpha-particle heating supplies a comparable amount of power, we can determine

TABLE IV-3. IMPURITY FRACTIONS CORRESPONDING TO 20 MW OF RADIATED POWER IN INTOR

	Fe	Mo	W
1 keV	10^{-4}	2×10^{-5}	10^{-5}
5 keV	10^{-3}	10^{-4}	2×10^{-5}
10 keV	10^{-3}	4×10^{-4}	5×10^{-5}
20 keV	10^{-3}	4×10^{-4}	10^{-4}

the maximum tolerable impurity level from the condition that the radiated power should not exceed 20 MW over this range of temperatures. Table IV-3 gives estimates based on this condition, for $\langle n_e \rangle = 1.4 \times 10^{14} cm^{-3}$. More details may be found in the report on Impurity Control, Fueling and Exhaust [5].

Several of the INTOR delegations have provided transport-code calculations of impurity effects in INTOR [1, 2, 3], considering both low-Z impurities (C, 0, etc.) and high-Z impurities (Fe, etc.). In the latter case, the impurity influx can be related to wall sputtering in a "self-consistent" way.

Almost all of the codes employ the same basic model for the transport of impurities within the plasma. In this model, the flux of impurity ions at every radius within the plasma is taken to be a sum of the radial flux due to neoclassical effects and a radial flux of the form $-D(\partial n_Z/\partial r)$, where D is the same (empirical) particle diffusion coefficient employed for hydrogenic ion transport. This model is in general agreement with results from a number of experiments on impurity transport (see Section 1.1.3). The neoclassical terms often take into account only the Pfirsch-Schlüter regime, but this assumption may be reasonably adequate for impurity transport. Sometimes, contributions to the neoclassical flux from ion temperature gradients are included, and sometimes they are not; when included, such terms may lead to a "temperature gradient screening effect", in which the impurity density can be peaked in the edge region of the plasma, even in steady state. As yet, there has been very little detailed experimental verification of impurity models of this type, although there is evidence from Alcator [1] that temperature-gradient screening may occur to some degree, at least for oxygen impurities.

Various models can be chosen to determine the influx of impurities into the plasma from the wall, namely:

(i) determine the influx from sputtering of the wall (usually Fe) by plasma ions and charge-exchange neutrals;
(ii) determine the impurity generation by charge-exchange sputtering only, and set the net impurity ion influx at the plasma boundary to this value, implying that all sputtered impurities are indefinitely retained within the plasma;
(iii) determine the impurity generation by charge-exchange sputtering only, as in (ii), but deposit the sputtered particles as neutrals distributed over some edge region of the plasma (typically ~5 cm wide), where they act as a source of impurity ions by ionization; in this case, some fraction of the impurity ions (depending sensitively on the edge-region conditions) are transported into the plasma, and some fraction diffuse out to the wall again, where they are assumed lost; eventually, a steady state may be reached in which the outflux of impurity ions balances the influx of neutral impurities.
(iv) specify the net impurity influx as a function of time, unrelated to sputtering yields;
(v) simply specify the fractional impurity content in the plasma.

None of the codes yet include sputtering by low-Z impurities or self-sputtering by high-Z impurities; however, in a divertor tokamak, such processes would be expected to occur on the divertor-plate, and not on the first wall. Moreover, it is expected that the efficiency of the divertor in preventing backflow of impurities from the divertor plate to the plasma will be high [5].

Essentially all of the codes estimate impurity radiation on the basis of the coronal equilibrium model. However, the characteristic time for diffusion of low-ionization-level impurity ions in the edge-region of the plasma is extremely short, with the result that the coronal model may not be reliable; no coronal effects are likely to increase the radiation by some factor, but quantiative estimates cannot yet be given.

At the present time, codes are in the final stages of development at various laboratories, which can treat separately the diffusion, ionization, and recombination of a large number of different ionization levels of a given impurity species (e.g., Fe), together with noncoronal treatments of the radiation. Such codes will represent improvements over any of the models (i)–(v), listed above, and will increase the realism with which problems of this type can be addressed. By including transport along a divertor scrape-off, these codes will also provide some estimate of the effectiveness of a divertor in screening out wall-sputtered impurities.

The results of a number of transport computations for INTOR including impurities are summarized in Table IV-4. In view of the differences in the models that have been employed, it is not surprising that there are significant differences in detailed aspects of the results. However, some general conclusions do seem to emerge.

If ignition is to be achieved, and a long burn to be sustained, it is clear from the various results with wall-material impurity species that a balance must quickly arise between the influx of sputtered impurity atoms and the diffusive outflux of impurity ions. The cooling of the edge region is also probably essential to diminish the sputtering rate.

Assuming that a steady state of this type can indeed be reached, the question then is: Can impurity radiation from the edge region be sufficient to remove a substantial fraction of the total power, while impurity radiation from the centre of the plasma remains sufficiently low that the burn is not quenched? A critical moment will occur just when the neutral-beam heating is switched off, when the temperature might be expected to drop over the outer part of the plasma (no longer supported by beam heating), resulting in an increase in impurity radiation, and a possible shrinking of the hot core.

The results summarized in Table IV-4 suggest that it may indeed be possible to radiate a substantial fraction of the power by means of edge-region impurity radiation in an ignited INTOR plasma. For Fe, Ni, and Mo impurities, we see that the radiated power density averaged over the edge region (arbitrarily taken to be 20 cm wide) can be of the order of 1 W/cm^3. (Note that W/cm^3 would correspond to a total radiated power of about 60 MW in INTOR). On the other hand, in the Garching calculations [1] this radiation loss was usually too high for ignition to be reached. In just one case, however, under strongly optimized conditions, ignition was obtained. In a somewhat larger sized INTOR plasma, excessive shrinkage of the hot core can be prevented more easily when neutral beam heating is switched off. Clearly, the parameters of INTOR are at best marginal for the achievement of ignition with impurities. At the minimum, one should make some allowance for the needed 20 cm of very-low-temperature edge plasma.

For carbon impurities, it does not seem possible to achieve an edge-region radiated power density significantly higher than the bremsstrahlung density from the centre of the plasma. The best case occurs with temperature gradient screening, but even here the edge-region radiated power density is only about 0.2 W/cm^3.

Neoclassical temperature gradient screening is, of course, always helpful in producing a desirable impurity density profile. However, at least in the USA calculations, the high relative level of impurity radiation from the edge region does not depend upon a sharp peak in the impurity density in this region. Such an outwardly peaked impurity density may be difficult to maintain over a long burn pulse, in the face of anomalous transport processes, MHD modes, minor disruptions, etc., that will tend to mix the different ion species more uniformly.

TABLE IV-4. COMPUTATIONAL RESULTS FOR INTOR INCLUDING IMPURITIES

Code	Impurity Species	Impurity influx Model	Temperature Gradient Screening	Impurity Fraction $\bar{n}_{imp}/\bar{n}_e$	Radiated Power Density (W/cm³) Centre	Edge Region	Edge Region Impurity Density (cm^{-3})	Impurity Profile Type	Comments
Culham Icarus*	Fe	(i)	No	0.12%	0.03	0.6	3×10^{11}	edge peaked	Ignition of the core is achieved, but overall energy balance positive only if beam heating of the edge is maintained.
Hermes*	Fe	(iii)	Yes	0.1%	0.05	0.5	$\cong 10^{11}$	weakly edge peaked	Ignition achieved. 50 MW beam switched off when $\langle n_e \rangle = 0.5 \times 10^{20}\ m^{-3}$; density then increased by gas puff.
Fontenay*	Fe	(v)	No	0.03%**	0.02	0.04	0.4×10^{11}	flat	No ignition for 0.05% Fe. Heating pulse only 4 s
Garching	Fe	(iii)	Yes	small		~1–2	$\sim 7 \times 10^{10}$	edge peaked	Ignition achieved in a larger INTOR plasma
JAERI+	C	(v)	Yes	6%	~0.2	~0.2	$\sim 10^{13}$	edge peaked	Sufficient radiation for some edge cooling
	C	(v)	No	6%	~0.4	~0.05	$\sim 4 \times 10^{12}$	centre-peaked	Central bremsstrahlung too large for ignition
Princeton	Ni (like Fe)	(iv)	No	0.15%	~0.1	~0.5	$\sim 1.5 \times 10^{11}$	flat	Prescribed impurity influx for 3 s during heating phase gives final steady fractions listed. Ignition occurs at these and lower fractions.
	Mo	(iv)	No	0.05%	~0.1	~1.2	$\sim 8 \times 10^{10}$	flat	

* These calculations were actually for JET rather than INTOR, but the results may be expected to be qualitatively similar for both devices.

** In this model, the fractional impurity density n_{imp}/ne is uniform over the plasma.

\+ The JAERI calculations do not include self-consistent evolution of the plasma profiles; the electron temperature profile is prescribed, and has an outer region about 20 cm wide at ~50 eV.

3.5. **Burn pulse scenarios** (see also Chapter VIII)

All of the INTOR delegations have reported long pulse calculations with helium build up. With complete retention of the helium within the plasma, the resulting depletion of the reacting fuel may be expected to limit the burn pulse to about 25–30 s if the temperature and $\langle\beta\rangle$-value are held fixed, and this is confirmed by the calculations.

If the DT fuel density can be continuously raised as the helium builds up, the burn pulse can of course be extended, provided the higher $\langle\beta\rangle$-values do not lead to instability. In one calculation [4], the burn pulse was extended to about 75 s but by this time the $\langle\beta\rangle$-value had clearly risen into what is believed to be an MHD-unstable regime. In this case, the plasma was operated in a weakly-driven sub-ignition mode in which there was no thermal runaway and the average temperature remained at about 10 keV. Similar calculations have been reported by the Garching group [1].

In an ignited plasma, thermal runaway will occur, i.e. the temperature will rise substantially and the $\langle\beta\rangle$-value will enter the MHD unstable regime, unless some form of burn control is provided. Similarly, some form of burn control must be incorporated into the computational treatments, if realistic calculations are to be made of long-pulse ignited plasmas.

A variety of possible burn control techniques are discussed in the report on Start-up, Burn, and Shutdown [6]. Some fairly detailed computations have been made of one of these techniques – namely, burn control by magnetic field ripple, by means of its enhancement of ion heat conduction. For good burn control, the field ripple must be adjustable to a value larger than that which permits ignition. In one calculation [7] it was found adequate to raise the ripple in the burn phase to ±1.25% at the edge of the plasma, from a much smaller value in the heating phase. In this case, the higher ripple was created by reducing currents in pull-back coils which, when activated, reduced the ripple in a 6-coil TF system to an acceptable level.

To treat the burn phase, field ripple contributions to ion thermal conductivity and helium build up have both been incorporated into many of the transport codes used to predict the performance of INTOR. However, realistic treatments are not yet available for the interesting case where a fraction of the helium recycling at the plasma edge is exhausted. One model that has been employed is to apply a boundary condition on the helium ion density n_{He} of the form $dn_{He}/dr = -n_{He}/\Delta$ at $r = a$, where Δ is typically taken to be about 10 cm [3]. (This model corresponds, in a very approximate way, to the case where about 90% of the outflux of helium ions are recycled as helium neutrals, penetrating about 1 cm into the plasma.) With this model, calculations [3] have demonstrated a steady state burn with $\bar{n}_e \cong 1.2 \times 10^{14}$ cm^{-3} in a case where the burn temperature is limited by field ripple.

4. OTHER TRANSPORT PROCESSES

4.1. Field ripple

4.1.1. General discussion and recommendations

It has been suggested that toroidal field ripple may be useful for a number of different purposes: thermal stability during burn, ejection of nearly thermalized alphas from the plasma core, and even (with top-bottom asymmetry) increased penetration of vertically injected neutral beams. The disadvantages of toroidal field ripple are enhanced thermal transport in collisionless plasmas, depletion of the high-energy tail of the thermal ion distribution, and spoiled containment of perpendicularly injected beam ions, or of banana-trapped alpha particles. It is our opinion, in surveying both the theoretical and the experimental literature, that the effects of toroidal field ripple on suprathermal particles are not yet well understood, and that even the calculations regarding the thermal plasma still have considerable uncertainties. Because of these uncertainties, we believe that the toroidal field system must be designed with extremely low ripple. If field ripple is desirable, for example for burn control or ripple-injection, it must be included as an add-on feature, capable of being adjusted to match plasma parameters, and also capable of being switched off altogether.

To illustrate the consequences of field ripple in INTOR, we have first calculated the field ripple for the case of 12 D-shaped TF coils of dimensions 6 m horizontally by 9 m vertically, a somewhat smaller bore than currently recommended for INTOR. We have located the inboard leg at R = 2.4 m and the outboard leg at R = 8.4 m, in the midplane. Each coil has current uniformly distributed over a 1 m × 1 m cross section, so that on the inboard side the coils are packed to fill about 80% of the available circumference. The plasma is centred at R = 4.8 m, and has a diameter of 2.4 m in the midplane. For this case, we find the peak-to-peak ripple to be 4.0% at the outer edge of the plasma, and 0.4% at the plasma centre. Fig. IV-7 gives the results of our calculations, showing contours of various values of the peak-to-peak ripple. The plasma is modeled as elliptical, with q = 3 at the surface, and q = 1 on axis. The shaded areas indicate the regions of the plasma cross section where $\alpha|\sin\theta|>1$, where

$$\alpha = r/(RNq\delta) \qquad (10)$$

with N the number of toroidal field coils, and $\pm\delta(r,\theta)$ the peak-to-average ripple magnitude. When $\alpha|\sin\theta|>|$ (the shaded regions in Fig. IV-7, the local ripple magnetic wells disappear, due to the variation in field strength along a field line arising from toroidal geometry. As the ripple magnitude is decreased, the regions where $\alpha|\sin\theta|>|$ increase in size. Note that in Fig. IV-7 the ripple magnitude

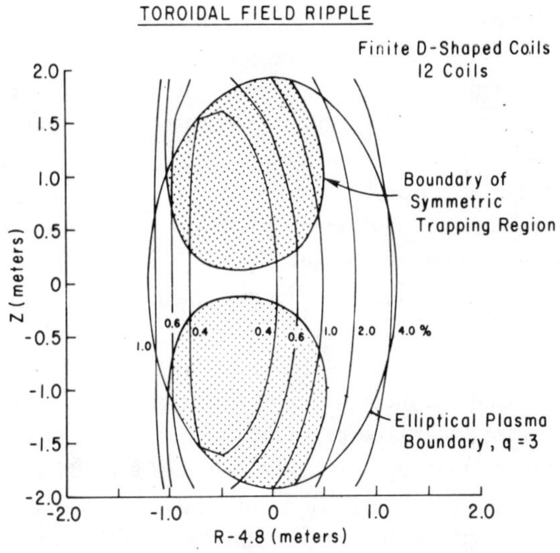

FIG.IV-7. Toroidal field ripple for INTOR with 12 TF coils of 6 m/9 m bore and 1 m² in cross section. The shaded regions indicate areas where local ripple trapping is eliminated. Note that the ripple values given are peak-to-peak values. The ripple rises again on the inboard side because the TF coils are only packed within 80% of the inboard circumference.

increases again with increasing r along $\theta = \pi$, owing to the fact that the toroidal coils are not tightly packed on the inboard side.

The contribution from toroidally trapped ions has been shown to arise mainly from those ions that pass through the ripple with such a small $v_{\|}$-value that they suffer about one effective collision as they traverse one phase of the ripple. The contribution from these ions to χ_i has been shown [3] to be independent of ν_{ii} (over a wide range of ν_{ii}), and has been termed the "ripple-plateau" contribution. It is usually comparable with or smaller than the ripple-trapping term given above, unless $\alpha \gg 1$. There is a third contribution to χ_i from the radial banana drifts of toroidally trapped ions whose turning points are scattered from one phase of the ripple to another, but this contribution is invariably smaller than the other two.

For the 12-coil case (6 m bore), for which the peak-to-peak ripple magnitudes are shown in Fig. IV-7, the ripple-trapping contribution to χ_i would be very large, approximately 50 times larger than the neoclassical χ_i within $0.5 < r/a < 0.9$, for parameters typical of an ignited INTOR [2, 3].

We conclude that the field ripple must be reduced to a much lower value (below 1% for the edge peak-to-average ripple) or else the thermal confinement of an ignited INTOR plasma will be seriously degraded.

4.1.2. Effect on suprathermal ions

While analytic calculations have been made of the effect of toroidal field ripple on ion thermal conductivity, no comprehensive treatment of the effect of ripple on the confinement of suprathermal ions is available. Simple estimates have been made of the reduction of plasma reactivity due to the radial broadening of the high energy tail of the ion distribution, Monte Carlo numerical simulations of perpendicular injection have been performed; and loss-cone-type calculations have been made of the slowing-down alpha distribution in a tokamak with substantial ripple. None of these, however, provides a quantitatively accurate basis on which to determine an acceptable ripple level for INTOR.

It appears to us that the most stringent requirement for low ripple will come from the need to confine nearly perpendicularly injected neutral beam ions. These fast ions are injected directly into the region of velocity space most susceptibl to ripple loss and, for shallow penetration, into the region of the plasma with the highest ripple. Alpha particles, by contrast, are "injected" isotropically and near the plasma centre; since they are highly collisionless, most will have largely thermalized before they sample the lossy $v_{||}/v \approx 0$ region of velocity space. The high-energy tail of the ion distribution function samples the range of pitch-angles more frequently, but at the temperatures anticipated in INTOR the loss of the extreme high-energy tail will not greatly affect the plasma reactivity, particularly if the ripple has been tailored to provide adequate confinement of 1–200 keV perpendicularly injected beam ions.

Monte Carlo orbit-following calculations for TFR showed that a substantial fraction (~30%) of the beam energy could be lost to ripple trapping in a case with 7% peak-to-peak ripple at the plasma edge. Similar calculations for TFTR indicate virtually no loss when the ripple at the plasma surface was ~1%. What distinguishes these two cases is the value of the quantity $\alpha(r)|\sin\theta|$. When $\alpha|\sin\theta|<1$, local ripple wells appear between the toroidal field coils, and fast ions can become locally trapped. A calculation of the ripple trapping time [3] shows that toroidally trapped fast ions with banana tips in regions where there are ripple wells are likely to become localized particles after only some hundreds of bounce times. Since the fast ions are highly collisionless, it is not so much the depth of the ripple wells, but rather only their presence which is of significance. When $\alpha|\sin\theta|>1$, the variation of B along the usual helical field line trajectories prevents local mirror wells from developing; thus no ripple-trapping effects are possible. Banana-drift diffusion effects may persist, but they are observed in the Monte Carlo codes not to be substantially greater than the usual neoclassical transport of injected ions, when $\alpha|\sin\theta|>1$. We tentatively conclude that the constraint that $\alpha|\sin\theta|$ be greater than unity over nearly all the plasma cross section is a necessary and sufficient condition for acceptable confinement of perpendicularly injected fast ions [3].

Referring to Fig. IV-7, we see that for the case with 12 coils and 6 m bore, near perpendicular injection with shallow penetration will put the banana tips of the injected ions into regions with local ripple wells, and so will expose the beam ions to a substantial danger of being lost from the plasma. On the other hand, a 12 coil system with 8 m bore has only a very small region with local ripple wells, and over most of that region, even if ions are locally trapped, they will drift vertically into regions without local wells and be detrapped onto confined banana orbits.

Monte Carlo orbit-following calculations by the Japanese delegation [2] show that the loss of beam ion energy due to field ripple during the slowing-down time could be as high as 15% for a 15° injection angle and 10% for a 20° injection angle, each with a ±0.75% edge ripple, for equal parts of co-injection and counter-injection. For a 15° injection angle, the loss is 10% even for co-injection alone. At the present time, however, these results might be regarded as preliminary, and further analysis of this matter is required.

In summary, if the ripple is low enough, the region where $\alpha|\sin\theta|>1$ is so extensive that the losses of beam ions and alpha-particles are small, even for near perpendicular injection. In particular, beam ion confinement should be adequate for injection at an angle of 20° to the perpendicular (in the co-injection sense, i.e. in the direction of the plasma current). However, either a 16-coil system, an extended outer-limb 12-coil system (i.e., 8 m bore), or ripple-correction coils seem needed to reduce the field ripple to the required level.

4.2. Alpha-particle confinement

Alpha particles are lost almost immediately after they are born if their "banana orbits" intersect the limiter, or if they are trapped in toroidal field ripples. Neglecting field ripple, confinement of alpha particles is a function only of plasma current and aspect ratio, and radial profile with which the alpha-particles are born. For different aspect ratios A, the current I necessary to achieve a given level of alpha-particle confinement scales such that IA = constant. Since alpha confinement depends mainly on poloidal flux, the "equivalent current" I_{eq} (see Section 3.1) should be used in assessing non-circular cross-sections such as INTOR; at the same time the actual aspect ratio A = R/a should be used, not A/a_{eff}, since aspect ratio enters via toroidal ∇ B drift effects on the alpha-particle orbits.

For I_{eq} = 3.8 MA and an aspect ratio A = 4, calculations show that the fraction of alpha-particles that are born on "prompt loss orbits" (i.e., orbits that intersect a "limiter" at r = a on the first transit of the torus) depends sensitively on the alpha-particle birth profile, and on the current profile within the plasma. For a "standard case" of parabolic density and temperature profiles, i.e. $n(r) = 2 \times 10^{14}(1-r^2/a^2)\ cm^{-3}$ and $T(r) = 20(1-r^2/a^2)\ keV$, we find that the

TABLE IV-5. ALPHA-PARTICLE PROMPT LOSS FRACTIONS

Birth Profile	Current Profile $j(r) \propto (1-r^2/a^2)^q$		
$P_\alpha(r) \propto (1-r^2/a^2)^p$	q = 2	q = 1	q = 0
p = 4	1.4%	2.0%	6.1%
p = 2	5.2%	6.2%	11.7%
p = 1	11.5%	12.6%	18.8%

alpha-particle birth profile is closely approximated by $(1-r^2/a^2)^{3.8}$. The alpha-particle prompt loss fraction for this and other cases, for a variety of current profiles, is given in Table IV-5.

An alpha-particle birth profile given by p = 3–4, and a current profile given by q = 1–2, probably represent realistic choices for INTOR. For such profiles, the prompt loss fraction should not exceed 4% and may be expected to be somewhat smaller. When alpha-particles produced by beam-plasma reactions during the heating phase are included, the loss fraction will of course be higher, due to the outwardly peaked beam deposition profile. However, the direct alpha-particle power to the wall due to beam-plasma reactions should not exceed a few MW.

Various mechanisms have been suggested for the loss of alpha particles by microturbulence before they have deposited their energy in the plasma. The most dangerous of these seems to be a mechanism of anomalous transport in the fluctuating electric and magnetic fields arising from unstable shear Alfvén waves*. The shear Alfvén waves are driven unstable by resonances between the wave and harmonics of the bounce frequency of magnetically trapped alpha particles. Either a spatial gradient in the alpha-particle density, or a region of velocity space where the distribution function f increases with increasing energy, is needed for instability. The latter seems to be more dangerous in practice, since the instability is predicted to cause cross-field transport of the alpha particles on a time scale somewhat shorter than their slowing-down time. For parameters close to those of INTOR, the theory predicts substantial cross-field loss of alpha-particle energy, although the loss decreases markedly as the plasma density increases above 10^{14} cm^{-}

* See the extensive published literature on this topic.

TABLE IV-6. R AND D NEEDS ASSESSMENT – ENERGY AND PARTICLE CONFINEMENT

	Status*	
	Adequate	Inadequate
Energy confinement scaling		
at high T_e	x[1]	
in noncircular cross-section	x	
Particle confinement and transport		
hydrogenic ions	x[2]	
helium ions		x
impurity ions	x[2]	
Particle and energy transport in divertor tokamaks	x[2]	
MHD effects on transport		
when $\beta \geqslant \beta$ crit	x[2]	
due to low m resistive modes	x	
Ripple effects on transport		
of thermal ions		x
of beam ions		x
of alpha-particles		x[3]
Demonstration of alpha-particle confinement and heating	x[3]	
Transport and diffusion of		
current density (including bootstrap current)	x[2]	
angular momentum (rotation)	x[2]	

[1] Assumes achievement of high temperatures ($\gtrsim$5 keV) in present-generation devices (e.g. ASDEX, DIII, PDX) by 1980/2, followed by a confirmation of scaling laws at reactor-like temperatures (~10 keV) in next-generation devices (JET, JFT-60, T-15, TFTR) in 1982/4.

[2] Can be carried out on existing experimental facilities but the programme in this area needs increased emphasis.

[3] Must rely on simulation studies of alpha-particle behaviour, using energetic beam ions, until initial D-T results are available from JET/TFTR in about 1984.

Adequate: Programmes are either planned or could be planned on existing or authorized facilities (with some upgrading if required) and using existing trained personnel.

Inadequate: Either the available and authorized manpower and facilities are inadequate or a substantial increase in effort is needed.

* The ratings are made on the world-wide fusion effort and not separately on the individual national programmes.

There are several approximations in the theory, especially in the treatment of the radial eigenfunctions, that might result in a serious overestimate of the dangers of this particular instability. Since the instability has a very low threshold alpha-particle density ($\sim 10^{-5}\, n_e$), it should arise in TFTR, and could be investigated experimentally in that device provided adequate alpha-particle diagnostics are available.

4.3. Synchrotron radiation

Calculations of synchrotron radiation to be expected from INTOR have been provided by the Japanese, European and USSR delegations. Estimates of synchrotron radiation from a tokamak geometry, for various reflection coefficients, that have been given in the literature were employed. At $\langle\beta\rangle \sim 5\%$, and assuming a reflection coefficient of 0.9, the power loss by synchrotron radiation is only about 2 MW at the ignition point in INTOR. Even without reflection, the synchrotron radiation is less than 5% of the alpha-particle heating power at the β-values at which ignition is achieved in INTOR.

5. R AND D NEEDS ASSESSMENT

A summary of the R and D needs assessment for energy and particle confinement is given in Table IV-6.

6. SUMMARY

This report on Energy and Particle Confinement contains an assessment of recent experimental results on energy and particle confinement in tokamaks, in particular those devices with strong auxiliary heating. The report also summarizes the results of many transport computations for INTOR carried out in each of the participating countries, and presents assessments of field ripple effects, synchrotron radiation, and alpha-particle confinement.

The major conclusions of the Energy and Particle Confinement Group are:

1. For the parameter range of present-day devices, there is a well-documented experimental data base on energy confinement. Electron energy confinement is found to obey an empirical scaling law, which provides a good basis for predictions of ignition in INTOR. Ion energy transport is found to be within a factor of about 3 of neoclassical predictions and, in any case, constitutes a relatively minor energy loss channel in INTOR. Uncertainties exist in the extrapolation of these scaling laws to the higher-temperature regime to be reached in INTOR, and an

experimentally-verified transport **theory** could provide a sounder basis for such extrapolations. However, the extrapolations in terms of dimensionless physical parameters are not substantial enough to prevent us from formulating, as a working basis, the requirements for reaching ignition in the INTOR plasma.

2. More data will be needed on the transport of hydrogenic and impurity ions in tokamaks in order to understand fully the requirements for D-T fueling and helium exhaust.

3. Computations for INTOR show that 60–75 MW of 150–200 keV D^0 neutral beam power applied for 4–6 seconds should be able to ignite the plasma core, provided that it is not excessively cooled by impurity radiation.

4. The volume average β at the time of ignition is in the range 4–5%, including contributions from beam ions and alpha particles; the value of the volume average β must rise to about 6% in the burn phase to achieve the desired neutron wall loading.

5. However, whether ignition is reached is very sensitive to many details of the physical effects included. A plasma that ignites slowly from the center outward, while the overall plasma energy density decreases slightly, is the type of ignition that occurs when the device parameters are such that ignition can just be achieved.

6. The D-T fuel mix must be carefully controlled during the heating phase, to ensure an adequate tritium density at the centre of the plasma to balance the substantial injection of deuterium from the beams. The operating scenario must involve supplying pure tritium by gas puffing or pellet injection during the heating phase, and it would be advantageous if some control could be exercised over the recycling rate (probably by a divertor).

7. With respect to impurity effects, the most **desirable** situation is when impurity radiation from the core of the plasma is negligible, while at the same time impurity radiation from the edge region of the plasma is large enough to account for a substantial fraction ($\gtrsim$50%) of the total power outflow. In this way, the edge region is cooled, and the plasma-wall interaction is reduced, so that a steady state would arise in which the influx of impurity atoms is balanced by outward transport of impurity ions. This situation occurs in most present-day tokamaks, often with carbon or oxygen as the dominant impurity. Computations indicate that this desirable situation may also be achievable in the burn phase of INTOR, the critical issue being to prevent the reacting core of the plasma from being excessively cooled by impurity radiation. In the heating phase, the critical issue will be to prevent excessive cooling of the plasma caused by a sharp increase

in impurity radiation, resulting from a lowering of the edge temperature when the beam heating is turned off. In the case of high-Z metallic impurities (Fe, Mo, etc.) the impurity particle density can be quite uniform, although it is always advantageous if the impurity density can be higher in the edge region than in the plasma core. The width of the radiating edge-region is estimated to be 10–20 cm. In the case of low-Z impurities (C, 0, etc.) sufficient in number to radiate a substantial fraction of the power, the impurity density must be peaked in the edge region, or else bremsstrahlung from the core will be excessive. Theoretically, the impurity density can become peaked in this way in the edge region by various neoclassical effects, but the lack of experimental verification of these impurity transport models makes quantitative conclusions in this area rather unreliable.

8. The fraction of alpha-particles in INTOR that are born on orbits that intersect the wall/limiter does not exceed 4% for all reasonable plasma profiles.

9. The magnetic field ripple must not exceed ± 0.75% at the plasma edge, ± 0.4% at $r/a = 0.7$, and ± 0.1% at the plasma centre, or else the ripple contributions to ion thermal transport will prevent ignition. For INTOR, this value of ripple corresponds to a horizontal bore of about 8 m, for a 12-coil design. If the ripple can be adjusted upwards to the range 1.0–1.5% at the plasma edge, then effective burn control is possible.

10. Ripple effects on beam ion confinement may be important. However, if the injection angle is set at 20° to the normal, the ripple losses of beam-ion energy will be diminished, and preliminary estimates indicate that they will certainly not exceed 10% of the beam power. Ripple effects on the alpha-particle confinement are not significant.

11. Synchrotron radiation will not be a significant power loss mechanism in INTOR, accounting for not more than 5% of the power.

REFERENCES TO CHAPTER IV

[1] European Contributions to the 2nd meeting of the INTOR Workshop, Energy and Particle Confinement, Chapter I, EUR FU/XII 501/79/EDV 50, June 1979 (2 Volumes). European Contributions to the 3rd meeting of the INTOR Workshop, Energy and Particle Confinement, EUR FU/XII 501/79/EDV 70, October 1979.

[2] Fusion Research Development Center, INTOR Workshop Report, Group 1, Energy and Particle Confinement, JAERI-M 8621 (1979).

[3] "Energy and Particle Confinement", Chapter in USA INTOR Report (see Appendix A.5.4).

[4] USSR Report to INTOR, "Energy and Particle Confinement" (see Appendix A.5.3 [7]).

[5] Chapter V, this report.

[6] Chapter VIII, this report.

[7] "Start-up, Burn and Shut-down", Chapter in USA INTOR Report (see Appendix A.5.4).

Chapter V

IMPURITY CONTROL, FUELING AND EXHAUST

T. TAZIMA – Japan, Chairman
J. BOHDANSKY – EC, Expert
F. ENGELMANN – EC
B. KADOMTSEV – USSR
V. PISTUNOVICH – USSR
J.R. PURCELL – USA, Expert
P. RUTHERFORD – USA
Y. SHIMOMURA – Japan, Expert
D. MEADE – USA, Expert

1. OUTLINE OF PLASMA-WALL INTERACTION, IMPURITY CONTROL, AND EXHAUST IN INTOR

1.1. Burn time

Impurity and helium accumulation will lead to serious limitations of the burn time in INTOR, if no effective impurity control and exhaust technique can be provided. In the worst case, the permissible impurity content of the plasma would be exceeded after a short time, of the order of a few particle confinement times, if impurities generated by the flux of plasma energy to the wall and limiter were allowed to diffuse freely into the plasma. Also, the helium concentration in an ignited plasma will necessarily reach the allowable limit of about 10% after about 20 s , unless some of the helium can be exhausted.

A burn time of *at least 100 s* has been prescribed for INTOR, to ensure a reactor-relevant mode of plasma operation, and to permit the main technological objectives of the device to be realized. Clearly then, such a burn time will require some active control of the impurity level and the helium content.

1.2. Requirements for ash exhaust

Very limited theoretical and experimental information is available yet on this topic. Certainly, the helium production rate over a volume V is given by

$$S_{He} = C\bar{n}_D \bar{n}_T \langle\sigma v\rangle_{DT} V$$

where $\langle\sigma v\rangle_{DT} \simeq 10^{-22}\ m^3/s$ at $\bar{T} \simeq 10$ keV, and where C is a numerical constant describing profile effects (e.g., $C \simeq 1.8$ for parabolic n and T profiles). For $\bar{n}_D \simeq \bar{n}_T \simeq 7 \times 10^{19}\ m^{-3}$, and $V \simeq 250\ m^3$, we obtain $S_{He} \simeq 2 \times 10^{20}\ s^{-1}$. If the allowable helium density is $\bar{n}_{He}$, and the helium confinement time, including recycling effects, is τ_{He} (the total helium flux to the wall being given by $(\Gamma_{He})_{wall} = \bar{n}_{He} V/\tau_{He} = \int (S_{He} + S^*_{He}) dV$ where S^*_{He} is the source of charged helium provided by recycling in the boundary layer), then the required exhaust efficiency ζ (i.e., the fraction of the helium flux to the wall that must be permanently removed from the system) is given by

$$\zeta(\bar{n}_{He}/\tau_{He})V = S_{He}$$

If the maximum allowable helium concentration is 10%, i.e. $\bar{n}_{He}/\bar{n} = 0.1$, from considerations of the contributions of helium and its associated electrons to the β-value, then for $\bar{n} \simeq 1.4 \times 10^{20}\ m^{-3}$ we obtain $\zeta \simeq 6 \times 10^{-2}\ \tau_{He}$. Thus, for $\tau_{He} \simeq 1-2$ s, we need exhaust only 6–12% of the helium escaping from the plasma. On the other hand, if $\tau_{He} \simeq 10-15$ s, we need to exhaust almost all of the escaping helium, and if $\tau_{He} > 15$ s, it will be impossible to maintain the helium concentration at the required 10% level.

It is not impossible that the helium recycling time at the edge of the plasma will be much shorter than the gross confinement time, just as it is for hydrogen ions. In this case, the effective τ_{He} will be short, and the fraction of escaping helium to be exhausted will be small. On the other hand, classical transport theory predicts some degree of central accumulation of the helium ions, in which case ζ might have to be larger than estimated above. At present, no reliable value for τ_{He} can be given, since this depends on transport processes not yet documented experimentally, and will depend also on the technique for helium exhaust that is employed.

The pumping requirements to achieve the above exhaust efficiency depend on the helium gas density in front of the pumping ports. If these ports are located in the main chamber, rather than in a divertor chamber, there seems no chance of achieving a high enough gas density, except perhaps in the immediate vicinity of a limiter at which the helium ions are recombining.

1.3. Plasma contamination by impurities sorbed on the wall surface

Appropriate surface conditioning of the first wall is the simplest and most reliable countermeasure to plasma contamination by absorbed and adsorbed impurities, as is demonstrated on present-day experiments.

In order to obtain a clean surface, many different kinds of surface treatments are necessary, both before and after assembly.

Furthermore, it will be important in INTOR to maintain sufficiently clean surfaces for a long time. To help in meeting this requirement, it will be necessary to reduce impurity flows from diagnostic instruments and neutral-beam sources, back-streaming from evacuating systems, and outgassing from walls not facing the plasma. Discharge cleaning between periods of experimental operation is necessary, and it may have to be employed during the dwell time between successive discharges.

1.4. Plasma contamination by first-wall materials

More severe problems will arise from contamination by first wall and limiter materials due to sputtering, self sputtering, evaporation, arcing, and other surface erosion processes.

In present-day tokamaks, several different processes are believed to be important mechanisms for metallic impurity production:

(a) sputtering by light impurity ions (carbon and oxygen) accelerated by the sheath potential near the limiter surface (e.g. PLT);

(b) self-sputtering by the metallic impurity ions themselves, also accelerated in the sheath as in (a) (e.g. DIVA);

(c) evaporation of the limiter surface by local deposition of high energy electrons in low density discharges;

(d) evaporation by arcing, especially in the case of an unstable discharge, or a dirty limiter surface (e.g. DITE).

In large tokamaks, with higher temperatures and longer discharge times, other mechanisms of impurity production are expected to become important:

(e) evaporation of the limiter surface due to the large energy flow from the plasma in cases of small plasma/limiter contact area would be so serious that such a design concept would be quite unacceptable;

(f) sputtering of the first wall by charge-exchange neutrals escaping from the plasma, mainly originating from the outer part of the plasma where the temperature will be in the 50 eV–1 eV range;

(g) sputtering of the first wall by unconfined beam ions in neutral-injection heating, and by high energy neutrals with a broad spectrum extending up to about 100 keV created by charge-exchanges between beam ions and the background neutral gas during slowing-down processes of beam ions;

(h) surface damage (blistering, etc.) and erosion of surfaces directly facing the neutral-beam injectors, which might receive ~10–20% of the neutral beam power in the initial phase of the heating pulse when plasma density is low.

1.5. Roles of a divertor

In attempting to assess the relative importance of various possible functions of a divertor, it is first reiterated that it has proved possible to realize a plasma essentially free from contamination by light impurities, such as oxygen and carbon, on many present-day devices without divertors, by maintaining clean wall surfaces, etc. Although, this is not yet definitely established, it is presumed that these techniques may be extrapolated to INTOR. Thus, removal of light impurities is *not* considered one of the essential functions of an INTOR divertor.

On the other hand, providing exhaust of helium is an *essential* function of an INTOR divertor, for reasons that have been discussed above (see Section 1.2).

The next most *desirable* function of a divertor in INTOR would be to remove some substantial fraction of the thermal power outflux from the plasma, in particular that part of the power transported by plasma conduction and convection processes which would otherwise go to a limiter. This power would now go to the divertor plate, or it might perhaps be deposited in the neutral gas in the divertor chamber, in which case it could be spread over the divertor chamber wall. As it seems likely that metallic impurities produced in the divertor chamber will be effectively prevented from passing to the plasma chamber, because of the screening action of the plasma scrape-off layer and their large sticking coefficient at the divertor chamber walls, the limiter-generated impurities are expected to be greatly reduced in the divertor case.

The exhaust of D-T particles is *not* an essential function of the INTOR divertor. Indeed, to increase the burn-up of tritium, a high degree of D-T recycling is desirable. Thus, the INTOR divertor should operate in a mode which permits the recycling of as many D-T particles as possible back to the main plasma chamber. A bypass providing a neutral gas flow from the divertor to the main discharge chamber might be useful for this.

In any case, the INTOR plasma must be continually fueled in some way (i.e., by gas puffing or pellet injection). This flux of neutral particles into the plasma will inevitably lead to an outflux of charge-exchange neutrals. These neutrals will bombard and sputter the first wall. Thus, a complete prevention of wall sputtering does *not* seem feasible in INTOR, even with a divertor. Therefore, a plasma scrape-off layer providing an efficient shielding of the plasma from the sputtered impurities, is desirable. This requires that the impurities be ionized in the scrape-off layer, and that the probability of their flowing into the divertor chamber be high.

To what extent these functions can be realized depends on the specific type of divertor considered and the regime of operation selected (see Section 5).

1.6. Particle and heat fluxes at the plasma boundary

The total thermal power from an ignited INTOR plasma will be about 100 MW. Of this, charge exchange and bremsstrahlung (for $Z_{eff} \simeq 1.5$) might be expected to account for about 10 MW each. The rest will be in conduction, convection, and impurity excitation and recombination radiation.

On the basis of a confinement time for an individual particle in the main body of the INTOR plasma of 2–3 s, one might estimate the total flux of hydrogenic ions from the plasma interior to be of order 10^{22} particles per s. The flux of particles crossing the plasma boundary (the "recycling flux") in INTOR is not well known but may be much larger than this, since it will be dominated by a high rate of particle recycling in the edge region.

The magnitude of the recycling flux is highly dependent on the conditions encountered at the edge of the plasma, in particular the local plasma parameters, the local density and temperature gradients, and the energy of the incoming neutrals. One way to estimate the recycling flux is to observe that the plasma energy must be transported across the separatrix (or limiter scrape-off) mainly in the form of convection, so that a given amount of power can be associated with a certain amount of particle transport. For example, if 50 MW of power is lost by electron and ion transport (i.e., not by radiation or charge exchange), and if the edge temperature is about 100 eV, then the ion recycling flux must be about 3×10^{23} particles per s when a heat transport enhancement (by electrons) $\gamma \sim 10$ is assumed. For $\bar{n}_e \sim 10^{20} m^{-3}$, with most of the density gradient occurring in the outermost 0.1 m, this flux corresponds to a local diffusion coefficient ~ 1 m²/s. In the case of deep pellet fueling, the recycling flux could be reduced, at the expense of a higher edge temperature. In the case where there is a divertor, the situation is quite complicated and requires careful consideration of the energetics and particle balance in the scrape-off plasma.

1.7. Erosion of the first wall

The first wall of a tokamak reactor, with or without a divertor, will be eroded by sputtering due to bombardment by charge-exchange neutrals escaping from the plasma. The limiter (or divertor-plate in the case of a divertor tokamak) will be similarly eroded due to bombardment by hydrogenic and light impurity ions, and due to self-sputtering by the metallic impurities themselves. In the case of bombardment of the limiter (or divertor plate) the impact energy will be increased due to the sheath potential required to repel electrons. In the present analysis, we will neglect this sheath potential, and will also ignore self sputtering and sputtering by light impurities. Thus, our discussion applies mainly to wall sputtering by charge exchange neutrals, but in a very approximate sense it also gives indications on limiter/divertor – plate sputtering by hydrogenic ions.

Given a certain sputtering rate, the total erosion that will occur will depend on the duty factor and availability of the reactor. For INTOR, we take these to be 70% and 25%, respectively, corresponding to a total discharge time of 5.5×10^6 s.

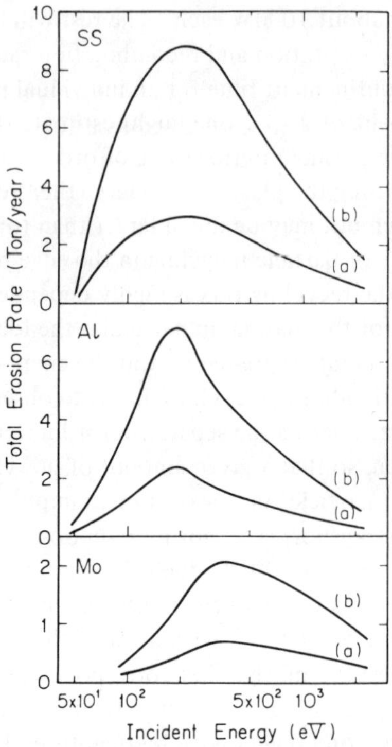

FIG. V-1.Erosion rate of the first wall sputtered by ions and charge-exchange neutrals with energy flux of 10 MW to the wall (a) or 30 MW (b).

The sputtering yield (atoms per incident particle) will depend on the energy of the incident particle, and the total sputtering rate (atoms per second) will depend on the flux of particles at various energies. As we have seen in Section 1.6, the magnitude of the recycling ion flux at the edge of the plasma is highly dependent on the conditions encountered at the edge of the plasma. Also, the flux and energy spectrum of escaping charge-exchange neutrals will depend strongly on the magnitude of the recycling neutral flux (which determines the number of neutrals available for the charge-exchange process), and the ion temperature in the outer part of the plasma (which determines the charge-exchange energy spectrum). Essentially, a certain fraction of the power outflux will be in the form of ion convection and charge-exchange neutrals, but the magnitude of the particle flux, and therefore the mean energy of the escaping particles, will depend sensitively on the edge conditions.

Accordingly, it seems most appropriate to define an erosion rate (atoms per MJ) at various incident particle energies (in the range 50 eV to 2 keV), and to use this erosion rate to estimate a total wall erosion (kg per year) for various assumptions about the total power outflux in hydrogenic particles that contribute to sputtering.

We have made such estimates for SS, Al, and Mo walls, and the results are shown in Fig.V-1. We have used sputtering coefficients (average of D and T values) provided in Ref. [1]. We have considered two cases: Case (a), in which the edge region is supposed to be sufficiently cooled by radiation so that only 10 MW of power is in sputtering particles; and Case (b), in which 30 MW of power is in such particles. If,in the worst case, all the sputtered atoms are lost from the first wall, this would imply that one would have an average erosion rate of 0.5 to 1.5 mm/year for stainless steel, 1.2 to 3.6 mm/year for Al, and up to 0.3 mm/year for Mo. Note that generally the highest erosion rates are encountered for edge temperatures between 200 and 300 eV. Therefore, to minimize wall erosion, one should cool the edge down to about 50 eV. Erosion is also reduced somewhat with edge temperatures in the keV range.

2. PERMISSIBLE LEVELS OF IMPURITIES [1–4]

Radiation from impurity ions may be an important loss mechanism. In particular, high-Z impurities, such as tungsten and molybdenum, not being fully ionized, irradiate strongly.

Whereas, the cross sections of the relevant atomic processes are reasonably well known for low-Z impurities, this is not the case for high-Z impurities. The theoretical results differ typically by a factor of 2 and experimental verifications are almost totally missing. A further uncertainty comes from the fact that impurity radiation from plasmas are normally estimated on the basis of the coronal model. However, for the core of a large plasma, like in INTOR, this is expected to be a good approximation. Typical radiation powers, estimated in this way, are shown in Fig. V-2 [3].

From these estimates, maximum permissible levels of impurity concentrations can be obtained for different impurity atoms, temperatures of the hot thermonuclear core, and Q values of the discharge. An example is given in Fig. V-3 [3], where nonradiative losses have been assumed to be negligible. As there is not a large margin for attaining ignition in INTOR without impurities, the radiation losses from the plasma core, induced by impurities should not exceed 10 to 20 MW. This implies that the impurity concentrations in the hot core of INTOR should not exceed about one tenth of the values given in Fig.V-3 for large Q.

It has to be noted that more stringent requirements must be fulfilled at the plasma edge and during start-up. In fact, here, the roughly inverse dependence of radiative losses on temperature, and stronger deviations from coronal equilibrium, lead to appreciably higher losses. On the other hand, if the plasma

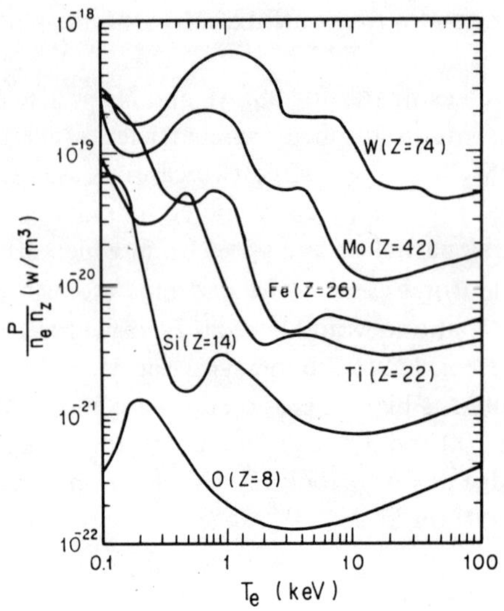

FIG. V-2. Comparison of total radiated power for a variety of impurities showing the increase with Z.

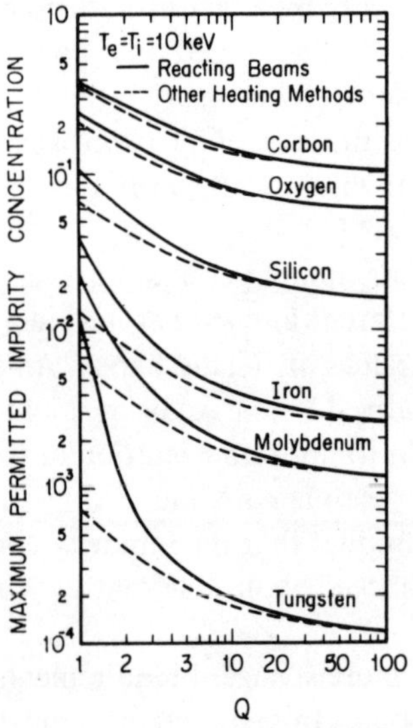

FIG. V-3. Maximum permitted impurity concentration.

core is sufficiently clean, an appropriate amount of impurities in the outside layers of the discharge is beneficial in leading to edge cooling and cold blanket formation. It is, at present, not clear whether such favourable impurity density profiles can exist and be sustained for the duration of an INTOR discharge (Chapter IV).

3. EXPERIMENTAL RESULTS ON PLASMA-WALL INTERACTION

3.1. Impurity production

Plasma contamination in present day tokamaks is related to either impurities sorbed on the wall surfaces (e.g. oxygen, carbon and nitrogen) or influx of first wall material, especially limiter material (metals or low-Z). In plasmas with large oxygen impurity concentrations (5–10%), most of the energy flow carried by plasma particles from the hot core region is converted into radiation due to oxygen impurities in the boundary region, so that strong interactions between the plasma and the limiter do not occur, and serious plasma contamination by metallic impurities is avoided.

A large effort has been made to reduce the light impurity concentrations, in order to improve the energy confinement time. Often these efforts resulted in an increase in the metallic impurity concentration, sometimes causing a hollow discharge owing to the increase of radiation from the hot core, and leading to extremely short energy confinement times. Therefore, in order to suppress large metallic impurity concentrations in plasmas with small light impurity concentrations, both gas puffing for edge cooling, and divertors for shielding the plasma from the influx of sputtered atoms, have been successfully applied in joule-heated plasmas.

However, in the presence of strong neutral beam heating (as in PLT, for example) gas puffing has not worked well for suppressing the strong interaction of the plasma with the limiter. In this case, a graphite limiter was more successful than a metallic one, but the energy confinement times were two times shorter than those in joule-heated plasmas at the same density with low light impurity concentrations. On the other hand, in ISX-B with a stainless-steel limiter, the energy confinement times for joule heating at about 200 kW and for beam heating up to 1 MW are similar.

(i) Absorbed and adsorbed impurity production

We can expect that the production mechanisms of light impurities in INTOR will not be much different from those in presentday tokamaks. Mainly these are particle induced desorption and/or thermal desorption of water and methane, which have been formed by chemical reactions between incident hydrogen and surface impurities.

However, further quantitative investigations of these desorption rates and of wall conditions in experiments such as PLT, T-10, PDX, ASDEX, etc, are necessary.

(ii) Metallic impurity production

Experimental results from a number of experiments have clarified the major mechanisms for metallic impurity production in tokamaks.

Experimental results in DIVA [2] show also that self-sputtering by the metallic impurity ions accelerated by the sheath potential near the limiter surface is a major mechanism of metallic impurity recycling. Evaporation of the limiter surface in low density discharges takes place due to local energy deposition by high energy electrons, and by arcing in an unstable discharge or in the case of a dirty limiter surface. These studies were done with the following plasma parameters in the scrape-off layer $n_e = (1.5-5) \times 10^{18} m^{-3}$, $T_e = 20-100$ eV, and $T_i = 20-80$ eV. The limiters are electrically biassed up to 500 V.

Results from PLT[3], T-10[4] and JFT-2 have indicated the important role played by limiter sputtering by light impurities.

Recently the metallic impurity production due to unipolar arcing has been investigated in detail in DITE [1], ISX [3] and T-12 [4]. Time resolved optical and electrical measurements in DITE and ISX show that arcing occurs predominantly at the beginning and end of the discharge when the confinement is poor. In disruptive discharges, arcing occurs also along with plasma disruptions. The arcing probability decreases with the distance from the plasma edge and is enhanced by surface oxide layers, oil-contamination layers or machining marks.

Erosion rates due to arcing are of the order of $10^{-5}-10^{-4}$ g/s or about $10^{16}-10^{17}$ atoms/s which are about the same number of particles as are released by sputtering throughout the discharge in a midsize tokamak. Therefore, unipolar arcing is potentially an important impurity source even in large tokamaks. However, the above experimental results show that use of a divertor, a clean surface, and a minimum surface area exposed to the plasma can greatly reduce impurities due to arcing.

(iii) Impurity production from low-Z limiters [1][2][3][4]

Low-Z limiters have been investigated in DIVA, JFT-2, PETULA, PLT, TFR, T-12 and T-10, and good results were obtained except in PETULA. Usually a careful positional control of the plasma, to avoid strong interactions with the limiter, and gas puffing were employed. Low T_e discharge cleaning and successive T_i deposition were also employed to reduce oxygen impurities. In T-10, krypton and argon discharges were used for wall conditioning.

The carbon concentrations in these experiments were several times larger than those in plasmas with large oxygen impurity concentrations, but these carbon levels did not significantly increase the effective ionic charge and radiation, except in PLT with a large auxiliary heating power by neutral beams.

The causes of the large carbon concentration in PLT have not been made clear. The results of JFT-2 [2] showed that the dominant production mechanisms of carbon were physical sputtering by protons and self sputtering by carbon ions, but not chemical sputtering. This conclusion was inferred from the observation that the methane pressure at the periphery, although increasing with increasing limiter temperature, was about the same in cases of Mo, SiC, and PyG limiters. On the other hand, the carbon concentration in the plasma was several times larger with PyG than with SiC or Mo limiters. Thus, there was no correlation between methane pressure (and its temporal evolution) and carbon contamination, suggesting that physical sputtering, rather than chemical sputtering, must have been dominant.

In T-10, the absolute Fe, C, and O content of the discharge was measured with a stainless steel and a graphite limiter. In the latter case, Fe is replaced by light impurities.

In TFR-400, oxygen was the major impurity present when a carbon limiter was used. The carbon line radiation (CIV) during current flat-top was the same as with Mo limiters; the total radiation loss was reduced.

In recent experiments in DIVA [2] it has been found that large chemical sputtering at the high temperature surface of a carbon limiter was easily suppressed by pre-bombardment at high surface temperature, and that carbon line radiation (CII) was almost the same in a temperature range of the limiter surface from 100°C to 650°C (the temperature was changed artificially under the same plasma conditions). Very low $q_a (< 2)$ stable discharges with good confinement and low-Z_{eff} were obtained in a plasma with a PyG limiter and a carbon wall in-situ coated by RF sputtering devices. Arcing was observed only in an unstable discharge.

Experiments in which an atomic hydrogen shower was incident on a carbon material surface [2] supported the results that had been obtained in DIVA and JFT-2, i.e. chemical sputtering was found to be reduced by a factor of 3–4 for PyG and nuclear grade graphite. One of the possible explanations is deactivation (passivation) of the surface by atomic hydrogen at a temperature of about 500°C.

Further investigations in actual devices are necessary to elucidate the carbon impurity production mechanisms in the case of a low-Z first wall.

3.2. Recycling of plasma particles

A recycling rate of at least 0.9 is found in normal operation on almost all present-day tokamaks with stainless steel walls and various metallic limiter materials. Sufficient Ti deposition can reduce the recycling rate down to 0.5. However, at low densities ($\bar{n}_e < 3 \times 10^{19} m^{-3}$) recycling rates down to 0.5 were also found in TFR and ALCATOR without Ti deposition.

Recycling mechanisms include reflection (backscattering), re-emission (including particle-induced desorption and thermal desorption) and trapping. The experimental results of DITE show that the dominant mechanism is re-emission (especially particle-induced desorption) and that most of the recycled particles are those implanted on the first wall surface in previous discharges.

4. SOME SPECIFIC MEANS FOR DECREASING PLASMA CONTAMINATION

4.1. Wall conditioning to reduce absorbed and adsorbed impurities (see also Chapter XVII).

4.1.1. Surface treatment before assembly

Before assembly, thick layers (several thousand angstroms) of the first wall contaminated by oxidation and machining should be removed by glow discharge cleaning, buffpolishing, or electropolishing, since such thick layers cannot be removed by in-situ cleaning after assembly. After once cleaning by such treatments the surface is contaminated no more than 200 Å in depth, even when it is exposed to air for a month.

4.1.2. Wall conditioning after assembly

(i) Baking and chemical surface treatment

In order to attain a high vacuum, baking is necessary not only for the first wall but for surfaces not facing the plasma which cannot be cleaned by in-situ methods other than baking.

For a molybdenum first wall, which can be baked up to 500°C, a chemical surface treatment in which the material is exposed to oxygen and hydrogen gases is a simple method, since it requires only 10–20 h to remove absorbed oxygen and carbon on the molybdenum surfaces. This treatment is more powerful than high-power discharge cleaning.

(ii) Discharge cleaning

Discharge cleaning has proven to be a very powerful means of decreasing plasma contamination. As observed in many experiments (T-10, T-11, T-12, JFT-2, PLT, TFR, ISR/CERN, W-VII, Pulsator, ORMAK, ISX, and PDX), discharge cleaning at low plasma temperature, including glow discharges, is faster than that with high-current discharges. Glow discharges are also suitable for cleaning surfaces remote from the first wall. Further optimization of the discharge cleaning regime is required.

The necessary current densities are in the range from 10^{-6} to 10^{-4} A/cm^2 and the cathode potential is 300 to 400 V depending on the gas used. The power required for INTOR is about 100 kW.

Further details on some of the discharge cleaning methods are given in Chapter XVII.

(iii) In-situ coating

Titanium coating onto the first wall is regarded as a good method to obtain plasmas free from gaseous impurities in existing tokamaks. In DITE, DIVA, ISX, JFT-2, and PLT oxygen impurities have been reduced down to 1–2% of the electron density by Ti deposition onto the torus wall and the limiter.

As a result of reduced impurity influx, broader T_e profiles were obtained, and the maximum attainable density in a given device was improved by a factor about 2. Also, it became somewhat easier to obtain low-q discharges. However, it has been found difficult to reproduce such conditions for many successive shots except for cases where sufficient Ti deposition (ISX, DIVA), or Ti deposition between discharges (JFT-2, PLT), was employed.

The necessary wall temperature to desorb the tritium strongly depends on the material in-situ coated on the wall surface, e.g. $\approx 300°C$ for Ni, $\sim 450°C$ for Mo, $\sim 800°C$ for C and higher than 1000°C for exothermic metals such as Nb, Ti, etc. We can imagine not only Ti but many sorts of materials for in-situ coated layers to obtain a clean wall surface. However, in any case in-situ coating will not be attractive because of exfoliation of coated layers after repeated reapplication.

4.2. Low-Z first wall

The advantage of low-Z materials is their relatively large permissible concentration in the plasma. However, there are many problems in using them for the first wall of large tokamaks with high temperatures and long discharge times.

In discussing such materials it is important to distinguish between graphite and other low-Z materials such as boron carbide, silicon carbide, etc.

(i) Thermal properties

The surface temperature rise is proportional to the temperature rise parameter $(\lambda c \rho)^{-0.5}$, where λ is the thermal conductivity, c is the specific heat, and ρ is the density. Since low-Z materials such as boron carbide have large temperature rise parameters, evaporation becomes more serious than with refractory materials such as Mo.

Low-Z coating or cladding onto the metal substrate such as Mo, W is ineffective, since these thin layers will be soon consumed by sputtering, unless the coating can be reproduced in situ.

Pyrolytic graphite (PyG) is one of the attractive materials from the viewpoint of thermal properties because of its high heat conductivity along the basal plane. However, considerable development is needed of a suitable structure of PyG limiters or divertor plates with cooling mechanisms.

(ii) Chemical sputtering

Laboratory experiments show that low-Z materials such as carbon have sputtering yields about one order of magnitude larger than those of high-Z metals in the temperature range of 400–800°C. Control of the first wall temperature, below 300°C or around about 1000°C, or suppression of chemical sputtering by plasma polymerization would be necessary to lower these values. However, strong chemical sputtering has not been observed in JFT-2 and DIVA experiments (see Section 3.1).

4.3. Impurity transport control [3]

A modification of impurity transport by poloidally asymmetric injection of hydrogen gas has been seen in Ti gettered discharges in ISX-A. The existence of this effect, which is predicted by neoclassical theory, coupled with some agreement between measured and neoclassically predicted transport times, raises the question of whether we can extend the concept to the edge of INTOR.

More gas driven flow reversal tests are planned on ISX-B and Doublet-III in 1980. In addition, tests are planned on ISX-B to attempt to verify the theoretically predicted effect of neutral-beam-produced momentum on impurity transport. RF-produced momentum might do the same, but no tests of this are planned yet. More work is required on means of implementing these concepts in a reactor environment.

5. DIVERTORS FOR IMPURITY CONTROL AND ASH EXHAUST

The basic function of a divertor is to exhaust particles and energy from the main discharge chamber to a secondary chamber, removing the contact between plasma and material walls from the limiter to the divertor plate. For INTOR, the most essential function of a divertor is to exhaust the helium produced in fusion processes. Heat unload and reduction of the impurity influx are important secondary functions of a divertor.

5.1. Divertor functions and exhaust requirements

5.1.1. Particle exhaust

In INTOR, the rate of He production during burn is typically $S_{He} = 2.10^{20}$ particles per second (see Section 1.2). In a steady-state, the helium exhaust flux has to be equal to S_{He}.

When He is exhausted, DT fuel is also inevitably pumped away. The value of the DT exhaust flux Γ_E and the DT fueling flux Γ_F (50%D + 50%T) are related to the He exhaust flux S_{He} via the ratio of the DT and He neutral densities at the pumping ports and the properties of the pumping system. As the former cannot be quantified, no number can be given. There is only a straightforward relation with the burn-up fraction f_B, namely

$$\Gamma_F = \frac{2}{f_B} S_{He}, \quad \Gamma_E = \Gamma_F\left(1 - \frac{f_B}{2}\right)$$

Of course, in order to keep the total tritium inventory low, one would aim at a high burn-up fraction, requiring to minimize the DT fueling flux. For example, if the latter could be kept at about 10 S_{He}, the fractional burn-up would amount to 20%. Therefore, optimum exhaust properties are achieved when the ratio of the DT and He neutral densities in the divertor chamber is as small as possible ("ash enrichment").

The gas densities in the divertor chamber depend on the flux of charged particles into the divertor. Therefore, the values of the particle confinement times in the outside layers of the discharge, and in particular in the scrape-off layer, are important for their determination: the shorter this time is, the higher in the gas density, and the larger is the degree of recycling acceptable for He (see Section 1.2). As, at present, no reliable estimate of the gas densities attainable in different divertors can be given, no statement can be made on the required pumping speed to ensure effective He exhaust. A He density of $n_{He} = 10^{18} m^{-3}$ would be satisfactory but the technical problems appear to be much less if the He gas density in the divertor chamber could be made to rise to

$n_{He} = 10^{19} m^{-3}$. Whereas in the former case the required pumping speed would be 2 X 10^5 ℓ/s, in the latter less than 2 X 10^4 ℓ/s would be needed.

It is very important to note that for the long pulses of INTOR, and in a situation where the main requirement on the divertor is to ensure He exhaust, one will have to work in a regime in which the flux pumped from the divertor chamber is small compared to the plasma influx into the divertor. This implies that in an INTOR divertor, like in a reactor divertor, the flux of particles (neutrals and/or ions) streaming back into the torus would be almost equal to the flux into the divertor.

5.1.2. Energy exhaust

Efficient energy exhaust provides edge cooling of the plasma and removes the bulk of the plasma – wall interaction into the divertor chamber where it is more easily controlled. Therefore, it constitutes an essential part of a divertor's capability to reduce the impurity influx. The thermal power to be exhausted from INTOR by convection and/or heat conduction from the outside layers can be estimated to be at most 80% of the total thermal power, i.e. less than 80 MW.

The energy exhaust efficiency of a divertor strongly depends on its working regime. The problem must be considered in three interactive regions:

(a) Energy transport in the toroidal scrape-off plasma.
(b) Energy transport through the throat of the divertor.
(c) Energy transport within the divertor chamber, either to a target or to a gas mantle that distributes energy by radiative and charge exchange mechanisms to the walls of the chamber.

The transport mechanisms lie between two limiting conditions, namely:

(i) Convective transport when the heat flow along the magnetic field lines can be expressed in the form

$$Q = \bar{n}_s \gamma T_s A_D v_f$$

Here $\bar{n}_s$ is the average density in the diverted plasma, T_s the temperature in the separatrix region, A_D the effective cross sectional area of the flow channel, and v_f is a characteristic flow velocity in the range 0.1–1.0 C_s (where C_s is the ion sound speed); γ is the energy coefficient associated with each electron-ion pair transported, and its value ranges from about 8 to 60, but the upper limit is reduced to about 20 if the main impedance to flow is imposed by a sheath at the divertor plate.

(ii) The alternative transport mechanism is electron conduction parallel to the magnetic field, and the heat flux over a transport length L is given by

$$Q = \frac{A_D}{L} \int \chi_{\|} \, dT_e$$

where $\chi_{\|} = v_{th,e} 2n/\nu_e \sim T_e^{5/2}$ is the electron heat conductivity parallel to the magnetic field, and $v_{th,e}$ and ν_e are the electron thermal velocity and collision frequency, respectively.

In general, convective (or free streaming) transport will occur at low values of $\bar{n}_s$ and small values of A_D and such transport results in a high exhaust temperature (i.e. $\geqslant$ 1 keV). A target sheath imposes somewhat comparable transport conditions and so it also causes high edge temperatures, unless there is substantial recycling of neutral gas at the target. Conversely, conduction at high $\bar{n}_s$ and larger values of A_D can provide substantial heat flow at moderate edge temperatures (see [1] where lower limits of 100 to 200 eV are quoted for 80 MW transport in INTOR). In practice, both forms of transport may be present depending upon the characteristics of the regions outlined in (a), (b) and (c). More detailed analysis of each contending divertor system is needed in order to extrapolate with confidence to INTOR.

The heat flow into the divertor must not exceed the maximum amount that can be safely dissipated within the divertor chamber. An acceptable power loading for a "conventional solid" target plate appears to be ~ 1 to 2 MW m^{-2} which implies a target area of 40 to 80 m^2 for INTOR. The poloidal system proposed in [2] aims to achieve this requirement by ergodization and repeatedly sweeping the plasma across the target plate. Target areas in bundle divertor concepts (even those in the plasma is expanded [1]) tend to be smaller than for the poloidal concept, but the plasma could also be swept across the target in this case and moreover, the target is more easily removable.

The presence of a sheath at a target presents an impedance to heat flow ($\gamma \approx 8$ to 20 in a sheath) and so causes the plasma temperature in the divertor to be high with the concomitant problems of a high erosion rate. The advantages of a cold gas target are evident, particularly in those divertors (such as the bundle) where the exhaust plasma acts as a powerful pump in the divertor throat and thereby ensures good retention of neutral gas within the divertor chamber (cf Sections 5.3.2 and 5.3.4). The chamber can be made large enough to accept the heat distributed to its walls. The target dissipates energy by the mechanisms of charge exchange and radiation and, even if a divertor plate is retained, the sheath potential is substantially reduced by the recycling action in the plasma/plate interface. Further analysis on whether a divertor with a cold gas target is viable is required.

5.1.3. Impurity control

A divertor has also to reduce the impurity influx efficiently, to keep the plasma contamination below the acceptable limits (see Section 5.2). This is achieved, in part, by removing the bulk of the plasma-wall interaction from the first wall of the discharge chamber into the divertor chamber, but also by a shielding effect of the scrape-off layer against impurities generated at the first wall. As the sticking coefficient of impurities on the divertor chamber walls is close to 1, no pumping of impurities is required.

The main requirements for efficient impurity control by a divertor, therefore, are a low probability for impurities to migrate from the divertor chamber into the main discharge chamber, and a high shielding efficiency of the scrape-off layer. A long divertor channel as well as a broad scrape-off layer with high density are, hence, favourable for reducing the impurity influx. Shielding efficiencies as high as 99% have been predicted for bundle divertor configurations [1], but it is to be noted that in the absence of a divertor, impurities from the wall appear to be appreciably shielded by the plasma in the edge region, so that the additional gain obtained by using a divertor, in this respect, may be limited. The effectiveness of a divertor in reducing the impurity influx, is, hence, mainly connected with the reduction of the interaction of the plasma with the first wall, and in particular with the limiter, and the possibility of preventing impurities generated in the divertor from entering the discharge chamber.

5.1.4. Recycling of unburnt fuel from the divertor to the torus

Recycling of unburnt fuel is accomplished when unburnt fuel (either as atoms or ions) flows back through the divertor throat as proposed in [2]. An alternative form of recycling can be achieved by providing a separate bypass for some neutral gas to flow back from the divertor chamber to the torus [1]. With such a device the recycling might also be appropriately changed, e.g., to provide some density profile control during start-up, and to increase the particle unload rate somewhat during shut-down. In addition, burn temperature control by recycling, if viable in principle, might be achieved in this way (cf. Chapter VIII). Recycling of unburnt fuel required the installation of a somewhat higher pumping speed to provide the necessary exhaust flux, unless He backflow through the bypass can be avoided. If it could, an increase of the burn-up fraction would also be achievable.

The benefits or these techniques are dependent upon the divertor chamber being able to be operated at high neutral He and DT density. Thus, the system that provides the maximum density gradient between the divertor throat and the vacuum pump is likely to have the maximum advantages in this respect. In this, localized divertors seem to have a better potential than poloidal divertors with external coils. On the other hand, using a bypass may be imposed if the neutral gas density in the divertor chamber would otherwise increase to the point that charge-exchange losses at the divertor throat become unbearable.

5.2. Results of divertor experiments [1][2][4]

5.2.1. Poloidal divertor

At the present time only DIVA and T-12 have provided results on the effectiveness of a poloidal divertor in a tokamak. The PDX and ASDEX devices will operate with poloidal divertors beginning in late 1979 and 1980, respectively, and will add to our understanding in this area. We can summarize the results mainly from DIVA [2] and T-12 [4], as follows:

(i) Regimes

Depending on the plasma current I_p and the divertor current I_D, with a copper shell, the following regimes are found:

$$\begin{cases} I_D/I_p < 0.7 & \text{Conventional discharge;} \\ 0.7 < I_D/I_p < 2.0 & \text{Divertor discharge;} \\ I_D/I_p > 2.0 & \text{Positionally unstable discharge.} \end{cases}$$

(ii) Stability

The separatrix magnetic surface stabilizes the surface mode but does not affect the m = 2/n = 1 tearing mode, which causes major disruptions and which can be triggered by impurity influx. The level of impurities leading to this disruption is twice as large with the divertor as without the divertor. No major disruption can be excited in divertor and non-divertor discharges with $q_a < 2$.

(iii) Confinement

Reduction of the impurity content by divertor action increases τ_E by increasing the radius of the hot column; the particle confinement time τ_p, and the diffusion and thermal conduction coefficients do not change. One has $\tau_p < \tau_E$ for low levels of radiation loss and $\tau_p > \tau_E$ for normal levels of radiation loss. A modified Alcator scaling describes τ_E both with and without the divertor, namely

$$\tau_E = 1.8 \times 10^{-20} \sqrt{q_a}\, \bar{n}_e\, a_{half}^2 \quad \text{(s)}$$

TABLE V-1. SHIELDING EFFECT OF DIVERTOR ON IMPURITIES

	No divertor, shielding by scrape-off only	Divertor
Shielding efficiency for impurities generated at the wall	~ 80%	~ 90%
Shielding efficiency for impurities generated at the divertor plate	–	99%

where q_a is the safety factor and a_{half} the radius where the electron temperature is one half its peak value.

(iv) Impurities

The shielding effect of the divertor on the impurity influx is shown in Table V-1.

The divertor action reduces the impurity radiation loss by a factor of 2–4. The largest reduction is observed in high current discharges. Impurity shielding from the divertor chamber to the main plasma column is good and no pumping is required for metallic impurities, as their sticking coefficient at the chamber wall is close to 1.

(v) Scrape-off layer plasma with and without a divertor

Energy and particle conservation laws give the following formulae for the electron temperature $\bar{T}_{es}$, the density $\bar{n}_{es}$, the particle flux density f_p and the heat flux density Q in the scrape-off layer, as well as for the scrape-off layer width d:

$$\bar{T}_{es} = \frac{3\tau_p(P_{in} - P_{cx} - P_r)(\bar{T}_e + \bar{T}_i)}{2\tau_E P_{in}\gamma} \quad (1)$$

$$d = \sqrt{D_\perp L/V_f};\ \bar{n}_{es} = L\bar{n}_e\ a/2V_f\ \tau_p\ d \quad (2)$$

$$f_p = \bar{n}_{es}\ v_f;\ Q = \gamma\ f_p\ \bar{T}_{es} \quad (3)$$

Here P_{in}, P_{cx} and P_r are the input power, charge exchange loss and radiation loss, $\bar{T}_e$ and $\bar{T}_i$ are the mean electron and ion temperatures, $\bar{n}_e$ is the mean electron density, γ, $D_\perp$, v_f and L are the heat transport rate, the diffusion coefficient, the particle flow velocity, and the half length of the magnetic field line between the plasma and the divertor plate in the scrape-off layer, respectively. $\bar{T}_{es}$, d, $\bar{n}_{es}$, f_p, and Q can be calculated if γ, v_f and $D_\perp$ are known. **γ, v_f and $D_\perp$ are thus essential for the understanding of the scrape-off layer plasma.**

The following results have been obtained with $B_T = 0.8–2.0$ T:

$\bar{n}_{es} = (1.5–5.0) \times 10^{18}\ m^{-3}$ ($(1–8) \times 10^{18}\ m^{-3}$ in T-12)

$\bar{T}_{es} = 20–100$ eV (10–40 eV in T-12), $\bar{T}_{is} = 20–80$ eV

$v_f \sim 0.3\ C_s$; $D_\perp \sim 0.1\ D_B$ ($0.05\ D_B$ in T-12), where D_B is the Bohm diffusion coefficient

$\gamma \geqslant 100$ in runaway discharges

$\gamma \sim 8–15$ in normal discharges

The observed values of γ agree well with the sheath model, including effects of secondary electrons and superthermal electrons.

From these results and Eq. (1) it follows that $P_{in} \gg P_{cx} + P_r$, $\bar{T}_{es}$ is extremely large and surface erosion and impurity production would be expected to be serious. Cooling of the plasma edge to several tens of eV is therefore required in conventional devices.

5.2.2. Bundle divertor

DITE is the only tokamak which has operated with a bundle divertor [1]. The shielding efficiencies found are 50–70% for carbon impurities and ~ 80% for iron coming from the wall. Only about 50% of hydrogen introduced by gas puffing is carried into the divertor. The unload efficiency of particles is about 30% and that for heat $\geqslant$ 60%. The following scrape-off layer parameters have also been determined:

$\bar{T}_{is} \sim 100–150$ eV (H_α doppler broadening)
$\bar{T}_{es} \sim 30$ eV
$\bar{n}_s \sim 1 \times 10^{18}\ m^{-3}$
$v_f \sim 0.25\ C_s$

The radiation loss is reduced considerably by the divertor action. In a diverted discharge $\tau_p > \tau_E$.

5.3. Divertor configurations for INTOR

The most promising divertors for adoption on INTOR are the poloidal divertor with exterior coils and some type of local divertor. A number of old and new configurations were considered and assessed. In the following, these configurations are described and their respective advantages and disadvantages are discussed. As a general remark it can be said that for none of these divertors has a full analysis yet been made of its potential capabilities in the working regime required for INTOR, nor of its technological implications, so that further design studies and experimental tests are needed. At the present state of knowledge, the least difficulties are seen in the poloidal divertor with external coils, so that near-term design activities for INTOR should be based on this concept. However, it seems improbable that a poloidal divertor could be installed in a commercial fusion reactor, because it requires too much space in the blanket area. This makes it desirable to develop a local divertor concept already for INTOR. The main problem with conventional bundle divertors lies in the possibly deleterious effects of the field ripple on confinement.

5.3.1. Poloidal divertor

Poloidal divertors can be conceived with coils situated inside and outside the toroidal field coils. Adopting internal coils would provide a divertor which would have a stiffer magnetic configuration and a narrower divertor throat, which are advantageous features. However, technologically such a concept leads to very severe difficulties. Either the coils must be linked with the toroidal field coils or the divertor coils have to be segmented, which leads to large forces in the cross-over leads as well as shielding problems. In either case, remote maintenance would be difficult. Therefore, only a poloidal divertor with coils outside the toroidal field coils is considered acceptable for INTOR. Conceptual designs of such a divertor have been made [2][3] and indicate that it may be feasible.

In fact, such a divertor is probably capable of reducing the back flow of neutral helium to small values [2], and should, therefore, provide a helium gas pressure in the divertor chamber sufficient for efficient ash exhaust. As back streaming of neutral DT gas is somewhat easier, a limited enrichment in the divertor chamber of helium with respect to the fuel gas may be possible ("ash enrichment" [2]). The heat load on the divertor plates is bearable, and the flow of impurities from the divertor to the main discharge chamber is expected to be very low. On the other hand, a strong shielding effect of the scrape-off layer against impurities emitted from the first wall is not certain. Moreover, in the absence of an experimental test of a divertor of this type in a working regime relevant to INTOR, nothing definite can be said on its efficiency. Furthermore,

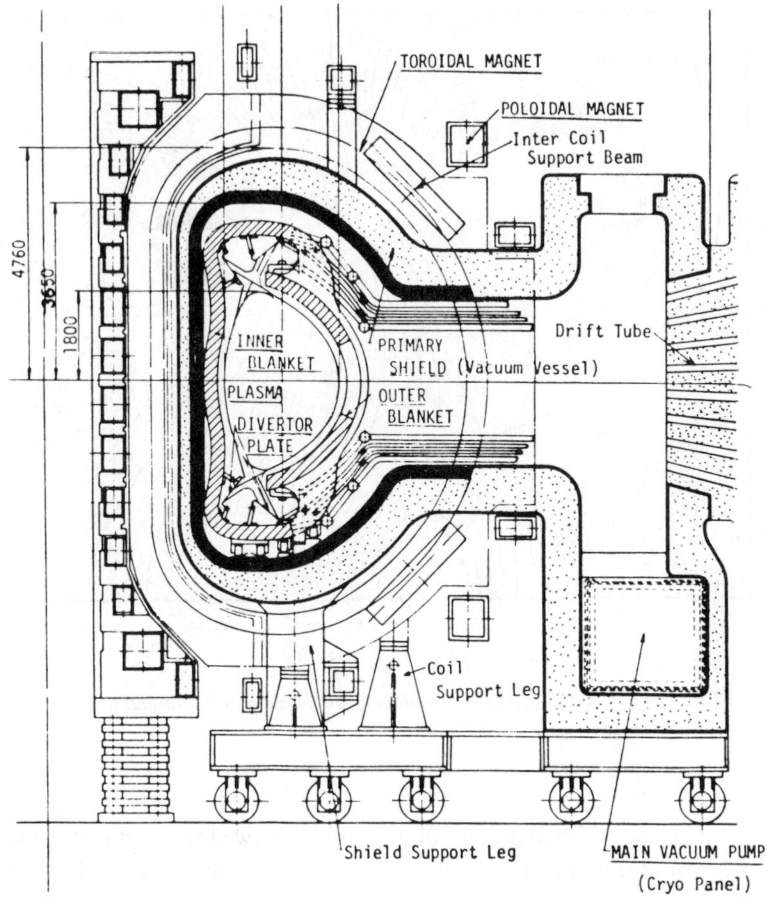

FIG. V-4 a. Design example of a poloidal divertor with exterior coils.

divertor action during start-up requires further study. In fact, keeping the scrape layer aligned with the exhaust ducts during this phase leads to stringent control requirements for the external currents [1], [3]. The main problem lies in the appreciable change in the plasma cross section when β_p is increased from small values to about 3 [1]. Technological problems related to this divertor concept are connected with the large volume, in the blanket region, required for it and with the remote assembly and maintenance of the divertor plates. A schematic drawing of an external coil poloidal divertor concept is shown in Fig. V-4a, and an example of a computed equilibrium is shown in Fig. V-4b.

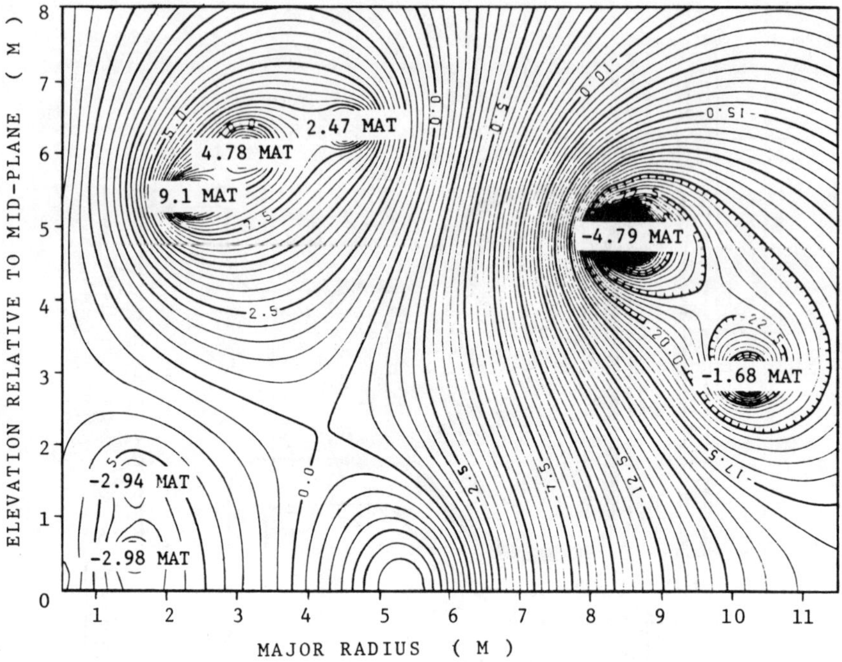

FIG. V-4 b. Computed free boundary equilibrium of a poloidal divertor.

One of the conceptual designs discussed [2] showed that a divertor throat having a width of 0.3 m and a length of 0.5 can be considered typical for a poloidal divertor with external coils. A pumping system with a speed of 5×10^5 l/s was adopted. Monte Carlo calculations were performed assuming that about 50 MW of power were diverted. These showed that some ash enrichment in the divertor was possible when the density in the scrape-off layer was low ($\lesssim 3.10^{18}$ m^{-3}). In this case the temperature in the scrape-off layer is expected to be larger than 200 eV. The heat load on the neutralizer plate can be reduced below 150 W/cm^2 by the following measures:

(1) the angle between divertor plate and scrape-off layer has to be small;

(2) the scrape-off layer has to be as wide as possible (mirror effects are useful, but ergodization of the field lines in the area around the separatrix by a small non-axisymmetric magnetic field, having typically a strength of 0.2% of the toroidal field, would also be required [2]);

(3) the scrape-off layer has to be moved sidewards by 5 to 10 cm, as employed in JT-60 (for this, in INTOR, an AC current of 400 k AT and about 1 Hz would be required; the peak power consumption of this system is about 250 MVA; and the average power 10 MW).

This way, the surface which has plasma contact can be increased to 40 to 60 m^2 and evaporation of the divertor plate is avoided. The flux of impurities from the divertor into the main discharge chamber should be low, as the divertor throat seems long enough to provide an effective barrier for particles having a sticking probability, at the chamber walls, close to 1. The total power required for a poloidal divertor with external coils remains in reasonable limits. In fact, comparing a poloidal field system with and without a divertor, the maximum power increases by 250 MVA, whereas the average power is increased just by the power necessary for swinging the scrape-off layer.

5.3.2. Bundle divertor and its improvements

The conventional bundle divertor incorporates two dipole coils opposing the toroidal field. A bundle of flux is pulled out and null points are created. The advantages of the bundle divertor lie in its simplicity and maintainability. The entrance area (measured perpendicular to the field lines) is an order of magnitude lower than for a poloidal divertor so the scrape-off layer is thicker. The area available for backflow (geometrical area) is also much less than for a poloidal divertor and, whereas the plasma provides powerful pumping, the gas conductance from the divertor chamber back into the torus is very low, especially in view of the convoluted shape of the entrance channel. Consequently, a high density, energy absorbing, gas target seems possible. The plasma recombines within the divertor and, because the density is high, the removal of ash may be carried out at low pumping speed. The installation of a duct to recycle exhaust gas into the scrape-off layer is conceivable. Moreover, if the desirable low edge temperatures below 50 eV are otherwise not achievable the high edge density associated with a bundle divertor is well suited to edge cooling by impurity radiation as the shielding efficiency (> 99%) should confine these impurities to the scrape-off layer.

The main disadvantage of bundle divertors stems from their perturbation of the toroidal field, ranging from 100% at the edge to a few per cent at the centre. The effect on confinement at high temperatures has not been exhaustively investigated for INTOR. For DITE it has been estimated at less than 10% [1]. There are indications that the main effects on confinement is related to the field maxima rather than the minimum on a field line [3]. Experimental investigation will be carried out on ISX-B and DITE.

The power required to oppose the toroidal field is large. There is a trade-off between shield thickness (i.e. lifetime), power consumption, field perturbation and the space needed. For INTOR parameters, a conventional bundle divertor would typically lead to a field ripple on the magnetic axis of about 2%, and would have a space for shielding of about 50 cm between the main plasma and the nearest coil. If the coil is larger and further from the plasma, to increase the shielding

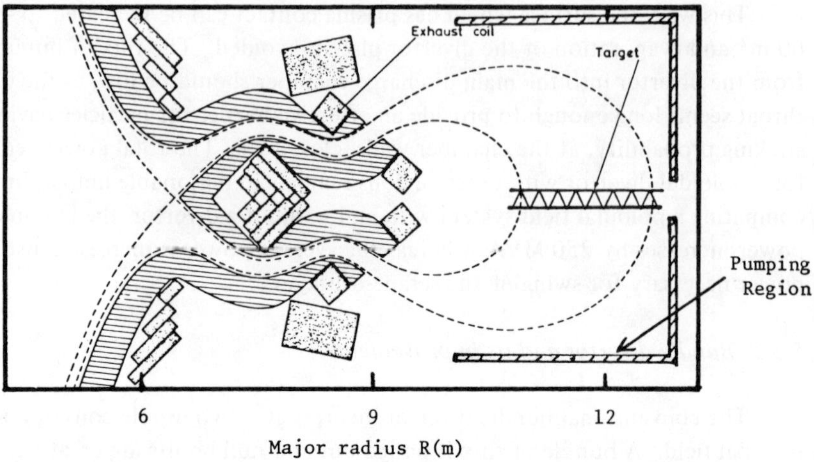

FIG. V-5. Section through the bundle divertor on the horizontal mid-plane showing the additions of two exhaust coils to expand the diverted flux bundle and a suitable location of cryopumping panels.

distance, the field ripple or the power consumption tends to increase. Optimization is possible only in the frame-work of a specific design which has not been undertaken for INTOR, yet.

Suggested improvements to the bundle divertor are as follows:

A 4 coil configuration (Fig.V-5) was proposed [1] which was based on the Culham Conceptual Tokamak CCTR and scaled down to INTOR. The purpose of the second pair of coils is to increase the area of the flux bundle, and thus to decrease the heat flux to the target, whether solid or gas.

An improved bundle divertor in which 4 coils are used for a different purpose, namely to reduce the field perturbation (Fig.V-6) was devised [3]. A value $\Delta B_T/B_T \approx 0.6\%$ on axis was achieved. There seems no reason, in principle, why a higher order null should not be used further to restrict the range of the perturbation to the toroidal field. However, there is a tendency for the space available for shielding to be reduced in such designs. In the case of 4 coil designs of this type for INTOR, shielding thicknesses are about 30 cm, but further optimization of the design is likely to yield better results. Moreover, if $\Delta B_T/B_T$ is very small on the magnetic axis, there will be a tendency for a local maximum in B_T to develop on the inboard side of the magnetic axis, which is believed to increase high-energy ion losses.

A bundle divertor having a T-shaped coil configuration (Fig.V-7) which is the larger divertor-angle limit of the conventional two-coil divertor was also suggested [3]. A preliminary design showed a field ripple of 1.6% on axis, but further improvement is possible.

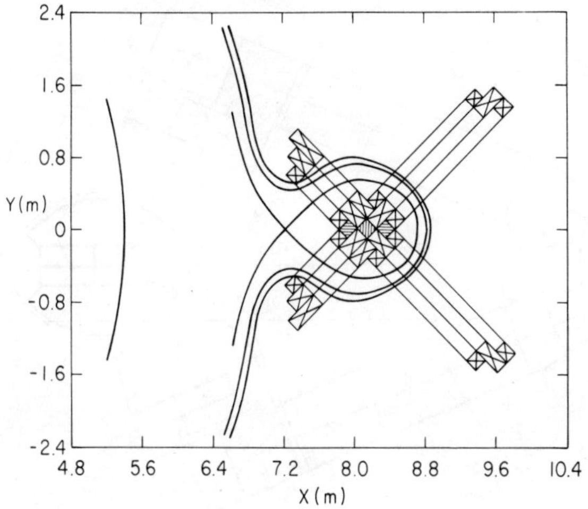

FIG. V-6. A 4-coil bundle divertor and magnetic flux pattern.

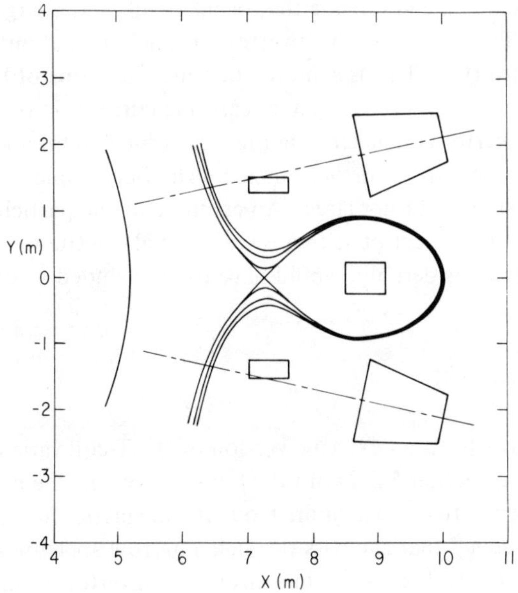

FIG. V-7. A T-shaped bundle divertor coil and magnetic flux pattern.

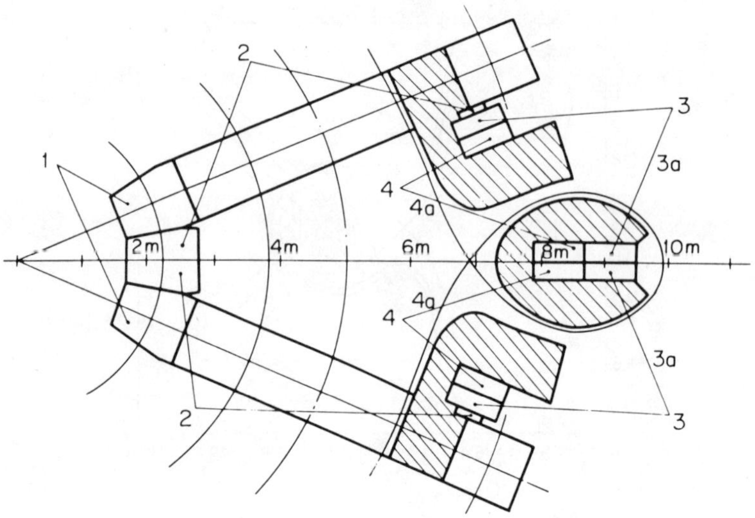

FIG.V-8. A 6-coil bundle divertor and magnetic flux pattern.

A 6-coil configuration (see Fig.V-8) has been suggested by the USSR [4], with two divertor coils located along the major radius, i.e. similar to the toroidal coils, and four coils almost perpendicular to the major radius. In this divertor, toroidal field lines are guided outside the toroidal coils and a target area of 20 m^2 may be attained. The current in the divertor coils might be about half that of the usual bundle divertor. This is achieved because, first, one of the toroidal field coils is just replaced by a coil-system carrying currents in opposite directions and, second, the divertor coil generating the field which counteracts the main toroidal field has been moved further outside. The field ripple on axis, produced by this divertor system, is rather large. An estimate of the particle losses induced has been made, but the effect of self-consistent radial electric fields, which may reduce the loss rate considerably, would have to be included to obtain a reliable result.

5.3.3. Hybrid divertor [*3*]

The hybrid divertor is an extreme version of the T-coil variant of the bundle divertor (see Section 5.3.2). In the hybrid divertor, the main role is played by local radial divertor fields, which draw out the magnetic flux surfaces into a thin horizontal "wedge" that can pass through a narrow aperture into a divertor chamber, as illustrated in Fig.V-9. Once inside the divertor chamber, there are two options as to the manner in which the scrape-off plasma is removed and neutralized: one possibility would be to attach a conventional bundle divertor at

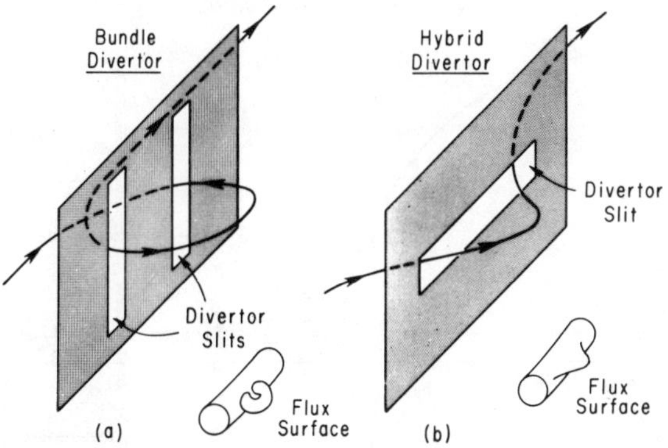

FIG. V-9. Schematic drawings of bundle and hybrid divertor concepts.

the point of maximum field line excursion, where the toroidal field is relatively low ($\simeq$ 0.35 T in one design for INTOR); a second possibility would be to utilize the horizontal hexapole null-trace, naturally created by the coils in the divertor chamber, to deflect the diverted field lines vertically upward or downward (as shown in Fig.V-10).

The virtues of the hybrid divertor concept arise from the relative ease with which flux surfaces can be "folded" along field lines as in Fig. V-9b, rather than across field lines as in Fig.V-9a. The resultant advantages include lower coil current densities, lower stresses, lower field ripple, and greater room for shielding. In particular, the requirements for low field ripple and adequate room for shielding the coil are not in conflict (as they are in conventional bundle divertors), and can be met by a variety of coil configurations and parameters. In one design, shown in Fig.V-10, a low field ripple ($\simeq$ 0.2% on axis, and $\leqslant$ 1% within $r/a \leqslant 0.5$) is achieved with considerable space for shielding ($\simeq$ 1.2 m), at the expense of rather large coil currents ($\simeq$5 MA).

Fig.V-10 shows a cross-sectional view at the central vertical plane ($\phi = 0$) of the divertor. The displacements of field lines (a, b–g, b'–g') in the scrape-off region in passing from $\phi = \pm 90°$ to $\phi = 0$ are shown. Note that there is no plasma current or rotational transform in this example. Field lines within about 20 cm of the midplane at $\phi = \pm 90°$ (a, b, b', c, c') are compressed to whithin a few cm of the midplane at $\phi = \pm 15°$, R = 7.5 m where they may enter the divertor chamber through a horizontal slit some 10 cm wide. The combination of a narrow divertor slit, followed by a broad spreading of the scrape-off field lines, seems promising from the point of view of removing the plasma thermal power.

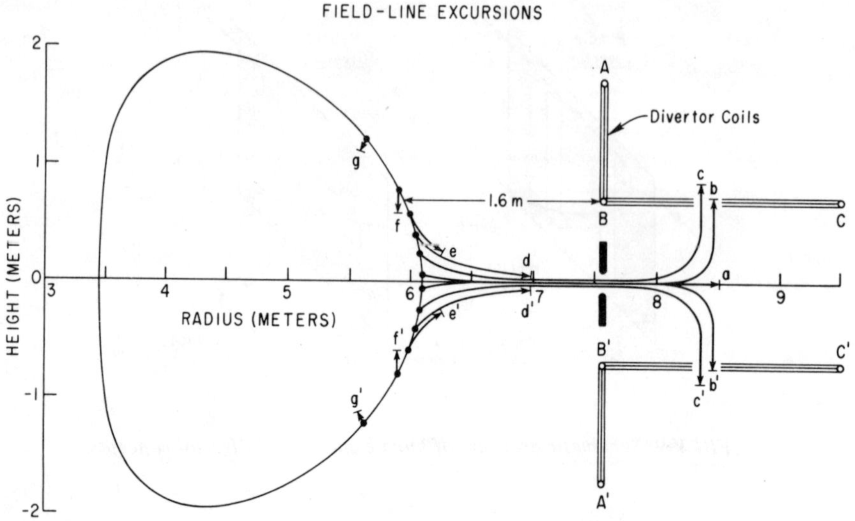

FIG. V-10. Side view showing the excursion of various field lines lying on an arbitrarily chosen "scrape-off surface". The excursions begin at $\phi = \pm 90°$ and are maximal at $\phi = 0°$, where ϕ is the toroidal angle relative to the central vertical "mirror plane" of the divertor system.

5.3.4. Branch (or poloidal bundle) divertor [4]

A "branch divertor", or "poloidal bundle divertor", has a magnetic configuration which resembles a magnetic tube joined to the main body of the plasma at its periphery. The magnetic tube is similar to a branch of a tree growing from a tree trunk (the main plasma); the simple nested configuration of magnetic surfaces will become slightly stochastic in the joining region, and the new configuration of magnetic surfaces in the plasma and the diverted branch will somewhat resemble the annual growth rings within the tree trunk and its branch.

Such a divertor can enhance plasma electron thermal conduction at the periphery, by providing a "shower" of cold electrons, created by ionization in the divertor chamber. To do this, it should have a high neutral density in the divertor chamber, to enhance recycling in the divertor chamber, and it should have a long narrow sleeve between the main plasma chamber and the divertor chamber, to prevent excessive back-flow of neutral gas to the plasma. The plasma physical processes in such a divertor chamber may be similar to those in the hollow cathode.

The branch divertor can be used for thermal unload, providing edge cooling in the form of (harmless) electron flow along the channel, and it can also provide for helium exhaust.

6. NON-DIVERTOR CASE FOR IMPURITY CONTROL AND ASH EXHAUST

In the absence of a divertor, for impurity control and ash exhaust one would have to rely on sufficiently favourable properties of the cool outermost layers ("cold blanket" or "cold mantle", generated by the presence of sufficient edge cooling, either by radiation of impurities in these layers or by an intense recycling of particles) of the discharge being in direct contact with the wall (limiter and, maybe, first wll). In principle, such non-divertor solutions [1], [2], [3], [4] are very attractive for their potential technological simplicity. However, as their viability has not been proven, yet, it would at present seem too risky to rely exclusively on them for INTOR. But if the effectiveness of a divertorless scheme could be shown in time, the implementation of a divertor might become unnecessary.

Whether a non-divertor approach is possible depends mainly on the plasma particle and energy transport properties which determine the density and temperature profiles of plasma and impurities. These are at present not sufficiently known (cf. Chapter IV), especially in the cool outside layers, to be able to make a reliable forecast for INTOR. As far as impurity control is concerned, a necessary condition to reduce sputtering at the wall is the presence of a layer with temperature below 50 eV having an average plasma density $\bar{n}_s$ and a width d such that $\bar{n}_s d \geq 10^{19}$ m^{-2}. This implies that, to have a practicable situation, the width d shoud be less than 0.2 to 0.3 m and the plasma density $\bar{n}_s$ larger than $(3-5) \times 10^{19}$ m^{-3}. Enhanced cooling of the plasma edge might be required to ensure the existence of such a layer. Furthermore, the impurity transport properties of the cold blanket would have to be such that the impurity penetration efficiency from outside is low, i.e., impurities would have to be required to accumulate within the cold blanket, possibly close to the wall, and their recycling back to the wall would have to be strong enough to keep the high-Z impurity content of the edge region within acceptable limits. Analogous properties with respect to helium might also allow a solution for the exhaust problem. The essential question here is whether it is possible to accumulate He gas sufficiently at the pumping ports to attain densities above about 10^{18} m^{-3} as required for efficient pumping. Moreover, since one would like to keep the quantity of DT gas pumped away together with the He as low as possible in order to minimize the external recycling of DT, one would like to have only a limited DT gas density at the pumping ports, possibly lower than 10 times the He gas density. On the other hand, no particular problem is expected to appear with fueling: gas puffing into the cold blanket and subsequent diffusion of the fuel into the reactor core seems to be a viable concept.

In order to be able to judge the viability of a non-divertor approach, further experimental and theoretical work is required. Although an exact simulation of the INTOR plasma is impossible, relevant experiments can be done with large machines in which the condition $\bar{n}a \gg 10^{19}$ m^{-2} is fulfilled. High-field tokamaks with a $\simeq$ 0.3 m operating at densities of the order of 3×10^{20} m^{-3} and higher seem particularly suited. The discharge duration needs to be long enough for a steady state to develop, i.e., longer than the particle confinement time in the core of the discharge. An essential parameter of the problem is also the energy flux through the cold layer which for Alcator-like confinement scales as T/a so that in compact high-field devices a temperature of a few keV is sufficient to attain typical INTOR values. JET, JT-60, TFTR and T-15 are also expected to come close to the relevant parameter range.

Three different versions of a divertorless approach have been proposed. The first relies on toroidal limiters for localizing the plasma wall interaction, the pumping port being situated adjacent to the limiters [3], [4]. In the second, termed "Wall Lapping Plasma", rotating resonant helical islands would be used to distribute the heat flux uniformly over the first wall and mechanical valves applied for exhaust [2]. In the third, a stream of small balls (perhaps titanium or graphite) is dropped at the edge of the plasma to provide limiter action at large heat load [4]. However, none of these schemes has been tested experimentally, yet.

6.1. Ports for ash exhaust adjacent to toroidal limiters [3, 4]

Experiments in T-11 and T-10 with rail limiters show that the effective length of thermal contact between the plasma and the limiter, in the poloidal direction, is 0.3 to 0.5 a. Using this estimate for the poloidal contact length, in the presence of two toroidal limiters of appropriate width, the area on which the plasma has wall contact is about 30 m^2, i.e. typically 10% of the first wall surface. For a total heat loss of the plasma of 100 MW, this implies a limiter heat load of about 3 MW/m^2. Such a load seems acceptable if cooled metallic limiters having a high heat conductivity, e.g. consisting of copper, aluminium and molybdenum alloys, could be used. As far as He removal is concerned, pumping ports adjacent to the limiter might be used, provided that concentrating the recycling at the limiter yields sufficiently high local He densities. As an alternative, implantation of He on the limiter and first wall surfaces was considered [3], [4], but at present it does not seem to be a workable scheme, since it requires a very high energy (above several keV for helium particles) which is inconsistent with a low edge temperature of about 50 eV.

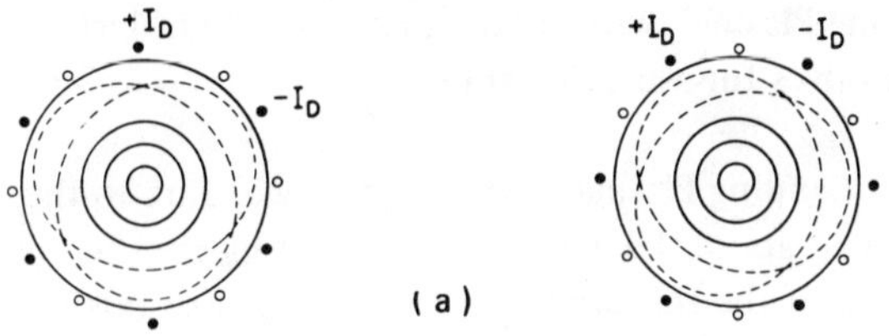

FIG. V-11(a). Schematic of helical islands with resonance at $q = 3$ in "wall-lapping" plasma. Two sets of helical coil currents are quickly alternated.

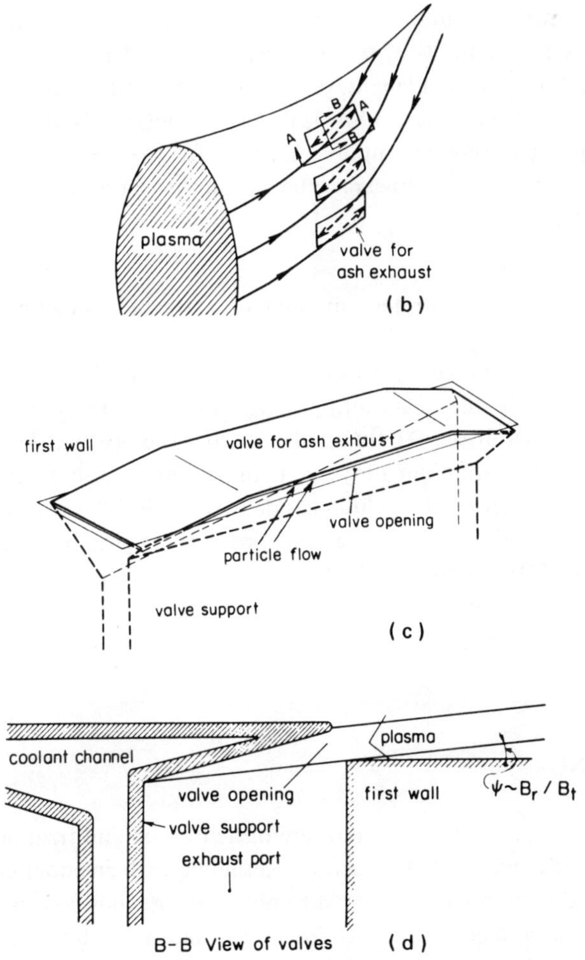

FIG. V-11 (b) (c) (d). Mechanical valves for ash exhaust in "wall-lapping" plasma.

6.2. "Wall Lapping Plasma" with rotating resonant helical islands and mechanical valves for ash exhaust [2]

In the "Wall Lapping Plasma" concept, the heat flux to the first wall would be kept uniformly at about 0.3 MW/m^2 by generating rotating resonant helical islands just in front of the first wall (see Fig. V-11). For this two sets of helical field windings behind the shield to carry an AC (1 to 10 Hz) current of about 50 kA when installed at about 1 m from the plasma edge, and generating additional fields having an amplitude of the order of a percent of the poloidal field, are required. Provided that for He exhaust only a small fraction of the He flux impinging on the first wall has to be guided out of the discharge chamber, simple valves (cf. Fig.V-11) might be used. For example, if this fraction were 5 to 10%, and for a favourable geometry of the port area, these valves might cover no more than 1 to 2% of the first wall surface. The size of one valve might typically be 1.2 m in length, 0.1 m in width and 1 cm in height. Valves of this kind are expected to have a low leakage conductance back to the plasma, and, hence, provide a good pumping efficiency for the He entering the valves. The heat load on the valve appears to be within the limits acceptable for a cooled metal structure.

6.3. Helium exhaust at a limiter consisting of a stream of small balls [4]

A stream of small balls dropped at the edge of the plasma might be used to provide a limiter capable of accepting a large heat load and simultaneously to exhaust a large part of the DT flux incident from the plasma. The ball material could be graphite or titanium alloys. The implantation of helium at low energy will be small. Therefore, the helium gas density behind the stream of balls might exceed the DT density by more than one order of magnitude, reaching about 10^{18} m^{-3} in INTOR. This would allow pumping of helium from ports behind the stream by turbomolecular pumps (cf. Chapter XVII).

7. FUELING

Fueling with a controllable mixture of deuterium and tritium will be needed in the density-rise phase of the startup. Refueling with an equal mixture of deuterium and tritium will be needed to replenish the ions consumed by the fusion reactions, and also to replace fuel exhausted through the divertor.

Gas puffing is the simplest method of fueling and refueling. However, there are uncertainties whether it will work satisfactorily for rapid fueling in the density-rise phase in a divertor case, because of the shielding action of the scrape-off layer and the pumping action of the divertor. Pellet injection will provide superior flexibility over the deposition profile. An attempt should be made to demonstrate experimentally whether pellet injection might also help to obtain an adequate target density for neutral beam injection or certain types of RF heating greater than the present empirical scaling law for the maximum density in joule-heated plasma (see Chapter VIII).

7.1. Gas puffing

Gas puffing is the simplest fueling technique and is used in all present-day tokamaks. The mechanisms by which gas puffing fuels the centre of a tokamak plasma is not well understood. However, there is no reason to believe that fueling by gas puffing will not work in a non-divertor INTOR in much the same way as it works in present machines, apart from the possibility that the inward plasma transport might be due to the neoclassical pinch. Since this effect depends on the parallel electric field, which will be small at the high electron temperatures of an ignited plasma, it will be much weaker in INTOR than in present-day machines.

In the presence of a divertor, gas puffing affects the physical parameters of the scrape-off layer and of the plasma and gas in the divertor chamber. Both the plasma density in the scrape-off layer, and the DT gas density in the divertor chamber, tend to be increased when gas puffing is applied. However, as back-streaming from the divertor chamber will be appreciable during steady burn, this effect will then be weak and, consequently, will not notably increase the quantity of DT to be pumped from the divertor and recycled back into the discharge chamber. Only little experience is, at present available on the use of gas puffing in the presence of a divertor.

7.2. Pellet injection [1, 3, 4]

7.2.1. Current experimental programme

The basic ablation experiments performed at Risø (Puffatron), Oak Ridge (ORMAK), Garching (Pulsator) and the recent fueling experiments in actual tokamak devices carried out on ISX-A and ISX-B have demonstrated the basic feasibility of the method. Based on these preliminary results, further experiments (some of them quite large-scale) are planned at Oak Ridge (fueling of ISX-B), Garching (injection of single pellets in W-VIIa, 1979–1980, fueling of ASDEX, 1980–1981), Princeton (PDX), and Risø (ablation experiment in DANTE, 1979–1980). Some activities are also carried on at the Lebedev Institute.

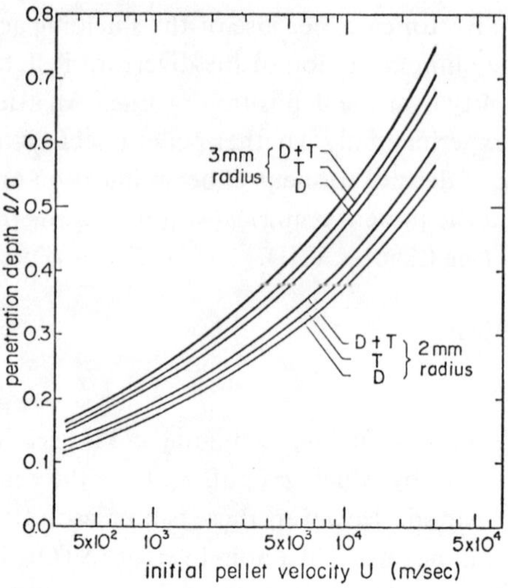

FIG. V-12. Penetration depth of pellet injection for mean density of 1.2×10^{20} /m³, mean temperature 10 keV and plasma radius 1.2 m, assuming parabolic distributions of the density and temperature.

These experiments involve pellet injection by light gas guns with injection velocities up to around 10^3 m/s (Oak Ridge, Garching, Risø, Princeton) and pellet acceleration and injection by centrifugal means (Garching, Oak Ridge) at lower velocities. The latter method seems to have advantages from the point of view of continuous fueling. Impact tests performed with a centrifuge at Garching showed that frozen deuterium pellets can withstand impact velocities as high as 50 m/s.

7.2.2. Injection requirements for INTOR

The calculated penetration depth of pellets in INTOR is shown in Fig.V-12 using a neutral shielding ablation model [3]. Based on the present state of the art, the following conclusions can be drawn:

(i) If wall effects and charge exchange collisions prove to be negligible in INTOR, injection velocities of the order of 10^3 m/s (f = 10 to 20 Hz) may be sufficient. Pellet velocities of this magnitude have already been obtained with single pellets at Oak Ridge and Garching by means of pneumatic accelerators. Such low injection velocity requirements, providing "shallow fueling" [3], can be obtained from transport-code calculations in which normal diffusive effects are adequate to feed the plasma centre.

(ii) In a less optimistic case, where charge exchange from pellet-generated neutrals seriously impairs particle confinement, deeper penetration is found to be required [1][4]. In this case, pellet injection velocities up to 10^4 m/s may be necessary. No injector systems capable of producing such pellet velocities are presently available, and a substantial development programme would be needed. It should be noted that, on the basis of the present state of knowledge, it is not possible to decide whether or not these larger velocities will be required for INTOR.

(iii) The results of pellet-plasma interaction calculations, as well as the experiments in Pulsator, ISX-A, and ISX-B, show that the plasma can survive the injection of rather large single pellets ($\Delta n/n \sim 1$). Assuming that a density perturbation of $\Delta n/n \sim 0.1$ is tolerable under conditions of continuous fueling, the pellet diameter corresponding to a total number of particles $N \sim 10^{22}$ in INTOR is of the order of 4 mm (spherical D_2 pellets, $n_{D_2} \sim 2.3 \times 10^{29}$ m^{-3}. None of the laboratories engaged in fuelling activities has experience in the production of such large pellets.

Additional experimental verification is needed in several areas germane to fueling by pellet injection. Of fundamental importance is a thorough understanding of the underlying physics of pellet ablation and penetration in the hot plasma of interest (1 keV $\lesssim T_e \lesssim$ 10 keV). Closely related to this is the extent to which pellet ablation is enhanced as a result of neutral beam injection heating, or by energetic alpha particles in an ignition device.

The principle of "shallow fueling" requires demonstration under high-temperature plasma conditions. Of equal importance is a determination of the practical upper bound on injected pellet mass, since it has been found experimentally that large pellets penetrate further in a given plasma discharge.

The development of cryogenic apparatus for the production of large size (in the mm-range) hydrogen isotope pellets should be commenced. The pellet producers should be designed for delivering a continuous stream of pellets.

Efforts should be undertaken towards the development of pellet injectors that are capable of accelerating large pellets to a velocity of at least 10^3 m/s (with potentials for further velocity increase) and at a frequency of at least 10 Hz (with a potential for a further increase in frequency).

8. R AND D NEEDS ASSESSMENT

A summary of the R and D needs assessment for impurity control, fueling and exhaust is given in Table V-2.

TABLE V-2. R AND D NEEDS ASSESSMENT – IMPURITY CONTROL, FUELING AND EXHAUST

	Status*	
	Adequate	Inadequate
– Experimental data on plasma-wall interactions, particle recycling, and edge plasma control in discharges with and without a divertor, particularly in INTOR-relevant discharge regimes	x [1]	
– Experimental and theoretical studies of physical processes in:		
a) poloidal divertors	x [1]	
b) local, e.g. bundle, divertors		x
– Design studies of divertors suitable for INTOR		x
– Experimental study of non-divertor techniques for ash exhaust and impurity removal		x
– Experimental studies of pellet injection both in ohmically heated plasmas and in the presence of strong auxiliary heating	x	
– Technology of D and T pellet injection at the required high repetition rate, at a minimum speed of 10^3 m/s, with a design goal of 5×10^3 m/s		x
– Surface physics studies of impurity generation and particle recycling, as well as studies of the atomic and molecular physics of impurities	x [1]	

[1] Programmes and facilities exist in these areas but there should be more emphasis on INTOR-specific problems.

Adequate: Programmes are either planned or could be planned on existing or authorized facilities (with some upgrading if required) and using existing trained personnel.

Inadequate: Either the available and authorized manpower and facilities are inadequate or a substantial increase in effort is needed.

*The ratings are made on the world-wide fusion effort and not separately on the individual national programmes.

9. SUMMARY

The control of the level, species, and distribution of impurities; and the provision for helium ash exhaust, are important and difficult problems in reactor-like tokamaks. We have concluded that provision of some technique for active control of impurities and helium ash is unavoidable in INTOR. Intensive theoretical and experimental studies on INTOR-relevant impurity control and exhaust techniques are, therefore, urgently required. On the basis of our present understanding, our major conclusions are:

1. To ensure that INTOR attains its main objectives in a reactor-relevant manner, a burn pulse of at least 100 s must be provided. Such a pulse length requires an effective technique for removal of the helium ash, and also some means for providing adequate control of the impurity influx.
2. Helium ions must be exhausted from the INTOR plasma at the rate of 2×10^{20} particles/s, i.e., the rate of helium production by fusion reactions. In order to be able to limit the helium concentration in the plasma to the required 10% of the plasma density, the helium particle confinement must not exceed 15 s. If the helium particle confinement time were about 2 s, it would be sufficient if about 10% of the helium transported from the plasma core to the plasma edge could be exhausted. The helium confinement time depends on a variety of transport processes in the plasma and should be investigated experimentally with some urgency.
3. Plasma contamination by impurities adsorbed and absorbed on the wall surfaces can be reduced to acceptable limits by providing proper treatment of the wall surfaces before assembly, and in-situ conditioning after assembly. Further work is needed to identify the optimum method for maintaining a clean surface for very long periods in a high-duty-factor device such as INTOR. The lack of adequate wall cleanliness may also result in the generation of wall-material impurities by unipolar arcs.
4. The prevention of plasma contamination by first wall and limiter/divertor-plate materials will impose severe design constraints on INTOR. The causes will be sputtering of the first wall by charge-exchange neutrals and by unconfined ions, and sputtering of the limiter/divertor-plate by plasma ions, by light impurity ions, and self-sputtering by heavy impurity ions. In fault conditions, impurities can also be produced by evaporation due to high heat loads and/or local heat deposition by runaway electrons.

5. Erosion of the first wall by sputtering may be a serious problem in INTOR. The sputtering erosion rates per power deposited on the wall by hydrogenic particles, for most materials, is maximum at edge temperatures between 200 and 300 eV. E.g., for stainless steel, erosion rates of 0.1 mm/MW-y, or 300 kg/MW-y, may be attained in this case.
6. Some form of magnetic divertor is probably required in INTOR to provide helium exhaust. Depending upon which concept is adopted, a divertor can also provide heat removal from the plasma to the divertor, protection against contamination by impurities generated at the limiter/divertor-plate, and some degree of shielding against impurities sputtered from the first wall. With respect to particle exhaust, the most desirable situation would be to pump the required quantity of helium gas from the divertor chamber, while allowing as much as possible of the hydrogenic gas to recycle to the main plasma chamber. The required pumping speed will depend upon the helium gas density in the divertor chamber. To diminish the exhaust of fuel ions, the hydrogenic gas density in the divertor chamber should be as low as possible. Several divertor concepts are under consideration for INTOR. At present, the poloidal divertor with exterior poloidal-field coils seems to offer a viable solution, and near-term design activity for INTOR should be based on this concept although it is improbable that it could be used in a commercial reactor. Therefore, local divertor concepts have to be developed further. The bundle divertor, in its conventional form, has the drawbacks of a large field perturbation, limited space for shielding, large power consumption, and high stresses. However, suggestions have been made for possible improvement. The magnetic configurations of proposed local poloidal divertors and hybrid poloidal/bundle divertors have been described, but engineering designs of these concepts are not yet available. Before any divertor concept can be finally specified for INTOR, detailed design studies and experimental tests are needed.
7. Non-divertor solutions are attractive for their potential simplicity, but their feasibility has not yet been demonstrated. Three preliminary concepts have been proposed:
   (i) ports for ash exhaust adjacent to a toroidal limiter;
   (ii) a limiter consisting of a stream of dropped titanium or graphite balls; and
   (iii) a "wall-lapping" plasma with rotating resonant helical islands to equalize the heat flux to the first wall, and with mechanical valves for ash exhaust. Further work in this field is required.
8. Fueling with a controllable mixture of deuterium and tritium will be needed in the density-rise phase of the startup. Refueling with an

equal mixture of deuterium and tritium will be needed in the burn phase, at a much slower rate, to replenish the ions consumed by the fusion reactions. Gas puffing is the simplest method of fueling and refueling. However, there are uncertainties whether it will work satisfactorily for rapid fueling in the density-rise phase in a divertor case, because of the shielding action of the scrape-off and the pumping action of the divertor. Pellet injection will provide much superior flexibility over the deposition profile. Pellet injection at 10^3 m/sec is technologically within reach, and is adequate at least for shallow penetration. For deep penetration, which would permit a further increase in the flexibility of the fueling scheme, about 5 times higher injection velocities will be needed. Assuming that the net pumping of hydrogenic gas from the divertor chamber is very small, so that most hydrogenic ions recycle back to the main plasma chamber, the small amount of refueling needed to replenish the spent fuel can probably be provided by gas puffing.

9. Further research on atomic and molecular data, on surface data, and on plasma-wall interactions would be useful for more reliable predictions of impurity behaviour, and to provide more information on which a selection of the first-wall material might be based.

REFERENCES TO CHAPTER V

[1] European Contributions to the 2nd meeting of the INTOR Workshop, Impurity Control, Chapter III, EUR FU/XII 501/79/EDV 50, June 79 (2 Volumes).
European Contributions to the 3rd meeting of the INTOR Workshop, Impurity Control, Fueling and Exhaust, EUR FU/XII 501/79/EDV 70, Oct. 79.

[2] Fusion Research Development Center, INTOR Workshop Report, Group 3, Impurity Control, Fueling and Exhaust, JAERI-M 8622 (1979).

[3] "Impurity Control, Fueling and Exhaust", Chapter in USA INTOR Report (see Appendix A.5.4).

[4] USSR Report to INTOR, "Control of Impurities, Fuel Maintenance, and Removal of Reaction Products in INTOR" (see Appendix 5.4.3 [9]).

Chapter VI

HEATING

P. RUTHERFORD – USA, Chairman
F. ENGELMANN – EC
V.I. PISTUNOVICH – USSR
T. TAZIMA – Japan

1. INTRODUCTION

1.1. Objectives

The purpose of this report is to compare the heating requirements of INTOR with the present state-of-the-art of tokamak plasma heating technology and demonstrated heating performance, and also with the technology expected by 1983–84 according to development and testing programmes in place. This comparison results in a set of recommendations for a heating technology development programme for the 1980's.

1.2. Possible heating methods for a tokamak reactor plasma

Table VI-1 lists the preferred methods for heating tokamak plasmas. To date, the most widely successful method has been energetic neutral beam injection. In addition, ion cyclotron wave heating has generated promising results in large tokamaks, wave heating in the lower-hybrid frequency range has had some success in a large number of experiments in smaller tokamaks, and electron cyclotron heating has proved effective in several very small tokamaks. Transit-time magnetic pumping has been effective in small tokamaks but our assessment does not support its serious consideration for INTOR. Adiabatic compression heating has also been successful in a number of relatively small tokamaks. Other methods, such as relativistic electron beam injection, have barely been attempted so far, and are not considered serious candidates for use in INTOR.

The application of neutral-beam injection heating to reactor plasmas seems feasible from the plasma physics point of view, but the method faces a host of technological problems which are summarized in this report. It is these problems, and the high capital cost of beam injectors, that provide the continuing incentive to investigate alternative heating techniques. But these alternatives, too, have problems. Adiabatic compression requires oversized vacuum chambers and rapid changes in large magnetic energies; ion cyclotron heating faces difficulties in

TABLE VI-1. PREFERRED PLASMA HEATING METHODS

Method	Advantages	Disadvantages
Neutral-Beam Injection	Demonstrated effectiveness	Significant development needed Neutron streaming problems High cost
Radio-Frequency Waves	Less development needed Reduced neutron problems May be cheaper	Heating effectiveness uncertain Launching structures near plasma (in some cases) in neutron environment
Major-Radius Compression*	Demonstrated effectiveness	Oversized vacuum vessel Very large EF power

* Used to multiply effectiveness of NBI or RF heating, but believed impractical for INTOR because of the disadvantages listed.

developing wave launching structures for the reactor environment; the physics of lower-hybrid wave heating in tokamak plasmas is still not understood and its effectiveness is uncertain; the power tubes for electron cyclotron heating face substantial development problems, and this heating method has undergone only limited testing; transit-time magnetic pumping would be lacking in flexibility with respect to changing plasma temperatures during start-up, and appears to have serious technological difficulties in a reactor situation.

The effectiveness of high-power neutral-beam injection heating has now been demonstrated in several tokamaks. Despite apparent disadvantages, this method offers, at the present time, the best assurance of providing adequate heating for the INTOR plasma. Consequently, neutral-beam heating has been adopted as the reference heating method. Whether beam heating will be displaced eventually by another technique as the reference method depends to a great extent on the results of heating experiments during the next four years. The INTOR Workshop recommends a more vigorous world-wide experimental effort in RF heating of tokamaks, to provide more definitive results on a time scale that could be compatible with the needs of INTOR.

2. HEATING REQUIREMENTS FOR INTOR

2.1. Power levels

The power required to heat the INTOR plasma to ignition is not known definitively, because of the continued uncertainty in high-temperature transport coefficients and impurity radiation losses, and because a variety of schemes for reaching ignition has been devised. These schemes are discussed in Chapters IV and VIII and result in power requirements in the range 60–75 MW for neutral beam heating. It seems advisable to specify a heating capacity at the upper end of this range, and even some additional margin will be useful when one or more heating stations are forced out of action for maintenance or by breakdown. However, for an "ideal" heating technique, in which the power can be deposited in the central region of the plasma throughout the start-up and heating-to-ignition phases, the total power requirements could probably be reduced significantly.

2.2. Pulse length

2.2.1. Ignition scenario

Depending on transport scaling and on the heating scenario adopted, a heating pulse length of 5–8 s is required for INTOR. However, provision for utilizing a heating system during the burn is required if an ignited plasma becomes extinguished. This might occur because of impurity (or alpha) buildup, or minor plasma disruptions that reduce the plasma temperature or density, and the plasma must then be reignited, or driven at high Q, to complete the 100 s burn pulse. The required heating capacity for this purpose could come from other than the primary heating system, but it is likely to be at least 10 MW. If reignition of an extinguished burn is all that is needed, it may be sufficient to have the capability to restart the original heating system during the pulse.

2.2.2. Driven systems

In the unlikely event that ignition cannot be attained at all, then the heating systems must be capable of long-pulse operation in order to provide the neutron fluence requirement of INTOR in high-Q operation ($Q \gtrsim 20$). Maintenance of the reacting plasma just below ignition conditions may actually turn out to be desirable for thermal stability control, and is favoured by the USSR delegation. Maintenance at $Q \cong 20$ would require a heating power of about 20 MW.

Thus, for any operating mode, the capability of long-pulse operation of at least a significant fraction of the INTOR heating system capacity is an important basic requirement.

For controlled operation of a subignition plasma, it would be extremely desirable for the long-pulse heating component to penetrate completely to the plasma centre, where there is a shortfall in alpha particle power. For example, negative-ion based beams at 400–500 keV could be used.

2.3. Neutral-beam heating status related to INTOR

Heating of the INTOR plasma to ignition conditions seems feasible using neutral-beam energies achievable with positive-ion-based technology at 175 keV per deuteron, but with significant advancements beyond TFTR/JET/JT-60/T-15 beam technology required to obtain long pulses and higher energy than has been currently demonstrated, and also a high duty factor. A 200 keV beam energy would be preferable if it could be achieved, but it is considered more important to provide a high fraction of monatomic ions, resulting in a high fraction of the neutral beam power in the full energy component. Other development requirements are an increase in overall efficiency (direct energy recovery is probably needed) and increased reliability and component lifetime.

Potentially serious disadvantages of the beam heating method are the exposure of the injector internals to the fusion neutron flux, and the expected high capital cost of beam systems. The difficulties incurred by the direct exposure of injector components to neutrons can probably be overcome, especially if the beam injectors can be closed off after start-up. Provision of effective neutral-beam heating during the density-rise phase in INTOR, probably needed, poses some difficulties.

2.4. Radio-frequency heating status related to INTOR

The continued attraction of pursuing the many different RF heating schemes derives in part from the perceived disadvantages of neutral-beam injection mentioned above. To date, only ion cyclotron wave heating (ICRF) has provided a convincing demonstration of substantial plasma heating in tokamaks of significant size. RF heating in the lower-hybrid frequency range has been successful in smaller tokamaks but the physical mechanisms of coupling wave power to the plasma are less well understood. The power sources for electron cyclotron heating face formidable problems of development and cost.

The greater success of ICRF heating, the reasonably straightforward extrapolation of the ICRF technique to INTOR, and the relative inexpensiveness of ICRF systems, made it the technique that won most overall endorsement from the INTOR Workshop. However, lower-hybrid heating has important technological advantages, and should be retained as a viable option. If the sources can be developed, electron cyclotron heating should be provided at least at small power (~ 10 MW), and possibly at a power level appropriate for bulk heating (~ 50 MW). Transit-time magnetic pumping does not seem extrapolatable to INTOR.

3. PHYSICS OF NEUTRAL-BEAM INJECTION

3.1. Beam penetration

The penetration of a neutron beam into a tokamak plasma is determined by impact-ionization collisions with the plasma electrons and ions, and by charge-exchange collisions. As far as the effect of impurity ions is concerned, the most recent measurements and theoretical analyses indicate that charge-exchange trapping on impurity ions dominates over impact ionization on impurities up to beam energies of at least 100 keV (D^0), and that the charge-exchange cross-section on impurities scales nearly as $Z^{1.4}$ [3]. Thus for the values of Z_{eff} which are permitted by the need to obtain ignition in INTOR (i.e., $Z_{eff} \lesssim 2$), ionization on impurities is expected to be relatively unimportant (see Fig. VI-1). Most experiments to date are consistent with "classical" beam penetration, but there is some indication that in high-density PLT plasmas the penetration of neutral beams is somewhat poorer than expected.

From Fig. VI-1, we can obtain a rough expression for the neutral-beam mean free path in the energy range of interest, namely

$$\lambda_{mfp}(m) = E_b\,(keV)/[180\; n_e\; (10^{20}\; m^{-3})\; A_b]$$

Where A_b is the atomic number of the beam particles. Writing the condition for effective penetration of a perpendicularly injected deuterium beam as $\lambda_{mfp} \sim a/2$ (although PLT data suggests that $\lambda_{mfp} \sim 2a/3$ might be more appropriate we obtain

$$E_b\;(keV) \cong 180\,\bar{n}_e\;(10^{20}\,m^{-3})\;a(m)$$

For an INTOR plasma with $\bar{n}_e = 1.4 \times 10^{20}\,m^{-3}$ and $a = 1.2$ m, the requirement becomes $E_b \cong 300$ keV. Several INTOR delegations [1, 4] have provided detailed calculations of beam deposition profiles for deuterium beams of various energies perpendicularly injected into plasmas with a = 1.5 m and various densities and density profiles. The results of these calculations are in good agreement with the rough formula given above. Specifically [1], for a **flat** density profile with $\bar{n}_e = 1.5 \times 10^{20}\,m^{-3}$, and three-species 200 keV beams with 50–80% of the power in the full-energy component, the penetration is quite unsatisfactory, with a peak in the energy deposition profile occurring at $r/a \cong 0.7$–0.8. For a parabolic density profile with $\bar{n}_e = 1.5 \times 10^{20}\,m^{-3}$, or for a flat profile with $\bar{n}_e = 1 \times 10^{20}\,m^{-3}$, the deposition profile is much more satisfactory, the energy being broadly distributed within $r/a \leqslant 0.8$.

Since positive-ion based neutral beams with $E_b = 200$ keV are not practically realizable, an ignition scenario for INTOR must be developed that involves a

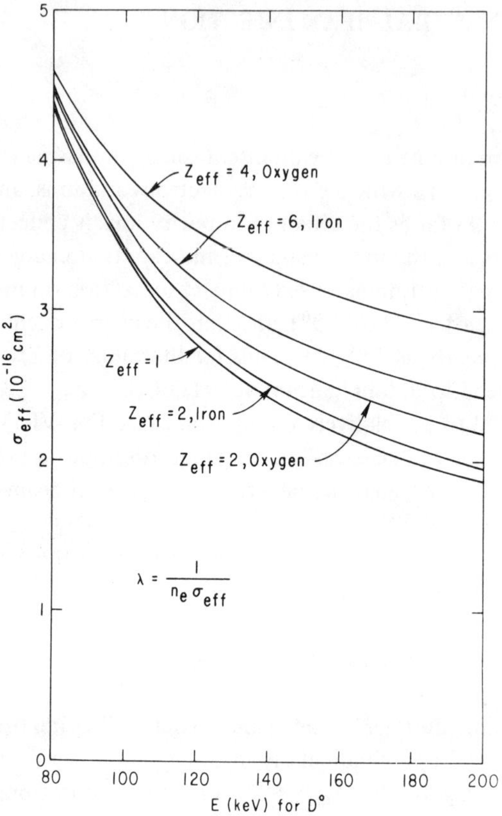

FIG. VI-1. Effective trapping cross-section versus beam energy, defined as

$$\sigma_{eff} = \frac{Z-Z_{eff}}{Z-1}\,(\sigma_{cx}^{H} + \sigma_{p\text{-}ion}^{H}) + \frac{\langle\sigma V\rangle_{e\text{-}ion}}{V_{beam}} + \frac{Z_{eff^{-1}}}{Z(Z-1)}\,(\sigma_{cx}^{imp} + \sigma_{ion}^{imp}),$$

for a deuterium neutral beam traversing a deuterium plasma with an impurity of charge Z.

plasma density that is lower at the beginning of the heating pulse ($\overline{n}_e \sim 7 \times 10^{19}$ m^{-3}) and reaches the final density ($\overline{n}_e \sim 1.4 \times 10^{20}$ m^{-3}) only at the end of the heating pulse, when the alpha-particle heating begins to play a role at the centre of the plasma.

Various self-consistent ignition scenarios of this type have been devised with near-perpendicular injection in the horizontal plane, and these indicate that penetration of neutral beams with energy E_b in the range 150 to 200 keV per deuteron should be sufficient, at least if the impurity level is very low (see Chapter IV)

Until further experimental data is available on penetration, however, one must be cautious about specifying an E_b in the lower end of this range. Accordingly, we recommend an E_b in the range 175–200 keV for INTOR.

A potential method of providing enhanced penetration of large reactor plasmas is the ripple-injection method which makes use of a vertically asymmetric ripple configuration. When neutral beams are injected vertically from the side of stronger ripple, energetic ions formed from the beam are trapped in the ripple magnetic well and drift upward to the central plasma region where the ripple is small and the fast ions are untrapped. This method has many potential problems, such as a possible enhanced loss of the tail of the bulk-ion distribution in the ripple field, and the need to provide space in the blanket for special ripple coils. A test of ripple injection will take place in the ISX-B tokamak in 1979–80. The INTOR Workshop considered it imprudent to specify ripple injection for INTOR, until its effectiveness has been demonstrated, and until calculations have shown that good confinement of the reacting tail of the ion distribution can be maintained.

3.2. Fast-ion loss channels

The efficiency of neutral-beam heating may be reduced by several loss processes for the fast ions:

3.2.1. Charge-exchange loss

For E_b = 120 to 200 keV, the concentration of neutrals introduced into the plasma by charge-exchange trapping of the beams is relatively small, and in fact negligible for $E_b > 150$ keV. However, the fast ions can exchange charge with the ambient neutral population and also with other injected beam neutrals; some of the resulting fast neutrals may escape the plasma. In present experiments this loss can be important at lower plasma densities, but for standard INTOR parameters an upper limit on the loss by this process (assuming that all re-charge-exchanged neutrals leave the plasma) has been put at 10 MW [3].

3.2.2. Orbit loss

When counter-injection or perpendicular injection is used, a fraction of the beam ions will be trapped in banana orbits that intercept the limiter or divertor scrape-off layer. Initially untrapped ions could also be lost by pitch-angle scattering into an escape orbit while slowing down, but this effect is expected to be unimportant at the high plasma currents needed in INTOR.

3.2.3. Ripple loss

The discrete toroidal-field coils set up a periodic toroidal field ripple which may result in local magnetic mirrors between TF coils. Fast ions injected or scattered into these mirrors experience a vertical drift which may carry them out of the plasma. This loss process was especially severe on the TFR device with near-perpendicular injection, and must be avoided on INTOR by careful choice of the injection angle, and by ensuring that the ripple magnitude $\delta(r, \theta)$ satisfies $\alpha|\sin\theta| > 1$ over most of the plasma cross-section, where $\alpha = r/(RN\delta)$ (See Chapter IV).

3.3. Microinstability effects

Microinstabilities that arise during neutral-beam injection can in principle cause an anomalously rapid thermalization of the fast ions or an enhanced radial diffusion rate. The first effect is unimportant for heating a thermonuclear plasma such as INTOR, and could even be desirable in the face of competing fast-ion loss processes, but the second effect must be avoided. Theoretical analysis predicts that ion cyclotron waves may arise during intense beam injection, especially for perpendicular injection, and such oscillations have indeed been observed in TFR. Also, the amplitude of drift-wave-type turbulence may be enhanced, but no adverse effect on plasma heating or on fast-ion confinement has been predicted.

In experiments to date, strong density fluctuations have often been observed by microwave scattering measurements, and were especially pronounced during high-power co- and counter-injection at low densities in the PLT device. However, the characteristic temperature scaling $\Delta T_i \propto P_{beam}$ observed at low and moderate beam powers continued throughout the higher-power regime where the intense fluctuations became evident. Thus these instabilities, still unidentified, had only a minor effect on the bulk-ion power loss. Examination of the fusion-neutron production rate showed also that the fast-ion slowing-down process continued to be classical in the presence of this instability.

Microinstabilities are also expected to arise when the beam velocity exceeds the Alfvén speed, but this condition will not occur in the range of INTOR injection energies.

3.4. Preferred injection mode for INTOR

As indicated above, ignition conditions in INTOR can apparently be attained with $E_b \cong 175$ keV, using near-perpendicular injection in the co-direction ($V_b \parallel I_b$) in the horizontal midplane. This injection mode results in greater beam penetration for a given E_b, and the co-injected trapped ions are carried farther inward on their banana orbits. Near-perpendicular injection is also more favourable for physical implementation, as less blanket need be penetrated by the beam duct, and the shielding around the beamline is more easily installed.

The direction of injection must be about 20° off normal to minimize losses due to ripple trapping and to limiter-intercepting orbits. The possibility of beam-induced instabilities arising due to intense perpendicular injection must still be investigated experimentally. Before the injection direction can be fixed, an analysis of the effect of the INTOR ripple is required, as well as stability and heating results from the perpendicular injection experiments on the PDX, TFR, and DIII tokamaks in 1980–81.

According to an analysis of the INTOR ripple effect [2], based on a Monte-Carlo particle code, the loss of fast ion energy to the wall during the slowing-down process is about 15% of the injected energy for the case of a combination of co- and counter-injection at 15° off-normal. In this calculation, the peak-to-average ripple magnitude was taken of the form

$$\delta(r, \theta) = 0.75 \times 10^{-2} (r/a)^2 \exp[-(0.50)^2]$$

in accordance with the recommended INTOR guiding parameters. The loss comes mainly from the ripple-enhanced banana drift motion of the ions, which is not strongly dependent on injection angle; thus, the energy loss remains about 10% even for injection at 20° off-normal. Further analysis of this matter is required.

Also, if beam heating is required during the start-up phase in which the density is raised from the ohmic limit ($\sim 3 \times 10^{19}$ m^{-3}) to the ignition requirement ($\sim 1.5 \times 10^{20}$ m^{-3}), then lower beam energies, or a different direction of injection, may have to be provided. A fuller discussion of this problem can be found in Chapter VIII. The various possibilities for using neutral-beam heating in this phase all involve severe technological difficulties. If 175 keV beams are used, the substantial "shine through" would require the installation of large beam dumps on the inboard first wall, a relatively inaccessible region. Alternatively, it has been suggested that it may be possible to change the beam energy, or the species mix, as a function of time during the injection pulse. A change in the acceleration voltage would seriously impair the beam optics, assuming this to be optimised for the full voltage mode, and it may prove easier to vary the species mix, thus changing the effective beam penetration. A third option, of course, would be to install a second set of beams at lower energy, or with a tangential direction, in either case involving some penalty with respect to decreased access and increased blanket penetrations. One interesting possibility would be to direct the beams so as to be tangential to the major circumference at the inboard edge of the plasma (a configuration to be employed in JET), corresponding to an injection angle of 38° off-normal in INTOR. In this case, compared with perpendicular injection, the path length to the magnetic axis is increased by only a factor 1.25 to $L_{axis} \cong 1.25$ a, whereas the total path length through the plasma to the beam exit is increased by a factor 4 to $L_{total} \cong 8$ a. If we

TABLE VI-2. RECENT EXPERIMENTAL RESULTS WITH NEUTRAL-BEAM INJECTION

Device	Year	Beam Power (MW)	Beam Energy (keV)	Pulse Length (ms)	Average Density (10^{13} cm^{-3})	max ΔT_i at r = 0 (keV)	max ΔT_e at r = 0 (keV)	Average Toroidal Beta
ORMAK	1976	0.36	30	50	2.8	1.0	0.4	0.01
TFR	1976	0.45	34	50	5.0	1.0	0.25	0.01
DITE	1978	1.0	30	50	1.6	0.5	–	0.018 (peak)
T-11	1979	0.65	25	20	3	0.8	–	0.025
PLT	1978	2.5	39	150	1.5	5.5	1.5	0.01
PLT	1978	2.0	39	150	4.0	1.0	1.0	0.01
ISX-B	1979	1.0	40	100	6	0.5	0.5	0.018

write the condition for effective beam penetration as $\frac{1}{3}\ L_{axis} < \lambda_{mfp} < \frac{1}{3}\ L_{total}$, then a density range given by

$$E_b\ (keV)/[100\ a(m)] < \bar{n}_e\ (10^{19}\ m^{-3}) < E_b\ (keV)/[15\ a(m)]$$

could be covered by a beam of energy E_b.

For $E_b \cong 175$ keV and a = 1.2 m, this corresponds to a density range $1.5 < \bar{n}_e\ (10^{19}\ m^{-3}) < 10$, which would be appropriate for INTOR start-up.

3.5. Experimental status

The slowing-down process of injected fast ions has been investigated on many tokamaks, using charge-exchange and neutron diagnostics, and found to be in accordance with classical theory. Heating has been more effective with the bulk ions than with electrons, as expected with the relatively low E_b ($\leqslant 40$ keV) used to date, where most of the beam energy is given up to the bulk ions. In smaller tokamaks, counter-injection has often resulted in an influx of impurities which clamped the electron temperature and thereby limited the ion temperature. In higher-power experiments with co-injection alone, and in PLT with co-and counter-injection, it has been possible to control impurity influx, and there has been apparently little, if any, deterioration of electron or ion-energy confinement time.

TABLE VI-3. PLANNED NEUTRAL-BEAM INJECTION EXPERIMENTS ON WORLD TOKAMAKS

Device	Year	Beam Power (MW)	Beam Energy (keV)	Pulse Length (ms)	Direction
PLT	1979–80	3.2	40	200	parallel
ISX-B	1979 1980	1.8 3.0	40 40	100 300	parallel
TFR	1979–80	1.2	40	100	perp.
DITE	1980	2.4	30	25	parallel
PDX	1980–81	6.0	50	500	perp.
ASDEX	1980–81	2.5	40	200	parallel
JFT-2	1980–81	1.5	30	50	parallel
DOUBLET III	1980–81 1982–84	7.0 20	80 80	500 1 000	perp.
TFTR	1982–83	35	120	1 500	parallel, perp.
T-15	1983–84	5	80	1 500	parallel
JET	1983–84	17	160	10 000	parallel
JT-60	1984	20	75	10 000	perp.

Table VI-2 summarizes the heating results obtained with neutral-beam injection during the last several years. The most spectacular results have so far been obtained in PLT, where at low density the central T_i was raised from 1.0 to 6.5 keV and the central T_e from 2.6 to 4.0 keV. Smaller temperature increases were measured at higher density, with $\Delta T_i \propto P_{beam}/n_e$. Significant electron heating could be obtained only with water-cooled graphite limiters and with titanium gettering of the vacuum vessel wall. On ISX-B, large electron heating has also been observed with stainless-steel limiters and without any gettering.

TABLE VI-4. PARAMETERS OF NEUTRAL-BEAM INJECTORS

	Example of Present Status on Tokamaks (PLT/ISX)	Systems Under Development						INTOR needs**
		PDX	DIII	TFTR	JET II	T-15	JT-60	
Current density (A/cm^2)	0.4	0.3	0.4	0.3	0.2	0.5	0.27	
Extracted current* (A)	60	100	85	65	100	35	35	
Monatomic fraction	0.85	0.85	0.8	0.8	0.8	0.7	0.75	0.85–0.9
Gas efficiency	0.5	0.5	0.4	0.4	0.5	0.5	0.4	0.5
Pulse length (s)	0.3	0.5	0.5	1.5	10	1.5	10	10
Lifetime (s)	$> 10^5$					10^4	10^5	3×10^5
Accel. potential (keV)	40	50	80	120	160	80	75	175
NB power^{+++} (MW)	0.85 (D^0)	1.5 (H^0)	3.6 (H^0)	8.4 (D^0)	11	2.5 (H^0)	1.5 (H^0)	15 (D^0)
NB current^{+++} (A)	30	40	70	90	80	40	30	120
Pumping speed^{++} (ℓ/s)	4×10^5	5×10^5	1×10^6	3×10^6	6×10^6	1.5×10^6	1.3×10^6	6×10^6

TABLE VI-4 (cont.)

	Example of Present Status on Tokamaks (PLT/ISX)	Systems Under Development						INTOR needs**
		PDX	DIII	TFTR	JET II	T-15	JT-60	
Transmission effic.	0.5	0.5	0.75	0.75	0.7	0.75	0.9	0.8
Direct recovery effic.	0	0	0	0	0	0	0	0.6
Overall power effic.⁺	0.4 (D^0)	0.3 (H^0)	0.3 (H^0)	0.35 (D^0)	0.25 (D)	0.25 (H)	0.3 (H^0)	0.3 (D^0)
Power density⁺ (kW/cm^2)	1.5	1.5	4.0	3.0	4–6	1.6	0.7–1.4	

* Per ion source
\+ Includes fractional energy components
‡ Per injector
** Assumes 5 of these injectors

In these experiments the relative fast-ion pressure has been at least as high as that expected in INTOR, with apparently no adverse affects. Average toroidal betas of 1.8% and peak betas of 8.5% have now been reached in beam-heated circular plasmas in ISX-B, and 2.5% and 8% respectively, in T-11.

Table VI-3 summarizes the beam-heating experiments that are planned for the next five years on world tokamaks. The results of these experiments should provide the more detailed data base necessary for the confident application of neutral-beam injection at longer pulse lengths to the heating of INTOR plasmas.

4. NEUTRAL-BEAM TECHNOLOGY

4.1. Status of neutral-beam systems

Table VI-4 gives the state-of-the-art parameters of neutral-beam injectors. On operating tokamaks, the PLT/ISX neutral-beam systems [3] have achieved the following parameters: E_b = 40 keV, $t_b \leqslant 0.3$ s, P_b = 600 kW injected H^0 per beamline with overall power efficiency η_I = 0.25 to 0.35 and P_b = 850 kW injected D^0 per beamline with η_I = 0.4. The effective transmission efficiency η_t = 0.50 ($\eta_t = \eta_I$/neutralization efficiency.) Test stand operation of the same systems has yielded slightly better parameters. The six-source 2XIIB injectors [3] have achieved E_b = 18 keV, t_b = 10 ms, P_b = 3.5 MW injected D^0 per beamline with η_I = 0.42 and $\eta_t \cong 0.50$.

4.2. Ion sources

Ion sources fall into several distinct classes [1]: the modified Duopigatron with magnetic bucket used in the ORNL injectors on PLT, ISX-B, and PDX; the magnetic-field-free LBL source to be used in Doublet III and TFTR; the Periplasmatron to be used on ASDEX and TFR 600; the bucket-type source being developed at ORNL and Culham and used on DITE. Ion sources for the JT-60 injectors will be of a new type with an electron beam dump developed at JAERI and called Lambdatrons [2], and those employed in the USSR on T-11 are magnetic-field-free sources, called IBM-5. For JET, a choice will be made between the Periplasmatron and the bucket-type source.

As an example of current beam development programmes, LBL and ORNL have operated ion sources at 110–120 kV and 15–20 A for $t_b \geqslant 0.25$ s. The TFTR prototype neutral beam system which is just beginning operation at LBL, will have one of three sources installed and powered by a 120 kV, 65 A, 0.5 s power supply. Projected overall efficiency η_I is 0.24 for 120 keV D^0 and η_I = 0.37 including fractional-energy components. By the end of 1982, three of these beamlines should be operating on TFTR, each delivering 6.5 MW of 120 keV D^0, and 8.4 MW including all D^0 components.

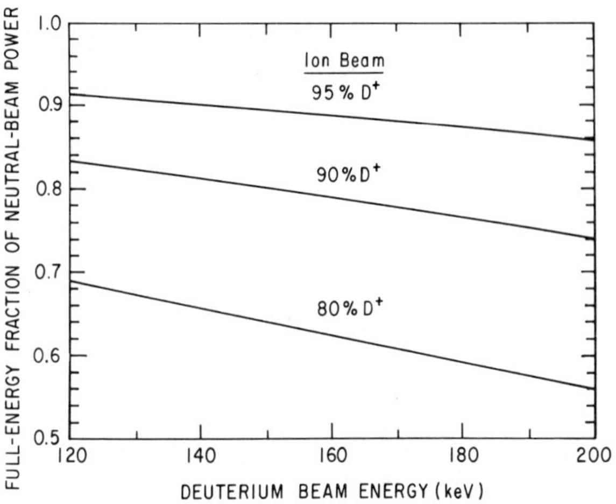

FIG. VI-2. Fraction of neutral-beam power in the full-energy component for various species mix in the ion beam.

4.3. Species mix

Fig. VI-2 shows the fraction of neutral-beam power at full energy for the energy range of E_b = 120 to 200 keV. The monatomic-ion fraction, f_{D^+}, of the extracted ion beam is as large as 0.80 to 0.85 in operating systems and test stands [3]. A reasonable design goal for systems operating in 1983 is $f_{D^+} = 0.90$. This value is desirable to insure adequate plasma penetration by most of the neutral beam, and to maximize the usefulness of direct energy recovery systems. For a 175 keV system, and a 90% fraction of monatomic D^+ in the ion beam, it should be possible, according to Fig. VI-2, to achieve a neutral beam with 78% of the power in the full energy component.

If this design goal $f_{D^+} = 0.90$ proves unrealistic, then the fraction of neutral beam power in the full-energy component will drop quite significantly. For example, for a 175 keV system with $f_{D^+} = 0.80$, the full-energy component of the neutral beam power drops to about 60%. With $f_{D^+} = 0.70$, the full-energy component of the neutral beam power is only about 45%.

4.4. Direct energy recovery systems

Instead of dumping the unneutralized ion beam, it can be advantageous to recover its energy by decelerating the ions electrostatically before dumping them on a target. Operating and planned beamlines on tokamaks do not yet include direct recovery systems. The results that have been obtained on test stands [1, 3] are shown in Table VI-5. There has so far been no demonstration of direct

TABLE VI-5. DIRECT ENERGY RECOVERY EXPERIMENTS

Laboratory	Beam Energy (keV)	Pulse Length (s)	Total Current* (A)	Total Current Density (mA/cm²)	Ion Current (A)	Ion Current Density (mA/cm²)	Recovery Efficiency
Livermore	100	0.5	2	7	1.5	5	0.6
Oak Ridge	37	0.1	20	200	7	70	0.6±0.2
Fontenay	70	0.2	20	250	13	150	0.45
Needed for INTOR	150 to 175	5–100	⩾ 20	> 200	⩾ 7	⩾ 75	⩾0.6

* Ions + neutrals.

recovery with the combination of beam energies, beam power densities, and pulse lengths characteristic of existing or future injector systems. The most severe requirement would seem to be the pulse length, which is limited to 1 s in present experiments.

Direct recovery is desirable for reducing power supply capacity (and cost), as well as for reducing power handling requirements in the beamline. The higher efficiency obtained with direct recovery may be an important step toward commercial reactor requirements, if NB heating is still to be used.

A direct recovery system is probably essential for the INTOR injectors. On the basis of estimates for the overall efficiency of a 175 keV system without direct recovery (see Section 4.5), the problems of power handling in the injectors themselves, i.e. the heat load on the beam dumps, etc., would be so severe as to call into question the viability of such a system, and would, at best, lead to very high capital cost of the beam dump and associated coolant systems. Direct recovery is also advantageous for reducing power supply costs. Here, the important cost factor is not so much that associated with the overall power consumption, but the capital cost associated with the needed power handling and power conversion equipment.

The efficiencies of the direct energy recovery system have been assumed by the various INTOR delegations [1, 2, 3] to lie in the range 60–80%. In all cases, it is assumed that only the energy of the D^+ ions can be recovered, and not the energy of the D_2^+ and D_3^+ ions.

4.5. Efficiencies of neutral-beam systems

Calculations have been made by several INTOR delegations of projected neutral beam efficiencies in the energy range of 150–200 keV, both with and without direct energy recovery. The calculations are generally in close agreement.

One representative set of [3] takes account of losses due to:

(i) power consumed in the plasma source;
(ii) grid gap ionizations, secondary electrons, and charge exchange;
(iii) incomplete neutralization;
(iv) beam divergence, assuming this loss to be 5% of the ion current emerging from the grids;
(v) unneutralized, unrecovered ions;
(vi) drift reionization; assuming this loss to be 5% of the neutral power;
(vii) cryopump refrigeration.

These particular calculations have assumed a source neutral gas density of 1×10^{14} cm^{-3}, a source gas efficiency of 0.5, a source power requirement of 1 kW/A, 95% equilibrium cell for all energies, 2-stage extraction with a 20 keV first stage, and a 1.0 cm grid gap per 100 keV. In addition it has been assumed that only full energy ions can be subject to direct energy recovery.

The results of these calculations for three beam systems that might be considered for INTOR are given in Table VI-6, both without direct energy recovery and with a direct energy recovery system assumed to be 60% efficient. In Table VI-6, we quote efficiencies based on the full energy neutrals only, as well as that based on all energy components of the neutral beam.

The overall efficiency of neutral beam systems (for all energy components, and for full energy component only) assuming various efficiencies of direct recovery, is plotted as a function of beam energy in Fig. VI-3 for the case of 90% D^+ ions.

Of the beam systems listed in Table VI-6, the 175 keV system with 90% D^+ ions, and 60% direct recovery efficiency, is strongly recommended for INTOR. The high monatomic ion fraction ensures a high mean energy in the neutral beam; the use of direct recovery is desirable for reducing total power requirements, and may be essential for solving beam dump problems.

Calculations by the Japanese delegation project somewhat lower efficiencies for the INTOR injectors [2], although their conclusions as to the desirability of a high monatomic ion fraction and direct recovery are the same. The lower efficiencies arise from higher estimates for the power losses in the grid and in the ion accelerating region.

Losses in the ion accelerating region arise from backstreaming electrons produced (i) by ion impact ionization of gas molecules in the grid gap; and (ii) in the form of secondary electrons created when slow ions strike the electrodes.

TABLE VI-6. EFFICIENCIES OF NB SYSTEMS WITH AND WITHOUT DIRECT ENERGY RECOVERY

Energy (keV)	200		175		150	
D^+/D_2^+ (%)	70/20		90/5		90/5	
NB power at full energy (%)	44		77		80	
NB mean energy (keV)	112		129		115	
Direct recovery effic. (%)	0	60	0	60	0	60
Power lost (%) by:						
Source and grid losses*	15.5	21.3	15.8	23.6	15.8	22.[illegible]
Beam divergence	4.2	5.8	4.2	6.2	4.2	6.[illegible]
unrecovered ions	56.2	39.9	57.4	36.9	52.3	32.[illegible]
drift, reionization	1.2	1.7	1.1	1.7	1.4	2.[illegible]
Power injected (%) in:						
full energy	10.1	13.8	16.4	24.4	21.0	29.[illegible]
half energy	7.6	10.5	2.2	3.2	2.4	3.[illegible]
third energy	5.2	7.1	2.7	4.0	2.8	4.[illegible]
Overall efficiency:						
all energies	23	31	21	32	26	37
full energy	10	14	16	24	21	30
Required primary power for a 75 MW system (MW)	325	240	355	235	290	200
Power to beam dump per 15 MW injector (MW)	37	19	41	17	30	13

* Includes power to plasma source, grid ionization, grid secondary electrons, grid charge exchang and cryopump refrigeration.

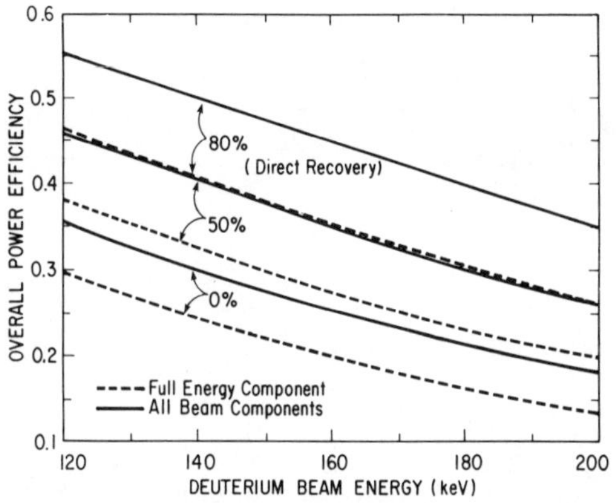

FIG. VI-3. Overall power efficiency, defined as neutral -beam power (in all energy components and in the full-energy component only) delivered to the torus divided by total electrical power into the injector, for various efficiencies of direct energy recovery, assuming 90% monatomic ions in the ion beam.

Losses in the ion accelerating region also arise from secondary ions and neutrals produced by ionization and charge exchange. Losses by these mechanisms have been included in all of the calculations, but their magnitude is somewhat higher in the Japanese estimates, mainly due to a somewhat higher value assumed for the gas density in the source.

The backstreaming of secondary electrons measured on the JAERI test stand for JT-60 will lead to substantial energy flux to be back plate of the extraction chamber, estimated to be as high as 900 W/cm^2 for an ion beam current density of 150 mA/cm^2 in INTOR [2]. To address this problem, a new type of ion source (Lambdatron) is being designed at JAERI for the JT-60 injector which has an electron beam dump in the source chamber. In this case, the ion beam current is 35 A at 75–100 keV for a pulse length of 10 s. The beam dump is made of a 1-shaped array of cooling pipes, and is capable of handling a total electron heat load up to 500 kW with a heat flux of 500 W/cm^2 maximum.

4.6. Negative-ion beams

Consideration of negative-ion-based neutral beams is motivated by potential advantages of power efficiency at $E_b \geqslant 160$ keV and by the potential option of efficient beam production at $E_b > 200$ keV (see Fig. VI-4). This advantage would be useful if higher energies are eventually thought to be desirable for penetrating to the centre of denser INTOR plasmas, or if the required number of beamlines is to be reduced by going to higher energies at the same beam current. However, the development of D^- systems [1, 3] is not certain to be sufficiently advanced by 1983 to allow a commitment to a D^- based neutral-beam system for INTOR.

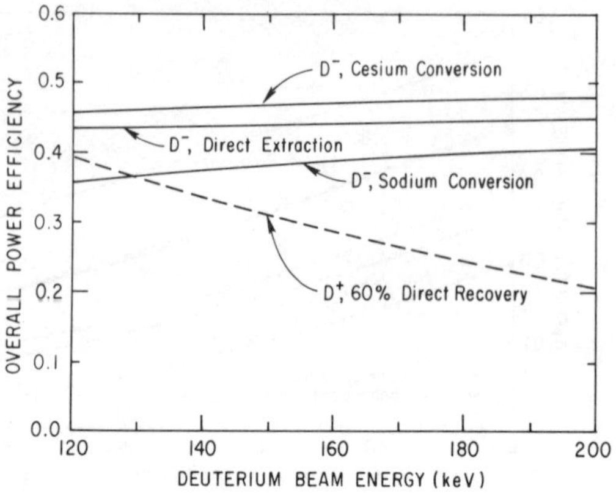

FIG. VI-4. Comparison of overall neutral-beam power efficiencies for various production method of negative-ion-based beams and a positive-ion-based beam with 60% efficiency of direct recovery.

Nevertheless, there remains the possibility of replacing one or more D^+-based systems with a D^- system at some later date during INTOR operation.

All four INTOR delegations agreed that it would be advantageous to have a small fraction of the heating power (~ 10 MW) from 400–500 keV negative-ion-based neutral beams with long pulse (~ 100 s) capability. As discussed in Section 2.2.2, this would be used for control of a slightly sub-ignition INTOR plasma, where the alpha-particle heating in the centre of the plasma must be supplemented by beam power at a modest level. All four INTOR delegations favoured continuation of a vigorous development effort on negative-ion beams.

4.7. Long-pulse beam development

While the reference pulse length for heating the INTOR plasma to ignition is 5 to 8 seconds, it is highly desirable (see Section 2.2) to have a 100 s pulse length capability (the duration of the burn) for part of the system. This requirement implies essentially steady-state operation for the INTOR duty factor of 0.7, but only a fraction of the heating capacity is needed for such an extended period. On present-day test stands, JAERI has operated a modified Duopigatron at 70 kV and 5 A for 10 s, and at 75 kV and 6 A for 6 s, the pulse lengths being limited by the capability of the beam dump and power supplies. ORNL has operated a PDX ion source at 40 kV and 100 A for 0.5 s, and also a relatively small ion source at 25 kV and 9 A for 10 s. The Lawrence Berkeley Laboratory has operated a ¼-scale TFTR ion source at 120 kV and 8 A for 1.5 s. Neutral-beam systems operating by 1983 on PDX, Doublet III, TFTR, and MFTF are expected to have 1–2 s pulse capability. No insuperable problems are foreseen in extending the pulse length as required, although the power density may come down, but a major development programme is needed.

The injection systems being developed for JET and for JT-60 will have a 10 s capability.

4.8. Deliverable power density

As examples of currently operating systems, the PLT and T-11 neutral-beam systems deliver 1.5 kW per cm^2 of torus port area. By 1983 the TFTR beam system is expected to be delivering 3.0 kW/cm^2 and the Doublet III system about 4 kW/cm^2 at the torus. These include all energy components.

Short pulse (50 ms) injectors are being constructed for JFT-2 with a power density of 5 kW/cm^2. For JT-60, the design value is 0.7 kW/cm^2, but an upgrade to 1.4 kW/cm^2 is being planned.

The JET injectors will deliver a mean power density of 4.3 kW/cm^2, with a higher power density planned for the JET II phase.

For INTOR, the power per unit area should be maximized in order to minimize the size of the injection ports, and thereby minimize the neutron streaming problem. A high power density is more difficult to achieve with the long duct necessary to penetrate the INTOR blanket and shield.

Because of the magnetic field, beam neutrals that are ionized by collisions with gas molecules in the beam duct are deflected to the duct walls. Bombardment of the wall may liberate more gas, leading to an enhanced rate of beam ionization and partial *blocking* of the beam. The duct pressure build-up observed in the PLT experiments must be further understood, because this phenomenon may impose design limits on power density and duct length.

The potential for significant duct reionization losses can be reduced markedly by designing high conductance ducts, insofar as such design is compatible with the need to minimize neutron streaming problems, and also by insuring a low gas evolution rate from the duct wall through the proper choice of wall materials and coatings. It is important that neutral beam test facilities have the capability to simulate tokamak magnetic fields in the duct region, so that the extent of reionization losses and the effectiveness of countermeasures can be determined.

Specific numbers of power densities, and the number of ports, on INTOR will require specific neutral-beam designs and total layouts, and will require more extensive prototype testing than in current development practice to establish relationships between power density and reliability/availability.

Nonetheless, each of the four INTOR delegations offered preliminary proposals in this regard, as outlined in Table VI-7.

The USSR configuration I has the ports in the midplane of the device. If this is not possible, for example because of a poloidal coil in the midplane, they recommend version II, in which the injector configuration at each port is divided into two parts, above and below the midplane.

TABLE VI-7. PROPOSED BEAMLINE CONFIGURATIONS

	EC*	Japan	USA*	USSR	
				I	II
Total power (MW)	75	75	75	75	75
Power density at port (kW/cm^2)	2.5	2.9	2.5	2.0	2.0
Number of ports	6	4	5	4	8
Size of port (m^2)	0.5	0.65	0.6	1.0	0.6
Ion Sources per port	4	7	3	4	2
Power per ion source (MW)	3.1	2.7	5	4.7	4.7

* The EC and USA emphasize that a specific design is needed before parameters such as these can be firmly specified.

4.9. Effects of neutron environment

There is no experimental data, and little analysis, on problems resulting from neutron streaming into neutral-beam systems. There will not be any on-line damage data even by 1983. There is no reason to believe, a priori, that any one of the different types of ion sources and extraction systems would be more resistant to radiation than others. As for the electrical circuits, semiconductor elements should be strongly shielded, as these are certainly very sensitive to radiation damage.

It has been speculated that there may be degradation in performance of filaments, grids, and insulators. However, Fig. VI-5 shows the fall-off in neutron flux, including backscattered neutrons, with distance up the beamline [3]. While a definitive analysis needs to be made, a comparison of the streaming neutron flux with the damage thresholds for various materials indicates that damage effects may be tolerable if the ion source and accelerator structure can be replaced each 3–6 months, for example.

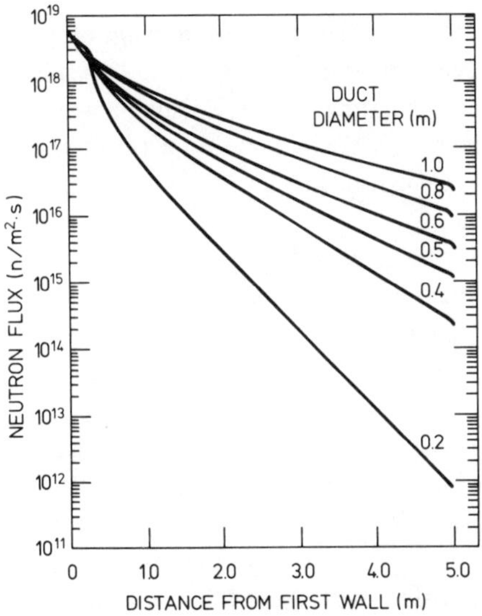

FIG. VI-5. Neutron flux including backscattered neutrons along the wall of a duct perpendicular to the vacuum vessel, for a fusion neutron wall loading of 1 MW/m^2.

The USSR INTOR delegation described a specific beamline layout in which the ion sources are as much as 10 m from the plasma [4]. For a 20 × 200 cm^2 injection window, they calculated that the 14 MeV neutron flux at the ion source would be about 2×10^{11} $cm^{-2}s^{-1}$, and the total neutron flux would be about 2×10^{12} cm^{-2} s^{-1}, in agreement with Fig. VI-5. Even for a 5–8 s pulse during start-up only, these levels would seriously limit the lifetime of organic insulators, but not of other filament and grid materials. However, organic insulators are usually only used in parts of the injector that are shielded from direct neutron irradiation.

An analysis also by the USSR [4] showed that, at this neutron flux, ion source materials would become sufficiently radioactive to require remote servicing. However, if the source is irradiated only during the injection pulse, and of course only when the device is operational, the radiation would be reduced to a level that might allow suitably-protected personnel to replace ion sources.

In principle, these radiation effects on the ion source and accelerating structure could be greatly reduced by the use of a bending magnet between the ion source and the neutralizer. It seems highly improbable, however, that such a configuration could be used without a considerable drop in overall power efficiency [1]. Indeed, very high neutral-beam power, combined with an acceptable divergence of the beams (0.5°–1°) are obtained by a very *close* coupling of the extraction and beam-forming system (which, itself, must be as short as possible) and the neutralizer, such that a very high degree of space-charge neutralization is immediately obtained.

To combine the electrostatic optics, the partial beam blow-up in the drift space (which must be strongly pumped in order to avoid losses by charge exchange), and the magnetic optics of the bending magnet, in such a way that a very low divergence beam results, seems hardly possible without losing a large part of the beam. All four INTOR delegations believed that a bending magnet should be avoided, if at all possible.

For neutral beam systems based on negative ions, the situation is different [1, 4 For the same unit power, the acceleration voltage is much higher and so the current considerably lower, which alleviates the space charge problem. The bending magnet could be placed after or before the full acceleration stage. In the latter case, the magnet could also serve to reduce considerably the electron contamination of the low energy negative ion flux and the magnet could be used to focus the negative ions on the acceleration system. However, with such a layout the acceleration system is directly exposed to the neutron flux emerging from the injection port.

A more attractive possible solution to the problem of damage caused by neutron streaming is provision of a 30-cm thick plug valve which is closed after ignition is attained. The plug valve would also lead to a large reduction in the heat load on the cryopumping system. Thus, this bulky valve would have to be opened and closed reliably once every 120 s, probably by rotation. However, the valve could not be used with any injectors that may have to operate for a substantial fraction of the burn pulse (see Section 2.2). All four INTOR delegations favoured the use of such a valve, if it could be designed.

In any event, it appears that the lifetime of ion source components may well be limited by the present-day problems of erosion, thermal cycling, and so forth, rather than by neutron bombardment. On the basis of very limited data, the probable source lifetime is estimated at about 3×10^5 s, corresponding to 3–6 months use at an assumed 25–50% device availability.

Neutron heating of cryo-panels may be much more serious, and the use of cryosorption pumps or getter pumping should be examined carefully.

4.10. Beamline pumping system

A PLT beamline cryopump has a speed of 4×10^5 ℓ/s and requires 2.5 hours for regeneration. The TFTR cryopump has a design speed of 3×10^6 ℓ/s and a regeneration time of eight hours. While the TFTR pumps could operate for 10000 one-second shots from the viewpoint of hydrogen explosion safety, the need to limit tritium accumulation results in more frequent regeneration.

As an example of the requirements for INTOR, we estimate that a total pumping speed of about 2×10^7 ℓ/s for D_2 gas is needed for the 75 MW system using differential pumping to maintain a pressure of 10^{-5} torr in the neutral beam drift tube. Such a 75 MW system would require about 250 m^2 of cryopanels.

The liquid He and liquid N_2 consumption by cryopanels of this size would be extremely high, especially if the system were under neutron radiation. Minimum estimates [1] put the liquid He consumption at 50 ℓ/h, with an instantaneous consumption during the heating pulse of 1000 ℓ/h.

For the INTOR beamlines, cryopanel pumping could provide sufficient pumping speed and acceptable regeneration intervals, but it has two potential disadvantages. Build-up of tritium inventory may require unacceptably frequent regeneration, and cryopanel heating by neutrons may be significant. Zirconium-aluminium getter pumping should be investigated as an alternative, as it would not release the tritium in the event of a vacuum leak and is not sensitive to small temperature rises. It is also possible that reliable cryosorption pumps operating at higher temperature will be developed.

If significant amounts of helium must be pumped in the beamlines, special cryosorption or cryotrapping pumps may have to be developed.

4.11. Gas influx from injectors

A neutral beam system injecting 75 MW of D^0 at 175 keV (with typical fractions of half and third energy components) injects deuterium atoms into the tokamak plasma at the rate of 3×10^{21} particles per second.

The gas flow from the injector has been estimated as follows. For a relatively short duct measuring 4 m × 1 m × 0.5 m, the conductance of D_2 gas is 3.5×10^4 ℓ/s. For a pressure at the inlet end of the injector of 10^{-5} torr, and six injectors, the total gas flow is about 2 torr ℓ/s, i.e. 3×10^{19} deuterium molecules per s which is only about 2% of the particle influx from injection. Even with somewhat larger-area ducts (2 m × 0.5 m), and with a duct length of 3 m (the shortest that could be considered possible for INTOR), the gas flow is found to be only 12% of the influx by injection [4].

An additional gas flow arises from gas liberated from the walls of the ducts when these are bombarded by ions formed by collisional ionization of energetic beam neutrals on gas molecules. This effect, known as beam blocking [3], has already been discussed in Section 4.8. To minimize the problem, one must have high conductance ducts, or ensure a low rate of gas evolution from the duct wall by choice of suitable materials [3], or one must pump directly from the drift tube either by surface or mechanical pumps [4]. In the latter case, space must be provided around the ducts for pumps and related equipment.

4.12. Impurity influx from injectors

Species analysis of the ion beams extracted from an ORMAK source and a ¼-scale TFTR source have indicated the presence of up to 2% oxygen. There is also evidence that neutral-beam injectors have introduced oxygen into 2XIIB and;

copper into PLT, but the injected impurity levels were too small to degrade plasma performance. Some Cu or Mo from sputtering on the extraction grids has to be expected; also some W can be evaporated from filaments of the source. The latter has still to be observed. Before a specific development programme in this area can be justified, more information on beam impurity levels is needed from test stands and tokamak heating experiments.

4.13. Magnetic shielding

The stray magnetic field at the ion source must be restricted to a few gauss. As the field outside the TF coils will be hundreds of gauss, magnetic shielding must be provided for the ion source, the beam acceleration region, and the beam extraction region. The neutralizer region must be shielded to reduce the stray field here to about 10 G. Some neutral-beam system designs include additional shielding between the neutralizer and the bending magnet to reduce the stray field in this region while allowing high pumping speeds. The INTOR beamlines may also require magnetic shielding of any direct recovery subsystem.

4.14. The injector cooling system

In a system without direct energy recovery, the most severely heat-stressed element of the injector is the ion beam dump. In the USSR proposal [4], in which each of 4 injectors receives 96 MW of electrical power to deliver 19 MW of neutral-beam power to the plasma, it is estimated that about 55 MW must be accepted by each beam dump. Assuming that the cross-section of the ion beam at the dump is about the same as that of the neutral beam at the tokamak inlet port (1 m^2), the average power at the dump will be 5.5 kW/cm^2, and the peak power may exceed 9 kW/cm^2 in the central part of the beam. The USSR proposes a collector plate tilted at an angle of about 7° to the direction of the beam, to reduce the power density to 1 kW/cm^2, and a pressurized (30 atm) water cooling cycle that would require about 5000 m^3/h of water (for all 4 injectors) during the heating pulse [4].

This coolant requirement is a powerful incentive for the development of an effective direct energy recovery system.

Development activities have been underway at JAERI in connection with the design of a beam dump for the 10 s JT-60 injectors. By using a long pulse ion source (10 s), a burn-out test has been made of a finned tube (copper with 0.2% silver added) to be used in the beam dump [2]. When the cooling pipe was exposed to a heat load at the level of 0.8–1.3 kW/cm^2, burn-out of the cooling tube occurred at a water pressure of 4–10 atm and a flow speed of 3–7 m/s. Based on this data, a value of 0.5 kW/cm^2 is being taken as the maximum design heat load for the JT-60 injector.

In one INTOR design with direct energy recovery, the heat load on the beam dump has been reduced to the same 0.5 kW/cm^2 [2]. Without direct energy recovery, the heat load would be a factor 2–3 higher.

4.15. Tritium injection

Tritium beams would be needed if penetration of tritons to the plasma centre could not be obtained by pellet injection or gas puffing, if certain beam-driven operating modes were to be adopted or in certain start-up scenarios. However, none of these is necessary to the reference INTOR operating scenario.

To give the same penetration as 200 keV D^0 beams, the T^0 beams energy must be 300 keV, but the power efficiencies of the T^0 and D^0 beams would be about the same. The principle problem encountered in the tritium injector is excessive build-up of tritium inventory in the ion source and elsewhere, which would hamper maintenance. Hydrogen could possibly be used as the neutralizer gas.

In view of the questionable need for T^0 beams in INTOR, there is no strong case at this time for a special tritium injection development programme. Only the USSR delegation saw a possibility of a need for tritium injection [3] for fueling purposes.

4.16. Availability, maintenance and reliability

4.16.1. Availability

The requirement that INTOR have a high availability (up to 50% is presently the goal) is a stringent requirement on systems such as the neutral beams which require routine maintenance. With a heating pulse of 5 to 8 seconds and as many as one pulse every 120 s each injector will be required to operate for up to 10^5 s per month. (Those injectors with longer on-times during each pulse would have a correspondingly greater requirement.)

4.16.2. Maintenance

Simple hot filaments used as cathodes generally have a lifetime of up to 3×10^5 seconds, and will need to be replaced each 3 months if INTOR runs at the full 50% availability. It may well be necessary to replace even more advanced cathodes and other components subject to erosion, such as the grids, on a monthly basis. Neutron irradiation experiments are required to determine the effect (if any) of neutron bombardment on normal replacement intervals.

4.16.3. Reliability

Each component must have very high reliability in producing the rated power during the rated lifetime (e.g. if there were 20 ion sources each with a 99.5% probability of successful operation on each pulse, then on average the system would provide only 90% of the maximum rated capability).

From the point of view of reliability, the weakest components of injectors are the cathode filaments, the extraction grids, the beam dumps, and the cryopumps. Accidents involving the injection of water into the vacuum chamber from the cooled extraction grids have to be foreseen. The power flux on the beam dump will be extremely high if direct recovery is not used and its temperature must be closely monitored. The cryopumps will probably involve ~ 250 m^2 of cryopanels, whose structures will undergo frequent thermal shocks.

A large number of diagnostics will be needed on each injector. The correct operation of the injector has to be monitored by electrical and thermal measurements on the source, the extraction system, and the beam dump. The beam itself must be controlled by thermal and optical means. The species mix must be measured regularly by the Doppler effect. A more complete list of needed diagnostics may be found in the USSR report [4].

4.16.4. Redundancy

It is important that components requiring regular maintenance (e.g., the ion source and recovery system) can be isolated, so that the rest of the system can continue to operate. For example, where the required power of 75 MW is produced by five injectors each with three ion sources, all to be working perfectly, it would be advisable to have a sixth injector since injector failure would lead to an extended downtime. In addition, to permit regular maintenance during operation, it would be desirable to have an extra ion source per injector. Thus we arrive at six injectors each with four ion sources rated for 120 MW total. To give 75 MW, there must be five injectors always operating (rated for 100 MW maximum), and requiring 98.5% reliability of each of the ion sources.

In an actual design, it may of course be difficult to provide for an additional port and associated injector. For example, in the USSR design with 4 injectors [4], a redundant port/injector is not allowed for, but 1–2 additional redundant ion sources per port are recommended.

4.17. Innovative power supply requirements

By 1983 several additional high-voltage power systems should be operational: 120 kV, 65 A for TFTR, 80 kV, 80 A for Doublet III and MFTF, and 100 kV, 80 A for a JT-60 prototype unit. Because of long lead times, these systems were specified at a time when little or no operational experience was available with very high voltage ion sources. Thus, in many respects, these systems are similar to those used on test stands, with good regulation and wide ranging voltage selection, and capability for frequent interrupts on all supplies. For the reactorlike conditions of INTOR, the sources will have to have evolved to a higher level of reliability, and certain simplifications which may be possible in the power system should be investigated: elimination of high-voltage switching, relying instead on star-point regulators and contactors in the primary lines; elimination of fast-restarting-after-a-fault requirement; elimination of wide-selection-range for accelerator voltage; etc.

None of these options has yet been tried. They will require interactive development of sources and power systems.

5. SUMMARY OF RESEARCH AND DEVELOPMENT NEEDS FOR NEUTRAL-BEAM HEATING AND TECHNOLOGY

5.1. Needs prior to INTOR authorization – major development requirements

Table VI-8 contains a list of the most important R and D needs listed in approximate order of urgency.

The most important task during the next few years is the further development of beam injector parameters, as described in Table VI-8. An investigation should be initiated immediately to determine whether present beam test stands are adequate and available in time to pursue most of the above development items. By 1983–84, the problems of achieving adequate reliability and ease of remote maintenance will become the most vital issues.

In connection with the provision of effective neutral-beam deposition during the start-up phase when the plasma density is low, there would be advantages in a neutral-beam system with variable energy or variable species mix, and the feasibility of such a system should be examined. However, since there are probably simpler ways of meeting this requirement, e.g. with an additional lower-energy beam, or with RF, development of a neutral-beam system with this capability is not an absolute requirement for INTOR.

In connection with the R and D needs in the area of radiation hardening, a neutron irradiation programme, for sensitive materials such as seals, insulation and accelerator grids, should be initiated. Beamline pumping systems (cryopanels or getter pumps) must be able to operate in the anticipated neutron flux. If it is determined that the ion source must remain out of the line-of-sight of fusion neutrons in the heating phase, or if a plug valve cannot be developed, then a development programme resulting in the demonstration of magnetic deflection of high-intensity beams would be required.

TABLE VI-8. R AND D NEEDS ASSESSMENT – NEUTRAL-BEAM HEATING

	Status*	
	Adequate	Inadequate
– Ion Beam. Development of high-current (~100 A) ion beam at 175 keV (preferably 200 keV if practicable)	x[1]	
– Pulse Length. Development of a 10 s pulse capability for the ion source and the beam dump, which must handle ~0.5 kW/cm^2 with direct recovery and ~2 kW/cm^2 without direct recovery. In addition, certain of the units should have a 100 s capability	x[1]	
– Monatomic Fraction. A 90% D^+ fraction in the ion beam should be a design goal, with 80% a minimum requirement	x[1]	
– Direct Recovery. Development of a direct recovery system for the unneutralized beam ions with at least 60% efficiency is almost certainly required		x
– Efficiency. Simultaneous achievement of high neutralization efficiency, high transmission efficiency, long ducts and low gas flow to the plasma		x
– Radiation Hardening. Adequate lifetime of all injector components against the expected neutron bombardment and other radiation must be insured. A plug valve shield should be developed for insertion in the beam ducts at the end of the heating pulse		x
– Reliability and Remote Maintenance		x
– Negative Ion Beams. At present, no absolute need is foreseen for this capability, but for purposes of control of a slightly subignited INTOR plasma, it would be desirable to have a portion of the beam power (~10 MW) at 400–500 keV, and with 100 s pulse capability		
– Power Supplies	x	

[1] An expansion of the present neutral-beam development programme, probably including upgrading of a present beam test stand or construction of a new test stand, is required to achieve these objectives.

Adequate: Programmes are either planned or could be planned on existing or authorized facilities (with some upgrading if required) and using existing trained personnel.

Inadequate: Either the available and authorized manpower and facilities are inadequate or a substantial increase in effort is needed.

* The ratings are made on the world-wide fusion effort and not separately on the individual national programmes.

5.2. Needs prior to INTOR operation

Maintenance. Intensified work on neutral-beam reliability, and design modifications to facilitate remote maintenance will be required. A detailed assessment is needed to determine the reliability and maintenance requirements.
Conditioning. To meet the reliability requirement, considerable effort will be required to determine the best materials and assembly techniques to minimize the need for time-consuming conditioning of the ion source.
Power Supplies. Significant impact might be made on injector costs and operational simplicity by investigating proposed schemes for elimination of high-voltage regulation, high-voltage switching, and fast-restarting after faults.

6. PHYSICS AND TECHNOLOGY OF RADIO-FREQUENCY HEATING

6.1. Summary assessment for RF heating in INTOR

Four RF methods for heating INTOR to ignition have been considered at the INTOR Workshop. All of these are based on the absorption of a wave in the vicinity of a resonance, and they can be classified according to the frequency range in which these processes occur.

Electron cyclotron heating (ECH) requires frequencies about 120–150 GHz. It was proposed by the EC and USSR for bulk heating at the 50 MW level, and by the USA for supplemental heating at the 10 MW level. The advantages of ECH are its probably simple physics, and the convenience of transmission through narrow waveguides. Its disadvantages are the limited practical experience presently available, the substantial requirements for gyrotron development, and its high probable cost.

Lower hybrid heating (LHH) requires frequencies about 1–3 GHz. It was proposed by the EC, Japan, and (as a second choice) by the USA. The advantages of LHH are convenient transmission through waveguides, well-developed power sources, and some degree of flexibility in the deposition profile within the plasma. Its disadvantages are the relatively poor physical understanding of present-day experimental results.

The ion cyclotron range of frequencies (ICRF) is about 50–70 MHz. ICRF was proposed by the EC, USA (as a first choice), and by the USSR. The advantages of ICRF are its relatively good proven effectiveness on present-day tokamaks, and its very high potential efficiency. Its main disadvantage is the need for internal launching structures (antennae, etc.) in the neutron environment near the plasma.

Very low-frequency heating has been considered only by the EC, in the form of transit-time magnetic pumping (TTMP) at about 30 kHz. At our present

TABLE VI-9. SUMMARY OF THE PHYSICS ASSESSMENT FOR RF

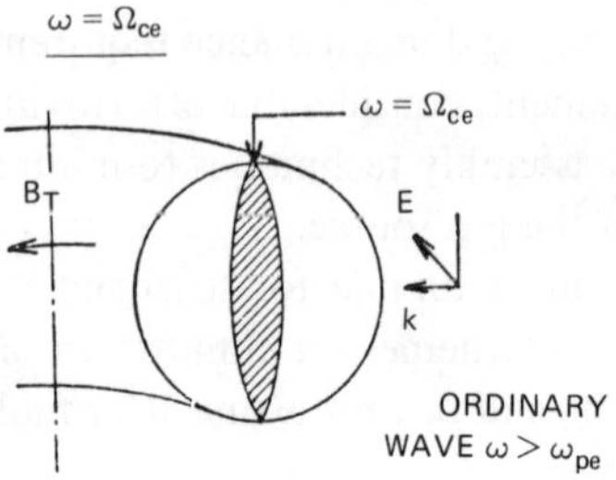

Both extraordinary ($E \perp B_T \perp k$) and ordinary ($E \parallel B_T \perp k$) modes are absorbed by direct cyclotron damping.

For INTOR size and temperature, the ordinary wave $\omega = \Omega_{ce} > \omega_{pe}$ appears to be much more promising.

Complete absorption is expected in a single pass. Absorption is essentially perpendicular to B but collisional effects are sufficient to keep f(v) Maxwellian.

Experimental evidence: Electron heating in TUMAN II, TM3, T-10.

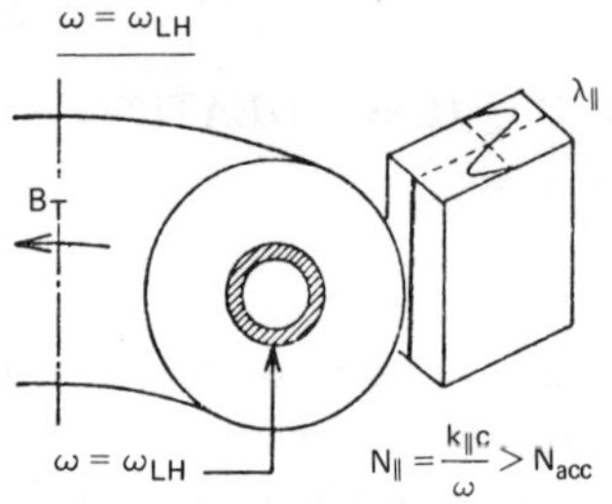

Near the lower hybrid resonance, the part of the power spectrum which is accessible $N_\parallel > N_{acc}$ can heat the ions by mode conversion near the linear turning point (ltp) if $N_{ltp} > N_\parallel > N_{acc}$ or the electrons by electron Landau damping (eld) if $N_{eld} > N_\parallel > N_{acc}$.

For INTOR conditions, these 3 inequalities fix the $N_\parallel$ spectrum of the waveguide grill, and the working frequency, and indicate that electron Landau damping will play a dominant role.

Experimental evidence: Plasma heating in ATC, WEGA, JFT-2, PETULA, ALCATOR A, FT-1

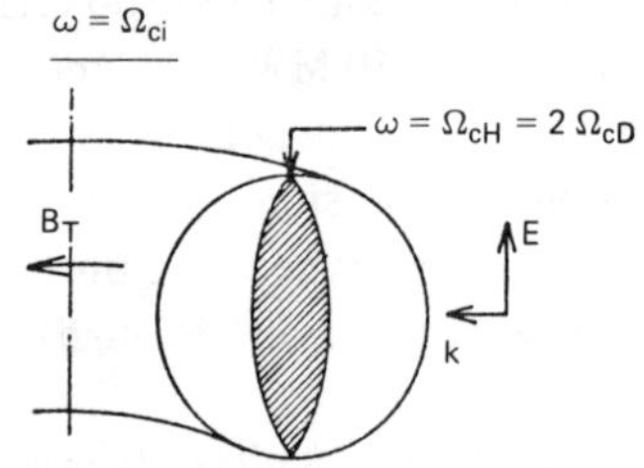

For INTOR conditions, direct cyclotron damping enhanced by ion-ion hybrid effects appears promising. As an example, $\omega = \Omega_{cH}$ in an H-D-T mixture with H as a minority species (~ 10%). Mode conversion is avoided by correct choice of $K_\parallel$. Complete absorption in a single pass is expected close to ignition. In the lower density and temperature ohmically-heated plasmas, about 25 passes would be required.

Experimental evidence: Ion-ion hybrid resonance – minority heating in TFR ST, ERASMUS, PLT, DIVA, T-4, TO-1

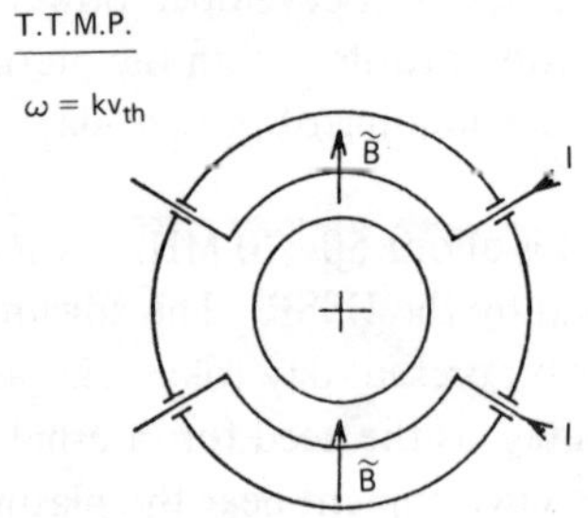

Both compressional ($\tilde{B}_{rf} \parallel B_T$) or torsional ($\tilde{B}_{rf} \perp B_p$) transit time magnetic pumping lead to a heating of the plasma if the resonant condition $\omega/k = v_{th}$ is satisfied. Heating occurs by ion Landau damping.

For INTOR conditions, the torsional mode is preferable because it saves room in the equatorial plane, absorption is higher, and the working frequency lower.

Experimental evidence: Ion heating in PETULA.

level of knowledge, TTMP would seem to offer no clear advantage over ICRF, and it shares its main disadvantage in that it, also, requires internal launching structures. Moreover, its efficiency is expected to be lower than that of ICRF.

The theoretical damping mechanisms for each of these cases are well-identified, and most of them are supported by some experimental results. These are summarized in Table VI-9. Some consideration has been given to the conceptual design of coupling structures for INTOR, and these are presented in Table VI-10. The 50 MW oscillator system needed for INTOR should be feasible in terms of 1982–83 technology, except in the case of ECH, where the gyrotrons that are needed for bulk heating would demand a prodigious development effort.

6.2. Electron cyclotron heating (ECH)

6.2.1. Theoretical considerations

The physics of absorption of wave energy at the electron cyclotron frequency has been extensively studied theoretically and now appears to be well understood.

To couple large amounts of power to a tokamak plasma, it is reasonable to assume that only access from the low field side of the plasma will be technically possible. Therefore, only heating at the electron cyclotron frequency by the ordinary mode (wave electric field parallel to toroidal magnetic field), with propagation along the major radius in the midplane, will be considered. In this case, the absorption is due to relativistic effects, and power is deposited in a narrow region where the wave frequency is just less than the local electron cyclotron frequency.

Plasma heating at the electron cyclotron frequency is attractive for the following reasons:

1. Power is completely absorbed by the plasma in a single pass in INTOR if $T_e \geqslant 1$ keV and $n_e \geqslant 10^{13}$ cm^{-3}. This is important for temperature profile shaping and for plasma heating at the beginning of the discharge when the density may be relatively low.

2. Power is absorbed almost exclusively in increasing the perpendicular energy of the electrons. Taking into account the details of the power deposition profile in the plasma, it can be shown that the electron-electron collision time is always sufficiently short to maintain a Maxwellian electron distribution. The possibility of generating run-away electrons with this heating method appears to be minimal.

3. Power deposition is well localized, and the wave frequency may be fixed so that power is deposited near the centre of the plasma. Edge heating effects which could cause an influx of impurities should be minimized.

4. Coupling the power to the plasma should be relatively simple. Oversized waveguides (10 cm diameter) with very low losses (1% per metre) can be used to transport power from the source to the tokamak, and a simple open

TABLE VI-10. SUMMARY OF THE TECHNICAL ASSESSMENT FOR RF

$\omega = \Omega_{ce}$

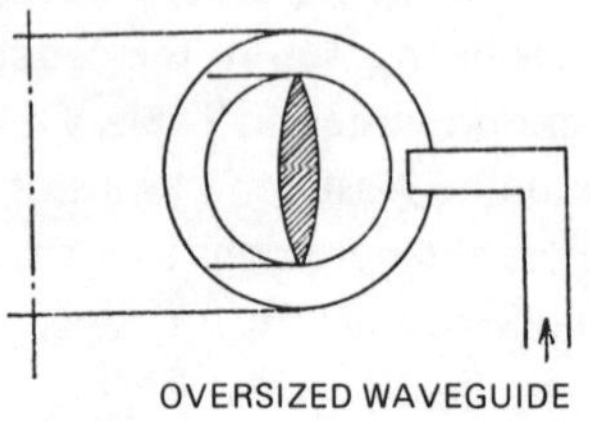

Coupling structure. Oversized waveguide – power density 5 kW/cm^2 which corresponds, for the 60 MW required for INTOR, to an access area of 10^4 cm^2.

Oscillator. The success of this promising method is directly related to the success of gyrotron development. Two 100 kW, 86 GHz, 100 ms gyrotrons are now in operation on T-10. Comparison between these values and the ones required for INTOR (500 kW, 130–140 GHz, 5 s) indicates the need for long-term development. However, an intermediate step (300 kW, 120 GHz, 1–2 s) should be interesting for current-profile shaping. An overall efficiency of 20–30% may be expected.

$\omega = \omega_{LH}$

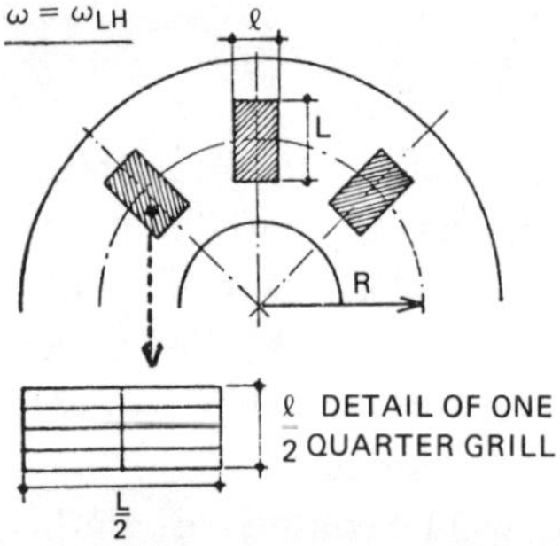

Coupling structure. Use of 10 waveguide grills has been proposed to couple 50 MW. Grills are equally spaced around the torus on the vertical ports. This allows a better accessibility and liberates the equatorial plane. Each grill is made of 8 waveguides along B_T to shape the power spectrum ($1.5 < N_\parallel < 3$) and 4 waveguides radially to handle the 8 MW. In these conditions, 5 MW are coupled to the plasma for an area of 760 cm^2. This power density of 10 kW/cm^2 has already been used in PETULA.

Oscillator. Klystron around 2.5 GHz. Even CW operation at the MW level with efficiency around 60% should not need long-term development. An overall efficiency of 30–40% may be expected.

$\omega = \Omega_{ci}$

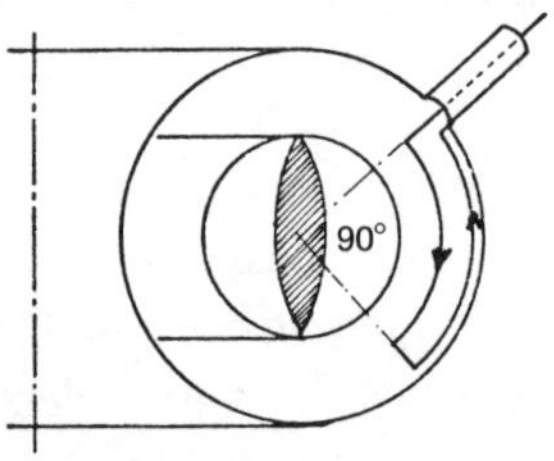

Coupling structure. Presently envisaged to be made of all-metal antennas. As an example, a loop, 2 metres long and 30 cm wide, can couple 3.3 MW to the plasma in the pessimistic case where no plasma exists between the limiter and the loop.

As much as 2–3 times this value can be expected if a low temperature plasma with density of 10^{13} extends all the way to the loop. Loops on top and bottom of the vacuum vessel can be considered.

Oscillator. At 76 MHz, tetrodes at several MW level and with efficiency of 70% do not require substantial long-term development. An overall efficiency of 30–50% may be expected.

T.T.M.P.

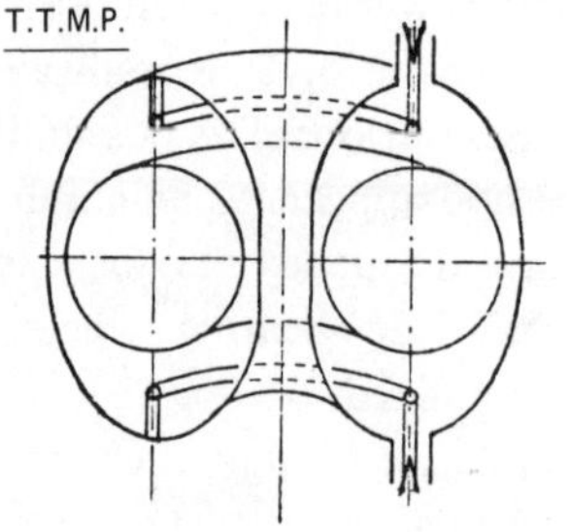

Coupling structure. Four loops located on top and bottom of the machine. Resistive losses in the metallic environment which are the crucial point vary rapidly with the distance between the loop and the first wall.

From numerical computations performed for INTOR plasma parameters, it appears that an overall efficiency of 15% can be obtained at the expense of an increase in the ellipticity of the vacuum chamber. As an example: chamber ellipticity 2.08, power coupled to the plasma 56 MW, overall efficiency 16%, assuming the rather pessimistic efficiency of 50% for the generator and 20% for the connection losses.

Oscillator. 5 MW, 100 kHz tetrode will be rapidly available. The overall efficiency is probably below 20%.

waveguide can beam the power into the plasma. The antenna-plasma coupling is independent of the plasma edge parameters. Refraction effects in the plasma should be tolerable with power generated by a gyrotron with a full beam width of 6° even for plasma density within 10% of the cut-off frequency.

The condition that the wave frequency be larger than the plasma frequency means that the central plasma density is limited to about 2.5×10^{14} cm^{-3} for 140 GHz during the period when the heating power is applied. This does not seem to be a major drawback for the heating method. The main problem has been the lack of a sufficiently powerful, reliable, and efficient source to verify existing theoretical work in a tokamak experiment.

The most extensive test of electron cyclotron heating (ECH) has been conducted on the TM3 tokamak. Heating experiments were carried out at the first and second harmonic of the electron cyclotron frequency using as sources gyrotrons with frequencies up to 60 GHz, power up to 60 kW and pulse length up to 1 ms. It should be noted that for the plasma parameters in the TM3 tokamak, many passes of the radiation through the plasma were required before the power was absorbed. Even under these unfavourable conditions, excellent results were obtained. Bulk heating of the electrons was observed, and a ΔT_e = 100–200 eV was found. It was also noted that the energy confinement time τ_e was proportional to T_e.

Table VI-11 compares some of the parameters of the ECH experiments on ISX-B, JFT-2 and T-10.

Initial results from the ECH experiment on T-10 have recently been reported (summer 1979). The ECH power was 200 kW for a pulse length of 0.1 s. Electron heating up to $\Delta T_e \sim 200$ eV was observed.

6.2.2. Technological considerations

The development of a suitable source presents the most severe difficulty for for ECH. However, recent progress in gyrotron development (Table VI-12) allows one to be optimistic in this regard. In particular, gyrotrons with a frequency of 86 GHz and at least 100 kW/unit with a pulse length of 100 ms have already been used on T-10. The sources needed to drive INTOR to ignition (see Table VI-12) represent a large but not unreasonable extension of present-day gyrotron technology.

The efficiency of gyrotrons in the required frequency range should be at least 30%, and for a 10 m transmission line, the overall system efficiency can be expected to be about 20%.

6.2.3. ECH proposals for INTOR

Detailed calculations of ECH for INTOR were presented by the USSR and EC delegations [1, 4] and these will be summarized. The USSR calculations were

TABLE VI-11. PLANNED ECH EXPERIMENTS ON T-10 AND ISX-B

T-10	ISX-B, JFT-2
$B_0 = 3T$, $\nu = 86$ GHz	B = IT, $\nu = 28$ GHz
Ordinary wave	Extraordinary, Ordinary wave
Launch at large R	Launch at large or small R
$P_{ECH} \sim P_{OH}$, $\tau_{ECH} \sim \tau_E$, $n_e \leqslant 8 \times 10^{13}$ cm^{-3}	$P_{ECH} \sim P_{OH}$, $\tau_{ECH} \gg \tau_E$, $n_e < 10^{13}$ cm^{-3}

TABLE VI-12. PARAMETERS OF GYROTRONS

	Experiments		Under Development			INTOR
	T-10	T-15	EC	USA	USSR	
Frequency (GHz)	86	86	94	120	150	130–140
Power (kW) per unit	100	200	300	200	200	500
Number of Units	2	24	–	–	–	<100
Pulse (s)	0.1	1.5	0.001	CW	1.5	5

carried out for B = 6T at the plasma centre, while the EC calculations were carried out for B = 5T, with the resonance zone displaced to r/a = 0.3, where a is the plasma radius.

According to the USSR calculations, the zone of energy release will be determined by the focussing of wave energy. Calculations of wave trajectories, for a source located in the midplane about 4 m from the centre of the plasma column, showed that refraction does not significantly distort the trajectories of the ordinary wave, so that local heating of the cyclotron zone of the plasma column is possible. In the case where the radiator has an angular divergency of 3°, with a Gaussian distribution of power density, a major part of the wave energy is released within $r/a \leqslant 0.33$. This was for $\bar{n}_e = 1.5 \times 10^{14}$ cm^{-3} and $\Omega_{ce}/\omega_p(0) = 1.1$.

The USSR also reported calculations on the heating to ignition of an INTOR plasma with $\bar{n}_e = 2 \times 10^{14}$ cm^{-3}. The wave energy was assumed to be distributed uniformly over the region $r/a < 0.5$. The power required was found to be 50 MW for 10 s in the frequency range 170 GHz.

In the EC calculations, the beam diameter was assumed much smaller than the plasma diameter, and refraction effects were ignored $\omega/\omega_{p,\,max} = 1.18$, $n_{max} = 1.4 \times 10^{14}$ cm^{-3}, $\nu = 126$ GHz, resonance zone located at $r/a = 0.33$). A toroidal shell of plasma was assumed to be heated by the beam, with the heat diffusing outward from this shell to heat the remainder of the plasma.

For the reference scaling law $\tau_E \propto n_e a^2$, it was found that approximately 35 MW of power are needed to ignite INTOR in 2.5 s. The EC group also employed the Mercier scaling law $\tau_E \sim n_e\, q\, T_e^{3/4}\, a^2$, and found that only 17 MW of power for 2.5 s was needed for ignition, due to the centrally-peaked deposition profile and the favourable T_e-scaling of energy confinement.

6.3. Lower hybrid heating (LHH)

6.3.1. Physics considerations

Lower hybrid heating (LHH) offers one of the most attractive outlooks of any of the present options for auxiliary heating of INTOR, although the viability of this choice is dependent upon a definitive experimental demonstration of its effectiveness. The anticipated flexibility of lower hybrid heating is derived from the ability to heat electrons or ions, and to control the spatial deposition profile of the energy.

The world tokamak programme includes many experiments on lower hybrid heating (see Table VI-13). In the near future several larger scale experiments will become operational which will hopefully provide the required demonstration of significant heating, and a confirmation of the physics basis, both of which are necessary before lower hybrid heating can be a viable choice for INTOR. Some of these planned experiments are listed in Table VI-14.

The lower-hybrid wave utilized for both ion and electron heating is an electrostatic wave which requires a slow wave structure for launching. Since the launched wave is absorbed in a single pass across the plasma the eigenmode structure is not involved. However, the choice of the parallel wavelength is restricted by accessibility and penetration requirements. In particular, the imposed parallel refractive index ($N_\parallel = k_\parallel c/\omega$) is bounded from below, in order to couple to a lower-hybrid mode which has access to the plasma. An upper limit on $N_\parallel$ arises from the need to ensure effective penetration of the lower-hybrid wave into the central region. More precisely, the linear turning point at which mode conversion leading to ion heating occurs, or the range where Landau damping by electrons is effective

TABLE VI-13. EXPERIMENTAL STATUS OF LHH

	Coupling Structure	Surface cm^2	P_{rf} kW	P_{ohmic} kW	τ_{rf} ms	B_T kG	f GHz	$\bar{n}_e$ 10^{13} cm^{-3}	W kW/cm^2	E kV/cm
DIII	slow wave structure	–	350 200	100	10	8	0.8	0.8	–	–
OCTOMAK	slow wave structure	–	10	10	1	0.85	0.4	0.04	–	–
ATC	2–4 waveguides	200	110	150–200	10	18	0.8	1–2	$\lesssim 0.5$	$\lesssim 1$
PT 1	slow wave structure	–	60	60	1	6–10	0.4	0.5–1	~1	~1
WEGA	2 loops	100	130	80–100	10	10–15	0.5	1–2	~2	~1.8
JFT – 2	4 wave guides	380 160	220 140	160–250	20	16	0.75	1–3	1	2
ALCATOR A	2 wave guides	16	90	300–400	20	60	2.5	10–30	~4.5	3.5
PETULA	2 wave guides	120 45	500 450	100 –150	6	16 27	1.2	1–5	4 10	4 6

TABLE VI-13 (cont.)

	Coupling Structure	Surface cm^2	P_{rf} kW	P_{ohmic} kW	τ_{rf} ms	B_T kG	f GHz	$\bar{n}_e$ $10^{13}\ cm^{-3}$	W kW/cm^2	E kV/cm
JIPP TII	2 wave guides	120	60	210	20	21 26	0.8	1.4 – 2.4	1.6	–
VERSATOR II	4 wave guides	300	200	–	10	15	0.8	≤ 5	0.7	–
FT – 1	4 wave guides	–	70	–	3	0.9	–	0.4	–	–

TABLE VI-14. PLANNED LHH EXPERIMENTS

	Year	Power (MW)	Pulse(s)	Power density (kW/cm^2)	Frequen[cy] (GHz)
JFT-2	1979 1980–1	 0.3	 0.1	2.0 3.5–4.0	 0.75
WEGA-2	1979–80	0.8	0.04	2.5	0.8
PETULA-2	1980	2.0	0.05	–	1.25
FT	1980	0.5	1	12	2.45
Alcator C	1980	4.0	0.2	10–12	4.5
PLT	1980–1	1.2	0.5	3.3	0.8
JFT-2 or JFT-2M	1979–81 1982–83	 1.0	 0.3	 2–4	 0.9
JT-60	1985	10	10	4.0	$\gtrsim$1.7 tentative

should lie in the central region. Which damping mechanism prevails – and hence whether ions or electrons are heated – depends on whether the linear turning point or the range of electron Landau damping is encountered first by a wave propagating into the plasma. A choice between the two cases is possible by adjusting the wave frequency appropriately.

The experiments performed so far have shown that the coupling efficiency of lower-hybrid waves to the plasma can be above 90%. Whereas the accessible fraction of the power has usually been about 50%, the heating efficiency of this fraction is close to 100%. Hence, by improving the wave spectrum to increase the accessible fraction of the power, a very good heating efficiency can be anticipated. The most efficient ion heating is obtained when the linear conversion point is situated at the plasma centre. The electron heating observed is consistent with electron Landau damping as the operating absorption mechanism. The wave propagation across the edge region can be explained by linear theory as long as the power flux is below 4.5 kW/cm^2. The main uncertainties regard the spectrum actually radiated, and the influence of the edge plasma on the coupling. In addition, increases in electron density are often observed during LH heating, perhaps due to contamination by light impurities desorbed by arcing at the waveguide or antenna. Recently, JFT-2 has succeeded in avoiding arcing in the wave-guide by improved surface treatments and improved insulators, with the result that no density increases were observed. Another uncertainty is whether run-away production would remain within an acceptable limit if electron heating were to be provided at densities as low as 2×10^{13} cm^{-3}.

6.3.2. *Technological considerations*

In the area of technology, lower hybrid heating appears to be the most attractive of all present heating options. The existing technology is capable of providing the required power in the appropriate frequency range with high conversion efficiency. The ability to remotely site the power supplies, transport power over large distances with waveguides through convenient paths with bends, and the simple, potentially radiation resistant, coupling structure provided by waveguide arrays considerably simplify the engineering requirements. An estimate of the overall DC to RF efficiency of the system is about 60% without thermal recovery. This would lead to an overall heating efficiency of 30–40%. The power flux available at the top of the launcher can be as high as 10 kW/cm^2 [1].

At present 500 kW klystron tubes at 2.4 and 2.1 GHz with conversion efficiency of 53% and 55 dB gain can be obtained. A modest development programme can provide a 1 MW tube around 2.0 GHz with conversion efficiency of 80% and 55 dB gain. It is estimated that five 1 MW tubes can be driven from a single high voltage power supply of conventional design rated at 75 kV, 83.3 A, and 6.25 MW. Also a higher heating efficiency should be developed before the construction of an INTOR LHH heating system. In the Japanese proposal [2] collector potential depression (CPD) for the collector of the klystron was adopted. From the disk model simulation, about 60% of the electron beam power passing through an output cavity is recovered using a three stage CPD collector. Therefore, if we assume 50% klystron efficiency, an 80% final efficiency will be obtained.

6.3.3. *Lower hybrid heating proposals for INTOR*

Specific proposals for LH heating of INTOR have been provided by the EC and Japanese delegations [1, 2]. In order to ensure plasma heating in the centre during heating to ignition, the frequency and $N_\parallel$ have to be suitably adapted to the change in temperature as well as to any change in density (see the Start up, Burn and Shutdown Report [6]). At ignition ($\bar{T}_e = 10$ keV, $\bar{n}_e = 1.4 \times 10^{14}$ cm^{-3}) a frequency of about 2.5 GHz and $N_\parallel = 2.0$–2.5 is appropriate. For lower temperatures and densities, lower frequencies and/or larger $N_\parallel$ would be required.

TABLE VI-15. EXPERIMENTAL STATUS OF ICRF

Device	f (MHz)	P_{rf} (kW)	Launching Structure	Coupling Efficiency (%)	Heating Efficiency (%)	ΔT_i (ev)	Pulse Length (ms)	Heating Mechanism
ST	25	< 450	Two ½-turn coils	90	20	200 D+	1	Fundamental two-ion hybrid
ATC	25	< 200	1/3-turn copper strip	90	< 40	200 D+	10	Two-ion hybrid
T-4	23	70	3/4-turn coil (all metal)	90	$\gtrsim 70$	100 D+	1	Direct minority heating, H^+
TFR	50–90	250	Tandem ½-turn coils (all metal)	80	70	150 D+	30	Mode conversion and direct minority
PLT	25	450	½-turn coils	80	$\gtrsim 70$	700 D+	100	Two-ion hybrid, D+ (direct minority heating, H^+)
TM-1	7–35	150	Two asymmetric coils	80–90	30–40	100	0.8	Two-ion hybrid
DIVA	25	180	½-turn coil	75	70–100	270 D+	3	Two-ion hybrid

TABLE VI-15 (cont.)

Device	f (MHz)	P_{rf} (kW)	Launching Structure	Coupling Efficiency (%)	Heating Efficiency (%)	ΔT_i (ev)	Pulse Length (ms)	Heating Mechanism
Macrotor	1–20	$\lesssim 200$	Rectangular loop			200	2–4	Off-resonant
TO-1	48	25	Coil		25	50 H+	20	Off-resonant
ERASMUS	5	40	½-turn coil (all metal)	50	12	30	3–6	Two-ion hybrid

One possible scenario would be to heat at fixed frequency, decreasing $N_\parallel$ from about 6 to 2.5 while the temperature rises, relying mainly on electron heating by Landau damping [1]. Another possibility is to start heating at a lower frequency, typically 1.5 GHz, when the plasma density is below $6 \times 10^{13}/cm^3$ and the temperature about 1 keV, and to change to 2.5 GHz when higher densities are reached, restricting the variation in $N_\parallel$ to a narrower range [2]. In this case, some ion heating might be achieved at an earlier stage. The power necessary for heating to ignition could be as low as 50 MW; the low-frequency power suggested in the second scheme for the early stage of heating might be 10 - 20 MW.

If it is considered that the power flux will be in the range of 4–10 kW/cm^2, for a 50 MW power input only up to 3 m^2 of launcher area is necessary.

In the case of the Japanese proposal, the power density of 4 kW/cm^2 and projected efficiency of 40% are the same as the values for the JT-60 lower-hybrid heating system.

6.4. Ion cyclotron range of frequencies (ICRF)

6.4.1. Physics considerations

Since ICRF heating was first introduced into tokamaks in 1973, experiments at power levels up to about 200 kW have been conducted on the ST, ATC, PLT, TFR, DIVA, T-4, TO-1, and TM-1 tokamaks. The present experimental status is summarized in Table VI-15. There are current plans to carry out megawatt-level experiments on PLT, Alcator, TFR, JFT-2 and T-10 tokamaks. The main purpose of these high power experiments is to determine the feasibility of employing ICRF heating in hot reactor-sized plasmas. Consequently, the emphasis of these studies is being placed on evaluating the physics of the wave-plasma interaction to permit extrapolation to reactor conditions.

In principle, a variety of heating schemes can be utilized in the ion-cyclotron frequency range. Primary heating regimes include (a) the two-ion regime for which minority fundamental cyclotron damping, majority second-harmonic cyclotron damping, and electron Landau damping contribute to the absorption of RF energy according to the dictates of the ion-ion hybrid resonance effect on the dispersion relation; (b) the pure second harmonic resonance regime; (c) possibly the pure fundamental resonance regime; and (d) under certain conditions the regime for which mode conversion into ion-Bernstein waves allows electron Landau and transit-time absorption mechanisms to become substantial.

The following practical scenarios appear attractive for INTOR:

(i) $\omega = \Omega_{CH} = 2\Omega_{cD}$ in a H-D-T mixture with H as minority species ($\sim$ 10%) which can be turned off after ignition. The RF power goes mainly to the hydrogen which, in turn, heats the other species by collisions.

(ii) $\omega = \Omega_{cD}$ in a D-T mixture with D as minority component (10 to 40%). The RF power goes to the deuterium.

(iii) $\omega = 2\Omega_{cT}$ in a T or D-T plasma. This scheme is based on harmonic damping by tritium.

(iv) $\omega = \Omega_{DT}$ is a D-T mixture of comparable concentration. Due to strong mode conversion, the power is absorbed mainly by the electrons.

The schemes (i), (ii) and (iv) allow complete wave absorption in a single pass through the plasma close to ignition. Scenarios (i), (ii) and (iii) can be excited by a coupler located on the lower or high field side, while (iv) needs an exciter situated on the high field side. All the schemes can be adapted to the density and temperature ranges foreseen for INTOR.

Minority hydrogen heating in a D-T plasma will require rather precise control of the hydrogen content of the discharge; some appropriate particle control techniques will have to be developed and tested.

In early heating experiments, heating efficiencies of about 20–40% were achieved but were hampered by substantial plasma density rise, indicating surface heating and loss of energetic ions. As the machine size and ohmic heating current have been increased, a systematic improvement in energetic ion confinement has resulted, as expected, and the mode spectrum has improved so as to increase the heating efficiency and considerably reduce the influence on recycling characteristics.

The best data presently available is for regime (a), the two-ion regime. Recent experiments have been performed for damping dominated by direct minority fundamental cyclotron-resonance damping in PLT, and for damping dominated by two-ion hybrid mode conversion in TFR, and for damping dominated by second harmonic cyclotron damping in DIVA. In the case of PLT, the minority distribution is made very energetic and serves as a conduit of the RF energy to the majority ion distribution. This regime is quite similar to ECH and scales very favourable to reactor conditions. In the case of TFR, sufficient minority concentration is maintained to permit mode conversion so that a large fraction of the power heats the electrons directly. In the case of DIVA, the minority concentration is controlled to explore the optimum heating conditions. The result indicates that the most favourable majority deuteron heating occurs for n_H/n_D = 5–10%, below which minority proton heating is observed, and above which mode conversion is dominant (although intense electron heating has not yet been observed.)

For optimization of ICRF heating in large tokamak systems, control of the RF power deposition profile is essential. Adequate control will assure that the RF power is delivered to the core plasma and that surface heating resulting in enhanced impurity influx is negligible. Since the strength of damping processes is approximately proportional to the beta of the species involved, considerable central peaking of the power deposition profile should ensue regardless of the wave k spectrum excited. However, selection of the k spectrum, also to avoid waves which have large fields

TABLE VI-16. PLANNED ICRF TOKAMAK EXPERIMENTS

Device	Expected RF Power	Launching Structure	f (MHz)	Pulse Length (ms)	Heating Mechanism
Macrotor	1 MW	Rectangular Loops	1–20	10	Stochastic Heating
Alcator A	200 kW	Tandem Coils	150–200	10	Second Harmonic
Alcator C	>4 MW	Tandem Coils	150–200	10	Second Harmonic
PLT	5 kW	Multiple ½-turn Coils	25 55	200 200	Two-ion Regime Second Harmonic
TFR	3 MW	Tandem Coils	50–90	100	Two-Ion Regime
T-10	1–5 MW	Loops	60	50	Two-Ion Regime
ERASMUS	>0.5 MW	Tandem Coils	5–15	2–30	Two-Ion Regime
JFT-2 or 2M	1 MW	½-turn coils	20	50	Two-Ion Regime

and field derivatives in the plasma surface, should serve to virtually eliminate surface heating.

The goals of larger planned experiments are given in Table VI-16. These plans are aimed at producing significant heating in the hottest plasmas available to assess the physics of the wave damping, the damping profile and means to control it, and the potential of the several heating schemes for extrapolation to reactor conditions. As indicated by the initial results on PLT and TFR, the physics of ICRF heating and its relevance for reactor conditions should become clear during the next few years from the experiments performed on the larger, higher-current tokamaks. With the recommended development programme, these results will be available in time to impact the choice of primary heating technique for INTOR.

6.4.2. Technological considerations

An ICRF heating system for startup and ignition of a device such as INTOR consists of a radio-frequency source, power transmission and matching system, and the wave launching structure or antenna.

External component technology, which embraces all components outside the neutron shield (i.e. excluding the wave launching structure and its power feed), is well in hand, but will benefit from some modest improvements. Megawatt tetrodes and coaxitrons suitable for ICRF heating sources currently exist with efficiencies of ~ 90%, but are now rated for pulse lengths of only 0.1–0.35 s at peak power. Although CW operation is now possible at the MW level for a single tube, development may be desirable to obtain the long pulse 5–100 s (essentially CW) operation envisioned for INTOR, at the highest possible power per tube. Power transmission can be accomplished easily with 10–20 MW capacity, 23 cm diameter lines which are commercially available. Furthermore in the ICRF frequency range, transmission over tens of metres incurs negligible loss of power, permitting the source to be located external to the reactor containment building, removing the need for remote handling maintenance of the source. Although radiative cooling of the line may suffice, liquid cooling with, for example, borated water should be investigated. Finally, double stub tuners are presently used to match the antenna feed impedance to the source and are likely to work well with minimal modification for the INTOR device. However, an optimal scheme should be developed during the next decade.

The one area for which the technology is only crudely developed is that of launching structures and feeds suitable for reactor operation. Efficient wave coupling > 90% is easily achieved with loop antenna systems located in the shadow of the plasma limiter, when such coupling systems are suitable. This includes all devices for which the heat load does not seriously increase the resistance of the loop elements. It is possible that loop systems can be developed for INTOR with adequate cooling of the system elements. The all-metal antenna recently employed on TFR [1] and T-4 [4] suggest that such cooling may be feasible. (All elements, including shields, are of sufficient size to permit liquid cooling.) Furthermore, insulators within the radiation shield may be eliminated, thereby avoiding unnecessary neutron damage of the loop system.

In addition to loop coupling, other wave launching structures need to be seriously explored for the reactor regime. Flush-mounted cavity apertures, dielectric filled waveguides, ridged waveguides, and strip line couplers are possible contenders. As the machine size is scaled up, these structures become more attractive, but remain virtually unexplored experimentally.

Power levels through ports on ICRF experiments have ranged from 1–10 kW/cm^2 without breakdown, and peak electric field intensities of 10 kV/cm have been obtained in the loop region. It is expected that these levels will remain at comparable values as scaling to INTOR is achieved, depending on the success of the development effort envisioned.

6.4.3. ICRF proposals for INTOR

The ICRF proposals from the EC [1] (see also Tables VI-9 and VI-10) and from the USA [3] emphasize the case $\omega = \Omega_{cH} = 2\Omega_{cD}$ (corresponding to a frequency of 76 MHz for $B_0 = 5$ T). In this case, direct cyclotron damping at the fundamental of a minority hydrogen species (~ 10%) dominated by ion-ion hybrid effects, will transfer most of the RF power to the minority hydrogen ions, which in turn heat the DT ions by Coulomb collisions. Complete absorption in a single pass is expected close to ignition. In the lower density and temperature ohmically-heated plasmas, about 25 passes would be required.

The ICRF proposal from the USSR [4], see also [1], emphasizes the case $\omega = 2\Omega_{cT}$, i.e. heating at the second harmonic of the tritium ions (corresponding to a frequency of 61 MHz for the USSR case of $B_0 = 6$ T). This mechanism will provide for wave absorption by one of the main components of the plasma, and no minority hydrogen species is needed. Since second harmonic damping is weaker than any fundamental resonance, the wave will require several transits through the plasma for complete absorption. The USSR calculations [4] show that a strong peaking of the deposited power can occur in the centre of the plasma. The tritium distribution will acquire a non-Maxwellian tail, which may increase the fusion reactivity. Experiments to test this method of heating, in non-reacting plasmas, could be done at the second harmonic of He_3 in an H-He_3 plasma. The USSR proposes [4] a 75 MW system with a 3 s pulse. The antenna is a set of coaxial segments located near the first wall. Neighbouring coaxials have antiphase connections, in order to ensure the necessary deceleration of waves. The length of the coaxials is about $\lambda/4 \cong 125$ cm. There are 4 antennae symmetrically arranged around the torus. The antennae are made of a vanadium alloy with 5% of titanium, as is the section of the coaxial transmission cable near the tokamak; the coaxial is water cooled. The overall electrical efficiency is estimated at 60%.

7. SUMMARY OF RESEARCH AND DEVELOPMENT NEEDS FOR RADIO-FREQUENCY HEATING

7.1. Physics research and development needs

Although the various RF techniques for heating plasmas involve physical processes that have all been verified, at least on small tokamaks, their feasibility for heating an INTOR plasma to ignition remains to be demonstrated. In particular, scaling laws for the regime where the RF power exceeds the ohmic

TABLE VI-17. DETAILED PHYSICS RESEARCH NEEDS FOR RF

Ω_{ce}	In these methods, the wave propagates ($\alpha \gg \lambda_{\perp}$) from the coupling structure to the resonance region where absorption occurs. The radial energy deposition is then the most important problem and has to be studied in a wide range of conditions. In particular, non-linear effects, impurity release mechanisms and their control, and parametric instabilities have to be determined.	The physics of absorption, in particular its dependence on polarization, should be verified experimentally at significant power levels.
ω_{LH}		Electron Landau damping should play a dominant role for INTOR conditions. Experimental evidence of the linear and quasi-linear theory is then necessary. The origin and location of the fast ions in the present heating experiments remain to be explained. Density increases need to be controlled by an INTOR-relevant technique.
Ω_{ci}		The choice of wave parameters is important in in the selection of the absorption mechanism. The exploration of these different situations with optimized launching structures has to be performed. The role of a Faraday shield remains to be clarified.
T T M P	As the vacuum wave-length is very large ($\lambda \gg \alpha$) the fields with or without the plasma are the same. Comparison between theory and experiment has been made only in the case of compressional T.T.M.P. The validity of the torsional version has to be checked.	The transition between the linear and nonlinear regime, i.e., the dependence of the critical density required to keep f(v) Maxwellian, must be confirmed for a larger range of plasma parameters.

power have to be established, and long-pulse high-power operation without excessive impurity influx has to be demonstrated. A broad programme of experimental research is needed, as outlined in Table VI-17.

The heating test programme at the megawatt level on the current generation of tokamaks should continue until the prospects for RF heating are fully established. Subsequent heating experiments for the attractive heating modes should then be performed on the next generation of tokamaks (i.e. JET, T-15, TFTR, JT-60), to demonstrate heating of larger plasmas to reactor temperatures where the problems of excessive heat loads and neutron deterioration of performance will have to be faced. The next step would be the application of RF heating to an ignition device such as INTOR.

In order to assure the timely development of the components needed both for the JET/T-15/TFTR/JT-60 generation of experiments, and for INTOR, it is considered that one or more dedicated "Radio-Frequency Test Facilities" (RFTF's) should be established as soon as the physics base warrants, probably within two years time.

For ICRF, the optimization of the loop coupling structures on present devices, will serve to delineate the degree of wave-spectrum selectivity required. This information can then guide the design of new types of coupling structures for testing on suitable devices, preferably on an RFTF. In this manner, all of the development required to optimize operation on much larger devices, and finally on INTOR, can be done off-site and can lead to highly reliable operation on these devices.

Clearly, the pressing need for the LHH method is a definitive demonstration of significant heating in a tokamak, and the corollary confidence in the physics understanding. It is also essential to determine any nonlinear limitations on the coupling and heating to ensure optimum choice of the required frequency, parallel wavelength, and port area for INTOR. The primary additional need for lower-hybrid heating before INTOR is the demonstration of long-pulse high power operation, without significant impurity influx, and to provide a test of the engineering reliability of the complete system. An additional need exists for engineering optimization of coupling structures, tests of phase tracking, and studies of nonlinear effects; these tests can be done more efficiently and economically on a smaller dedicated RF facility, such as the proposed RFTF.

For ECH, the theoretical picture of linear absorption needs experimental verification, in particular also the dependence of absorption on wave polarization. Furthermore, an intense development programme for gyrotrons is required, as well as development of oversized waveguides for power transmission, as discussed below.

7.2. Technology research and development needs

For a coupling structure sustaining a power density of 5–10 kW/cm^2 a total area of 5×10^3–10^4 cm^2 of loops, antenne, or waveguides will be subject to neutron bombardment. From this point of view, the problems of the coupling structures are materials problems very similar to those of the first wall. Table VI-18 lists some aspects of these problems which call for a strong, dedicated research and development programme.

For ICRF, with internal antennae, the problems of the neutron environment are very severe. In addition to continued testing of structures in the plasma

TABLE VI-18. DETAILED TECHNOLOGICAL DEVELOPMENT NEEDS IN RF LAUNCHING STRUCTURES

As has been already discussed, waveguides, antennae, or loops are required to match the wave coming from the generator to the boundary conditions of the plasma. In the radiation environment expected in INTOR, a proper RF coupling design is a major problem. Such a structure must be resistant to radiation and heat damage and, from this point of view, the constraints on the choice of the materials are very similar to those for the first wall. However, specific RF requirements such as low resistivity dimensional stability, arcing and multipactor effects, also constrain the choice of material. A dedicated development programme on these topics will be necessary to reach the INTOR aim.	$\omega = \Omega_{ce}$ – Power handling capabilities of the oversized waveguides and of the associated elements – Windows $\omega = \omega_{LH}$ – Shielding and location of the vacuum window – Base pressure allowed in the waveguide grill between the vacuum window and the plasma – Conditioning of the waveguide grill before operation $\omega = \Omega_{ci}$ – Improvement of the antenna power handling capabilities – Shielding and location of the vacuum window – Possible use of ridged or loaded waveguide T.T.M.P. – Possible use of insulating material such as SiC to protect the RF structure from particle bombardment.

environment, materials-testing relevant to ICRF heating is needed. Radio-frequency conductivity and the effects of surface blistering should be measured for candidate conductors (copper, vanadium, molybdenum) at neutron and plasma flux levels comparable to those expected in INTOR. Methods must be found so that all insulators (sintered beryllia, alumina, yttria) can be located out of the direct view of the neutron flux, their compatibility with conductors, thermal stress and neutron bombardment should be studied.

Power supply requirements for RF, other than ECH, would involve only modest development of present technologies. For ICRF, 1–5 MW sources in the frequency range 25–75 MHz which are capable of long-pulse operation (practically CW) should be developed. The selection of LHH for INTOR would make modest development work on klystron sources beneficial. The development of long-pulse klystrons of 1 MW or larger and improved efficiency would be the key objectives.

TABLE VI-19. R AND D NEEDS ASSESSMENT – RF HEATING

	Status*	
	Adequate	Inadequate
– MW-level experiments to prove heating effectiveness, followed by demonstrations of heating larger plasmas to reactor temperatures (ICRF, LHH)	x	
– Experimental verification of absorption and heating by ECH at the 200–400 kW level, followed by MW level experiments	x	
– ICRF and LHH launching structures resistant to thermal loading and neutron radiation		x
– Vacuum windows resistant to thermal loading, neutron radiation and high RF power		x
– Gyrotron development at 120–150 GHz, 500 kW, 5 s pulse	x	
– ICRF, LHH power supplies	x	

Adequate: Programmes are either planned or could be planned on existing or authorized facilities (with some upgrading if required) and using existing trained personnel.

Inadequate: Either the available and authorized manpower and facilities are inadequate or a substantial increase in effort is needed.

* The ratings are made on the world-wide fusion effort and not separately on the individual national programmes.

For ECH, research is required into achieving high-efficiency, (>20%) high-power gyrotron operation at frequencies in the 120–150 GHz range. Such research will require new techniques for optimizing gyrotron resonator design, and techniques for eliminating mode competition in high mode resonators. The requirement of a 10^6 pulse lifetime in 5 s pulse operation has not been tested even in low frequency gyrotrons. Techniques for resonator and collector cooling and for gun operation under long-pulse operation must be investigated. Improved techniques for efficiently transforming the working mode of the gyrotron into a linearly polarized mode for plasma heating should be developed. Systems, including quasi-optical systems, for efficiently transmitting gyrotron radiation from the tube to the reactor should be studied. Window materials for the gyrotron and ECH port must be developed.

Table VI-19 provides a brief summary of R and D needs in RF heating.

8. SUMMARY

8.1. Summary of neutral-beam heating

Heating of the INTOR plasma to ignition conditions seems feasible using neutral-beam energies achievable with positive-ion-based technology (150–200 keV per deuteron), but with significant advancements beyond TFTR/JET/JT-60/T-15 beam technology required to obtain long pulses at high duty factor, as well as high currents at the upper end of this energy range. Other development requirements are a high monatomic ion fraction in the ion beam, and possibly an increase in overall efficiency to be obtained by direct energy recovery, and increased reliability and component lifetimes.

Potentially serious disadvantages of the beam heating method are the exposure of the injector internals to the fusion neutron flux, and the expected high capital cost of beam systems. The difficulties incurred by the direct exposure of injector components to neutrons can probably be overcome, especially if the beam injectors can be closed off after start-up.

The following are probable technical parameters for the INTOR neutral beam injection system:

Beam energy	175 keV (the beam development programme should attempt 200 keV)
Beam power	75 MW
Pulse length capability	10 s (the capability should also exist to restart injectors during the burn to achieve an effective 100 s pulse length at 10 MW)
Injection direction	20° to normal
Monatomic ion fraction	About 90%
Bending magnet at ion source	None
Fraction of neutral beam power at full energy	About 80%
Direct energy recovery	Probably required (efficiency ⩾ 60%)
Power density at port	2–3 kW/cm² (depends on detailed design)
Number of injection ports	4–5 (depends on detailed design)
Cross-section of port	0.5–1.0 m² (depends on detailed design)
Neutron protection	Plug valve closed after ignition

Lifetime of source	3×10^5 s
Pumping speed per port	5–8 Mℓ/s
Redundancy	Additional ion sources will be needed to provide required reliability
Negative ion beam	Advantageous for the long pulse capability (10 MW for 100 s); useful energy range 400–500 keV.

The provision of effective heating (if needed) during the start-up phase when the plasma density is low poses some difficulties, due to beam shine through, if the main injector system is to be used. A lower-energy injector may have to be provided, or a larger angle of injection, or perhaps a system capable of variable energy or variable species mix.

There remain some uncertainties about the absolute need for a direct energy recovery system for the INTOR injectors. The advantages of direct recovery (60% efficient) would be an increase in the overall efficiencies of the injectors from an estimated 20% to an estimated 30%, and a reduction in the power to be handled on the beam dumps from about 200 MW to about 85 MW for a 75 MW system. However, to date, there has been very little development effort on direct recovery systems for neutral beam injectors.

8.2. Summary of radio-frequency heating

The continued attraction of pursuing the many different RF heating schemes derives in part from the perceived disadvantages of neutral-beam injection mentioned above. To date, only ion cyclotron wave heating (ICRF) has provided a convincing demonstration of substantial plasma heating in tokamaks of significant size. RF heating in the lower-hybrid frequency range has been successful in smaller tokamaks, but the physical mechanisms of coupling the wave power to the plasma are less well understood. Electron cyclotron heating has demonstrated its heating effectiveness in recent experiments on quite large tokamaks. Transit-time-magnetic-pumping has been found effective on small tokamaks but its extrapolation to INTOR conditions seems technologically too difficult.

The experimental success of ion cyclotron heating, and the potential inexpensiveness of ICRF systems make it an RF technique that presently seems attractive for INTOR. However, lower-hybrid heating has important technological advantages, and should be retained as a viable option. However, before any RF method can displace neutral-beam injection as the preferred heating technique for bulk-heating INTOR, this method must demonstrate the multi-keV temperature increases already obtained with neutral-beam injection, and convincing designs for reasonably efficient and neutron-resistant wave launching structures must be devised and tested.

TABLE VI-20. ADVANTAGES AND DISADVANTAGES OF RF HEATING METHODS

	Advantages	Disadvantages
Very low frequency (TTMP) 100 kHz	Physics understood	No access to coils. Neutron effects on internal coils. Low efficiency of heating
Ion cyclotron 50–70 MHz	Proven effectiveness in large tokamaks. High efficiency	Neutron effects on internal antennae
Lower hybrid 2–3 GHz	Very attractive transmission (waveguides) Efficient sources available (klystrons)	Effectiveness in large tokamaks yet to be proved
Electron cyclotron 120–150 GHz	Very attractive transmission (waveguides) Simple physics.	Substantial tube development needed. High probable cost

The technological problems associated with electron cyclotron heating (ECH) lie primarily in gyrotron development, rather than in transmission of the ECH power to the plasma (which is technologically very attractive). Gyrotrons at 120–150 GHz, delivering about 500 kW per unit for a 5 s pulse, should be developed for INTOR, for a total ECH power level of at least 10 MW. Such an ECH system would be an attractive option for providing the needed start-up power at low plasma density.

Table VI-20 summarizes the main advantages and disadvantages of the various RF methods.

REFERENCES TO CHAPTER VI

[1] European Contributions to the 2nd meeting of the INTOR Workshop, Heating, Chapter IV, EUR FU/XII 501/79/EDV 50, June 79 (2 Volumes).
[2] Fusion Research Development Center, INTOR Workshop Report, Group 4, Heating, JAERI-M 8624 (1979).
[3] "Heating", Chapter in USA INTOR Report (see Appendix A.5.4).
[4] USSR Reports to INTOR, "Plasma Heating in INTOR. Injection System"; and "Plasma Heating in INTOR. RF-System" (see Appendix A.5.3 [14, 15]).
[5] Chapter IV, this report.
[6] Chapter VIII, this report.

Chapter VII

STABILITY CONTROL

F. ENGELMANN – EC, Chairman
B. KADOMTSEV – USSR
H. MOMOTA – Japan, Expert
P. RUTHERFORD – USA
T. TAZIMA – Japan

1. MHD STABILITY LIMITS ON β

1.1. Theoretical predictions

The possible MHD perturbations of a toroidal plasma may be classified as

1) high toroidal mode number (n) perturbations occurring in a plasma with fixed boundary ("ballooning modes"),
2) low toroidal mode number perturbations occurring in a plasma with fixed boundary ("low-n internal modes"),
3) low toroidal mode number perturbations occurring in a plasma with free boundary ("kink modes").

The axisymmetric instability (n = 0) could be taken as a subcase of 3). It will, however, be treated under "Equilibrium and Stability Control" (Section 3), as it corresponds to an almost rigid shift of the whole plasma column, mainly in the vertical direction and, hence, leads to a position control requirement but not to a β limit. Only if this control of position were not feasible, would there be an indirect β limit, by excluding practically any sizeable elongation of the plasma cross section in the vertical direction.

In the following discussion only such equilibria will be considered which are accessible from states with $\beta = 0$.

1.1.1. High-n, fixed boundary modes

The behaviour of high-n, fixed boundary modes can be analyzed by asymptotic analytical methods, supplemented by numerical calculations for relatively large n typically $n \approx 5$. It has been found that stability at high β, in the presence of large pressure gradients, requires low shear (rq'/q) and the presence of a magnetic well.

Anticipating that an ignited INTOR plasma would have pressure profiles such that $\alpha \equiv 8\pi\, p'R/B_\varphi^2 \lesssim 1$, the shear parameter rq′/q would be required to be smaller than about 0.1 for a circular cross section, and smaller than about 0.5 taking the beneficial effect of elongation and triangularity (D-shape) into account [1]. In terms of a limit on the volume average $\langle\beta\rangle$ of β, for INTOR (A = 4, b/a = 1.6) this means that

$$\langle\beta\rangle_{max} \approx 5 \text{ to } 6\%$$

can be anticipated, provided that the plasma shape is appropriately tailored and the current profile is not too strongly peaked. In fact, considering ballooning modes only, still higher values of $\langle\beta\rangle$ are possible, even in a circular cross section. However, in practice the shear cannot be arbitrarily small, in particular in the edge region; moreover, if it were very small, low-n modes (see below) would tend to be unstable.

It is important to note that it is an open question if high-n modes are dangerous at all. In ISX-B and T-11, discharges theoretically expected to be ballooning unstable have recently been made, and no deleterious effects have been seen, at least as far as the macroscopic behaviour of the discharge is concerned [2], [8] (see Section 1.2). From a theoretical point of view, non-ideal and non-MHD effects are expected to affect the ballooning mode stability properties but it is not clear in which way (finite resistivity has been reported to be destabilizing). The consequence of ballooning instability is likely to be an enhancement of transport, but it is impossible to say how large this could be compared to transport effects appearing in present-day machines.

1.1.2. Low-n modes

The investigation of low-n modes requires a numerical approach, which has the consequence that general trends in the dependence of the stability limits on aspect ratio, q(a)*, plasma shape, and profiles is much more difficult to display and understand. However, the extensive work done in recent years has begun to clarify the situation.

---

* In the following, the safety factor q(a), defined as the inverse rotational transform at the plasma edge, and the quantity

$$q_J(a) \equiv \frac{1}{2}\left(1+\frac{b^2}{a^2}\right)\frac{2\pi\, a^2 B}{\mu_0 R I_p} < q(a)$$

which is related to the plasma current, will be distinguished, as these coincide only when the poloidal magnetic field is independent of the poloidal angle.

As far as **low-n, fixed boundary modes** are concerned the results obtained indicate that for high shear the limits of $\langle\beta\rangle$ are similar to those of ballooning modes, whereas for low shear they may be higher by up to a factor of 2, provided that q on axis is larger than 1. The data available are consistent [2], [4] with a β-limit that scales with aspect ratio, q(a), and elongation, according to

$$\langle\beta\rangle_{max} \sim \frac{a}{R}\, q(a)^{-2}\, \frac{b}{a} \left\{1 + \delta\left(\frac{b-a}{a}\right)^2 \times f\left(\frac{b}{a}\right)\right\} \qquad (1)$$

where δ describes the triangularity of the plasma cross section and f(b/a) is a function of elongation. The benefit due to elongation tends to saturate for b/a about 2 [4]. The favourable effect of low shear, in Eq.(1), appears via the dependence on q(a). For D-shaped INTOR equilibria again

$$\langle\beta\rangle_{max} \approx 5 \text{ to } 6\%$$

is estimated for reasonably flat current profiles. It has been noted, however, that for equilibria having low shear in the plasma interior and consequently high shear in the edge region (that is, just the profiles which seem most favourable otherwise), perturbations with azimuthal mode numbers m = 2, 3, . . . may impose stringent limitations on β, if the positions where q = m are in the inner low-shear range [1].

The most favourable situation with respect to **low-n, free boundary modes** is encountered for large shear, at least in the edge region [1], [4], and $\beta_J \leqslant a/2R$ where β_J is the poloidal β defined by $\beta_J = \beta[2R^2 q_J(a)^2/(a^2 + b^2)]$. If $\beta_J \gtrsim R/a$ these modes tend to be of the ballooning type and their stability is reduced. For n = 1 modes, the scaling (1) again seems to apply [2] with a 30 to 50% uncertainty in the absolute magnitude. In particular, from calculations [9] and from scalings, including those from JET [4], there is fairly extensive evidence that in INTOR, without a conducting wall, kink modes and **also** all internal modes will be stable when $\langle\beta\rangle$ does not exceed 3 to 4%. Lowering $q_J(a)$, and hence β_J, has a beneficial effect. As a preliminary conclusion it can be said that in INTOR, without the need for a conducting wall, stability of kink modes can be anticipated for

$$\langle\beta\rangle_{max} \approx 4\%,$$

provided that $q_J(a)$ is reduced to values between 2 and 2.5. The question whether the guiding value of

$$\langle\beta\rangle_{max} \approx 5 \text{ to } 6\%$$

can be reached without some sort of MHD instability arising, requires further research to be answered. Anyway, careful control of the plasma shape and current profile will be required.

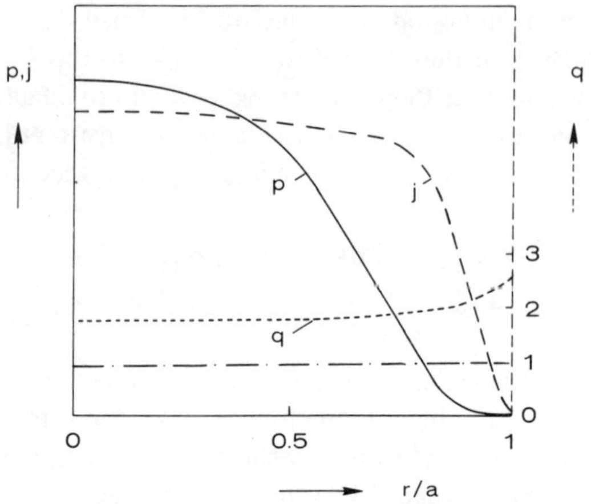

FIG. VII-1. Guess of the most favourable profiles of pressure (p) and current (j), averaged over the poloidal angle, to reach high ⟨β⟩; a typical $q_J(r)$ is also shown.

Low-n free boundary modes can be stabilized, or their growth rate strongly reduced, by a **current-carrying pressureless plasma** around the main plasma column or by a **conducting wall.** For the latter to be effective its radius (a_w) should be fairly close to the plasma, typically $a_w/a < 1.2$ to 1.4. The growth rate of any unstable kink mode would then be of the order of the current decay time in the conductor; for full stabilization a feedback control system would be required. The technological complication connected with using such a device would have to be traded off against the possible gain in $\langle\beta\rangle$.

1.1.3. Optimum profiles

From the requirements discussed in Sect. 1.1.1 and 1.1.2 profiles of pressure and current can be devised which are expected to maximize the β limits (see Fig. VII-1). Anyway, it is advisable to keep q on axis somewhat above 1 and $q_J(a)$ not higher than 2.5. These values also recognize the constraints imposed by the disruptive instability (see Section 2) as well as the desire to lower β_J. As an example, for $q_J(a) = 2.5$ one would have, for INTOR guiding parameters, $\beta_J \approx 3$. For this value the evolution of the equilibrium by current diffusion will be appreciably reduced, in particular at the edge where, for high β_J, the magnetic configuration tends towards a smaller rotational transform.

1.2. Experimental status

The highest values of $\langle\beta\rangle$ as well as of β in the discharge centre ($\beta(0)$), attained in tokamaks [2], [4], [8], are

$\langle\beta\rangle = 2.5\%$ and $\beta(0) = 8\%$ in T-11 *

$\langle\beta\rangle = 1.8\%$ and $\beta(0) = 8\%$ in ISX-B

$\langle\beta\rangle = 2\%$ in TOSCA.

All these devices have a circular cross section. In both T-11 and ISX-B, the theoretical limit for the onset of ballooning instability seems to have been exceeded. Degradation of gross plasma confinement has not been observed, but local effects on the pressure profile cannot be excluded. In ISX-B, significant changes in the characteristics of the MHD oscillations are, however, observed. The result quoted for TOSCA has been obtained in a decompression experiment.

The highest values of β_J, in a steady state, have been reached in

ISX-B; $\beta_J = 2.8$ at $R/a = 3.5$,

TM-3; $\beta_J = 2.3$,

T-11; $\beta_J = 2.1$,

and many other experiments have attained values of β_J somewhat above 1, again without causing any deleterious effect on confinement.

Within the next two years, the β-value in a number of tokamaks will be raised to the range pertinent for studies of β limits. Particularly relevant are the studies on noncircular plasmas in Doublet III, PDX, and ISX. Work will also be done on Alcator C, ASDEX, DITE, JFT-2, JFT-2M, PLT, TFR, T-10, T-11 and others.

1.3. Conclusions

Whereas from the **theoretical and numerical analysis** performed so far it can be concluded that $\langle\beta\rangle = 4\%$ should be possible in INTOR, **experimentally** $\langle\beta\rangle = 2.5\%$, a value somewhat above the theoretical stability limit, has been reached in a machine with circular cross section (T-11). **Further research** is required to answer the question whether $\langle\beta\rangle = 5$ to 6% is attainable in the geometry of INTOR.

---

* Since this report was written, ISX-B has reported $\langle\beta\rangle \approx 2.3\%$.

The **requirements** for being able to reach high $\langle\beta\rangle$, following from theory, include:

1) control of the plasma shape;
2) control of the current profile (see also Section 3);
3) operation at low q(a) and $q_J(a)$; ($q_J(a) \approx 2.5$ is advisable and, if not excluded for an increased disruption frequency, even lower values of $q_J(a)$ should be considered).

The possibility of feedback control of low-n free boundary modes should be further analyzed.

1.4. R and D needs

1) It is most urgent to perform experimental studies on the β-limits and on the impact of increased β on plasma transport, including the dependence on aspect ratio, q(a), plasma shape, and profiles of pressure and current density (see also Section 3). The ability to produce $\langle\beta\rangle \approx 6\%$ in a plasma column of large aspect ratio ($R/a \approx 4$) and appropriate shape should be demonstrated experimentally.
2) Theoretical investigations should be carried out to determine the class of pressure and current profiles for which $\langle\beta\rangle \approx 6\%$ can be reached for $R/a \approx 4$ and elongation $b/a \approx 1.6$. Furthermore the accessibility of these equilibria from low β situations and their consistency with transport effects (changes in the pressure profile and diffusion of the current in the edge region) have to be checked.

2. DISRUPTIONS

2.1. Disruptions in present-day machines

All the experience in present-day tokamaks shows that the prerequisite for a major or minor disruption is a growing $m = 2/n = 1$ mode usually preceded by some contraction of the current channel. Other modes, in particular the "sawtooth" mode ($m = 1/n = 1$) and the $m = 3/n = 2$ mode, are believed to play an important role in creating a major disruption. However, it is **not** really **understood** under which circumstances a dangerous growth of the $m = 2/n = 1$ tearing mode is induced so that a major disruption appears.

Major disruptions can be triggered by impurity influx, bad positioning, exceeding an upper limit for the density, a too rapid density increase, or a too rapid change in current.

There are **two regimes** of operation in which the **occurrence** of major disruptions is **minimized.** The **first** is typically given by

$$\langle n \rangle < (2 \text{ to } 4).\ 10^{19}\, B_{[T]}/R_{[m]} \equiv n_{crit,\Omega}, \quad (2)$$

and good control of position,

$\dot{I}_p < \dot{I}_{crit}$,

$\dot{n} < \dot{n}_{crit}$,

$q(a) \gtrsim 2.2$ to 2.5.

The numerical coefficient in $n_{crit,\Omega}$, Eq.(2), strongly depends on the impurity content of the discharge. In fact, the largest values of $n_{crit,\Omega}$ are obtained in very clean discharges using Ti gettering. Furthermore, $n_{crit,\Omega}$ tends to increase with increasing current [5]. For $\dot{I}_{crit}$, on the basis of experimental findings in different tokamaks and, in particular, in Alcator and T-11, it was suggested [1] that taking

$$\dot{I}_{crit} = I_p/\tau_{skin} \quad (3)$$

with τ_{skin} the skin time of the plasma, is a good estimate. For $\dot{n}_{crit}$, no scaling is known, but it is certain that there is a correlation between $\dot{n}_{crit}$ and $\dot{I}_p$. Similarly, there seems to be a correlation between the density limit and $\dot{I}_p$. In particular, during shutdown $n > n_{crit,\Omega}$ seems possible, perhaps because the decrease of $\dot{I}_p$ has a stabilizing effect on tearing modes. Even within this regime, tokamak operation is not really disruption-free. The frequency of major disruptions varies widely from one machine to the other. At best it is of the order of 10^{-2} in present-day devices.

The **second** regime has been mainly investigated in DIVA [3]. Here it has been shown that operating at

$q(a) < 2$

suppresses major disruptions completely. The accessibility of this regime may be connected with the presence of a conducting wall close to the plasma and/or of a separatrix at the plasma boundary. Careful control of impurities to keep the plasma clean is essential. Preliminary results from T-11 seem to confirm some of these findings [1].

In a disruption the width of the negative voltage spike seems to be about equal to the growth time of the m = 2/n = 1 mode [2] which in present-day tokamaks is of the order of milliseconds. The typical **current decay** and **energy deposition times** can, however, sometimes be much shorter, namely

current decay time:	0.5 to 5 ms,
energy deposition time:	>0.05 ms (the lower limit is reported from PLT, [2]).

On the **spatial distribution** of the **energy** deposited not much is known, but a maximum around the equatorial plane is seen in all machines.

It is to be noted that this picture refers to **ohmically** heated discharges. Only very little is known so far on disruptions in the presence of intense (non-ohmic) heating.

Minor disruptions correspond to a limited growth of the m = 2/n = 1 mode which does not significantly affect the current distribution.

2.2. Extrapolation to INTOR

In INTOR it will not be possible to stay strictly within the first regime of low occurrence frequency of disruptions, described in Section 2.1. In order to be able to reach the working density of INTOR, one must be able to exceed the **critical density** (cf. Eq.2) appreciably, typically by a factor 4 to 5. If the density limit were related to the available heating power, one might infer that in the presence of a non-ohmic heating power P

$$n_{crit} \approx n_{crit,\,\Omega} \left(1 + \frac{P}{P_\Omega}\right)^{\frac{1}{2}}$$

is valid where P_Ω is the ohmic heating power. Such a scaling would be sufficient to be able to make up for this factor. Note that the density limitation might lead to a particularly serious problem during shutdown when the non-ohmic heating power is P = 0, unless smoothly running down the discharge current is sufficient to avoid growth to tearing modes. During start-up and perhaps shut-down the condition of Eq.(3) can be satisfied, limiting the rate of **current change** to

$$\dot{I}_p < 10^7 \text{ to } 3.10^5 \text{ A/s} \qquad (5)$$

where the larger value corresponds to low temperatures ($T < 100$ eV), $\dot{I}_p$ having to decrease monotonically with increasing temperatures to reach the smaller value at $T \approx 1$ keV. On the limitations on the permissible rate of **density increase,** nothing can be forecast.

Whether working in the second regime ($q(a) < 2$) described in Section 2.1 is a viable choice to minimize disruptions in INTOR is presently uncertain.

Anyway, at the present state of knowledge, the occurrence of major disruptions in INTOR cannot be excluded. The growth time of the $m = 2/n = 1$ mode is expected to be about 20 ms in INTOR. Therefore, it might be inferred that the **current decay** and **energy deposition times** might be 10 times longer than in present machines, i.e. of the order of 10 ms, but this estimate is based on scalings that have not been definitely established.

As far as the **consequences of disruptions** are concerned, these fall into two categories, **thermal** and **electromechanical.** The dimenison of the **first** problem is given by the fact that 200 to 300 MJ are deposited on the first wall in a time t_d which could be quite short. If one assumed that the energy deposition were uniform, melting of a first wall made from stainless steel [1] would not occur if the energy deposition time were **longer** than about 10 ms. Otherwise a surface layer, whose width might somewhat decrease when t_d becomes shorter, would be melted (and perhaps partially evaporated). The typical thickness of this layer would be about 0.2 mm, implying that tens of litres, i.e. hundreds of kilograms, of wall material are melted. For non-uniform deposition, about the same volume would be affected. However, for short deposition times, localized evaporation of wall (or limiter) material can be expected to form a radiating mantle which tends to render the deposition more uniform. As a tendency, non uniform deposition may lead to melting of a correspondingly thicker layer. As a result, holes might be burnt into the first wall after a very low number of disruptions. If the first wall were made of pyrolithic graphite, and for uniform deposition, the critical deposition time below which evaporation would occur is only about 0.5 ms; the affected layer thickness is about 0.1 mm, corresponding to about 50 kg of carbon. It is, hence, necessary to install sacrificial elements against fast heat deposition in INTOR, at least if the first wall is in stainless steel.

The **electromechanical consequences** are connected with the large voltage induced by current disruption (which tends to drive eddy currents wherever possible and in particular in the area adjacent to the plasma) as well as with the forces resulting from these currents. A further concern is the heating of superconducting coils when a large $\dot{B}$ appears. These effects will have to be kept within acceptable limits by an appropriate (electrical and mechanical) design of the structures (blanket/shield, toroidal field coil casing, etc.) surrounding the plasma.

The main consequence of a **minor disruption** [2] is probably a change of the poloidal flux of the order of 10%. The problem here lies not so much in the effects of a single event, but in their high frequency which might cause failures due to fatigue. Furthermore, it might somewhat modulate the heat and particle flux to the wall.

2.3. Conclusions

Due to its thermal and electromechanical consequences (see Section 2.2) the disruption phenomenon is a crucial problem for INTOR. All **measures against the occurrence** of disruption and to **control** them when they appear, must be taken. Those presently known include:

1) very careful position control;
2) control of the current density profile (see also Section 3) to avoid the m = 2/n = 1 mode;
3) use of an appropriate primary OH circuit to counteract any decrease of the plasma current (as, for example, in T-10); and
4) operation in a regime where disruptions are improbable (see Sections 2.1 and 2.2).

Feedback control of the m = 2/n = 1 mode would require coils close to the plasma [1], [8] and must therefore be traded-off with the technological difficulties which it causes. Such a system would require [1], e.g., four helical conductors of finite extension as well as suitable diagnostics; the response time has to be of the order of milliseconds, the necessary AC currents are estimated to be of the order of 10 kA with a frequency of about 10^3 Hz, and the power required is of the order of 10 MVA.

Helical windings behind the blanket/shield carrying a constant current (like the ones used in Pulsator) might also be applied, but their effectiveness is uncertain, and they are not recommended for INTOR.

Even when all precautions are taken, at present the probability of disruptions in INTOR may be foreseen to be at most decreased to 10^{-3} per shot. A considerably higher frequency of disruptions can be expected to occur in the early phase of operation of INTOR in which the boundaries of the optimum working regime are explored. This implies that INTOR would have to withstand **more than 1000 disruptions** during its lifetime of 10^6 shots.

2.4. R and D needs

The investigation of disruptions and their consequences requires much more emphasis:

1) More intense investigations on disruptions and the class of equilibria developing into a disruption are urgently needed to come to a better **understanding** of the circumstances that lead to a disruption. This could lead to better methods of control, in particular, if it were possible to recognize a disrupting discharge early.

2) Operation at q(a) < 2 and with a separatrix limiting the plasma should be further investigated and extended to larger machines. An important issue is the accessibility of this regime.
3) Information is needed on disruptions in discharges with strong non-ohmic heating, in particular also during the shut-down phase. Methods for smooth run-down have to be developed.
4) The methods presently known of reducing the occurrence of disruptions and controlling them should be further developed and tested on large non-circular machines.
This includes the development of appropriate diagnostics for the feedback system as well as the use of partial turn helical windings.
5) Transport phenomena during disruption and the space and time dependence of energy deposition must be studied.

3. EQUILIBRIUM AND STABILITY CONTROL

3.1. Control of position and shape

In addition to the specific requirements of divertor operation, control of position and shape requires keeping the plasma column centred in the discharge chamber and with a D-shape. For centering the discharge horizontally and vertically, two dipole fields perpendicular to each other (the "vertical" field B_V and the "radial" field B_R, respectively) are needed, whereas shape control also asks for quadrupole and hexapole components. Since quadrupole and hexapole components can be easily superposed on the dipole fields B_V and B_R, using the same coil set (part of the hexapole component required is readily generated by toroidicity) shape control effectively can be performed using the same coils as for centering the discharge. The quadrupole currents [4] are appreciably smaller than the dipole currents for B_V when the poloidal field control coils are situated **between blanket/shield and toroidal field coils,** and they are of the same order as the dipole currents when these coils are placed **outside** the toroidal field coils, which are the **two principal options** possible. **The total number of ampere-turns** required for the control system is twice as large when the coils are **outside** the toroidal field coils, as when they are **between** blanket and toroidal field coils [2], [3], [4], [8], [9], [10], [11]. For larger β_J the dipole currents for B_V and the quadrupole currents have to be somewhat increased.

While the control of the horizontal position, via B_V, and of the shape, involve only equilibrium properties, the vertical position control has to handle an instability (the axisymmetric instability mentioned in Section 1.1) when the plasma elongation is above $b/a \approx 1.2$. This has important consequences for the **time scales** on which the different components of the control system have to react. For the B_V and

shape control components, this is somewhat **shorter** than the time of current rise and decrease as well as of the change in plasma pressure, leading typically to times <1 s.

For the **radial field** B_R the response time would have to be of the order of typical MHD times, unless the growth rate of the instability is reduced by using a **passive conducting "shell"** (i.e. conducting elements in which dipole currents generating a radial field can be induced) at a distance not too far from the plasma. In this case the growth time of the instability becomes of the order of the resistive decay time of the "shell", its value depending also on the coupling of the "shell" with the plasma (i.e. on the structure and distance of the "shell" from the plasma edge). The response time of a feedback system ensuring full stability can then be of that same order. The maximum distance of the "shell" from the plasma edge beyond which a "shell" is not effective any more is dependent on its structure. For an ideal closed shell this maximum distance is, in the case of INTOR, about one minor plasma radius [11]. For INTOR guiding parameters this implies that the "shell" has to be **part** of, or placed in, the **blanket/shield** structure. It does not seem difficult, at least as long as INTOR is not equipped with a full breeding blanket, to provide such conducting elements with a resistive decay time [10], [11] of the order of 100 ms.
Of course, these conducting elements must not be shielded from the active coils of the feedback system.

In the plasma, the additional **poloidal field** required **for control** is dominated by the vertical field B_V which, for $I_p = 5$ MA and $\beta_J = 2.5$, is about 0.5 T. Approximately one half of this value is related to the finite value of β_J. The typical rate of change of this control field during start-up and shut-down [2], [4] is, hence, smaller than 0.2 T/s, apart from the first 0.1 s during which the current rise is anticipated to be $\dot{I}_p \approx 10^7$ A/s (see Chapter VIII), so that a value of 0.5 T/s will be necessary. Close to the coils the values are, of course, somewhat higher, depending on the configuration chosen (cf. [11]). Since 50% of the control fields required have to be added during the heating phase, it is necessary to guide the **total field change by preprogramming and feedback** [4].

The **power** involved in the control system for plasma position and shape are considerable. During start-up the power for horizontal position and shape control is estimated to be less than 100 MVA when the poloidal field coils are **between blanket/shield and toroidal field coils**, and [2], [3], [4], [8], [9], [10], [11] about 400 MVA when they are **outside** the toroidal field coils. During burn, the feedback power required is smaller than these values [4], [10], typically by a factor 10. However, in the former case there are appreciable additional ohmic losses.
The feedback power for vertical position control, in the case of active coils outside the toroidal field coils and an intervention time of the order of 100 ms, is less than 100 MVA, when maximum excursions of 10 cm are tolerated [9], [11]. For active coils inside the toroidal field coils this power would decrease by a factor 3

An adequate design of a poloidal field coil system **outside** the main field coils has been shown to be possible [3], [4], [9], [10]. As avoiding linked poloidal field coils has great advantages in simplifying access and maintenance, this is the preferred option for INTOR. No compelling reason is, at present, seen which would lead to requiring internal poloidal field coils for equilibrium and shape control for INTOR. Only if a more **detailed** control of the plasma **shape** than provided by outside coils turned out to be necessary, could the situation change. In this case, however, the installation of special shaping coils even close to the first wall might have to be considered.

A **hybrid** solution between the two alternative placements of the coils, inside and outside the toroidal field coils, was also discussed. In this, poloidal coils inside the toroidal field coils would provide the necessary short response time, but on the average carry little current, whereas coils outside the toroidal field coils respond slowly (over several seconds), so that the power requirements are reduced. However, since this solution does not avoid coil linking, it conserves the main disadvantage of an internal poloidal field coil system.

The position and shape control system will have to be guided by **diagnosing** the position of the outermost magnetic surface, corresponding to the plasma edge, by measuring the poloidal flux as well as the poloidal field in front of the first wall [4]. Also, monitoring the heat flux to the first wall seems indicated.

3.2. Control of the scrape-off layer of a divertor

The requirements for the control of the scrape-off layer of a divertor obviously depend strongly on the type of divertor applied. Scenarios for this will have to be developed (see Chapter V). As far as the magnetic topology of the scrape-off layer is concerned, a poloidal divertor leads to a more sophisticated poloidal field coil and control system, but the total number of ampere-turns and the power necessary are estimated [2], [3] not to increase substantially. For a conventional bundle divertor no particular magnetic control problems are anticipated.

3.3. Control of the current profile

The control of the current profile necessary for reaching high values of $\langle\beta\rangle$ (see Section 1.3), and for minimizing the occurrence of disruptions (see Section 2.3), can be best performed by applying an additional heating pulse which has an appropriate radial and time dependence, **during the ohmic heating phase.** This should be effective because, in INTOR, the typical skin times will be long enough that during heating to ignition the discharge will go through a sequence of flux-conserving equilibria. During the burn period of 100 to 200 s no appreciable diffusion of the current will take place in the main body of the plasma column. If the **power deposition profile,** during heating to ignition could also be controlled as a function of radius and time, one would gain additional flexibility. For controlling the current profile during the ohmic heating phase, the **heating power** required [6] is a fraction of the ohmic heating power, typically of the order of 10 MW. The **duration** of the heating pulse will have to be of the order of a second. Electron cyclotron heating or, alternatively, lower hybrid heating would seem to be attractive possible techniques.

An **alternative way** of controlling the current profile is by expansion of the minor radius of the discharge during current rise [7]. This requires appropriate control of the plasma position and shape, the density, and the loop voltage.

During the burn phase, there will be some adjustment of the current profile, by diffusion, close to the plasma edge. This is probably tolerable, and no means to counteract this diffusion are at present believed to be necessary. A simple means of counteracting this diffusion would be to keep the loop voltage slightly above the resistive value.

3.4. Induced time dependence of the wall loading

An effective control system for plasma equilibrium and stability will prevent appreciable excursions of the wall loading. **Variations** of no more than 10% over times of the order of 1 s might be considered typical.

3.5. Conclusions

1) For control of the horizontal plasma position and shape a poloidal control field of about 0.5 T is necessary in the plasma whose rate of change during start-up and shut-down does not exceed 0.2 T/s, except for the first 0.1 s, during which 0.5 T/s will be attained. This field can be produced by coils situated either between blanket/shield and main field coils or outside the main field coils or both. For poloidal coils inside the main field coils, the maximum power required during start-up is less than 100 MVA. If they are situated outside the main field coils, the maximum power required during start-up will be about 400 MVA. Furthermore, larger values of B must be accepted at the toroidal field coils. However, since the design problems of such a system can be solved (see Chapter IX), it is preferred for technological reasons. The control field has to be preprogrammed and in part adjusted by feedback. The response time has to be smaller than 1 s, leading to a power requirement of ≅40 MVA. For control of the vertical plasma position, for the guiding parameters of INTOR, passive conducting elements within the blanket/shield structure, supplemented by a feedback system whose coils can be outside the toroidal field coils, are required. The resistive decay time of the passive elements should be about 100 ms, to keep the feedback power below 100 MVA. No appreciable time dependence of the wall loading

due to plasma equilibrium and stability control is to be expected. If a detailed shape control turned out to be necessary, coils close to the first wall might also be needed. Satisfying the partially **conflicting** requirements of plasma **position** and shape **control,** plasma **current build-up** and **protection against disruptions** is the key problem of the design of the poloidal field system.

2) It is anticipated that current profile control can be satisfactorily performed by using selective auxiliary electron heating (electron cyclotron or lower hybrid) during the ohmic heating phase. About 10 MW over about 1 s will be required. An alternative method is expansion of the discharge radius during current rise.

3.6. R and D needs

1) Experience which is extrapolatable to INTOR has to be gained on the control of plasma position and shape in large non-circular tokamaks. This includes the development of appropriate diagnostics.
2) Control schemes for suppressing the unstable growth of low-n free boundary modes will have to be devised and tested.
3) Control of the current profile by selective additional electron heating as well as by minor radius expansion will have to be tested.

4. SUMMARY

The problem area "Stability Control" comprises three main topics, namely **β limitations, disruptions,** and **equilibrium and stability control.**

1. In present-day tokamaks the discharge is magnetohydrodynamically stable. From theoretical considerations it is, however, inferred that going to **higher values of the ratio** β of the kinetic and magnetic pressures, as required for INTOR, at some point **stability limits** will be encountered. From the **theoretical and numerical analysis** performed so far, it can be concluded that in INTOR discharges with a volume average $\langle\beta\rangle$ up to 4% will be stable without a need for wall stabilization. Higher values may be possible, but have not been assessed, yet. They will certainly require careful plasma shape and profile control. **Experimentally,** stable discharges at a value of $\langle\beta\rangle = 2.5\%$ have been obtained in a tokamak with circular cross section (T-11). This value is probably higher than theoretically expected for the plasma shape and profile present in this device. Experiments with still larger heating power, already underway, are needed to explore the actual β limitations, in tokamaks.

2. The **requirements** for being able to reach high $\langle\beta\rangle$, following from theory, include
   1) a D-shaped plasma cross section of strong triangularity;
   2) an appropriate, relatively flat current profile;
   3) operation at low q_a and β_J ($q_J(a) \approx 2.5$, corresponding, for a = 1.2 m, B = 5 T, R/a = 4, and b/a = 1.6, to a plasma current of 5.3 MA, is advisable and, if not excluded for an increased disruption frequency, even lower values of $q_J(a)$ should be considered).

3. During the life time of INTOR, at least 10^3 disruptions are expected to occur. Hence, INTOR must be designed with the necessary protections against the thermal and electromechanical consequences of disruptions. Against fast heat deposition sacrificial elements will have to be installed.

4. On the other hand, all **measures to reduce the occurrence** of disruptions and to **control** them when they appear must be taken. Those presently known include
   1) very careful position control;
   2) control of the current profile;
   3) use of an appropriate OH transformer system counteracting current decrease;
   4) operation in a regime where disruptions are improbable.

   Feedback control of disruptions would require coils close to the plasma and must, hence, be traded-off with the technological complications which it causes.

5. For **horizontal position** and **shape control** mainly a vertical and a quadrupole magnetic field with admixtures of higher multiple components are required. The strength of this control field in the plasma is about 0.5 T. During start-up and shut-down its rate of change is below 2 T/s except for the first 0.1 s where 5 T/s are reached. The maximum power requirement during start-up is about 400 MVA for a poloidal coil system situated **outside** the toroidal field coils. Poloidal coil configurations of this kind, able to generate the required plasma equilibrium, have been found.

6. For **vertical position control** appropriate **passive conducting elements,** supplemented by a **feedback system** whose **coils** may also be situated **outside** the toroidal field coils, are required. The distance of the conducting elements from the plasma adge has to be smaller than about the minor plasma radius. For INTOR guiding parameters this implies that these elements have to be **part** of, or placed in, the **blanket shield** structure. This current decay time

TABLE VII-1. R AND D NEEDS ASSESSMENT — STABILITY CONTROL

	Status*	
	Adequate	Inadequate
β limits and transport at high β with the aim of demonstrating $\langle\beta\rangle \approx 6\%$ at $R/a \approx 4$	x	
Occurrence, early detection and avoidance of disruptions, including low-q operation, discharges with intense heating, and shut-down	x[1,4]	
Space and time dependence of energy deposition as well as transport during disruption	x[1]	
Position and shape control in large noncircular tokamaks in reactor-relevant conditions	x[1,2]	
Feedback control of kink modes and disruptions[3]		x
Current profile control		x

1 More specific emphasis required.
2 Extrapolability to INTOR to be analyzed.
3 Such Feedback control systems might not be practicable in INTOR.
4 Early detection of disruptions, on a time scale on which operating a mechanical protective system is possible, is essential.

Adequate: Programmes are either planned or could be planned on existing or authorized facilities (with some upgrading if required) and using existing trained personnel.

Inadequate: Either the available and authorized manpower and facilities are inadequate or a substantial increase in effort is needed.

* The ratings are made on the world-wide fusion effort and not separately on the invidual national programmes.

should be of the order of 100 s, which does not seem difficult to realize at least as long as INTOR is not equipped with a full breeding blanket. The maximum feedback power then required is less then 100 MVA.

7. Although no experiments have been done so far, it is anticipated that current profile control can be satisfactorily performed either by selective electron heating by RF during the ohmic heating phase or, alternatively, by controlled minor radius expansion during current rise. In the former case about 10 MW heating power over about 1 s will be required.

5. R AND D NEEDS ASSESSMENT

The R and D requirements related to the area "Stability Control" are summarized in Table VII-1. From the world tokamak programme, an adequate data base can be expected to be available in time for INTOR check points, when due emphasis is given immediately to investigation of the critical issues identified.

REFERENCES TO CHAPTER VII

[1] USSR Contribution to the 2nd meeting of the INTOR Workshop, Group 11, 1979.

[2] "Stability Control", Chapter in USA INTOR Report (see Appendix A.5.4).

[3] Fusion Research Development Center, INTOR Workshop Report, Group 11, Stability Control, JAERI-M 8625 (1979).

[4] European Contributions to the 2nd meeting of the INTOR Workshop, Stability Control, Chapter XI, EUR FU/XII 501/79/EDV 50, June 79 (2 Volumes).

[5] European Contributions to the 2nd meeting of the INTOR Workshop, Energy and Particle Confinement, Chapter I, EUR FU/XII 501/79/EDV 50, June 79 (2 Volumes).

[6] European Contributions to the 2nd meeting of the INTOR Workshop, Heating, Chapter IV, and Start-up and Shut-down, Chapter XII, EUR FU/XII 501/79/EDV 50, June 79 (2 Volume

[7] Fusion Research Development Center, INTOR Workshop Report, Group 12, Start-up, Burn and Shut-down Control, JAERI-M 8625 (1979).

[8] USSR Report to INTOR, "Stability Control in INTOR" (see Appendix A.5.3 [28]).

[9] "Poloidal Field Coil", Chapter in USA INTOR Report (see Appendix A.5.4).

[10] Fusion Research Development Center, INTOR Workshop Report, Group 4, Heating, JAERI-M 8624 (1979).

[11] European Contributions to the 3rd meeting of the INTOR Workshop, Homework and Special Supplementary Information, EUR/FU/XII 501/79/EDV 70, Oct. 79.

Chapter VIII

START-UP, BURN AND SHUT-DOWN

F. ENGELMANN – EC, Chairman
V.I. PISTUNOVICH – USSR
P. RUTHERFORD – USA
T. TAZIMA – Japan

1. START-UP

Start-up of INTOR, analogously to the situation in any tokamak, can be subdivided into four sequential phases (see Fig. VIII-1), namely

1) the ionization and current initiation phase,
2) the phase of current rise at low electron temperature ($T_e \lesssim 100$ eV),
3) the phase of current rise at medium electron temperature (100 eV $\lesssim T_e \lesssim 1$ keV) in which the current reaches its maximum value and the main ohmic heating takes place, and
4) the phase of "auxiliary" heating to ignition.

Phases 1 and 2 can be grouped together; during this period the typical tokamak configuration is built up and the physics is strongly influenced by atomic processes. Also phases 3 and 4 have many features in common, since during this period the bulk of the plasma heating occurs. Therefore, the following discussion of the start-up will be structured according to these two main time intervals.

1.1. Build-up of the tokamak configuration

Extrapolating from present-day tokamaks [1–4, 9, 10], one would anticipate that **ionization** and **current initiation** in INTOR will take about 10 ms (see Fig. VIII-1). As, for technological reasons, one would prefer to limit the initial loop voltage as far as possible, one would try working with the lowest electric fields which ensure satisfactory ionization and current initiation. Recent experimental [9] and theoretical results [9, 10] show that an electric field of 1 to 2 V/m is sufficient for break-down, provided that the filling pressure is taken correspondingly low (10^{-4} torr or somewhat below), the walls of the discharge chamber are kept clean from impurities, and the poloidal stray magnetic fields are less than 0.2% of the toroidal field. This implies for INTOR that a loop

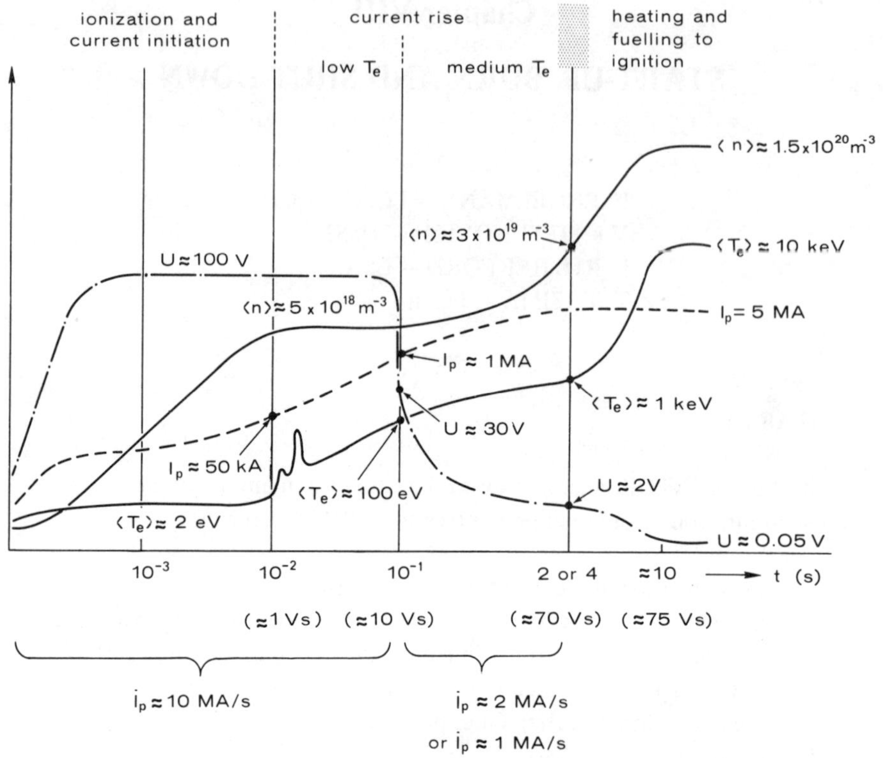

FIG. VIII-1. Schematic plot of a typical start-up scenario as extrapolated from present-day machines to INTOR;

I_p	=	*plasma current*
U	=	*loop voltage*
$\langle n \rangle$	=	*average density*
$\langle T_e \rangle$	=	*average electron temperature*
t	=	*time*

The volt-second requirements up to different moments are also indicated.

voltage of about 100 V would seem satisfactory. A voltage ramp rate of 10^4 to 10^5 V/s is probably sufficient. The density would rise to about $5.10^{18}\ m^{-3}$ during the ionization and current initiation phase, and the plasma current I_p may reach the order of 100 kA, the electron temperature remaining low, typically just about 1 eV. This scenario is not far from the situation in some of the largest present-day tokamaks, and has been closely modelled in T-10 [9]. It must be noted, however, that in a larger machine the inductive fraction of the loop voltage tends to be larger, implying a decrease in the ohmic power available. On the other hand, with a loop voltage of 100 V, a current rise rate of $I_p = 10^7$ A/s cannot be reached before the end of the ionization phase, as this requires the resistive voltage to drop significantly below 100 V which will happen only when electron heating has started. A consequence of this is that the value of q(a) will decrease in time much more slowly than in present-day devices.

During the **following phase** [1–5], lasting for about 100 ms, the plasma is heated through the **radiation barrier** reaching about 100 eV at the end, while the current rises to about 1 MA (see Fig. VIII-1). In the beginning, **charge exchange** also provides a sizeable loss mechanism, whereas later, when the current density tends to build up a skin profile, **MHD activity** starts favouring current penetration, but also contributing appreciably to the losses [5]. It is during this phase that both the ohmic power input and the plasma losses reach their maximum. Provided that the current extends essentially over the whole cross section of the discharge chamber, one would have, for a given electric field and at a fixed instant, the following scalings with the size of the device:

loop voltage	$U \sim R$,
plasma current	$I_p = \text{const.}$,
ohmic power	$P_\Omega \sim R$,
current density	$\langle j \rangle \sim (ab)^{-1}$,
ohmic power density	$\langle W \rangle \sim (ab)^{-1}$,
safety factor	$q_y(a) \sim \frac{1}{2}(1 + b^2/a^2)(a^2/R)$.

With the scenario of Fig.VIII-1 the ohmic power in INTOR would increase from about 3 MW to 10 MW during the phase of current rise at low temperature. Hence, the average power density is extremely small, namely some 10^{-2} W/cm^3. This shows immediately the stringent requirements imposed in a large machine like INTOR, similar to those in JET [4], to keep the energy balance positive during this phase (see, for example, the losses as estimated for PLT [2] at a density of 3.10^{19} m^{-3} and an oxygen content of 5%, shown in Fig. VIII-2, which reach 4 W/cm^3; scaling the radiation losses down to a density of 5.10^{18} m^{-3} and 1% oxygen content would still yield a loss rate of 2.10^{-2} W/cm^3). These were also recovered in 0–D calculations performed for INTOR [2] and JET [4]. The calculation performed for INTOR [2], assuming a density of 2.10^{19} m^{-3}, leads to a limitation of the oxygen content of about 2% when no other impurities are present and the loop voltage is raised to 400 V. However, 1–D simulations of the start-up phase performed by the Kurchatov group [5] (see Figs VIII-3 and VIII-4) with a maximum loop voltage of only 220 V, the same plasma density, and losses due either to

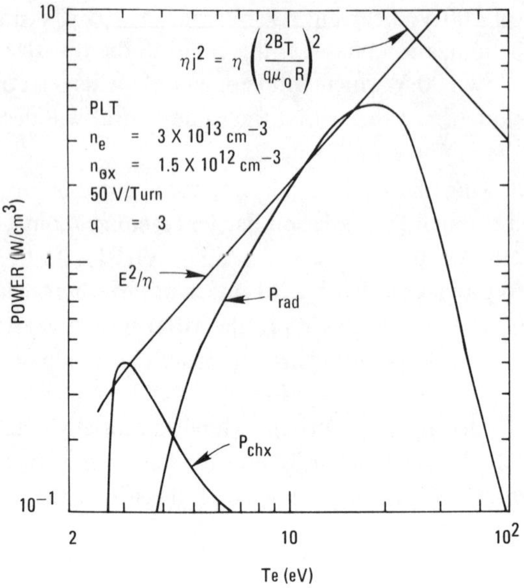

FIG. VIII-2. Power loss by charge exchange and oxygen line radiation as estimated for PLT, compared with the constraints imposed by ohmic heating in this machine.

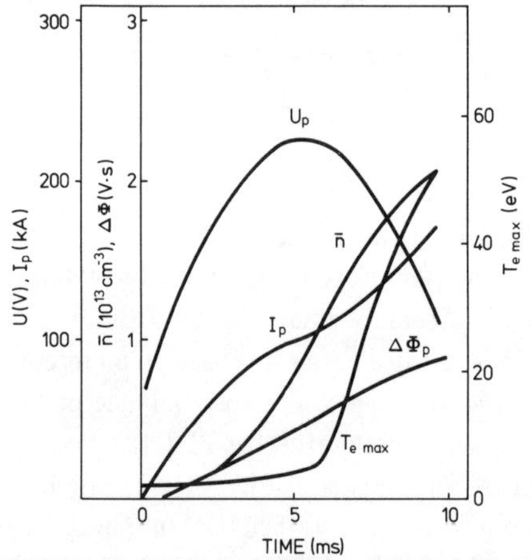

FIG. VIII-3. 1-D simulation of start-up for $t < 10$ ms (ionization phase), assuming a maximum loop voltage of 220 V:

I_p = plasma current,
U_p = loop voltage,
$\bar{n}$ = average density,
$T_e max$ = peak electron temperature,
$\Delta\phi_p$ = flux swing.

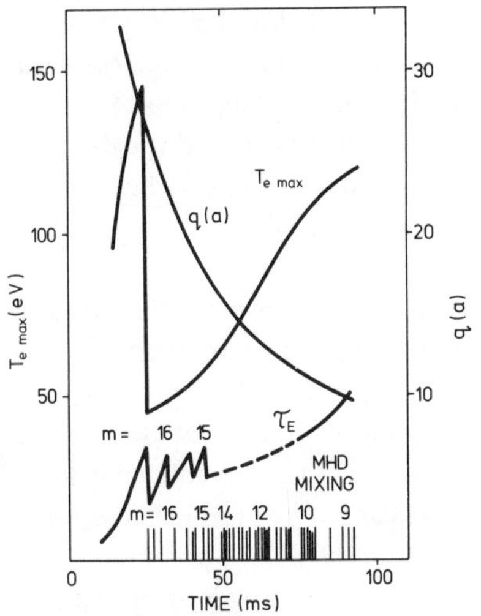

FIG. VIII-4. 1-D simulation of start-up for 10 ms $< t <$ 100 ms in a plasma without impurities, taking account of MHD activity induced by skin forming of the current:

$T_e max$	=	*peak electron temperature,*
$q(a)$	=	*safety factor at edge,*
τ_E	=	*energy confinement time;*

the dashes indicate the times where tearing modes with mode number m located on the two sides of the current maximum in the skin overlap.

MHD activity or to impurities, show more favourable results. In fact, with skin-like impurity profiles (both oxygen and carbon: 2% at the edge and 0.5% at the centre; as well as iron: 0.2% at the edge and 0.05% at the centre) no problem was encountered. This is believed to be related with the non-uniformity of the current profile which locally leads to a more favourable energy balance, permitting more easily to burn through the radiation barrier [9]. At a density of 5.10^{18} m^{-3} and a loop voltage of 100 V the situation is somewhat better than in this example, provided that the absolute plasma contamination is not higher. As far as the losses due to MHD activity are concerned, these depend essentially on the formation of a current skin [9]. To minimize losses, the current rise rate, therefore, has to be chosen appropriately, probably not much higher than I_p/τ_{skin} where τ_{skin} is the skin time of the plasma at the temperature reached. This implies that $\dot{I}_p$ should not exceed 10^7 A/s in this stage. Note that the example reported in Fig. VIII-4 does satisfy this requirement.

Decreasing the loop voltage-below 100 V, e.g. to 50 V, seems possible when pre-ionization as well as some auxiliary heating, typically 2 to 5 MW for about 10 ms, is provided [9]. The latter would improve the marginal energy balance during ionization as well as during heating through the radiation barrier. Both electron cyclotron [2, 9] and lower hybrid heating have been proposed for this purpose. In the FT-1 device [10], electron cyclotron heating was effective in reducing the loop voltage by a factor of 3. In T-12, electron cyclotron heating was used for pre-ionization [9]. This permitted to localize the break-down area around the region of electron cyclotron resonance. Of course, applying large RF heating powers in an early stage of start-up and at very low density might favour the generation of run-away electrons and should, therefore, be considered with caution. Furthermore, lowering the loop voltage below 100 eV implies that a current rise rate of 10^7 A/s cannot be attained anymore during the first 0.1 s of the discharge, but this might very well be acceptable.

A way to make working with a low loop voltage easier is to start the discharge in **smaller dimensions** [2], [3], [4]. Such a scheme leads to more severe control requirements for the plasma density [3] and position.

Note that for a current rise rate of 10^7 A/s during the whole time interval considered ($t \lesssim 0.1$ s) the vertical field in the plasma has to increase at a rate of 0.5 T/s (see Chapter VII).

On the basis of the above simulations [2, 5] and other estimates [1, 9 – 11], the **flux swing** necessary for the entire phase under discussion (up to 100 ms) can be estimated, for the scenario described, to be about 10 Vs.

More information on the optimum conditions during the early start-up phase in a tokamak of large size will be obtained from JET, JT-60, T-15, and TFTR.

1.2. Ohmic and auxiliary heating

1.2.1. Current rise and ohmic heating

The **duration** of the **main ohmic heating** phase in INTOR will be determined by the limitations imposed on the current rise rate $\dot{I}_p$ and by the requirements of plasma position and shape control (see Chapter VII). The former, if imposed by the necessity to avoid excessive skin formation (see Section 1.1) are more severe. They would lead to a limitation of the current rise rate to less than 10^6 A/s for temperatures in the range above 100 eV. Assuming that a rise rate of (1 to 2) × 10^6 A/s is still acceptable, a duration of 2 – 4 s would follow for the main ohmic heating phase (Fig. VIII-1). During this phase, the **current** will reach its **maximum** value, $I_p \approx 5$ MA, and in order to keep the q profile acceptable, the **shape** of the plasma has to go **from circular to elongated.** Furthermore, the profile of the current will have to acquire the shape appropriate for the later high-β burn phase (see Chapter VII). The electron **temperature** will rise to about 1 keV, provided that losses due to impurity radiation can be kept low, and the ohmic heating power drops to about 7 MW. The plasma **density** has to be increased as far as possible. Since there is, in present-day tokamaks, a limitation on the density of ohmic discharges, following the scaling (see also Eq.(2) of Chapter VII)

$$\langle n \rangle_{crit,\Omega} \approx (2 \text{ to } 4) \times 10^{19} B_{[T]}/R_{[m]}$$

the coefficient depending sensitively on how well the plasma is kept free from impurities (the highest densities are reached employing Ti gettering) and also on the value of q(a), in INTOR one might, however, not be able to increase the density above 3×10^{19} m^{-3}. Gas puffing is expected to be suitable for this density increase. During the main ohmic heating phase the **flux swing** requirement is largest, namely [9–11] 50–60 Vs, of which about 20% is resistive. This yields, as an upper limit, 70 Vs to be necessary in total during the first few seconds of the discharge.

1.2.2. Auxiliary heating and fuelling to ignition

In the auxiliary heating phase, the **plasma parameters** have to vary over a very wide range (see Table VIII-1). In fact, not only heating, but also fuelling has to be provided during this phase.

This implies that the **auxiliary heating** method adopted must be very **flexible** and that both the plasma **position** and **shape** must be carefully **controlled** (see Chapter VII).

The **duration** of the auxiliary heating phase is about 5 s (see Chapter IV), depending also on the heating power applied and on the energy deposition profile. The necessary **power** for neutral beam heating at 170 to 200 keV is of the order of 75 MW, while for heating schemes providing an optimum deposition profile, the power requirements might be decreased by almost a factor 2 (see Chapters IV and VI).

Less than 5 Vs are required to **drive the current** during this phase against the plasma resistance.

Fuelling by gas puffing and/or by pellet injection is anticipated to serve the purpose for all heating schemes.

TABLE VIII-1. CHANGE OF PLASMA PARAMETERS DURING AUXILIARY HEATING AND FUELLING

	Initial	Final
Density ⟨n⟩ $[m^{-3}]$	$\approx 3 \times 10^{19}$	$\approx 1.5 \times 10^{20}$
Electron temperature $\langle T_e \rangle$ [keV]	≈1	≈10
Ion temperature $\langle T_i \rangle$ [keV]	smaller than $\langle T_e \rangle$	$\approx \langle T_e \rangle$
β_J	≪1	2.5 to 3
β [%]	≈0	4 to 5

1.2.3. Problem areas

From the preceding discussion, several points are apparent where problems may arise:

1) The **duration of the current rise** phase (2 to 4 s), while reasonable from the point of view of position and shape control, and probably acceptable from the point of view of current penetration, might be too long from the point of view of impurity accumulation, if no effective impurity control is present during this phase. In fact, in 1–D simulations of start-up including sputtering due to neutrals on the wall, the ohmic heating phase had to be reduced to 1 s to avoid plasma cooling by impurity radiation [6].

2) One may ask whether this problem cannot be eased by **overlapping** the **current rise** phase with the **auxiliary heating** phase. However, at least when the plasma current fills the full bore of the discharge chamber, this is anticipated to lead to difficulties with current penetration resulting in longer current rise times than are acceptable (Chapter VII). If the current is raised at reduced discharge radius, [2, 3], some overlap of the two phases will be possible, but the control requirements become more severe, since an excessive local heat load must be avoided. In particular, shape control is necessary early during current rise in this case. Furthermore, the range of variation of the penetration parameter na during auxiliary heating increases which is unfavourable for neutral beam heating [3].

TABLE VIII-2. RELATION BETWEEN NEUTRAL BEAM PARTICLE ENERGY E_0 AND PLASMA DENSITY $\langle n \rangle$ FOR FLAT PROFILES, $Z_{eff} = 1$ AND a = 1.2 m ($a/\lambda_0 \approx 3.10^{-18} \langle n \rangle a/E_0 = 1.6$)

E_0	≈50	100	200	400	keV
$\langle n \rangle$	0.2	0.5	1	2	10^{20} m^{-3}

3) A **crucial point** is how to **ensure** the required **flexibility** of the additional **heating** system to cover the parameter range of Table VIII-1. The flexibility of **neutral beam heating** is limited, due to the requirements on penetration which have to be respected. In fact, both a too long and a too short mean free-path λ_0 of the neutrals injected would lead to problems, the former because the neutrals hitting the opposite wall produce a high wall load, leading to an increased impurity influx, the latter because only the plasma edge is heated, which implies a low heating efficiency and again enhanced impurity production. For almost perpendicular-injection and flat density profiles the optimum penetration is given by $a/\lambda_0 \approx 1.6$. The heating of the plasma core is substantially impaired when a/λ_0 differs from this optimum value by more than 50% (see Chapters IV and VI). On the other hand, to avoid excessive "shine-through" and edge heating, a/λ_0 should stay between about 1.5 and 3. Table VIII-2 illustrates the density for optimum penetration in INTOR for a range of beam energies and $Z_{eff} \approx 1$. Hence, with a neutral beam energy of 170–200 keV, heating is satisfactorily possible only above a density of about 7×10^{19} m^{-3}. Thus a problem arises **if** the density in the ohmically heated discharge cannot be higher than 3×10^{19} m^{-3} (see Section 1.2.1). In this case a more sophisticated, and hence more complex, heating scheme would have to be devised. One possibility would be to inject at least 20 MW at about 40° off-normal [11], in which case a large margin with respect to shine-through is gained and very little penetration is lost at high densities, the bounds on α/λ_0 becoming typically 0.4 and 2.4, respectively. This may, however, be difficult for topological reasons. Another possibility would be to start injection at reduced particle energy [11] and/or degraded species mix [2] at the largest (off-normal) injection angle topologically acceptable, if flexible enough injectors can be built. The simplest solution might, however, be to apply a mixed approach, providing at least 20 MW of neutral injection at an energy between 70 and 100 keV, or preferably a somewhat smaller power of RF heating (see below), for about 2s. In such a case the low-energy neutral beam or RF heating would have to precede the main auxiliary heating phase, raising the plasma temperature to about 3 keV and providing the possibility to increase the density

to about 7×10^{19} m^{-3} by fuelling, before neutral injection at 170 to 200 keV starts. To determine the optimum heating scenario, based on neutral beam injection, further work is necessary.

The potential of **RF heating** methods to provide the flexibility required during heating and fuelling to ignition seems much **better.** **Electron cyclotron** heating, using the ordinary mode incident radially from the low field side, is expected to cover the whole parameter range required. In fact, for $T_e \gtrsim 1$ keV, the optical thickness of the plasma is larger than 1, ensuring single-pass absorption, for $n > 10^{19}$ m^{-3}, and the accessibility condition is $n \lesssim 2 \times 10^{20}$ m^{-3}. The localization of the absorption is good, so that heating of the plasma centre is possible. The ions are efficiently heated when $n > 5 \times 10^{19}$ m^{-3}.

Also **ion cyclotron** heating is expected to be able to cover the parameter range of Table VIII-1. Depending on the heating scheme chosen (see Chapter VI), however, different heating regimes would appear, because the absorption increases with the pressure of the species to be heated. In the starting phase of heating and fuelling, i.e. at low pressure, as many as 25 reflections will be necessary for absorption, and the energy deposition profile will be broad. With increasing pressure, the situation will gradually evolve to a single-path absorption with localized energy deposition in the resonant layer [11].

Lower hybrid heating, based on electron Landau damping, would be only subject to the constraint $T_e \lesssim 10$ keV, provided that the parallel refractive index could be appropriately lowered, from about 6 to 2.5, during heating to ignition (see Chapter VI). This implies that a limited outward shift of the deposition profile is unavoidable towards the end of the heating period. However, as the plasma centre then is heated mainly by α-particles, this seems acceptable. As in the case of electron cyclotron heating, the ions are heated efficiently only if $n > 5 \times 10^{19}$ m^{-} If the heating is started with a somewhat lower frequency, some direct ion heating might be achieved in this early stage (see Chapter VI).

4) Accurate control of the **fuel mix** is another important issue when ignition is approached, particularly in the case of neutral beam heating [2]. Furthermore, keeping the **density profile** somewhat peaked [4] (see also Chapter VI) would be favourable for reaching ignition. Gas puffing and pellet injection are considered to be satisfactory fuelling methods, yielding also some flexibility in varying the particle deposition profile. Gas puffing can also provide some **edge cooling.** However, only very limited flexibility in influencing the density profile is available, as long as there is no possibility of controlling recycling by an appropriate particle unload device (see Chapters IV and V). In the case of neutral beam heating, the discharge would have to be started in a tritium-rich gas mixture into which D beams are injected (Chapter IV), the additional fuelling required being ensured by puffing of pure tritium gas and/or tritium pellet injection. A favourable by-product of this scenario is that beam-plasma fusion effects assist the heating in its early stage. Appropriate wall conditioning will be a pre-requisite for this mode of operation.

5) The plasma **position and shape** control requirements are particularly severe during the phase under consideration. They are discussed in Chapter VII. Also the requirements for controlling **divertor** action are rather extreme in this phase of rapid variation of the plasma parameters (see Chapter V).

6) During the heating phase the **current profile** necessary for reaching high-β operation will have to be generated. Both auxiliary heating (electron cyclotron or lower hybrid) during the ohmic heating phase, and appropriate build-up of the discharge with expanding minor radius, have been proposed for this purpose (see Chapter VII).

2. THE BURN PHASE

The main aspects of the burn phase to be considered here are

1) the duration of burn;
2) the methods for burn temperature control, and their reactor relevance.

The questions of impurity control, fuelling and exhaust, as well as stability and equilibrium control, are not dealt with, as these are discussed in Chapters V and VII, respectively.

2.1. Duration of burn

There are four factors which may potentially limit the duration of the burn phase on different time scales, namely

1) the necessity to **drive the plasma current** against the finite plasma resistivity;
2) **helium** accumulation;
3) influx of **impurities**; and
4) **diffusion** of the plasma **current**, leading to a decay of the poloidal magnetic field configuration.

Driving the plasma current by the **transformer** is **not a problem** in INTOR, during the burn. In fact, the loop voltage during burn being estimated to be about 0.05 V, as few as 10 Vs of flux swing are sufficient for a burn pulse of 200 s. Hence, no other method for driving the current is required.

How far **helium accumulation** [1 – 5] is important in providing the limit to the pulse length of INTOR depends both on the maximum allowable $\langle\beta\rangle$, to be denoted β_{max}, and on the efficiency of He exhaust (see also Chapter V). More precisely, if β_Q denotes the value of β necessary to produce a required wall loading Q, then, for complete helium accumulation, the pulse length is limited to

$$t \approx 100\,[(\beta_{max}/\beta_Q - 1)]\,s \tag{1}$$

for standard INTOR burn conditions (T ≈ 10 keV, $\langle n \rangle \approx 1.5 \times 10^{20}\ m^{-3}$).
If one wants to operate in the ignited regime, $\beta_Q \geqslant \beta_{ign}$ must also be valid, when β_{ign} is the value of the plasma $\langle\beta\rangle$, excluding the beam contribution, at ignition. Eq. (1) shows that a 100 s pulse would be possible without exhaust if $\beta_{max}/\beta_Q = 2$. On the other hand, if the α-particle transport is such that with the help of an appropriate exhaust system the total **confinement time** of He can be reduced to 20 s or shorter, then the relative α-particle concentration remains below 10%, requiring only an increase of $\langle\beta\rangle$ by less than 15% above β_Q to make an indefinitely long burn possible. These numbers define the framework for the **conditions** to be provided **for effective exhaust.** For example, if the confinement time of He in the boundary layer of the discharge were 2 s, only about 10% of the He arriving at the wall (including limiter and divertor, if present) would have to be pumped away or gettered by the wall [2]. This fraction could still be much smaller, if the confinement time of He ions in the boundary layer were much smaller than the above value, due to rapid recycling (see Chapter V).

Pulse length limitations due to **impurity influx** are **severe** [1–4], as long as no effective way of impurity control is found and applied (see Chapter V). It may be noted that, under the condition that the lifetimes of impurities and hydrogenic ions are actually about equal, indefinitely long burn would be possible if, for all impurity species generated at the wall, the allowable (relative) concentration divided by the effective yield coefficient (sputtering etc., corrected for recycling in the edge layer) is larger than 1. This does, of course, only refer to the plasma interior, and more stringent conditions might be imposed to be able to ensure a proper energy balance in the boundary layer.

The **diffusion** of the plasma **current** profile is not an important problem in INTOR. In fact, the typical time scale of this process is the **skin time** which under burn conditions is close to 10^3 s. Thus, no limitation of the pulse length is expected due to the decay of the overall poloidal field configuration. Some adjustment of the current profile will, however, take place close to the plasma edge, in particular if β_J were above 3, but this is probably tolerable (see Chapter VII).

It is apparent from the above that operating INTOR with a burn pulse of 100 to 200 s is **reactor relevant** in the **sense** that full impurity control and exhaust, as well as refuelling are required. On the other hand, phenomena appearing on the time scale of the **skin time** are not addressed. In a commercial reactor, this same situation would pertain, if the burn time were kept shorter than about 30 min.

Whether this is acceptable for a commercial reactor will have to be assessed [2–4]. If not, the problems related with burn pulses longer than the skin time will have to be solved in a later demonstration reactor. The same is true for alternative methods for driving the current, since for pulses up to about 30 min a transformer would be suitable.

2.2. Control of burn temperature

Burn temperature control is necessary because ignition in INTOR and later reactors will occur in a **thermally unstable** situation. Three different **modes** of operation are possible, which require different means for burn control:

1) operation **in** the ignited state, that is, in a situation where the α-particle heating exceeds the "normal" thermal losses; here two methods might be applicable, namely
   1.1) control by a variable field ripple;
   1.2) operation at the β limit for high-n MHD modes;

2) operation in a **marginally** ignited state, that is, just at ignition; for this case, again, two methods might be considered, namely
   2.1) major radius compression and decompression;
   2.2) fuelling control;

3) **subignition** operation, where the plasma burns just below ignition, the proper energy balance being ensured by a small amount of auxiliary heating (i.e. $Q \gg 1$); if desired, an effective barrier against reaching ignition can be provided by the methods mentioned under 1).

2.2.1. Field ripple

Burn temperature control using a variable **field ripple** [1 – 3], seems to be the **most flexible** control method among the ones considered. However, neither the physical problems (see also Chapter IV) nor the technical feasibility are analysed sufficiently to guarantee its functioning as yet. The open questions regard the effect of the ripple on the ion and α-particle distribution functions at different locations, non-uniform thermal fluxes to the wall, as well as the technical realization, including the time necessary for changing the field ripple. An example of a 1-D simulation of plasma burn controlled by increasing the field ripple from ± 0.5% to ± 1.25% after reaching ignition was reported [2] (see Fig. VIII-5). This case was based on a 6-coil toroidal field system with ripple correction coils which, when activated, reduced the ripple to the low values

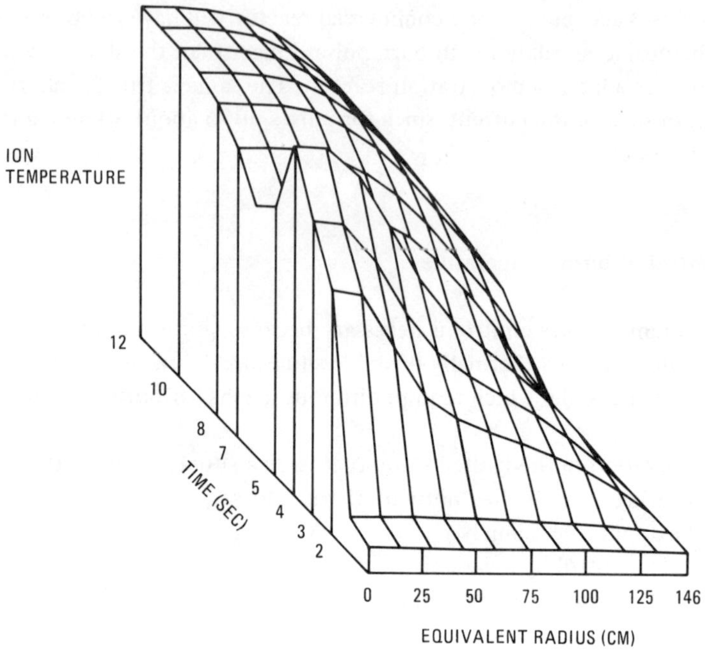

FIG. VIII-5. 1-D simulation of burn temperature control by a variable field ripple (± 0.5% during heating to ignition, ±1.25% during burn).

necessary for reaching ignition. If the discharge chamber is large enough, major radius compression and decompression could be used to move the plasma in and out of the high ripple region rather than changing the magnetic field ripple itself. The effect of the increased ripple on the density profile, when fuelling is performed by gas puffing, has also been addressed [9, 10]. Furthermore, the loss of fusion α-particles which can be induced by a variable asymmetric ripple was calculated [10].

An advantage of this method is that the effect of a variable field ripple on ion particle and heat transport can be tested in reactor-grade hydrogen or deuterium plasma, i.e. without requiring α-particle heating.

2.2.2. β limits

Controlling the burn temperature by operating the discharge at the **β limit** for the stability of high-n MHD modes [1, 2, 5], is at present just a speculation.

There is no self-consistent theory of plasma transport induced by these modes, nor have experiments reached operation at high enough β. In T-11 and ISX-B, where the highest $\langle\beta\rangle$ and β_J were reached, and which should have been unstable for ballooning modes, no obvious effect on transport could be identified (see also Chapter VII). If this method worked, it would be simple to implement because, to get a required power density and wall loading, just the toroidal field would have to be pre-set at an appropriate value. The prospects for changing or adapting the burn parameters during the pulse rests on the possibility that the β limits can be modified by changing the plasma shape via the poloidal field system. This would be needed to avoid changes in power density and wall loading caused by a limited influx of low-Z impurities or α-particle accumulation. Changing the density profile by altering the fuelling particle source might also change the β limit.

A 1-D simulation of plasma burn in INTOR with the temperature excursion limited by β (more precisely, by the value of the pressure gradient), using a phenomenological enhancement of transport when the stability limit was reached, was performed by the Kurchatov Group [5]. It was found that ignition is reached before the ballooning stability limit comes into play, but that the thermally stable burning points lie very close to the ignition curve.

For this method tests are possible in (non-burning) reactor-grade H or D plasmas by investigating if and how plasma transport is being changed when the β limit is reached.

2.2.3. Compression – decompression

In the compression-decompression scheme [2, 4, 11], the plasma is periodically shifted back and forth between an ignited and a non-ignited state, leading to burn stabilization. The required β, to obtain the desired power density and wall loading, has to be adjusted by appropriate fuelling. A special feature of this scheme is that it leads to **direct conversion** of a fraction of the thermal power generated by the burning plasma. This fraction is, however, small in a burn control mode where small excursions are required ($\approx$0.3 MW for INTOR).

The Garching Group [4] has reported extensive 0-D calculations on this mode of burn control. A change in **major** radius $\Delta R/R \approx 4\%$ is necessary, leading to an additional space requirement in **minor** radius of about 10 cm. For the duration of one burn cycle about 1 s is found to be satisfactory (see Fig. VIII-6). In the example considered, the power density varies by 20% peak-to-peak during the cycle. To what extent problems with time-dependent heat fluxes to the wall can be minimized, needs further analysis. The system must be feedback controlled and requires powers in the range of tens of MVA during compression. Compatibility with a divertor depends on the stiffness of the magnetic configuration at, and the width of the divertor throat and needs further study.

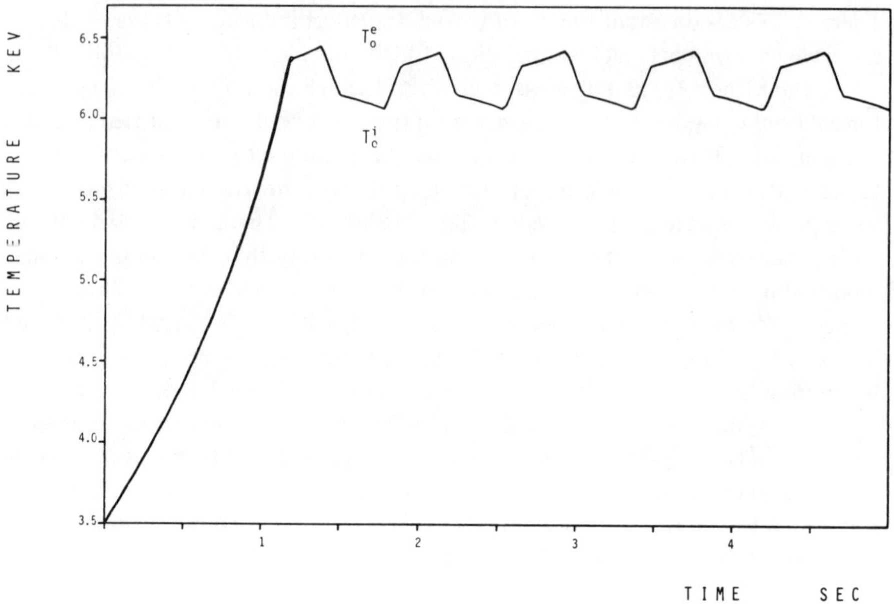

FIG. VIII-6. 0-D simulation of feedback-controlled burn by major radius compression and decompression ($\langle n \rangle = 0.8 \times 10^{-20}$ p/m^3, strongly peaked profiles).

2.2.4. Fuelling

Keeping the plasma close to the ignition point would also be possible by time-varying, feedback-controlled **fuelling**, if the particle transport times were shorter than the time constant of temperature run-away [1, 2, 4, 6]. It is anticipated that this is, at best, marginally the case.

An example of a 1-D simulation of temperature control by fuelling over a limited time has been given by the Culham Group [6]. Due to the unfavourable ratio of the characteristic transport and thermal runaway times, large excursions (about 50% peak-to-peak) seem unavoidable.

Thus, fuelling provides more a means for **adjusting the reactivity** than an efficient method for burn control.

2.2.5. Subignition operation

Subignition operation [1, 2, 4, 5, 9], can be looked at from two different points of view. It may be necessary because no other control technique is viable, or it may be unavoidable because ignition cannot quite be reached.

However, in both cases the operational margin is very narrow because, if $Q \gtrsim 50$ cannot be attained, the auxiliary heating power, necessary during the whole burn pulse of at least 100 s, would rapidly become prohibitive. Furthermore, the heat flux to the wall, at given neutron flux, would appreciably increase. At $Q \gtrsim 50$, the required **heating power** would be $\lesssim 10$ MW, and the heat flux to the wall would be less than 10% higher than in the ignited regime. This implies that one would have to approach very close to ignition anyway. On the other hand, if there is no physical barrier to reach ignition, such as a strong increase of lossess, e.g. due to a β limit **just** below the value at which ignition is attained, **diagnostic** and **instrumental requirements** to **avoid** ignition probably impose [4] $Q \lesssim 50$. Of course, there is a **trade-off** between the value of Q, the efficiency and accuracy of diagnostics, and the performance of the heating technique applied. Therefore, the latter should be able to affect the plasma interior, which at $Q \approx 50$ would already be "ignited", that is, the local α-particle power deposited would exceed the power lost. This implies that either **neutral beam injection** at high energy, typically $E_0 \sim 400$ keV, or **electron cyclotron** heating would be required. The conditions are less stringent when ignition **cannot** be attained, but high Q operation turned out to be possible. However, also in this case heating of the core of the plasma would be almost imposed because of its higher efficiency and to avoid enhanced impurity production at the wall.

The variation of the heat flux to the wall is smaller than 10% peak-to-peak if $Q \geqslant 50$.

The Kurchatov Group has performed a 1-D simulation of a feedback-controlled burn [1,5], applying the heating power with a peak value of 10 MW (see Fig.VIII-7). The control is performed on the difference between the heat losses and α-particle power (measurable via the neutron flux), which is controlled every 20 ms and kept between 5 and 10 MW. The dynamics of feedback-controlled high Q operation [4] and typical working regimes [1] have been explored by 0-D calculations.

2.2.6. Other methods assessed

Three further potential methods for burn temperature control were considered and found not to be viable, namely

1) the hot ion mode,
2) control by cyclotron radiation,
3) control by high-Z impurities.

The **hot ion mode** [2, 4], implies working at a temperature high enough that the ignition point is thermally **stable.** The difficulty is that, on the basis of Alcator scaling for the electron heat conduction, a 2 to 3 times larger thermal

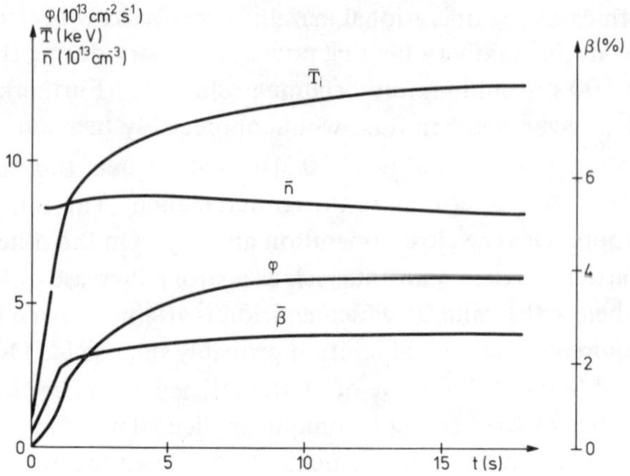

FIG. VIII-7. 1-D simulation of feedback-controlled burn in the subignition regime with 10 MW additional heating and control at time intervals of 20 ms:

$\overline{n}$ = *average density*
$\overline{T_i}$ = *average ion density*
β = *average* β
φ = *neutron flux* (5×10^{13} cm^{-2} s^{-1} *correspond to a wall loading of 1.1* MW/m^2*).*

power and wall loading results than for low-temperature burn of INTOR. The corresponding thermal flux appears to be unacceptable. Furthermore, higher heating powers would be required to reach ignition. It is also to be noted that the existence of the hot ion mode relies on the absence of enhanced ion thermal losses, an assumption which does not apply in practice, due to the bounds on the value of the field ripple imposed by the coil design.

Control by varying the **cyclotron radiation** losses [1, 3, 4] is not feasible, because these are only about 5% of the total power loss. Thus, apart from the technical difficulties, no effective intervention is possible (see also Chapter IV).

High-Z impurities [1,4] are unsuitable for burn temperature control, because there seems to be no ion that, in the relevant temperature range (10 to 25 keV), has a radiation loss which increases more rapidly with temperature than the power generated by DT fusion.

2.2.7. Reactor relevance of burn temperature control methods

In a commercial reactor both the power density and the wall loading will have to be as high as technologically possible in an economic manner. Furthermore, the total power will have to be larger than in INTOR, perhaps by a factor 5 to 10.

This implies that the size of the plasma will have to be somewhat larger than in INTOR, the plasma density remaining about the same or being slightly increased also. Therefore, the thermal insulation properties of the plasma will be better. As a consequence, the steady state burning temperatures of such a plasma will be lower than in INTOR, for the thermally unstable working point, and higher, for the thermally stable one. On the other hand, for given $\langle\beta\rangle$ and magnetic field, the highest power density is attained for temperatures between 10 and 20 keV. **Working on, or close to, the ignition curve, hence, is inappropriate for a commercial reactor.**

The problem can be made more quantitative by adopting the Alcator scaling for the plasma heat conductivity and considering that, in a reactor, radiation is unlikely to be able to provide the dominant energy transport from the hot thermonuclear core. In fact, in this case, on the ignition curve, the wall loading scales as T_e/a, and the total reactor power as $T_e R$. Therefrom it is obvious that in a larger machine, keeping the wall loading at the level of INTOR, or higher, requires high working temperatures, in the range typical for the hot ion mode, for which the power density tends to decrease, impairing reactor economy. Note also that for given $\langle\beta\rangle$ and $\langle T_e\rangle \gtrsim 30$ keV, the temperature, on the ignition curve, grows less than linearly with the minor radius.

It can, therefore, be anticipated that a **commercial reactor** will have to work **well in the ignited regime**. Adopting this mode of plasma operation in INTOR thus corresponds to the most reactor-relevant choice. Accordingly, the burn temperature control methods suitable for this mode (ripple, β limits) are to be considered the most reactor-relevant ones.

3. SHUT-DOWN

Shut-down consists of an **initiation phase** and **shut-down proper** during which the plasma cools down and the plasma current goes to zero. During this latter stage the essential requirements are that the heat flux to the wall does not increase above the values during burn, and that the rate of current change is not larger than during start-up, which implies that shut-down has to be **smooth**. This yields a minimum length of the shut-down phase of about 5 s. On the other hand, as the dwell time should not be longer than necessary, the shut-down time should not be much longer than that value either.

3.1. Initiation of shut-down

The **best way** of initiating shut-down [2] **depends** on the plasma **burning regime** adopted for the burn pulse and on the **burn control method** applied.

TABLE VIII-3. BURN TERMINATION TECHNIQUES

Technique	Time scale (seconds)	Control mechanism	Main advantages	Main disadvantages
Ripple enhanced transport	2+	Reduced alpha heating and increased ion transport	Maintains reactivity	Significant additiona equipment required, alpha losses may be highly localized
Fueling control	3++	Reduced reactivity	Minimum additional equipment if recycling control is available	Potentially long uncertain timescale, reduced reactivity
Impurity injection (pellets?)	0.5–2	Increased loss	Minimum additional equipment, losses uniform on wall	Potential degradatio of operation, increas likelihood of disruption
Decompression	<1	Reduced reactivity	Fast	Large confinement volume required

If the discharge is operated in the subignition regime, no special provision, apart from switching off the auxiliary heating, is necessary. In the other regimes, if control by ripple or by compression-decompression is used, the discharge can be brought below ignition by these same means. Normally, one would simultaneously stop fuelling. Only in the case of burn control via the β limit is a special technique necessary for initiating shut-down. This could be either stopping or increasing fuelling, which, however, is slow, or alternatively again an increase of the field ripple. A summary of possible techniques for initiating shut-down [2] is given in Table VIII-3.

3.2. Cooling and current run-down

The cooling of the plasma and the run-down of the current [1 – 3], starting from a state just below ignition, should be about the time-inverse of start-up. This implies that **first** one would let the plasma **cool down** at constant, or weakly decreasing, current, simultaneously reducing the density, if necessary.

Impurity injection may be useful in this phase to accelerate cooling, but this increases the risk of disruption. Running down the density over a few seconds also seems difficult, unless an efficient unload divertor is available.

After the plasma has been cooled into the ohmic heating regime (T $\approx$ 1 keV), the **current** has to be **decreased** on a time scale of about 2 s, analogously to what is done in present-day machines, accompanied by further plasma cooling. Again adding impurities might help to reduce the run-down time [1]. The main problem is again that a disruption may be induced in a fairly early stage.

The means necessary to avoid **disruptions** are expected to be essentially the same as in the earlier phases of the discharge (see Chapter VII), namely careful position and shape control, keeping the plasma in the regime of minimum disruption occurrence, and possibly measures against the development of precursors to the disruption. A particular problem is that during shut-down the density limit of ohmically heated discharges will have to be exceeded, but it is not clear if this is necessarily unacceptable, since the decrease of the current has a stabilizing effect on tearing modes. If there were a difficulty with the density limit at some stage, perhaps applying some additional heating would have to be considered [8].

4. NON-INDUCTIVE CURRENT DRIVE

Non-inductive current drive is **unnecessary** for INTOR, both for sustaining the current and for current profile control (see Section 2.1).

The assessment carried out [1, 4] has led to the conclusion that, on the basis of trapped-particle effects [4], the power necessary to drive a current of about 5 MA in INTOR by injection of **neutral beams** of 200 keV or higher would be of the order of 100 MW. The underlying theory is in agreement with experimental findings in the Culham Levitron. If this mechanism were not effective, still larger beam particle energies and powers would be necessary [1]. Current-drive by **travelling waves** has potentially more favourable energetics, because the momentum can be transferred to faster particles, but this approach has not been subject to experimental tests at this time.

5. CONCLUSIONS

As the operation regime of INTOR will deviate appreciably from that in present-day machines, there is a number of unresolved questions with respect to its start-up, burn and shut-down scenarios. Many of them are related to building up, controlling and quenching a burning plasma. However, significant work is being, and can be done in the near future, on present-day devices and the schemes developed can be tested in JET, T-60, T-15, and TFTR. This should provide a satisfactory data base for INTOR.

A. Among the three phases **start-up** is the closest to the situation present in the largest devices now in operation. Nevertheless, there are several new problems which require attention:

i) For technological reasons it is preferable to apply in INTOR, during break-down and heating through the radiation barrier, **low loop voltages** like 100 or even 50 V. Only very recently work on start-up with low loop voltage has been done. It seems now evident that electric fields of the order of 1 to 2 V/m may be sufficient, at least in sufficiently clean devices of present-day size and at low stray magnetic fields. This implies for INTOR that, indeed, a loop voltage of 100 V seems satisfactory, if the stray field level is kept below 100 G. Providing **preionization** and **auxiliary RF heating** (2 to 5 MW for 10 ms) might allow lowering the loop voltage to 50 V.

ii) The rate of **current rise** will have to be chosen appropriately (not above $\dot{I}_p = 10^7$ A/s when the plasma is cold and, perhaps, by a factor of 10 smaller at $T_e > 100$ eV) to minimize plasma-wall interaction by MHD activity. At least towards the end of the ohmic heating phase efficient control of the plasma shape and position will have to be provided, and the required **current density profile** will have to be established. As the ohmic heating phase may extend up to 5 s **impurity control** is essential.

iii) **The heating and fuelling** phase to ignition is accompanied by a very **strong change in all plasma parameters,** namely

density:	from perhaps	3×10^{19}	to 1.5×10^{20} m^{-3}
temperature:	from	1	to 10 keV
β_J:	from	$\ll 1$	to 2.5 or 3
$\langle\beta\rangle$:	from	≈ 0	to about 5%

If the density, during the ohmic heating phase, has to be kept as low as 3.10^{19} m^{-3}, quasiperpendicular **neutral injection** at fixed energy will not be satisfactory. **RF heating** methods seem to have a **better** potential in this respect. Therefore, if for INTOR a full RF heating scheme cannot be developed, either a somewhat more complex neutral injection scheme or a mixed RF/neutral injection scenario might have to be adopted. Furthermore, in particular in the case of neutral beam heating, controlling the fuel mix will probably require both **puffing of pure T gas** and **T pellet** injection. Finally, efficient plasma position and shape control is needed.

iv) The flux swing required for start-up until ignition is reached is estimated to be not larger than 75 Vs for INTOR as to guiding parameters.

B. The most serious limitations to the **burn pulse** length are expected to arise from **impurity** influx and **He accumulation.** The latter problem can be much eased if there is a "reserve" in β between the value necessary to attain the required wall loading and the maximum value allowed. Alternatively, if the total lifetime of He can be reduced to 20 s or less, by an efficient exhaust system (see Chapter 4.1.2), indefinitely long burn at a low He-particle concentration ($\lesssim$10%) would be possible.

C. The flux swing needed for a **burn duration** of 200 s is only 10 Vs. This constitutes a small part of the **total flux swing** requirements, amounting to about 90 Vs. Non-inductive current drive is not necessary.

D. There are several possible methods for **controlling the burn temperature,**

i) in the ignited regime:
   a) variable field ripple (± 1 to 2%),
   b) self-regulation by enhanced transport at the β limit,

ii) in the marginally ignited regime:
   a) compression-decompression,
   b) fuelling (if not too slow)

iii) subignition operation at $Q \approx 50$, implying about 10 MW from an external heating source during the entire pulse.

None of these has been tested to the point that its viability can be guaranteed. The most reactor-relevant, and probably simplest, burn control methods are those of D i) above. A variable field ripple, providing an active control method if practicable at all, would appear most attractive. For INTOR, also subignition operation, perhaps combined with one of the methods of D i), would be satisfactory. The time variation of the thermal wall loading will depend on the method adopted (ripple: small; at β limit: depending on impurity and He-particle accumulation; compression-decompression: 20% (preliminary estimate); fuelling: large; subignition operation: <10%). Using the "hot ion mode", or control by cyclotron radiation or high-Z impurities is not considered feasible for INTOR.

E. **Shut-down** can be **initiated** by the **same means** as used for burn control, perhaps combined with a change in the **fuelling rate.** Its minimum duration is about 5 s. Shut-down should be essentially the **time-inverse of start-up.** The main concern in shut-down is the possibility of early **disruption,** in particular if **forced cooling** is to be applied and in case a high plasma **density** would be deleterious.

6. R AND D NEEDS ASSESSMENT

The areas where R and D is urgently required comprise

1) continuing studies and tests to further develop methods for ionization and current rise at low voltage, including the use of preionization and RF auxiliary heating;
2) research into the optimum current-rise scenarios, in particular in large noncircular tokamaks;
3) investigation of the maximum density attainable in clean ohmic discharges;
4) demonstration of density rise and profile control in discharges with intense heating, in particular by neutral beams;
5) development of sufficiently flexible heating schemes (neutral beams, RF, or mixed) compatible with the parameter variations during heating to ignition;
6) development of sufficiently flexible fuelling techniques, including gas puffing and pellet injection for D and T separately (see also Chapter V);
7) development and test of impurity control and He exhaust methods (see also Chapter V);
8) development of INTOR relevant burn temperature control methods, including diagnostics (the effect of a variable field ripple as well as enhanced transport at high β can and must be tested in reactor-grade H or D plasmas);
9) investigation of smooth shut-down of tokamak discharges after intense heating (see also Chapter VII).

The R and D needs assessment is summarized in Table VIII-4.

7. SUMMARY

The questions related to scenarios for the main part of start-up and for shut-down in INTOR can be studied, to a large extent, on present-day devices, although the operating regime of INTOR will deviate appreciably from that of tokamaks in operation by now. On the other hand, only limited work is possible on burn scenarios as well as on the last stage of the start-up phase, as reaching, or at least coming close to, ignition is required for this. A possibility of testing the schemes being developed now and in the near future on smaller devices, and extending them to heating a tokamak plasma into a regime where sizeable α-particle heating is present, will be provided by the large devices JET, JT-60, T-15, and TFTR.

TABLE VIII-4. R AND D NEEDS ASSESSMENT – START-UP, BURN AND SHUT-DOWN

	Status*	
	Adequate	Inadequate
Ionization and current rise at low loop voltage	x	
Optimum current-rise scenario (experiments in large, in particular noncircular tokamaks)	x[1, 2]	
Maximum density in clean ohmic discharges	x	
Density rise and profile control during heating to ignition	x	
Compatibility of heating schemes (neutral beam, RF or mixed) with parameter variation during heating to ignition	x[1]	
Burn temperature control, including diagnostics		x
Smooth shut-down, in particular after intense heating	x[1]	

[1] More specific emphasis required.
[2] JET, JT-60, T-15 and TFTR will provide full-size information from 1982 on.

Adequate: Programmes are either planned or could be planned on existing or authorized facilities (with some upgrading if required) and using existing trained personnel.

Inadequate: Either the available and authorized manpower and facilities are inadequate or a substantial increase in effort is needed.

* The ratings are made on the world-wide fusion effort and not separately on the individual national programmes.

Items that require particular attention include:

1. There is now evidence that for **ionization** and **current initiation** electric fields of 1 to 2 V/m may be sufficient, provided that the discharge chamber walls are kept clean from impurities and the stray magnetic fields are small. For INTOR this implies that in this phase a **loop voltage** of 100 V seems satisfactory, if the stray fields are less than 100 G. Providing **preionization** and **auxiliary RF heating** (2 to 5 MW for 10 s) might allow to lower the loop voltage to 50 V.
2. During the **ohmic heating** phase, and in particular in its early stages, the rate of current rise will have to be chosen appropriately to minimize plasma-wall interaction. Impurity, plasma equilibrium and current profile control will have to be provided.

3. During the subsequent phase of **intense heating** and **fuelling** to reach **ignition** large changes of all plasma parameters will be encountered (e.g. the temperature increases by a factor of 10, the density perhaps by a factor of 5). Hence, the heating scheme adopted has to have enough **flexibility** to cope with these variations. Potentially **RF heating** is better suited in this respect than **neutral beam injection**. Therefore, if for INTOR a full RF heating scheme cannot be developed, either a somewhat more complex neutral injection scheme or a mixed RF/neutral injection scenario might have to be considered. Furthermore, in particular in the case of neutral beam heating, for controlling the species mix of deuterium and tritium a flexible **fueling** scheme is required, implying probably the use of both puffing of pure tritium gas and tritium pellet injection. Finally, efficient impurity, plasma position and shape control are needed.
4. A **burn time** of 100 to 200 s requires efficient impurity control and helium exhaust (see Chapter V).
5. The total flux swing necessary for INTOR, as specified by the guiding parameters, is about 90 Vs.
6. There are several conceivable methods for **controlling the burn temperature,** such as
   a) variation of the field ripple up to ±1 or 2%,
   b) self-regulation by enhanced transport at the β limit,
   c) operation just below ignition, applying a limited auxiliary heating power, e.g. 10 MW, during the entire burn pulse,
   d) periodic adiabatic compression and decompression,
   e) variable fueling (if it is not too slow).

   The most reactor-relevant, and probably simplest, burn control methods are the first three of the above, or combinations of them. Only the methods (a) and (b) can (and must) be tested experimentally in reactor-grade H and D plasmas.
7. Shut-down has probably to be approximately the time-inverse of start-up. It will have to be initiated by reducing the reactivity below the ignition point, using the same means as applied for burn control, combined with stopping fueling. Then the plasma will cool down to the ohmic discharge regime. Finally, the current will have to be brought to zero smoothly. The time necessary for shut-down is expected to be at least 5 s. The main concern is that a disruption might occur in an early stage.
8. The R and D requirements related to the area "Start-up, Burn and Shut-down" are summarized in Table VIII-4. From the world tokamak programm an adequate data base can be expected to be available in time for INTOR check points, when due emphasis is given immediately to investigation of the critical issues identified.

REFERENCES TO CHAPTER VIII

[1] USSR Contribution to the 2nd meeting of the INTOR Workshop , Group 12, Start-up and Shut-down (1979).

[2] "Start-up, Burn and Shut-down", Chapter in USA INTOR Report (see Appendix A.5.4).

[3] Fusion Research Development Center, INTOR Workshop Report, Group 12, Start-up, Burn and Shut-down Control, JAERI-M 8625 (1979).

[4] European Contributions to the 2nd meeting of the INTOR Workshop, Start-up, Burn and Shut-down, Chapter XII, EUR FU/XII 501/79/EDV 50, June 79 (2 Volumes).

[5] USSR Contribution to the 2nd meeting of the INTOR Workshop, Group 1, Energy and Particle Confinement (1979).

[6] European Contributions to the 2nd meeting of the INTOR Workshop, Energy and Particle Confinement, Chapter I, EUR FU/XII 501/79/EDV 50, June 79 (2 Volumes).

[7] USSR Contribution to the 2nd meeting of the INTOR Workshop, Group 4, Heating (1979).

[8] European Contributions to the 2nd meeting of the INTOR Workshop, Stability Control, Chapter XI, EUR FU/XII 501/79/EDV 50, June 79 (2 Volumes).

[9] USSR Report to INTOR, "Start-up, Burn and Shut-down in INTOR" (see Appendix A.5.3 [29]).
USSR Report to INTOR, "On Current Programming in INTOR" (see Appendix A.5.3 [30]).
USSR Report to INTOR, "Some Aspects of Start-up, Burn and Shut-down in INTOR" (see Appendix A.5.3 [32]).

[10] Fusion Research Development Center, INTOR Workshop Report, Japanese Contribution to the 3rd meeting of the INTOR Workshop, JAERI-M 8625 (1979).

[11] European Contributions to the 3rd meeting of the INTOR Workshop, Homework and Special Supplementary Information, EUR/FU/XII 501/79/EDV 70, Oct. 79.

Chapter IX

MAGNETICS

J. GILLELAND – USA, Chairman
C. BAKER – USA, Expert
G. CHURAKOV – USSR, Expert
J.R. PURCELL – USA, Expert
P. REYNOLDS – EC
K. SAKO – Japan

1. INTRODUCTION

This report provides an assessment of the magnetic systems required for INTOR. Included in this study are the toroidal field (TF) coils, induction coils, shaping and control coils, and divertor coils. Both the existing data base and the additions to the data base anticipated in the early 1980's were analyzed in order to identify any major technological uncertainties. Recommended development activities to eliminate these uncertainties are also presented.

2. SUMMARY

2.1. Contents

The main conclusions of the Workshop participants are summarized in this Section. The observations do not present a complete picture of magnetics status. Rather, they represent the principal points of interest and discussion at the workshops in Vienna. Further information can be found in the various national reports referenced at the end of this report, which describe in detail the supporting studies carried out during 1979.

2.2. TF coil

1. The various development programmes now being conducted are thought to be sufficient for the necessary technology bases to feed INTOR. The programmes appear to explore essentially all reasonable conductor and cooling options.

2. INTOR has a central field of approximately 5 T, leading to 10–11 T maximum at the coil. This leaves open both the Nb_3Sn and the NbTi options. The existing programmes and better definition of the INTOR mission will clarify which conductor and cooling method will work best.

3. The superconducting toroidal field coils are expected to be very reliable over the anticipated lifetime of INTOR. However, our experience is limited and studies with regard to failure modes, repair and replacement must be conducted in order to provide a basis for major decisions on machine design. These studies should attempt to determine the most probable modes of coil failure and the corresponding repair requirements as well as the disadvantages due to a design which allows easy coil removal.

4. Plasma current disruptions change the poloidal field at TF locations resulting in rapid increases in overturning moments and rapid field changes. This will be a major, but solvable, design problem for the SC/TF coils.

5. Shielding is required to protect the insulators and stabilizer from radiation damage and to reduce the radiation-induced heating of the toroidal field coil. The limiting value of radiation to the toroidal field coils is determined by the resistivity increase of the stabilizer. Copper is more resistant to radiation damage than aluminium. However, its recovery when annealed at room temperature is not as good as that of aluminium. Annealing of the toroidal coil takes a great deal of time and therefore is to be avoided if possible. The toroidal field coil can be designed to operate satisfactorily without the need for annealing the stabilizer, provided the fluence is kept below 10^{-4} dpa or 2×10^{17} n/cm^2 (fission neutron spectrum) for copper stabilizer material.

6. The toroidal field coils can be of sufficient size to permit space for proper access to the blanket or the inclusion of a bundle divertor. However, an upper limit on the size is imposed by the requirement of plasma shaping by PF coils placed outside the TF coils and the available power supply for these coils.

7. Poloidal coils outside the TF coils increase the pulse field problems and losses in the TF coils; however, an improved access to the PF coils is more important than the disadvantages of increased refrigeration loads and increased mechanical loads.

8. TF coils that can be split through the use of demountable joints are assessed as impractical for INTOR. This results from considerations of stress and dewar complication, not from considerations of contact resistance.

2.3. Induction coil

1. Small scale experiments show that a superconducting induction coil is feasible. The operation of such a coil is less reliable than that of toroidal field coils. The development programmes are limited at present and must be expanded to become a viable option for INTOR. It must be noted that the induction coil probably cannot be used for plasma initiation if the required voltage per turn is greater than 50–100 V. This question requires further conductor and coil development.

2.4. Shaping and control coils

1. Designs should favour the placement of poloidal coils outside the TF coil despite increased complexity of structure, higher pulse losses in the TF coils and larger poloidal system power supplies. This results partly from considerations of coil reliability, but mainly from the requirements of access to the blanket.

2. Coils inside the TF coil should be normal conducting coils and must be designed to be replaceable. If coils must be located inside the shield (for fast stabilization), the insulators may have short lifetimes and must be remotely replaceable. It is recommended that design concepts should avoid inside-inboard coils.

3. Present estimates indicate a one-turn voltage for plasma initiation of ~100 V/turn. It may be possible to provide this breakdown voltage with the superconducting OH coil. The use of a special coil for providing this voltage, located near the plasma, is undesirable from the viewpoint of engineering. Another option, suggested by the USSR, is to use a "hybrid" OH coil consisting of a superconducting coil for the slower ramp and a room temperature coil for the faster ramp.

2.5. Divertors

1. For INTOR it is almost impossible to design poloidal divertor concepts with internal coils such as described previously; the design of bundle divertors also entails extreme difficulties. Several new concepts, including poloidal divertors with coils exterior to the TF coils, were discussed at the Workshop, which may allow reasonable engineering practice and still permit acceptable divertor action. Further review of these new ideas is recommended.

2. Divertor designs have received very little attention compared with other magnetic systems. If it is considered necessary to place a divertor coil within the TF coil, then a major new development programme will be required to bring divertors anywhere near the stage of reliability of the rest of the magnetics.

3. A poloidal divertor with external coils and bundle divertors are clearly more accessible and replaceable than an internal poloidal divertor. An overriding consideration in divertor choice must be the ability to provide remote maintenance.

2.6. R and D needs assessment

A summary of the R and D needs assessment for magnetics is shown in Table IX-1. These items are discussed in further detail in this chapter and in the national reports listed in the references.

3. TOROIDAL FIELD SYSTEM

3.1. Description

Tokamaks require a strong toroidal field (TF) for stable plasma confinement. The fusion output power of the reactor is proportional to B_t^4 for a given β. It is essential for the toroidal field coils to be superconducting in the case of a reactor, such as INTOR, to reduce the parasitic power.

3.2. Trade-off considerations

3.2.1. Superconducting materials

If the peak field is < 9 T, it is generally recognized that NbTi is the best superconductor to use because it is cheaper than Nb_3Sn; it is very ductile and has been tested in many large magnets. Furthermore, fabrication of NbTi is a well-estabished state-of-the-art and high quality conductors can be readily produced; this is a very important aspect for a project such as INTOR.

If the peak field is > 9 T, three options must be considered: (1) a large amount of NbTi could be used and the coil operated at lower temperatures, or (2) Nb_3Sn could be employed in the high field region with NbTi in the low field region, or (3) Nb_3Sn could be used for the entire coil. It is not clear at this time which of the three options is best. The present development programme will answer this question.

TABLE IX-1. R AND D NEEDS ASSESSMENT – MAGNETICS[1]

	Status *	
	Adequate	Inadequate
Superconducting toroidal coil development and 12 tesla (Nb_3Sn) conductor development	x[2]	
Superconducting induction coil development		x
Divertor development		x
Internal poloidal field coil development, if required		x
Radiation effects on superconducting TF coils components	x	
Superconducting magnetic systems integration experience	x	
Development of large, reliable cryogenic systems	x	

[1] It is assumed that a prototype INTOR TF coil will be built as part of the project.
[2] Additional R and D for high field coil demonstration, similar to Large Coil Task at ORNL (USA), is required by Japan.

Adequate: Programmes are either planned or could be planned on existing or authorized facilities (with some upgrading if required) and using existing trained personnel.

Inadequate: Either the available and authorized manpower and facilities are inadequate or a substantial increase in effort is needed.

* The ratings are made on the world-wide fusion effort and not separately on the individual national programmes.

3.2.2. Stabilizer material (Cu versus Al)

The stabilizer for the TF-coil conductor should have a low overall resistivity and, ideally, high mechanical strength. Both high purity copper and high purity aluminium are considered as potential candidates for the INTOR TF coil conductor stabilizer.

The low-temperature resistivity of high purity copper is determined by the radiation-induced resistivity, strain-induced resistivity and the magnetoresistivity. The intrinsic resistivity of copper is negligibly small. The magnetoresistivity of copper is about $0.455\ B \times 10^{-8}\ \Omega cm$, where B is the magnetic field in tesla.

The rather large magnetoresistivity at high field strengths significantly increases the overall resistivity of copper. Radiation produces lattice imperfections (atomic displacements). The radiation-induced resistivity of copper increases with dose and eventually becomes saturated at an induced resistivity of 3×10^{-7} Ωcm.

The low-temperature resistivity of high-purity aluminium is mainly determined by the radiation-induced resistivity, the magnetoresistivity and the strain-induced resistivity. The latter is important because high purity aluminium is rather soft with poor mechanical properties (low modulus and low yield strength.) For strains less than 10%, the strain-induced resistivity of aluminium is about $1.1\ \xi^{1.19} \times 10^{-7}$ Ωcm, where ξ is the strain in the aluminium. The radiation-induced resistivity of aluminium also increases with dose and eventually saturates at a resistivity of 8×10^{-7} Ωcm. However, aluminium has a rather small magnetoresistivity that saturates at a low field (~0.5 T).

A total neutron fluence of about 10^{17} n/cm^2 will increase the total resistivity of copper by 25% and aluminium by 100%. A magnet design that will tolerate these resistivity changes during its lifetime is not difficult and the coil thickness is not greatly increased. Annealing to reduce the resistivity is a more complex problem. Apart from the long time necessary to warm and then recool the coils, the combined effects of radiation and mechanical strain do not anneal out as well as either effect separately. When warmed to room temperature, copper will only recover about 60% of its induced resistivity and aluminium will recover about 80%. When aluminium is heated up to 375 K, still within the thermal capabilities of the coil, it will amost completely recover its original conductivity.

Radiation damage to the superconductor occurs only at a fluence of an order of magnitude above that for the stabilizing material and is not a problem for INTOR.

3.2.3. Cooling mode

The TF-coil can be bath cooled (4.2 K), forced flow cooled, or cooled at subatmospheric pressure (temperature less than 4.2 K).

It is generally recognized that bath cooling is the simplest and most natural cooling mode and is thus highly reliable. The operation of the TF-coil system in a bath cooling mode will be independent of the continuous operation of the refrigeration system provided that some safety margin exists for the refrigeration capacity.

Forced cooling, however, simplifies the cryostat structure, and removes some of the problems involved in providing sufficient electrical and mechanical strengths. Additionally, in the case of forced cooling, the cryostat contains less helium.

Operation at a temperature below 4.2 K may be desirable as a means of improving the high field characteristics of NbTi.

3.3. Machine interaction

3.3.1. Tolerable neutron exposure

Neutron irradiation of the superconducting TF coil causes radiation damage and nuclear heating. One of the functions of the blanket and shield is to protect the TF coil from excessive radiation damage and nuclear heating. However, the inner blanket/shield occupies the high toroidal magnetic field region where space is at a premium. Therefore, the inner blanket/shield should be designed with efficient compositions to minimize its thickness. The cost of a high efficiency shield system must be traded off with the cost of space required for a less efficient shield and with nuclear heating and the radiation damage to the components within the TF coil.

The questions of radiation-induced resistivity in the stabilizer and damage to the superconductor were addressed in Section 2.2.2. Radiation damage in the TF coil insulator is irreversible. Therefore the insulators in the TF coils must be designed with regard to proper operation for the lifetime of the INTOR. Organic insulators are much less resistant to radiation damage than inorganic insulators. However, organic insulators are desirable for large magnetis since inorganic insulators are brittle and difficult to use. It appears that several organic insulators are available that will be suitable for INTOR from a radiation point of view.

3.3.2. Pulsed poloidal field

The superimposing field from poloidal coils (the plasma current, the control coils, the induction coil, the divertor, and the initiation coil) is found to be the most troublesome factor in the TF-coil design. The out-of-plane load is large and cyclic in nature; thus it creates very challenging problems for the structural design. It also complicates the cryogenic stability of the coil due to the frictional motion induced by bending moments and shear stresses in the coil structure. Large a.c. losses will occur in the TF coil unless the TF conductor is made of cable. The sponginess of the cable conductor adds further to the problem of coil structure integrity. Finally, additional losses occur due to the pulsed parallel field coupling between twisted filaments at different distances from the conductor axis. This coupling effect must also be addressed in the conductor design.

3.3.3. Toroidal field strength

Higher toroidal field strength improves energy confinement and will increase the INTOR power performance. However, high field coils require more space, make more difficult the provision of sufficient space at the centre for the induction coil, and result in a larger, more expensive machine. Thus, the field strength at the plasma centre is one of the most important trade-off parameters of the entire reactor.

3.3.4. Maintenance provisions

The TF coils can be built as large as required to provide access to the machine interior, within the limitations imposed by the poloidal coil system located outside the TF coils. This access will be required for neutral beam ducts, vacuum pumping, assembly and disassembly of blanket and shield, and other maintenance provisions.

3.3.5. Disruptions

The tokamak magnet system consists of the TF coil system, the OH-induction coil, the control coil (field-shaping coil), the plasma current ring, the divertor and the initiation coil. All of these magnet systems are very large and are strongly coupled together. Disruption of one system will trigger very strong interactions not only with the rest of the magnet system but also with the other metallic components within the reactor such as blanket and shield, first wall, and torque frame. These interactions must be fully analyzed during the magnet design.

3.3.6. Design guidelines

In the early phases of the INTOR Project, it will be necessary to establish the design guidelines, safety factors and appropriate design codes for the magnetic systems.

3.4. Development programmes

The technical outlook for superconducting toroidal field coils in INTOR is very good. Although at present we do not know the optimum design, it is clear that the system can be built and the various problems can be solved. The development programmes, summarized here, indicate the broad range of information that will become available within the next few years. The builders of INTOR will have many technical options to choose. The information required for construction of reliable toroidal field coils will be in hand.

In the USA, the Large Coil Test (LCT) is the major development programme for toroidal field coils. This facility, at Oak Ridge National Laboratory, will ultimately consist of six subsize coils, about 1/3 INTOR scale, operating with a peak field of about 8 T. Three of the coils will be manufactured in the United States of America, one in Japan, and two in Europe. Both bath cooling and supercritical forced flow cooling will be applied, as well as Nb_3Sn and NbTi.

In the tests the coils will be exposed to pulsed external fields and simulated neutron heat loads very much like those experienced by the coils of INTOR. The coils can be directly compared in terms of performance and manufacturing difficulty, providing valuable input to the INTOR design options. Testing of the first coils is scheduled for 1982.

A second United States programme is the 12 T programme. This programme is aimed at producing small test coils (about 1m) that operate at 12 T under reactor-like conditions. The programme will test NbTi at reduced temperature and Nb_3Sn with both bath cooling and supercritical forced cooling.

The Japanese development programme consists primarily of a cluster test facility with a test module coil and a coil for the LCT. The cluster test coil will have a peak field of 8 T, an operating current of 10 000 to 12 000 amperes with bath cooled, copper stabilized NbTi. It will be about 1/2 the size of the LCT coil and reactor conditions will be simulated. The Japanese LCT coil will also be bath cooled NbTi with copper stabilizer. After the coil has been first tested in Japan, it will be shipped to Oak Ridge for final testing at the end of 1982.

The LCT programme is to be followed by the STTA programme in Japan. Carried out as a part of the R and D effort for PETF, the effort is aimed at integral tests of the toroidal magnet system, with particular emphasis on plasma-magnet interactions.

In the USSR, the primary goal of the development programme is to utilize Nb_3Sn superconductor with forced cooling. To this end, the T-7 machine has been built, using coils of this construction, and now a larger device, T-15, is being designed. This machine, which will be stabilized against disruptions as well as normal pulsed field loads, is designed for a nominal peak field of 5.8 T, giving 3.5 T on the machine axis. Engineering margin is such that the field on the axis may be taken as high as 5 T. Both single phase and two phase forced flow will be used on T-15 to determine the best method of cooling. The ultimate goal of the whole superconductor development programme is to create the current-carrying elements for the toroidal and poloidal magnetic fields.

In the European community, development work is being done with Nb_3Sn and small, high field coils will be built in the next few years. They are also building a coil for the LCT. This coil will be NbTi stabilized with copper and forced cooled with supercritical helium at an inlet temperature of 3.8 K.

Torus II is a tokamak in the design stage that will have a peak field of 9 T produced with copper stabilized NbTi bath cooled with superfluid helium. The advantages of superfluid helium are the lower temperature, which improves the superconducting properties of NbTi, and the high thermal conductivity, which distributes concentrated heat loads throughout the bath. Development work on the Torus II coils has already begun and this work will provide important information for the design of INTOR, in particular the performance of superconducting coils under tokamak operating conditions.

4. INDUCTION COILS

4.1. Description

The induction coil system consists mainly of a solenoid coil located in the central core of the machine. There are some additional induction coils whose purpose is to control the location of flux return from the central solenoid; but the field, stress, and space problems are less severe with these coils than with the central solenoid. Consequently, the discussion will be limited to the central solenoid. It is generally believed that the induction coils of INTOR-sized tokamaks should be superconducting because the machines can be made more compact with reduced power supply requirements and lower costs. A superconducting induction coil eliminates the need for a warm support structure and thermal insulation for the toroidal field coils along the inside of the torus.

Magnetic iron can be utilized in the induction coil system to reduce the ampere-turn and stored energy requirements. A full iron return path would give the most reduction in ampere-turns but is very bulky and would interefere with other machine components. Iron only in the central solenoid section of the induction coil may provide some reduction in ampere-turns while not interfering with the rest of the machine. In the design of INTOR, consideration should be given to both induction coil systems, i.e. with iron in the central solenoid region, and an air core system.

During a machine pulse, the field in the induction coil undergoes a complete reversal. The time rate of change of the field is non-uniform, with most of the field change occurring during plasma production and the remainder during burn to make up for losses due to resistivity. The peak voltage of the induction coil may be limited to ~50–100 volts per turn. If higher voltages are required, the plasma initiation pulse may have to be provided by a separate system.

4.2. Trade-off considerations

4.2.1. Superconductor

The preferred superconductor is NbTi because of the well developed and simpler fabrication technology for this material.

4.2.2. Operating field

For the bipolar field swing a maximum value from −8 T to +8 T uniformly in one second appears reasonable for NbTi-based superconductors. Larger field changes, e.g. ±9 T in one second, lead to heat transfer problems which can be solved only by using unreasonably thin insulation or by cooling below 4 K. The proposed INTOR startup after current initiation requires a field change of only ≈2–3 T/s.

4.2.3. *Losses*

Because of the large amount of stored energy and the high value of dB/dt characterizing the induction coils, copper to superconductor ratios as high as 10:1 should be considered. To reduce the resulting eddy current losses in the copper, a mixed Cu-CuNi matrix can be used with the CuNi forming barriers around the NbTi and Cu, or forming radial webs. The fabrication problems of extruding and drawing high copper ratio mixed matrix superconductors are not completely understood. The technology for Cu-CuNi matrix conductor requires considerable additional development. An alternate way of producing high copper ratio conductor is to bundle together pure copper wire with superconducting wire of lower copper to superconductor ratio. Energy losses in the induction coil should typically amount to 0.1 to 0.3% of the total stored energy during a bipolar swing of the magnetic field.

4.2.4. *Cooling methods*

The three general methods of cooling considered are forced flow, natural convection, and forced convection cooling. In forced flow cooling, the cryogen is forced through conduit enclosing the conductor by pumps. In natural convection the flow of helium is driven by buoyant forces. The form of natural convection known as pool boiling implies a change of phase from liquid to gas at the conductor produced by heat transfer from the conductor. In forced convection, the flow is moved across the conductor through channels and driven by pumps or by a level difference in the helium at the entrance and exit reservoirs. The simplicity of forced convection or natural convection is preferred for the induction coils; pump-driven forced convection is the least desirable of the alternatives.

4.2.5. *Winding methods*

The two methods of winding are helical multilayer and spiral pancake windings. The spiral pancake type winding requires more joints in the conductor, but it is easier to wind and much easier to repair. The pancake coil also lends itself more readily to modular fabrication.

4.2.6. *Cabling methods*

Cabling methods considered are flat braid, three-dimensional lattice braid, Rutherford cable, and conduit. Conduit is appropriate only with forced flow cooling. Flat braids tend to have high aspect ratios which can result in mechanical support and cooling problems. The three-dimensional lattice braid has certain cooling advantages, but requires development of equipment for its manufacture and more work on mechanical support. The Rutherford cable is essentially a Roebel transposition of subcables. If the subcables have complete transposition, the final cable will have complete transposition. In addition, the Roebel transposition can be done around a stainless steel strap to give extra support to the cable. The Rutherford cable is the preferred method of cabling.

4.2.7. *Insulation*

Possible insulation methods are plastic film, wrapped tape, and chemical coatings. The physical integrity of wrapped tape is difficult to maintain if it must be applied at an early stage of the cabling because repeated cabling operations tend to tear the tape. However, at times tape is the only practical way of insulating a conductor. This is the case if the cable being insulated is made of pure copper strands and superconducting strands with electrical contact between the strands required for stability of the cable. Chemical coatings such as Ebanol or CuS have the advantage of higher thermal conductivity than plastic films, but voltage strength has caused chemical coatings to fall into disuse, although they may be acceptable if voltage insulation provided by the coating is only between parallel strands in the braid and not layer-to-layer in the pancake coil. Plastic films have the advantage of high dielectric strength and resistance to cutting or abrasion and are the preferred method of insulation.

4.2.8. *Structure*

Axial forces can be taken up by end flanges of the individual coil form for each pancake coil and passed to the coil form. Radial forces are supported by the internal strap of the Rutherford cable or by some combination of internal strap and banding. Plastic impregnation is not considered practical because of the pulsed duty of the induction coils. Neither the banding nor the support structure for the internal strap can be allowed to produce a shorted turn.

4.2.9. *Current*

Current in individual cables should be limited to 50 kA; as an example, a 50 kA, 8 T cable has a cross section of 12 cm by 2 cm; a larger cable becomes increasingly difficult to fabricate and wind into a coil. Several pancakes can be connected in parallel if higher currents are required to reduce voltage levels or to simplify current programming.

4.3. Machine interaction

4.3.1. Helium vessel and structure

To eliminate induced currents in dewar walls and the resulting heating and force problems, the helium vessel should preferably be made of an insulator such as epoxy-fiberglass. If stainless steel is used for dewar walls and support structure, care must be taken to eliminate shorted turns around the induction coil and to subdivide the structure within the induction coil to reduce eddy currents. A single structure concept in which the coil form for the induction coil supports the TF coils seems possible. Some designs have found that a G-10 support structure could not support the resulting compressive strain and have thus resorted to stainless steel. Graphite-epoxy coil forms have also been proposed because of the high stress and the need to reduce eddy currents with a high resistivity material. Technology development in reactor-like nonconducting dewars and related structure is needed.

4.3.2. Refrigerator

Several trade-offs exist between the refrigerator size and the design of the induction coil. If simpler but higher loss superconductor is used for the induction coils, the coils may operate satisfactorily but require more refrigeration. The method of electrically connecting the individual pancakes or groups of pancakes also affects the amount of refrigeration required. Parallel connections imply larger current leads into the dewars and consequently larger heat leaks. The possible inability to make Cu-CuNi matrix superconductor must be considered as a burden on the refrigeration system.

4.3.3. Power supplies

The electrical connections of the individual pancakes also affect the peak current and voltage levels to the power supply. Unusually high current supplies will be more costly because of the larger bus work required. If high current must be switched, novel switch gear to handle the current will also increase the cost. Power supplies requiring homopolar generators or commutated d.c. machines still require development, and alternator-rectifier systems are to be preferred.

4.3.4. Intercoil coupling

There will be electromagnetic forces between the induction coils and the field-shaping and control coils and also between the main induction coil in the core of the machine and the outer induction coils. These forces on the outer induction coils can be comparable in magnitude to the forces produced by the self-field and can act in directions different from the self-field forces. This may affect the support structure for the coils and the manner in which the conductor is braced against the coil form. There is also a loading of the outer turns of the induction coil due to the interaction of the current of the induction coil with the stray field of the toroidal field coils.

Inductive coupling between the induction coil and the equilibrium and shaping coils can be significant in some machine designs. Usually the problem shows up in the equilibrium or shaping coil circuits rather than in the induction coil circuit. The options available are to let the power supply buck out the induced voltage or to insert an extra decoupling winding.

4.4. Technology assessment: present and expected

Both the induction coil and its supporting equipment are very large, and in some cases larger than any similar item built today. In principle, the power supply and refrigerator required by the induction coil could be built as straightforward extensions of present technology. The use of salient pole alternators at their pulsed ratings in conjunction with thyristor power supplies is a well established technology for powering tokamak experiments. The alternator technology required exists and is well developed, and the production of cheaper higher-rating thyristors would be welcomed by other larger markets than that composed of fusion experiments. In these two areas there is little to be done which will not be done anyway, with or without the encouragement of the fusion community.

The power supply switch gear required by large tokamaks has tended toward higher current ratings and more operations before maintenance than is typical of the power industry. This requires continued development.

Plastic dewars are made at present, but not of the size required for the induction coils. This is a technology area requiring more development.

Test cables capable of carrying 50 kA at 7 T are available but no coils of INTOR size have been built yet. Several coils capable of carrying 10 to 12 kA at fields of 2.5 to 4.5 T have been made. These coils have been charged or swung in a bipolar fashion at rates of 5 to 10 T/s. The superconductor used in both coils was copper matrix multifilamentary NbTi. It is possible that this material is adequate for the induction coils and mixed matrix material will not be needed. There is a trade-off of the cost of the refrigerator against the cost and difficulties of producing mixed matrix material, and some substantial advantage arises if Cu-CuNi matrix NbTi conductor could be readily available.

4.5. Development programmes

The induction coil development programme is small compared to the developmen programme for toroidal field coils. One demonstration programme to produce a ± 7 T,

14 T/s, 50 kA coil is in place. This is part of the 20 MJ coil programme at LASL in the USA. Several smaller programmes are contributing to the development of high-field, high-current pulsed superconducting coils. Examples are the 10 to 12 kA coils at LASL and ANL and a 2.5 T, 400 kJ, 25 kA coil is being built by Westinghouse for testing at LASL. This latter coil is not a cryostable coil intended to solve the problems of the induction coils, although some of the technology will be transferable.

Some plastic dewar development is expected in connection with the 20 MJ coil.

5. SHAPING AND CONTROL COILS

5.1. Description

A moderately high β tokamak such as INTOR may require some poloidal field coils trapped within the toroidal field coils for plasma shaping and stabilization. Studies carried out during the Workshop indicate that the shaping and stabilization coils could be located outside the TF coils provided there is a conducting shell (perhaps part of the blanket or shield) inside the TF coils.

Vertical field coils located outside the TF coils can be either superconducting or copper. They will present few design problems since they are well shielded from neutrons, are accessible, and can be moved out of the way. However, their sheer physical size will make them difficult to fabricate.

If poloidal coils are trapped within the TF coils, they represent a more serious problem. These coils must be of copper and may be quite close to the plasma with a resulting high neutron flux. Access to these coils will be poor at best and the closer they are to the plasma, the more difficult access becomes. Maintenance on the coils may have to be done remotely, once the machine has had a few D-T shots and has become activated. Electrical insulation must be accomplished utilizing inorganic materials such as aluminium oxide, if located in intense radiation zones. The coils will be water-cooled and operate near room temperature.

5.2. Trade-off considerations

Complicated trade-offs must be made in the overall design of the machine, regarding the coil positions. The coil position affects the obtainable β, the amplitude of a disruptive voltage on other components, neutron flux and heating, access to the machine interior, pulsed field on the TF coils, and a host of other interactions.

The method of construction of the poloidal coils located inside the TF coils strongly affects the access and maintainability of the machine. The coils can be formed of segments that will slip between the TF coils with each segment having a current path into and out of the machine. When installed, the segments effectively form a circular current loop but the coils are not captive in the TF coils. This method requires a minimum number of TF coils to reduce the number of segments. Forces on the leads may represent a significant problem in this case.

A second method of construction would be to assemble preformed rigid circular sements with bolted joints. These joints would have to be disassembled and reassembled remotely once the machine is in service.

A third possible construction technique would be to form the coils from a flexible conductor that is wound in place inside the machine. These coils would have to be unwound and then rewound remotely, for any repairs or maintenance.

Operating currents will depend somewhat upon the method of construction, although conflicting requirements make the choice of coil current uncertain. In general, a high current is desirable from the point of view of minimizing the number of joints, and reducing total voltage on the coil. These coils must have a current shaped waveform similar to the induction coils. This waveform can be generated either by very large power supplies or by decoupling them from the induction coil by proper selection of number of turns. In either case, the waveform generation favours a smaller current, conflicting with the desire for few joints.

5.3. Machine interactions

The position and construction of the poloidal coils have a large impact on the design and operation of the machine. Remote maintenance and machine interior access are strongly affected by the poloidal coils. A plasma disruption will be attenuated by the poloidal coils and the amount will depend on the design and placement. An additional consideration is the unusually high momentary voltage that may appear within these coils due to plasma disruptions. The pulsed field imposed on the TF coils is also a function of the interior poloidal coil design.

5.4. Technology assessment

A great deal of experience exists in the construction and maintenance of ordinary copper coils, however, the radiation environment and remote maintenance requirements bring new and large problems into the design of INTOR coils. At present no technology programmes exist for the development of reliable, remotely maintainable joints. This may be a major problem in the design of INTOR. (See Chapter V).

The copper conductor should not present a material problem for the INTOR coils; however, the electrical insulation is less clear. Aluminium oxide or a similar ceramic might tolerate the neutron flux if it is not too close to the plasma.

The existing data may be misleading in that the samples were irradiated without the presence of an electric field. However, an electric field will exist in the coil insulation during neutron irradiation. This field may cause displaced atoms to migrate and damage the insulator at a lower dose than without the field. The problem can be somewhat relieved by placing the coils farther away from the plasma, behind the blanket and some shielding.

6. PLASMA INITIATION

In order to make the induction coil drive sufficient plasma current for tokamak confinement and ohmic heating, a relatively low plasma resistance is required at the start of the induction coil current swing. In existing machines, the initial plasma breakdown is usually achieved with a rapid induction coil current swing at the beginning of the cycle to produce the toroidal electric field required for breakdown. This produces sufficient ionization so that the required plasma current can be produced with a slower induction coil current swing during the remainder of the discharge.

This approach to plasma initiation may not be practicable with INTOR for two reasons. First, the induction coil and the plasma are poorly coupled due to their relative positions. In order to produce the initial breakdown voltage spike, an energy transfer of $\approx$ 100 MJ over a period of $\approx$ 10 ms would be required, resulting in excessively large power transfer. Second, the toroidal L/R time constant of all material within the toroidal field coils must be less than a few milliseconds in order to generate the fast breakdown pulse at the induction coil. This requirement might impose severe restrictions on the blanket and shield designs.

An alternative to utilizing the main induction coil for producing the breakdown pulse is to use a separate coil designed especially for this purpose. The breakdown coil would be located near the plasma (for good coupling) and would be energized early in the startup cycle by discharging a capacitor bank through it, with a diode to prevent current reversal. This method is preferable to utilization of the main induction coil because the breakdown coil only requires small currents for short times, the energy transfer required for breakdown is reduced to about 10 MJ, and the breakdown coil need only deliver about 3 V-s of flux change However, the breakdown coil will need to be well supported mechanically, because plasma disruptions could force a large fraction of the full plasma current to flow in the poloidal circuit for the time required to transfer the current to the outer poloidal coils. A further disadvantage of this approach arises from the difficult access and remote maintenance problems inherent in a startup coil close to the plasma.

The most desirable method of producing the initial plasma ionization may be the use of a radio-frequency electric field. The RF-induced breakdown approach entails none of the disadvantages of the separate breakdown coil, and would be easier to control than inductive breakdown using either the main induction coil or a separate coil. The main disadvantage at present is a lack of experience with this type of startup in tokamaks.

No particular technology programmes are currently in place to develop plasma startup techniques. If an internal breakdown coil is used, then the associated design problems will be primarily those generic to all internal poloidal coils, such as joints, maintenance, radiation damage to insulators, and so on. A development programme is needed to address these problems, but the same programme would also serve the needs of the startup coil.

The preferred startup method, using RF waves, does require demonstration on a present-day tokamak, and possibly further physics R and D. However, this work could be accomplished as part of the existing R and D programmes on RF auxiliary heating, requiring a small enlargement of existing programmes.

7. TECHNOLOGY ASSESSMENT OF DIVERTORS FOR THE INTOR MISSION

7.1. Introduction

Some of the earlier tokamak reactor studies have included a divertor as a means of impurity control and spent-fuel pumpout. As these studies have progressed, however, and as the next step reactor relevant machine rapidly approaches, it becomes increasingly clear that inclusion of a divertor represents a major complication. Unfortunately, there will probably exist no data base which can assure successful operation of an INTOR device without a divertor. It is therefore of vital importance to carry out a programme leading to a divertor of acceptable reliability and minimum complexity, so that the divertor option can be weighed against the risk of omission. It is also necessary to expand research efforts on non-divertor concepts for impurity control and helium ash pumpout. There is no adequate programme directed toward this goal at present.

7.2. Description

Divertors fall into two general classes; poloidal divertors that divert poloidal flux, and toroidal divertors that divert toroidal flux. In both cases flux lines at the outside of the plasma column are diverted out of the main vacuum chamber to a region where the particle flux can be pumped with minimum back-streaming to the main chamber. The pumping requirements in an INTOR scale device are severe, being of the order of 10^7 l/s at a particle thermal power as large as approximately 100 MW. Obviously, it is desirable to divert the flux to a region where there is room for sufficient pumping area and sufficient expansion to spread the thermal power to an acceptable thermal flux. (It may be desirable to reduce the heat load in the divertor by radiating a substantial portion of the alpha particle energy to the first wall.)

Poloidal divertors utilize coils carrying current parallel or antiparallel to the plasma current to produce a poloidal field null outside the plasma. The position of the null is dependent on plasma current and position. Both single and double null versions are described, and particularly in reactor designs several sets of coils are generally used to channel and expand the flux to regions inside or outside the TF-coils.

The majority of poloidal divertor designs place the divertor coils inside vacuum cans, and the main vacuum wall outside the coils and blankets, and sometimes outside the TF coils as well. In several designs the poloidal divertor coil sets carry no net current and hence can be made in sectors with current cross-overs.

Poloidal divertor designs [4–6] were presented at the Workshop that utilize coils outside the toroidal field coils. In this case the coils could be made superconducting and the engineering problems are thus greatly reduced. Vacuum pumps and shielding might still remain a problem with a poloidal divertor outside the toroidal field coils.

Toroidal divertors generally provide a null field at the outside of the plasma, but they must do so without producing a damaging toroidal field ripple in the confined regions of the plasma. A "blister" toroidal divertor was proposed, however, with one toroidal field coil sufficiently enlarged to create just such a ripple, causing particles to be trapped and drift out into a suitable pumped region. The effects of such a ripple on general confinement are thought to be too severe, however. The "bundle" divertor is a toroidal divertor which produces a local null in the toroidal field by means of coils whose dimensions are small compared to the TF coils. The null location is not dependent on the plasma current as it is in the poloidal divertor. Typical designs produce a null at the plasma edge with a 2% ripple on the plasma axis. Once diverted out of the main chamber, the flux is again channelled and expanded to a region outside the toroidal field coils. The bundle divertor coils need not be interlocked with the TF coils, and hence offer increased potential for maintenance access.

New divertor concepts were presented at the workshop. The USSR described a new "toroidal divertor" concept [7], which has the advantage that it does not require the cancellation of the toroidal field, and forces are much smaller. The disadvantage is a rather complicated coil topology and the presence of some coils near the plasma. The coils would be more accessible than most poloidal designs and further evaluation is advisable.

Another new divertor concept, called "hybrid divertor" [6], was also presented at the Workshop by the USA delegation. This concept utilizes local radial divertor fields to draw out magnetic flux surfaces into a thin horizontal wedge that can pass through a narrow aperture into a divertor chamber. Since the hybrid divertor affects mainly the poloidal field, the coil currents and field ripple tend to be less than for a bundle divertor. Detailed engineering designs have not been carried out.

7.3. Trade-off considerations

For each of the divertor concepts, there are major trade-off considerations. These trade-offs involve factors such as shielding versus lifetime, resistive versus superconducting coils, and first cost versus maintainability. There is a strong opinion among engineers that any device of the INTOR time scale and mission should allow only resistive coils internal to the TF system. It is further the strong sentiment of the magnetics group that use of internal coils should be restricted to absolutely necessary functions. When external coils, despite increased power are suitable, they should be used. On this basis preference should be given to poloidal divertors with external coils, and any internal poloidal divertor system should utilize copper conductors. Furthermore, if an acceptable bundle divertor can be developed, its external nature will be a prime factor in its favour.

The amount of shielding can be traded off against lifetime and construction materials. Internal poloidal divertors restricted to copper coils can trade shielding against organic or inorganic insulation. They can clearly also trade lifetime versus shielding. It is the engineering sentiment that increased shielding, while resulting in increased first cost because of increased scale and power, may allow dependence on organic insulation, and will therefore greatly increase reliability and lifetime.

Bundle divertors and external poloidal divertors, being in principle external to the TF coil system, would not be required to utilize copper coils. Here trade-offs could be done, after firm engineering designs, on the basis of the degree of complexity and maintainability. Peak toroidal divertor fields are equivalent to peak toroidal fields, and thus in principle, the divertors could utilize the same superconductors; provided they can handle the current density, stress and heat load requirements. On the other hand, if copper coils can be built which can handle the cooling load, it may be preferable to invest in external power supplies in favour of simpler divertor coils. In addition the reduced shielding requirement of copper versus superconductors can be a factor. Copper coils for an INTOR mission require a minimum of 50 cm of shielding, while superconducting coils require even thicker shields. A copper bundle divertor might require 100 MW of power, however, versus 1 MW for a superconductor coil.

One additional trade-off is of great importance to the bundle divertor, namely divertor coil requirements versus the need to limit ripple in the plasma region. Where this is not a factor, the "blister" divertor, or a relatively simple single local coil, could serve as a divertor. Unfortunately, relatively little theory and few experimental results are available concerning the effect of a single ripple disturbance. Multicoil designs which minimize ripple in the plasma region suffer from increased complexity.

7.4. Machine interaction

It is clear that divertors interact in a major way with the physics and engineering of the balance of the device. Since it is prudent to postulate an INTOR scale device with some provision for a divertor, then preliminary design concepts should reserve space for such an option.

In the case of the poloidal divertor, with coils internal to the TF coils, there is a particularly strong interaction with the machine. Most previous poloidal divertor designs have placed the divertor coils inside the vacuum system in order to postulate a reasonable shape of vacuum tank. The blanket modules and shields would thus also be within the primary vacuum enclosure. Some of the more recent poloidal divertor concepts with coils external to the TF coils would significantly improve reactor maintenance and assembly.

Studies have pointed out the leakage of neutron flux to the TF coils through the poloidal divertor. Shaping of the channels to prevent leakage must be traded against effectiveness of the divertor.

The bundle divertor has relatively little interaction with vacuum and shield, but does interact strongly with the structure required for the TF coils which are near the divertor. Designs incorporating bundle divertors have generally required enlarged TF coils to accommodate the divertor.

7.5. Technology assessment

7.5.1. Current experimental divertors

1. Bundle divertor. Much more information is currently available from poloidal divertor experiments than from bundle divertor experiments. The only bundle divertor in operation is the DITE experiment at Culham. Also significant information will result from experience with the TEXTOR bundle divertor now being fabricated.

The DITE divertor has shown successful divertor action without any adverse effects on the plasma. Some difficulties arose in connection with the coils because of high current densities and difficult support problems. The divertor channel is 15 cm in diameter, pumping speeds of 5×10^4 l/s are provided, and ion trapping is achieved with a moving 300 cm^2 molybdenum target.

A bundle divertor is planned for ISX-B during FY 1981. Improved coil shape and available space make it possible for the ISX-B divertor to operate at approximately one-third of the current density of DITE.

These two divertor experiments are of rather small scale and thermal loads are very modest compared with an INTOR scale experiment.

2. *Poloidal divertor.* Several major poloidal divertor experiments are coming on line in the near term, namely PDX, ASDEX, and JT60. Smaller experiments like DIVA (Japan) and Finger Ring T-12 (USSR) have demonstrated proof of principle, in that diversion can be accomplished without adverse effect on plasma performance. PDX and ASDEX are generally similar in that both utilize a number of vacuum encased coils internal to a vacuum tank. Pumping speeds in the divertor sections of the order of 2×10^6 l/s are planned. The ASDEX experiment has 180° joints in the divertor coils, which are inside demountable vacuum enclosures within the main vacuum tank. The PDX has joints in the TF coils, allowing full 360° coils in welded enclosures, again within the main vacuum tank.

The scale of the current programmes for poloidal divertors is considered adequate to provide a physics base for INTOR. The major problem seems to be in the engineering area in that the requirement for internal vacuum encased coils is not compatible with remote maintenance requirements. Additional development of external poloidal divertor concepts is required.

7.5.2. Current INTOR-relevant studies

1. *Bundle divertor studies.* A 1978 study by Westinghouse in the USA of a Demonstration Tokamak Hybrid Reactor, DTHR, has included a bundle divertor. This machine, with major radius of 5.2 m and central field of 5.5 T is of the INTOR scale. The divertor coils have a minimum of 60 cm of shielding and are designed to use Nb_3Sn superconductors at a peak field of 10 T and average current density of 4×10^3 A/cm^2, not unreasonable parameters for Nb_3Sn conductors. Structural support of the windings is accomplished by embedding the conductor in grooved plates. The divertor also includes a pair of auxiliary coils above and below the main divertor pair, which help reduce the demands of the main coils and improve the characteristics of the separatrix. Rather simple expander coils are used to take the flux well outside the TF coils where the 280 MW of thermal power are reduced to an average of 0.30 kW/cm^2 and a peak local flux about 10 times higher. The design shows considerable promise for maintenance accessibility and possible replacement. Trade-off of a copper coil for the superconductor could be examined. A direct substitution of copper for the Nb_3Sn results in a power consumption of approximately 300 MW. Increased copper cross-section would, of course, reduce this value.

Oak Ridge National Laboratory studies show a bundle divertor in their 1978 baseline TNS design, a change from the 1977 baseline design which carried a compact poloidal divertor (see next Section). The change was made to move the divertor coils outside the TF coils to improve overall access and to provide more room for particle collection. Their divertor coils call for 0.4 metre shielding, and copper coils with inorganic insulation. Expander coils are postulated.

Princeton Plasma Physics Laboratory (PPPL) studied the use of a bundle divertor for use in their proposed copper coil ignition test experiment. This ignition machine, while smaller than an INTOR scale, must face the same activation problems, and the bundle divertor was chosen over the poloidal divertor with internal coils for maintenance reasons. The copper divertor coils were designed at 4.5×10^3 A/cm^2 and a total power consumption of 52 MW. This device was designed with 30 cm of shielding and assumes use of specially formulated organic insulators.

Culham Laboratory has done a 1978 study of a bundle divertor for their MK II Fusion Reactor concept. The reactor has a 7.4 m major radius and a 4.1 T central field. Their divertor coils have a peak field of 7 T and an overall current density of 1.5×10^3 A/cm^2. They have chosen a shield of 0.75 m and postulate copper coils with inorganic insulators, requiring a total power consumption of 92 MW. These values are within the state-of-the-art of continuous water-cooled coils, although there is essentially no experience with high stress inorganically insulated coils.

Exploratory studies of advanced fuel fusion reactors done at the University of Illinois in the USA have included bundle divertors in their designs. Trade-off studies have been done of copper, cryogenic and superconducting coils, with the conclusion that all are possible, but that cryogenic coils (aluminium conductor and liquid hydrogen cooling) are a particularly attractive trade-off of shielding needs, current density limits, and power consumption. It is our assessment that cryogenic coils will not prove as attractive when more detailed designs are undertaken.

2. *Poloidal divertors studies.* Poloidal divertors have been included in several reactor designs.

A 1976 Workshop study at PPPL, involving participants from PPPL and three USSR laboratories, conceptualized a poloidal divertor for an EPR scale device. The internal coils were 180° loops with no net current, and hence in principle demountable, and utilized superconducting coils. The diverted flux is bent and channelled horizontally between the TF-coils. These same basic concepts are also utilized in the UWMAK-III reactor design. It is postulated that the diverted flux will pass between the TF-coils due to the fringe field components of the TF field near the windings.

A poloidal divertor was designed as an option for the 1977 ORNL TNS baseline design. This Westinghouse designed divertor included space for 30 cm of shielding and was located inside a specially shaped vacuum chamber. Special coils were designed to compress the separatrix into an elongated channel to minimize height and avoid a necessary increase in the height of the TF coils. Solid metal plates backed by highly complex fins were designed to provide a surface area of 320 m^2. Several problems with this compact poloidal divertor were the extreme difficulty of reaching the divertor and target, of providing for pumping surface regeneration, and the impossibility of adding adequate shielding for an INTOR mission machine. Increasing the TF coil size would have provided some design flexibility, however.

The proposed Princeton SLPX device was designed with a single null poloidal divertor. The study is of particular interest in that it outlines tooling requirements for remote winding the divertor coils in place inside the activated vacuum chamber, and for adding welded enclosures and supports. Removal of a coil would require destruction of the coil and production of a duplicate coil. Design details for the 60 cm of shielding for the divertor for the SLPX III option were postulated, but not carried out. The SLPX divertor target design utilized swirl tube collectors at a peak heat flux of 2.9 kW/cm^2.

7.5.3. Relative feasibility of poloidal and bundle divertors for the INTOR mission

Given the strong motivation to limit the internal coils in the INTOR device, strong preference should be given to poloidal divertors with external coils and bundle divertors. Studies of shielded bundle divertors for the INTOR scale device suggest fields and current densities which are somewhat more ambitious than those in the throat of the TF coils. Conceptual designs suggest that it will be possible to remove and replace the bundle divertors without disturbing the vacuum chamber and TF coils provided the TF coils are built sufficiently oversized. Use of expander coils should allow the bundle divertor flux to be expanded to an arbitrarily low thermal flux to permit reasonable target design. The targets would be fully external to the TF system.

7.5.4. Materials requirements

Poloidal divertors with external coils and bundle divertors for INTOR must trade-off superconducting coils, with their increased support complication and increased shielding, with copper coils which require reduced shielding but increased power supply costs.

It is highly desirable to utilize sufficient shielding to allow radiation resistant organic insulators to be used. Nevertheless, development programmes should be carried out on materials and coil construction methods for inorganic insulations as well as for radiation resistant organic insulators.

7.6. Development and technology programmes

7.6.1. Expectation of current divertor programmes

Poloidal divertors are being explored on an adequate physics scale to study effects on stability and impurity control of modest energy content plasmas. Bundle divertors are not being explored on an adequate scale.

7.6.2. Design studies needed

Major design activities should be initiated to detail alternate divertor designs suitable for the INTOR mission. It is our assessment that the most acceptable divertor will be a bundle, new poloidal, or hybrid divertor not linking the TF coils. Such designs should therefore be pursued most vigorously with the aim of physics and engineering selfconsistency. These designs should pay particular attention to expansion of the flux to a sufficient degree (possibly taking advantage of induced instabilities in the channel), and to engineering considerations such as replaceability, maintenance and the trade-off of resistive versus superconducting coils.

7.6.3. Development programmes needed

The bundle divertor programmes lag behind the poloidal divertor programmes. Major bundle divertor designs should be initiated, and model coil programmes which can simulate current densities and asymmetric stress loads should be initiated. A magnetics test stand simulating a major divertor experiment could be undertaken, and could involve tests of superconducting coils and copper coils. Copper coils with organic and inorganic insulators should be tested.

It is quite possible that such a test stand could also investigate flux expansion, thermal loads, and pumping. Injection plasmas perhaps need not be from a tokamak, as basic questions of stability under divertor action have been, and are being addressed on machines such as DITE, ISX-B, and TEXTOR.

REFERENCES TO CHAPTER IX

[1] European Contributions to the 2nd meeting of the INTOR Workshop, Magnets, Chapter V, EUR FU/XII 501/79/EDV 50, June 79 (2 Volumes).
[2] European Contributions to the 3rd meeting of the INTOR Workshop, Design Options and Alternatives, EUR FU/XII/501/79/EDV 60, Oct. 79.
[3] European Contributions to the 3rd meeting of the INTOR Workshop, Homework and Special Supplementary Information, EUR FU/XII/501/79/EDV 70, Oct. 79.
[4] Fusion Research Development Center, INTOR Workshop Report, JAERI-M 8711 (1979).
[5] Fusion Research Development Center, Engineering Aspects of JAERI Proposal for INTOR (II), JAERI-M 8518 (1980).
[6] "Magnetics and Cryogenics", Chapter in USA INTOR Report (see Appendix A.5.4).
[7] USSR Report to INTOR, "INTOR Electromagnetic System" (see Appendix A.5.3 [16]).
[8] USSR Contribution to the 3rd meeting of the INTOR Workshop.

Chapter X

POWER SUPPLY AND TRANSFER

P. REYNOLDS – EC, Chairman
C. BAKER – USA, Expert
G. CHURAKOV – USSR, Expert
T. HIRAOKA – Japan
J. GILLELAND – USA

1. INTRODUCTION

This summary is based on group discussions at the second and third sessions of the INTOR Workshop; see reports listed in the references.

There was substantial agreement on the amount of power required for INTOR and on ways of supplying the required power. There was, however, some difference in emphasis. For example, the reports from Japan [3, 4] concentrated on the poloidal magnetic field supply for the JAERI conceptual INTOR designs and that from the United States [5] included details of neutral injection supplies, protection circuits, RF supplies and alternate methods. The reports from the USSR [6, 7] described a system using combined liquid nitrogen cooled coils and superconducting coils for the ohmic heating inductor with superconducting coils for control and also the protection of the superconducting toroidal field system under quench conditions. The European report [1] included a discussion of the characteristics of the European electricity supply network relating to INTOR.

A common feature of all the reports was the opinion that the power supplies of INTOR could be mostly designed on the basis of present technology. This was particularly the case for the motor generator flywheel sets for supplying the poloidal magnetic field coils and neutral beam injectors, as such machines are already under construction for the new large tokamaks. The neutral injection supplies pose some development problems because the voltage is higher than that at present in use and the power greater.

Because of the relatively few new problems with power supplies, this report is kept very brief. However, there is extensive material contained in the references and these reports should therefore be consulted for further details of the work.

A summary of the amount of power required for INTOR based on preliminary estimates is given at the end of this report.

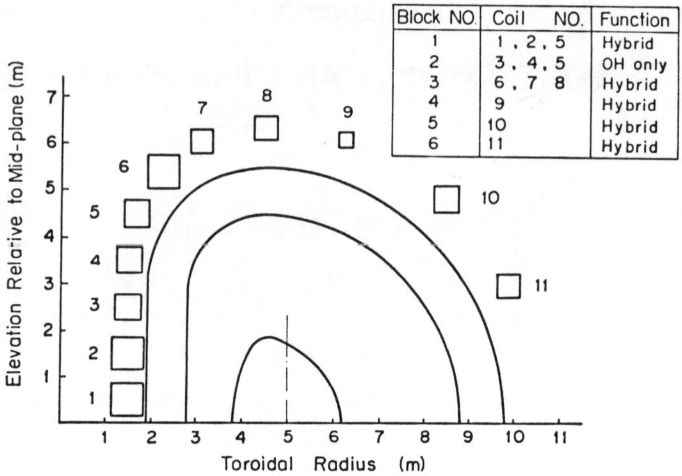

Block NO.	Coil NO.	Function
1	1 , 2 , 5	Hybrid
2	3 , 4 , 5	OH only
3	6 , 7 8	Hybrid
4	9	Hybrid
5	10	Hybrid
6	11	Hybrid

FIG.X-1. INTOR hybrid poloidal coil locations.

2. THE POLOIDAL MAGNETIC FIELD SYSTEM

Systems consisting entirely of superconducting coils and those with a mixture of superconducting and resistive coils have been considered for INTOR. The most detailed study is that made by the Japanese team [3, 4]. As an example, therefore, this system will be described below in more detail than the others and will be used in the estimation of the INTOR power requirements.

2.1. All superconducting poloidal magnetic field coils outside the TF coils

2.1.1. Coil arrangement

Fig.X-1 shows the arrangement of poloidal magnetic field coils for the JAERI conceptual design of INTOR. The coils are all superconducting and are placed outside the toroidal magnetic field coils, which is consistent with the current guiding parameters for INTOR. The plasma loop voltage is limited to 100 V [3] and 50 V [4] to limit the rate of change of magnetic field on the superconducting inductor. The assumed plasma current waveform is shown in Fig.X-2 The coil system is divided into six blocks with functions as given in the table of Fig.X-1 . Poloidal divertor action can be obtained at the top and bottom of the plasma by suitable adjustment of the coil currents and, by swinging the current in coil number 9 at a frequency of 1 Hz, the diverted plasma can be swung across the collector plate to distribute the heat load.

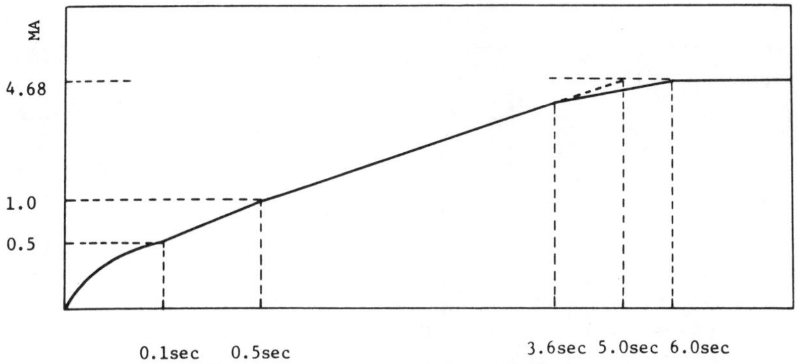

FIG.X-2. Plasma current in the build-up phase.

2.1.2. System description of power supply [*3, 4*]

The system configuration of the poloidal field power supply is shown in Fig.X-3. All the electric power and energy required for operation of the poloidal field system are supplied from two sets of motor generators and the utility power network. The utility power network supplies the average electric power needed for operation of the entire poloidal field system and the motor generators have flywheels which store energy taken from the power network to supply the peak electric power during the plasma current build-up stage. The collapsing magnetic field energy at the end of each tokamak pulse is recovered and used to accelerate the motor generators, except in the case of a pulse terminating in a plasma disruption.

As shown in Fig.X-3, each block of coils is subdivided into two identical sub-blocks (for example B1 and B1'), since the arrangement of the coils is symmetric with respect to the median plane and the coil of each sub-divided block has its own DC power supply (thyristor convertor) E_1 and E_1 respectively as described above (except blocks B2 and B2').

The thyristor convertors E_{OH} and E_{OH}' and the DC vacuum circuit breakers are used to build up the plasma current. The DC circuit breakers are essential components to make a fast build-up of the plasma current in the first 0.1s or so. The second slow rise up to prescribed flat top values is made during a few seconds by using the thyristor convertors E_{OH} and E_{OH}'.

The thyristor convertors E_1, E_3 -E_6 (and also E_1', E_3' -E_6') are used to adjust and control the currents in the poloidal field coils to produce a required magnetic field configuration to give the desired plasma position and shape and to sweep the plasma across the divertor target plate.

Details of the operation sequence are given in [3] and [4].

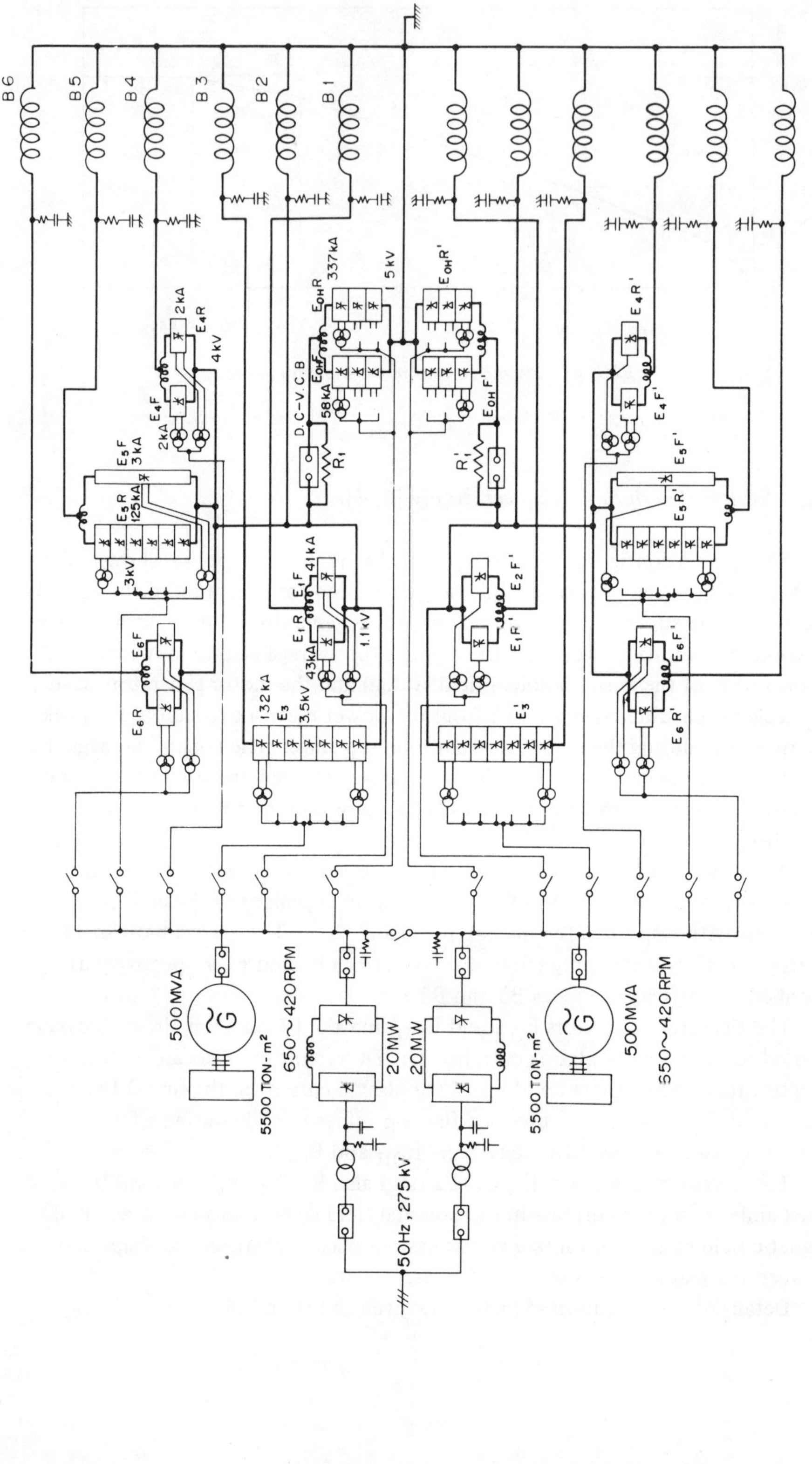

FIG.X-3. Schematic diagram of poloidal field power supply system.

2.2. Poloidal magnetic field coils which are not all superconducting

Although superconducting coils are, at present, the preferred choice for INTOR, the power requirements for poloidal coils which are not superconducting are within the capabilities of the motor generator and supply network and they were therefore considered as an alternative to all superconducting coils. The rating of the generators is little affected during the current rise since most of the power is reactive.

2.2.1. All water-cooled copper coils [1]

Extrapolation from JET gives a value of power of ~200 MW in the flat-top phase for water cooled PF coils positioned outside the TF coils.

2.2.2. A combination of superconducting coils and water-cooled copper coils [5]

The superconducting coils are placed outside the toroidal magnetic field coils and take care of the slowly rising field; the copper coils are placed inside the toroidal coils and are used to control the plasma position and shape. Being nearer to the plasma, they can do this more effectively and as they are water-cooled copper coils, it is conceivable that demountable current joints suitable for remote handling could be made in them to allow removal of the coil for maintenance or repair. Such joints will, however, need development.

2.2.3. Ohmic heating induction coil: part superconducting and part liquid nitrogen cooled

It is hoped that it will be possible to initiate plasma current in INTOR with an applied voltage of as low as 50 V/turn, which will be within the capabilities of superconducting inductors. Should much higher voltages be required, the rate of change of magnetic field at the inductor may be too high for superconducting material in the absence of further successful development. It was proposed therefore [5], but not considered in detail, to make the induction coil in two parts: one part made of aluminium cooled to a temperature between 20–70 K for the fast change of field and a superconducting part for the slow change.

2.2.4. A separate plasma initiation coil

As an alternative to the system described in Section 2.2.3., the high electric field required to initiate the plasma current could be provided by a water-cooled copper coil placed as near to the plasma as possible to give good coupling to the plasma but poor coupling to the other coils in the poloidal magnetic field system. High rates of change of field on these coils would thus be avoided, but at the expense of the complication of placing a coil within the toroidal field system and perhaps within the blanket/shield region. It should be noted that this arrangement is not the one normally used in present tokamak experiments.

3. DIVERTOR

3.1. Poloidal divertor

The poloidal coil system shown in Fig.X-1 is such that poloidal divertor action should be obtainable by a suitable distribution of current in the coils without the necessity of employing coils inside the toroidal field coils [3, 4]. An increase of reactive power of approximately 60% would then be required during the current rise, with a mean power increase of ~30% both for steady and swept operation of the poloidal divertor [4]. The increase in mean power is much less than for the water-cooled poloidal or bundle divertors.

For a water-cooled poloidal divertor inside the vaccum vessel, an estimated power requirement of approximately 40 MW during the flat-top phase is obtained which could be taken from the supply network.

3.2. Bundle divertor

The power required will depend critically on the distance of the divertor from the plasma dictated by blanket and shield requirements. It seems unlikely that a superconducting system will be possible and a normal water-cooled copper system will need to be compact due to the small space available for conductor and support, and it will, therefore, consume a large power estimated at about 100 MW in the flat-top phase.

4. TOROIDAL MAGNETIC FIELD COIL SUPPLY [5]

The supply of the toroidal magnetic field coils is expected to be a straightforward extension of present systems for supplying large superconducting magnets.

During the current rise in the coil (10 hours), a 5 MW DC power supply (thyristor convertor) is required with automatic current control accurate to within ~1%.

A superconducting magnet needs a rapid energy release in the case of the appearance of a normal phase. Provision must therefore be made for a highly reliable protective system consisting of special dump resistors shunted by high-voltage DC breakers. A possible system is given in [5].

5. THE NEUTRAL BEAM INJECTION SYSTEM [4, 5]

5.1. Power system

The supply of power to neutral beam ion sources is an extension beyond existing equipment. In general, there will be required an AC to DC conversion unit, some type of switching scheme to transfer the power from the DC power supply to the ion source and equipment protection systems.

AC to DC conversion units which can deliver up to 200 kV at pulse widths of 5 s and currents of 100 A are within the state-of-the-art. Present ion source design requires a direct relationship between the arc current of an ion source and the accelerating grid voltage. Active regulators can be designed to correlate the accelerating voltage with the arc current.

The beam accelerating voltage of 175 kV proposed for INTOR is somewhat higher than that planned for the TFTR (120 kV) and for JET (160 kV). Switch tubes in the DC side of the power system presently have a voltage hold-off capability in the region of 120 kV. This parameter would have to be extended in additional development programmes; however, it is felt that this would be an extension, rather than the creation of the state-of-the-art in vacuum tube technology. There is also the possibility of connecting switch tubes in series to increase the hold-off voltage, and solid state switches may eventually be a viable alternative.

If positive ion sources are used, the neutralization efficiency will decrease at higher accelerating voltages. 120 kilovolt sources have a power conversion efficiency of approximately 30% (full and partial energy neutrals). If energy recovery schemes are neglected, the overall power conversion efficiency (neutrals/power input) will be of the order of 20%. Therefore, total pulsed power input required for 75 MW of delivered beam power will be of the order of 375 MW. Approximately 10 to 15% additional auxiliary power will be required for filament, arc, and other miscellaneous supporting electronics.

Due to the nature of the pulsed loading of the neutral beam system, as well as the other power systems of INTOR, it may be well to consider the use of a dedicated motor-generated system just for neutral beam injectors. This approach would reduce interpower system reaction and might provide a higher overall tokamak system reliability. Assuming that the neutral injectors are pulsed for only 10 s at 400 MW total power, a driving motor on the flywheel of a power of about 25 MW would be required.

5.2. Neutral beam injector protection system

The neutral beam ion sources must be protected from energy stored in transmission lines, as well as follow-on current from the power system in the event of a source fault. In present systems, this is accomplished through the use of current-limiting, energy absorbing reactors, as well as fast opening switches (vacuum tubes) and shunting crowbars (ignitrons or SCR's). Sources presently operate at 120 kV and appear to suffer damage when more than 5 J of stored energy are dumped into a fault. This poses an even more severe problem when the voltage is increased to 175 kV. A device must be developed to absorb this stored energy, as it will probably be physically impossible to construct transmission lines which will store less than 5 J at this voltage (including supporting equipment which is tied onto these lines).

6. RADIO-FREQUENCY HEATING POWER SUPPLIES

Consideration was given to power supplies for Electron Cyclotron Resonance Heating (ECRH), Ion Cyclotron Resonance Heating (ICRH) and Lower Hybrid Heating (LHH).

Power output requirements for RF would range from 25 MW to possibly as high as 100 MW, dependant upon losses and plasma coupling efficiencies. Lower hybrid and electron cyclotron heating would be accomplished through the use of either existing electronic tubes or upgraded versions of existing microwave tubes.

Lower hybrid heating would lie in the 1–2 GHz region and would require negative power supply voltages of the order of 75–125 kV. This is based on the assumption that power modules would be created in the 1–5 MW class. Prime power for these systems would require similar power generating sources as neutral beam power systems require AC switchgear, fast interrupters, voltage controllers, AC to DC power conversion units, capacitor banks, crowbars, as well as some type of modulator or switch to turn the RF source on or off. Lower hybrid efficiencies (of power out/input power) would probably be of the order of 50%, provided extensive transmission line runs were not required. Coupling structures would have to be developed to efficiently transfer the energy from the wave guides to the plasma.

Electron-cyclotron RF equipment would operate at lesser efficiencies due to inherent losses within existing or proposed developmental RF tubes. In addition, transmission problems would be much more severe at frequencies of 50–100 times that of the lower hybrid frequency. Because of this, ECRH modules would probably be similar to LH modules, but possibly of the order of hundreds of kilowatts, rather than megawatts.

ICRH heating would probably use super power tetrodes or triodes to generate RF in the area of 50–200 MHz. These devices require positive voltages of the order of 30–60 kV at an appropriate current corresponding to their power output. Efficiencies for an ICRH system would probably lie in the range of 75% (RF output/input power).Transmission problems become somewhat more difficult due to voltage breakdown problems which would exist in coaxial transmission lines at the MW power level. Transmission losses would be minimal compared to the other two RF systems previously discussed. The primary power system would again be similar to that used in ECRH and LHH RF systems. Coupling structures would have to be developed to transfer the energy efficiently from the transmission lines to the plasma.

Much of the power system constructed for neutral beam heating (60–70%) could be used for RF heating. Some loss in efficiencies would be encountered in both operation and cost due to the different voltages involved in all of the systems. This decision should be based on a cost study which should take place during the initial conceptual design stage of this project.

7. POWER FOR COOLING

7.1. Machine cooling

The thermal output of INTOR of ~600 MW and the thermal load arising from coil losses must be removed by several coolant systems.

It is estimated that a pumping power of 25 MW will be required for the whole system.

7.2. Power for cryogenic refrigeration

A considerable amount of power will be required for cryogenic refrigeration equipment for vacuum pumping (mainly for the neutral beam injectors) and for the superconducting coils. Estimates of the total required vary over the rather wide range of 30–85 MW depending upon the coil design adopted, the AC losses in the coils and the type of coolant (bath or forced cooling).

7.3. Neutral beam loss cooling power

Based on the assumption of 20% neutral beam efficiency (neglecting energy recovery) more than 300 MW of dumped beam power will have to be removed. Thermal inertial concepts may not prove viable, resulting in considerable water flows and associated pump power in the region of 10 MW.

8. ELECTRIC POWER GENERATION SYSTEM

A demonstration of electric power is a necessary step in the development of practical fusion power. The most significant areas are related to the development of a recovery blanket and the conversion of this to steam. A 2–10 MW_e steam turbine-driven electrical system could be made to deliver electrical power back to the utility system.

Cooling will be by a pressurized water-cooling system operating at approximately 300°C. An electric motor-driven coolant pump will circulate the water through the reactor and the steam generator. Energy storage will be incorporated into the design of the steam generator via the mass of the system and the latent heat of vaporization of the water.

The steam turbine would be a conventional industrial turbine operating with a condenser. The turbine will be geared down to the proper generator speed.

The generator will be a standard salient pole machine with a rotating brushless exciter, designed for top air discharge to a cooling duct leading outdoors.

The control of the steam system must be an integrated control, sensing temperatures and liquid level in the steam generator, power from the electrical generator and making appropriate adjustments in the speeds of the coolant pump, the feedwater pump, and the turbine steam admission valves.

It is not proposed to govern the generator speed as a "stand-alone" power system where generator frequency is regulated constant. Instead, the generator frequency will remain constant because it is synchronized to a large electric utilitysystem. Power output from the generator to the power system will be controlled by the steam throttle valve.

9. ALTERNATIVE CONCEPTS

9.1. Superconducting energy storage system [4]

An alternative energy storage and transfer system for the ohmic heating coil, equilibrium field coil, divertor coil system and auxiliary heating power supplies can be based on the use of a superconducting energy storage coil and thyristor converters. This system was not selected as the primary design option because development work is still required on the energy storage coil and thyristor networks and, because of the lack of any sizeable demonstration units, cost information is too inaccurate to determine whether any apparent capital cost advantages are likely to be real or not.

9.2. High-frequency solid-state neutral beam power supplies [5]

Conventional power systems for Neutral Beam Sources (NBS) use rectifier transformers to produce HV DC. The HV is applied to the NBS via a switch

TABLE X-1. R AND D NEEDS ASSESSMENT – POWER SUPPLY AND TRANSFER RESEARCH AND DEVELOPMENT

	Status*	
	Adequate	Inadequate
– Energy storage		
– flywheel generators	x	
– superconducting storage systems	x	
– Switches		
– HV switch tubes	x	
– vacuum circuit breakers	x	
– SCR switches	x	
– superconducting switches	x	
– Neutral beam injectors		
– transmission lines	x	
– ion source protection	x	
– Radio-frequency system		
– tubes	x[1]	
– replaceable antennae	x	

[1] Only the ECRH tube needs development

Adequate: Programmes are either planned or could be planned on existing or authorized facilities (with some upgrading if required) and using existing trained personnel.
Inadequate: Either the available and authorized manpower and facilities are inadequate or a substantial increase in effort is needed.

*The ratings are made on the world-wide fusion effort and not separately on the individual national programmes.

tube, which serves both as an on-off switch and a voltage regulator. Presently, switch tubes limit neutral beam sources to $\leqslant$120 kV, and dissipate <2.1 MW while regulating. The lifetime of a switch tube is thousands of hours. At the INTOR energy level, switch tubes may be subject to damage from sparking. The development of a solid state HV power system may give a more reliable, more compact and less expensive system than the conventional one.

10. R AND D NEEDS ASSESSMENT

A summary of the research and development needs assessment for power supply and transfer is shown in Table X-1.

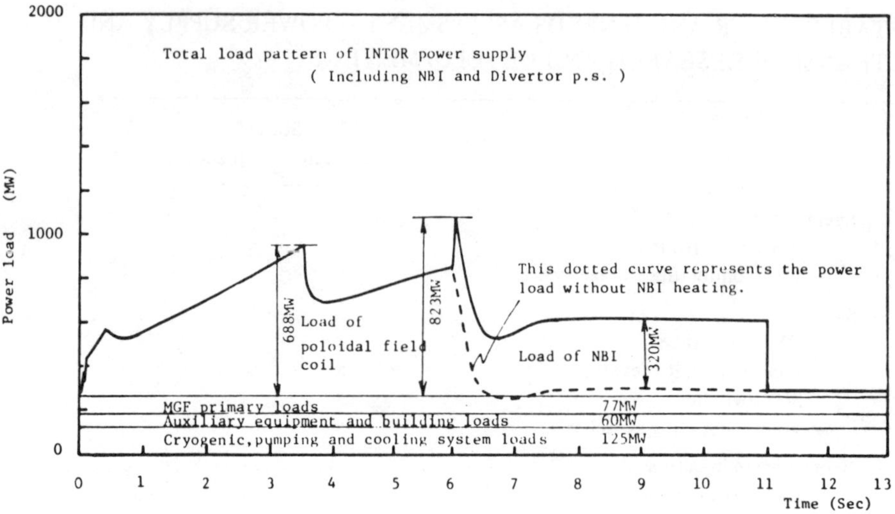

FIG.X-4. Total load pattern of INTOR.

11. SUMMARY OF POWER REQUIREMENTS AND CONCLUSIONS

Table X-2 lists the power required for the various systems in INTOR. It must be stressed that these values are only preliminary estimates: more accurate figures can be given only after much more detailed work on a specific design.

Noteworthy features are the great power savings obtainable by the use of superconducting poloidal magnetic field coils in place of copper water-cooled coils and by the use of a superconducting poloidal divertor employing coils external to the toroidal magnetic field coils in place of a water-cooled internal poloidal divertor or a bundle divertor.

Total power requirements for three systems are given at the end of the table. These are for the lowest estimates of power for the cryogenic systems and for the miscellaneous systems: if the higher estimates of power for both these systems were used, the totals would need to be increased by 90 MW. Block diagrams for the power supply system are given. Fig.X-4 shows the power load as a function of time [2, 4].

From this preliminary review of ways of providing the power for INTOR it is concluded that a level of technology only slightly more advanced than that being used for the new large tokamaks is required. Further work is required in the following fields:

(i) More detailed design of the main power supplies when definite requirements are supplied by the INTOR team;

TABLE X-2. POWER REQUIREMENTS FOR INTOR

Assuming:

Burn time = 200 s

Dwell + start-up + shut-down time = 40 s

		Peak	*Mean* (MW)
(i)	Poloidal Magnetic Field (PF coils outside TF coils)	*Peak*	*Mean* (MW)
	(a) all superconducting coils	440 MVA	*26*
	(b) all copper water-cooled coils	440 MVA	200
(ii)	Feedback system for plasma control, of the order of	100 MVA	*10*
(iii)	Divertor		
	(a) poloidal using the superconducting coils outside TF coils of (1)a and sweeping	250 MVA	*9*
	(b) Water cooled copper poloidal inside vacuum vessel		40
	(c) Bundle with water cooled coils.		100
(iv)	Toroidal magnetic field coil		
	(a) charging		5
	(b) operation		*1*
(v)	Neutral beam (10s pulse, 20% efficiency, no direct recovery	400 MW	*25*
(vi)	Pumped cooling of NB system		*10*
(vii)	RF heating. Efficiency not sufficiently known		
(viii)	Cryogenic systems*		*30*
(ix)	Pumping power for water cooling of shield.		*5*
(x)	Pumping power for helium cooling of blanket.		*20*
(xi)	Vacuum pumps.		*5*
(xii)	Miscellaneous** systems (control systems, diagnostics, buildings)		*25*
	TOTAL		
	(a) All SC coils, swept external poloidal divertor		170
	(b) SC Coils, water-cooled poloidal divertor		200
	(c) SC coils, water-cooled bundle divertor.		260

* Lowest estimate: highest estimate is 85 MW

** Lowest estimate: highest estimate is 60 MW

(ii) Development of design for large feedback-control circuits for stabilizing the plasma when coil positions and magnetic fields have been specified;
(iii) Development of design for higher voltage neutral beam power supplies and protection circuits;
(iv) Development of high power RF systems particularly for ECRH;
(v) Development of methods of plasma initiation with low loop voltage;
(vi) Study of alternate designs for power supplies and energy storage as possible replacements for the present well established methods.

REFERENCES TO CHAPTER X

[1] European Contributions to the 2nd meeting of the INTOR Workshop, Power Supply and Transfer, Chapter VIII, EUR FU/XII 501/79/EDV, 50 June 79 (2 Volumes).
[2] European Contributions to the 3rd meeting of the INTOR Workshop, Power Supply and Transfer, EUR FU/XII 501/79/EDV 70, Oct. 79.
[3] Fusion Research Development Center, INTOR Workshop Report, Group 8, Power Supply and Transfer, JAERI-M 8511 (1979).
[4] "Power Supply and Transfer", Chapter in USA INTOR Report (see Appendix A.5.4).
[5] USSR Report to INTOR, "INTOR Power Supply" (see Appendix A.5.3 [24]).

Chapter XI

FIRST WALL, BLANKET AND SHIELD

R. HANCOX – EC, Chairman
G. CASINI – EC, Expert
J. DARVAS – EC, Expert
B. KOLBASOV – USSR
G. KULCINSKI – USA
K. SAKO – Japan
G. SHATALOV – USSR

1. INTRODUCTION

The design of the first wall, blanket, and shield is central to the design of any fusion reactor. In INTOR, these items are of particular importance since they are closely coupled to the objectives of the reactor. Thus the report of this Group includes a discussion of alternative objectives for INTOR as well as descriptions of various design alternatives and a data base assessment. The final choice of the objectives will depend on the assessment of whether or not it is judged possible to achieve them in a machine of acceptable cost and reliability.

In conformity with the decisions of the INTOR Workshop, a Reference Design has been proposed which is the minimum step to fulfil the requirements of a next generation tokamak. This Reference Design is consistent with the suggested Parameters adopted by the Workshop, and is to be understood in the same way – a design which can be used as the basis of further study, and which is currently thought to represent the most likely form to be chosen for INTOR, but which does not imply any final decision. It is the basis for further discussion. It is also a norm from which to measure the advantages and disadvantages of alternative proposals, and three such alternative designs are presented in this report which represent larger steps toward the eventual development of commercial fusion reactors. Consideration of these alternatives is consistent with the wish of the International Fusion Research Council to know what is "the maximum reasonable step beyond the present generation of experiments ... ".

As in all the groups of the Workshop, the work of the First Wall/Blanket/Shield group interacts strongly with that of others. In particular, the construction of the first wall, blanket and shield must be consistent with the remote maintenance procedures described in Chapter XV, with the choice of the location of the vacuum boundary proposed in Chapter XVII, and with the possible form of divertor suggested in Chapter V. The mechanical design of the primary shielding is included in the work of this group since it interacts strongly with the design of the blanket; the neutronic design of the primary shielding and of shielding around penetrations is dealt with in Chapter XVI. The discussion of most materials problems, especially bulk radiation damage, has been dealt with in Chapter XIII.

2. SUMMARY

The main conclusions from the data base assessment were:

1. In general, a wide range of design options has been explored in past fusion reactor studies and wide experience relative to this area is available from fission reactor design and construction. An adequate data base exists for many components of the Reference Design and the only area which could seriously affect the decision to construct INTOR in the mid-1980's is the design of components which interact directly with the plasma, such as the limiter, first wall or liner, or divertor plate.

2. The data base with regard to structural, thermal hydraulic and vacuum vessel design is considered adequate at this time. Codes exist and experiments verify that first walls can be designed and built to withstand the normal operation heat flux in INTOR. However, the situation is not so promising for limiters or divertor collector plates, and further development work is required. It is not clear that adequate design solutions exist to handle the heat in all cases.

3. The data base required to design the first wall is acceptable, but the exact form of the first wall (i.e. water- or helium-cooled) must await a more detailed design. The data base for coatings to reduce tritium permeation needs to be expanded in this regard. It appears that it will be necessary to protect the first wall from fast plasma disruptions, but, if liners are used, the data base on which to choose liner materials is rather sparse. Perhaps the most critical feature is the lack of information on how many disruptions will occur during the lifetime of INTOR or the magnitude of the power density deposited during a disruption.

4. Adequate codes exist for analyzing the expected life of the first wall and blanket, but irradiated materials data is inadequate, particularly for materials other than steels containing nickel. The results of the analysis will be very design-dependent.

5. The composition and necessary thickness of the shield can be decided on the basis of known data. The form of the structure is not yet clear, but no serious design problems are foreseen.

6. An area of uncertainty in the data base for the first wall, blanket and shield design is the implication of high voltages at start-up and during a disruption. The introduction of conductors in these regions for equilibrium or stability control would create design problems. A better appreciation of this problem will probably only come from further detailed design studies.

3. INTOR GUIDING PARAMETERS

For completeness, the suggested INTOR Parameters relevant to the first wall, blanket, and shield are given below.

Vessel half-width	a_W	= 1.4 m
Vessel half-height	b_W	= 2.2 m
Major radius	R	= 5.2 m
Plasma elongation	b/a	= 1.6 m
Plasma aspect ratio	R/a	= 4.0
Fusion power during burn		= 620 MW
Neutron wall loading during burn, Pn		= 1.3 MW/m^2 (1.6)*
Burn time	τ_b	$\geqslant$ 100 s
Duty cycle	τ_b/τ_c	= 70%
Maximum availability (final phase)		= 50%
Number of pulses in lifetime		= 10^6
Blanket and shield thicknesses, inboard		= 0.85 m
Distance between first wall and toroidal field coil conductor, inboard		= 1.1 m
Number of toroidal field coils		= 12
Toroidal loop resistance		$\geqslant$ 0.2 mΩ
Poloidal field response time		= 10 ms
Plasma breakdown voltage		= 100 V
First wall heat loading during burn		= 0.33 MW/m^2 (0.40)*
Particle flux to wall, if no divertor		= 10^{23}/s
Particle flux to beam dumps		= 2 × 10^{20}/s at 200 keV for 5 s if 90% plasma trapping

* Alternative value for comparison.

4. DESIGN CHOICES AND CONSTRAINTS

In designing the first wall, blanket, and shield of INTOR, there are several areas in which free choices may be made, which mainly relate to the objectives of the device. Other design options only allow limited choice because of interaction with other parts of the machine, and in some cases design constraints are imposed from elsewhere. The most important of these choices are considered here.

4.1. Free choices

It is generally accepted that the main function of INTOR will be as an engineering test facility, and that, as such, major functions will be to investigate the essential technologies required for the generation of electricity and breeding of tritium. The extent to which these technologies are investigated and demonstrate is a matter for discussion. This discussion also includes the possibility of achieving these objectives in a staged manner, which implies that the device is upgraded or significantly modified during its life to achieve some objectives at a later date. The possibility of staged operation will depend, however, on suitable provision having been made during the initial design to allow for later modifications.

4.1.1. Generation of electricity

A variety of views have been expressed as to the extent to which the fusion power released in INTOR should be used to generate electricity. A first reaction might be that most of the heat should be so used, producing up to 150 MW of electricity, since the technology of electricity generation is relatively well understood and the power would otherwise go to waste. Against this view, there is the argument that the generation of electricity imposes constraints upon the flexible experimental operation of the device, so that the generation of electricity should only be considered for a limited period. The provision of extensive capital equipment such as steam generators, turbines, and alternators to deal with the full output of INTOR for a short period would not be sensible, so that in this case only a fraction of the heat would be used for this objective. It would then be argued that experience in fission reactors will allow the system to be scaled up without difficulty.

A quite different view is that the development of fusion is so urgent that the largest possible step towards a demonstration reactor should be made, and that this would be the generation of net electricity from INTOR. Such a choice would place constraints on many components beyond the first wall and blanket requiring, for example, improvements in the efficiency of auxiliary plasma heating and of the energy transfer to and from the poloidal field windings, so that the electricity generated would exceed that required to maintain the system. This choice might also influence materials choices in the blanket if higher coolant temperatures were required, or lead to a larger overall device. Such an objective would be a greater challenge and would reveal more clearly the areas in which further major development is required before useful fusion power can be achieved, but is not thought to be achievable at the present time.

4.1.2. Tritium breeding

Similar arguments have been made to those concerning the generation of electricity. In the case of tritium breeding, however, the required technologies are not well established, so that there are good reasons for starting with small-scale investigations even if it is considered desirable to provide a more extensive demonstration later. Another major difference is that tritium which is burnt in the plasma but not replaced in a breeding blanket must be provided from elsewhere, at a cost of around 3500 to 7500 $ per gram. If only a negligible proportion of the tritium were replaced by breeding it is estimated that the total cost of supplying tritium over the life of INTOR might range from 300 to 500 M$.

A major reason for not breeding significant quantities of tritium is that the required breeding blanket complicates the machine design, and might affect machine reliability. If a breeding ratio of unity were required, it might be necessary to use a blanket on the inboard side of the torus, increasing the thickness of the blanket and shield and probably the overall size of the machine as a whole.

However, if lithium-lead alloys were used it would be possible to achieve a breeding ratio of unity with 80% of the outboard first wall area. Since the required tritium technology is not yet developed for such alloys, and much more development is required before the best breeding material can be chosen, the decision to obtain significant tritium breeding might delay the design of INTOR. This problem might be overcome by staged operation, although the installation of a breeding blanket in the active machine at a later stage would be a major undertaking.

The objective of a tritium breeding ratio of unity would be, like net electricity generation, a major challenge. Reasons for adopting such an objective would be that it is an essential step in the development of fusion, or that it is not clear that a smaller demonstration will uncover all the problems or yield a solution which will scale directly to a larger reactor. Since a major claim for fusion reactors is that they are safe and environmentally preferable to fission reactors, it might be advantageous to demonstrate the full-scale operation of a tritium breeding cycle as soon as possible.

4.1.3. Staged operation

The cheapest and most reliable construction for INTOR would be the one seeking to achieve the minimum objectives. More ambitious objectives might delay construction of the machine, and the desire to fulfil several different objectives might introduce incompatible design requirements. One way of overcoming such difficulties might be to introduce staged operation, with an initially simpler device which can be modified later. As has already been mentioned, this could be of interest if initial studies of different tritium breeding systems were required before installing a full breeding blanket. The same approach could be used in other areas of uncertainty, for example the installation of a higher temperature blanket for more efficient electricity generation after suitable materials testing. More generally, it could be claimed that a machine designed initially with the capability for staged operation would be more flexible and have a greater development potential. However, the cost would be higher, and between 1 and 2 years might be lost during major changes.

4.2. Limited choices

Many of the design decisions relating to the first wall, blanket, and shield interact strongly with other parts of the reactor, and hence with the work of other groups. Some of the main alternatives are discussed briefly.

The extent of the fixed structure will have a major impact on the design of the blanket and shield, and also on the possibilities for remote maintenance. A blanket and shield which formed a permanent structure would be the cheapest to provide and would be acceptable if known to be very reliable. If INTOR consisted entirely of a well- cooled shield, inboard and outboard, with all engineering development in removable test modules, this approach could be considered. However, the high power loading of the first wall with the possibility of a very uneven distribution of energy deposition during plasma disruptions throws doubt on the ability of such a structure to have a sufficient life expectancy, unless effective protective devices are developed. The difficulty of remotely repairing any damage suggests that a segmented approach would be preferable, in which the structure is sub-divided into segments which may be withdrawn radially from the torus. In a partially segmented system only the first wall and blanket need be withdrawn leaving the toroidal field coils in place; some shield could be withdrawn with the blanket or remain in place as a permanent structure. In a fully segmented system a complete unit composed of first wall, blanket, shield, and toroidal field coil would be removable as a unit.

Interacting with the form of segmentation is the question of the location of the vacuum boundary. The use of the first wall as the vacuum boundary is possible, but difficult because it is already one of the most highly stressed components and is least accessible. The use of a vacuum building is attractive in that it removes the problem from the region of the blanket and shield, overcoming the need to provide good vacuum joints in a segmented structure. The complications introduced elsewhere, however, are so extensive that this solution is not recommended. Thus, the vacuum boundary is generally taken to be in the shield region, preferably with the vacuum joints on the outside where they are most accessible.

If the first wall is not the vacuum boundary but part of a segmented system, it should not be continuous around the major circumference of the torus. This is in contrast to many present generation experiments. In this situation the first wall may be a separate component, or simply the front part of the blanket or cold shield. Several alternative designs are possible, depending on whether the majority of the alpha-particle power is expected to go to the wall or to other components such as limiters or divertor dumps, and on whether it is considered necessary to provide coatings of low-Z materials. If the power loading on the first wall is expected to be high, it may be constructed with a thermally isolated but unstressed liner to smooth out thermal cycling and reduce fatigue damage. Alternatively, it may be possible to reduce thermal cycling by coolant flow rate control.

The design of the first wall, blanket, and shield all introduce extensive choices of materials. In the case of blankets for electricity production and tritium breeding, it will certainly be necessary to test different combinations of structural, breeding, and coolant materials. For the general structure it is generally accepted that stainless steel will be used, although the choice of several materials such as aluminium with lower levels of long-lived induced activity appears attractive. In the shield, the choice of conventional shielding materials is accepted, although in some cases the use of special materials to reduce the shielding thickness even with considerably increased costs may be desirable.

4.3. Design constraints

For completeness, several design options are listed here which will be dictated by considerations outside the region of the first wall, blanket and shield.

Firstly, the mean power loading of the first wall, corresponding to a neutron loading of 1.3 MW/m^2, has been fixed in the Suggested Parameters. This level was chosen to ensure that the neutron loading and energy fluxes were reactor-relevant. Whilst this neutron loading and the maximum machine availability are too low to allow the lifetime neutron dose in INTOR to approach the levels expected in a power reactor, it will be possible to undertake some limited materials testing.

The possibility of maintaining the plasma purity and of exhausting helium during the burn without a divertor is doubtful. On the other hand, there is insufficient understanding of the physics or engineering implications of either a poloidal or a bundle divertor to make a definite choice between them. Although in neither case does the inclusion of a divertor strongly affect the first wall and blanket design, nevertheless it is important in considering the various design possibilities in the following Sections to take into account the possible implications of having to incorporate a divertor in INTOR.

The implications of different forms of auxiliary plasma heating must also be taken into consideration. At the present time it is thought that neutral beam injection will be used, requiring 4 to 6 injection ports of 0.5 to 1.0 m^2 area each. If radio frequency heating methods are sufficiently developed, they may have several advantages including the elimination of the straight injection ports which allow the streaming of 14 MeV neutrons through the shield. On the other hand, a variety of wave-guides or antennae will be required.

Finally, the design of the first wall, blanket, and shield will all be constrained by the need to admit pulsed magnetic fields to the plasma with time constants of 1 s and low distortion due to induced currents. This may require sub-division of the blanket and shield and may introduce the need for electrical insulators. The electrical resistance around the torus should be greater than 0.2 mΩ to limit induced currents.

4.4. Design examples

The above considerations have led to particular design examples being considered by three of the four delegations participating in the INTOR Workshop. Such design examples are valuable in illustrating the variety of choices possible, and form a necessary step in the detailed study of the possibilities and implications of each design choice. They are in some cases derived from several previous studies of TNS or EPR devices or built on power reactor studies. Further details are given in the reports prepared by the delegations, and only a brief summary is given here.

The design example studied by Euratom consisted of a partially segmented structure with the vacuum boundary at the back of the shield. Staged operation was proposed, with an initial medium temperature blanket to generate about 50 MW of electricity. Alternative blankets with water or helium cooling were considered. Later, alternative segments would be tested which bred tritium, and finally a complete set of tritium breeding segments, with or without electricity generation, would be installed.

A systematic design was studied in Japan, with a fully segmented structure divided into six sections with two toroidal field coils on each. A cellular breeding blanket covered all but the inboard surface, using lithium oxide, stainless steel, and helium coolant, to give a breeding ratio of 0.6. Separate sections of the first wall were covered with a refractory alloy such as TZM. The vacuum boundary was at the inner surface of the shield. Designs with and without a poloidal divertor were considered. As an alternative, water or helium cooled tube-in-shell type blankets were proposed.

No detailed design was proposed by the USA since extensive EPR and TNS design studies are available. Separate demonstrations of electricity generation and tritium breeding from small test modules were preferred. A possible system would be as in TETR, with a cold shield structure and 12 test modules of about 1 m^2 area. Alternatively, a hot blanket segment for electricity generation could be inserted between two toroidal field coils.

A fully segmented structure with 12 sections of simple water cooled shield was considered by the USSR. Four were equipped with experimental test modules of 3 to 6 m^2 area. The possibility exists to replace these four segments during the third stage of operation. The remaining segments are fixed and contain test channels of 100 to 200 mm diameter. All test modules and channels have vertical access, and in most cases do not penetrate the primary vacuum vessel. The vacuum boundary coincides with the first wall and side walls of the segments, with seals located in the shield region.

5. REFERENCE DESIGN AND ALTERNATIVES

The group discussed the many choices outlined in the previous section and suggested a possible reference design and three alternatives which span the whole range of objectives for INTOR. These possibilities are illustrated in Fig. XI-1, in which the extent of tritium breeding increases along the horizontal axis and the extent of electricity generation down the vertical axis. In either case, four main possibilities exist:

(a) The use of test modules. Each 1 m^2 of module would provide heat to generate continuously about 0.3 MW_e or breed about 20 g of tritium per year at 50% availability and 70% duty cycle.

(b) The use of test segments. Each segment of about 25 m^2 would provide heat to continuously generate 8 MW_e or breed 500 g of tritium per year at 50% availability.

(c) The use of all outboard regions of the torus. The generation of about 100 MW_e or a breeding ratio of up to 0.8 would be possible.

(d) The use of the whole surface of the torus. This would be necessary to achieve the maximum generation of electricity or a breeding ratio of unity.

Combinations of these possibilities which appear of interest form the reference and alternative designs.

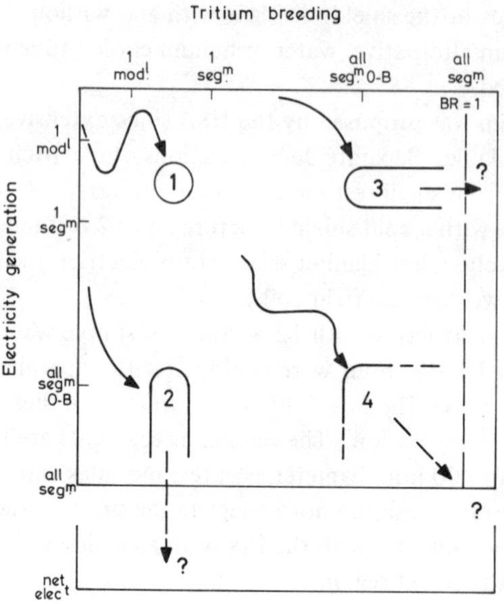

FIG.XI-1. Schematic plot of electricity generation versus tritium breeding.

5.1. Reference design

The reference design is indicated as Area 1 in Fig. XI-1. It is a device with a cold (<100°C) shield inboard and outboard with several test modules. The necessary number of test modules is probably in the range of 4 to 12, and their size 1 to 6 m^2. These test modules would be used for demonstrating the generation of electricity up to a level of 5 or 10 MW_e, for testing several alternatives of breeding materials, and finally, for demonstrating at least one breeding system which operates under conditions suitable for electricity generation.

5.2. First alternative design

An alternative which is only a modest advance on the reference design is a device with a hot (250–350°C) blanket in which at least half of the energy released in fusion reactions is recovered for electricity generation. It is indicated by Area 2 in Fig. XI-1. The blanket would have a similar composition to the shield of the reference design and would only slightly increase the overall thickness of the first wall, blanket and shield region. Whether it covered only the inboard or outboard regions of the torus or the whole surface is left open; in the extreme case, this device could aim for the generation of about 150 MW of electricity. Test modules would still be provided for testing of tritium breeding and for the demonstration of tritium breeding in a system suitable for electricity generation.

5.3. Second alternative

A more significant advance might be a device in which at least 50% of the tritium consumed is recovered by breeding in a cold blanket. It is indicated by Area 3 in the Fig. XI-1. In this case the increase in thickness of the first wall, blanket, and shield region would be greater and the effect on machine design greater if the blanket were extended beyond the outboard region; the extreme situation would be a breeding ratio of unity. Again, test modules would be provided to generate electricity, to test alternative breeding systems, and to demonstrate tritium breeding in a system suitable for electricity generation.

5.4. Third alternative

The largest step which could be contemplated would be a device which demonstrated simultaneously the generation of electricity from more than half the fusion energy released and a breeding ratio of more than 50%. This would require breeding in a hot blanket. For such an objective, it would probably be desirable to introduce staged operation, testing alternatives in test modules, trying the most likely solutions in single segments, and then modifying the device to include a full set of new segments. Such a machine would obviously be more expensive but would have considerable flexibility, would demonstrate the technologies on the maximum scale possible, and could lead directly to the construction of a demonstration reactor.

5.5. Choice between alternatives

The final choice between the various alternatives for the objectives and form of INTOR must depend on a technical assessment of the available technology to construct the device and the relevance of tasks to be performed. It is the purpose of the data base assessments in the following Sections to provide the technical basis for the answers to these questions.

6. BLANKET TEST FACILITIES

One of the objectives of INTOR is to serve as an engineering test facility for tritium breeding, electricity production, and materials testing, and therefore there is a clear need for test facilities in the blanket region of the device. These facilities could range from “test channels” suitable for materials test specimens to complete segments for "full engineering" tests of structures in preparation for the construction of future reactors. Intermediate between these extremes are "test modules", of order of 1 m^2 area, in which a wide range of materials and engineering tests can be performed. The characteristics and uses of these various facilities are discussed in this Section.

6.1. Test channels

Holes of 100 to 300 mm diameter, could be made available in the blanket and shield region with vertical, horizontal, or tangential access. They would be roughly equivalent to the experimental channels of a fission reactor. Many would lie behind the first wall and not break the primary vacuum of the torus so that the introduction of test specimens did not affect the operation of the tokamak, but others might be allowed with access to the plasma for plasma-wall interaction experiments.

Typical uses of these test stations include small scale materials tests on structural materials and electrical insulators such as radiation damage, corrosion, or tritium permeation. Both static and dynamic tests would be possible, the latter including irradiation creep, crack propagation, and fracture. In general, a controlled environment and constant temperature would be required.

6.2. Test modules

The most useful test facility availaible in INTOR would be the test module, and between 4 and 12 have been proposed. Modules could be used to study alternative concepts for the breeding of tritium, to allow the demonstration of electricity generation, and to allow the engineering testing of mock-ups of blanket cells. Such modules would also allow realistic studies to be made of questions such as heat transport, corrosion, and tritium containment. They would in many cases be associated with experimental loops external to the reactor, for example for tritium recovery. Important considerations are that such modules should be accessible without disturbing the toroidal or poloidal magnetic field coils, and that they be designed so that failure of one module does not represent a risk to other modules or the reactor as a whole. Various means of inserting the modules either vertically or horizontally have been proposed, but in all cases they would be in the outboard region and close to the position of peak neutron flux.

The preferred size of test modules depends on several factors, and proposed solutions have areas which range between 1 and 3 m^2. A minimum requirement is that they be large enough to reduce boundary effects on the neutron flux and spectrum. In practice, the size is determined by the level of electricity generation or tritium breeding required or by the space required for a realistic engineering mock-up, and this usually leads to areas of the order of 2 m^2. This larger size is also compatible with the estimated need for about one hundred litres of space for materials testing since in this case only the first 100 mm depth of the module will experience a high neutron flux.

The use of test modules for engineering testing and development will be the main activity in the second stage of INTOR operation, and will continue in the third stage where higher machine availability will also allow higher neutron doses to be accumulated.

6.3. Test segments

The use of complete reactor segments, i.e. 1/6 or 1/12 of the reactor, for engineering testing during the third stage of operation has been suggested. The ability to change segments remotely must be built into INTOR, but this would be a time consuming activity and therefore only possible on a few occasions. An argument in favour of using complete segments is that a full engineering test could be undertaken of the construction of a subsequent Demonstration Fusion Reactor which was large enough to expose all the problems and demonstrate the necessary reliability. Arguments against the use of segments include the cost, lost time during the change, and the very small number of alternatives which could be tested in this way. The question must remain open.

6.4. Other considerations

A range of designs have been proposed for test channels and modules. No concensus on the best approach has emerged, but it is concluded that the provision of these facilities is possible and that considerable volumes of testing space could be made available within the concept of the reference design for INTOR. In the case of some of the alternative designs, however, in which extensive electricity generation or tritium breeding is required, the space for test facilities might be limited.

7. DESIGN CONCEPTS AND DATA BASE ASSESSMENT FOR THE FIRST WALL

The first wall in INTOR will be very different from those in existing tokamak experiments, due to the high heat loading. Various designs have been considered for reactors, including

(a) Simple panels with water or helium cooling
(b) Panels constructed entirely from tubes cooled by water or helium, capable of withstanding higher heat loads
(c) Walls protected by carbon or refractory liners which are radiation cooled.

In all cases, further design options are open. The wall may be designed to accept the full heat load from the plasma, or may be protected by limiters or the divertor. The first wall may be an identifiably separate component, or be integral with the blanket or shield. In either case it may be continuous, possibly forming a vacuum boundary, or sub-divided for easier remote maintenance. Finally it may include a surface coating of low-Z material to reduce plasma contamination or of a material which reduces tritium diffusion.

Designs with both helium and water cooling have been considered, and both appear practical. Common problems with helium cooled designs include the extent of manifolding required, the difficulty of fabricating an engineering system which is sufficiently free of helium leaks that plasma performance is not affected, and the higher structural temperatures that arise for a given wall loading due to the poor (relative to other coolants) heat transfer characteristics of a gaseous coolant. It is of importance to take into consideration the non-symmetrical nature of the energy deposition. Attractive features of using helium as a coolant are its compatibility with austenitic stainless steels and nickel-base alloys, its inherent lack of chemical reactivity with either tritium breeding materials or air, and the relatively minor consequences for the system structural integrity of a coolant line failure.

Water cooling permits the structural material temperature to be decreased to 150 to 200°C, in which case aluminium alloys may be considered as well as stainless steel. Water cooling also allows the use of more conventional equipment and engineering. The main problem is the tritium diffusion into the water and the danger of an accident with the radioactive coolant.

A major concern in the design of a tokamak reactor first wall must be to find ways, through both design and material selection, to minimize the tendency of the first wall to fail by cyclic loading. Radiation enhanced stress relaxation would be expected to reduce the thermal stresses imposed on the first wall during the burn period of the cycle, but these stresses are reimposed, with an oppossite sign, during the dwell period of the cycle. Thus, what is initially a creep-fatigue problem becomes translated into a fatigue or crack growth problem. The ability of inspection techniques to detect small flaws in the as-fabricated structure becomes very important to the expected lifetime of the structure.

Further problem areas which must be considered in the design have been identified. The high voltages around the torus during start-up or discruptions lead to arcing in present tokamak experiments, and must be reliably withstood in a reactor. Similarly the possibility of large currents flowing in the first wall during a discruption must be avoided, or else the mechanical structure must be capable of withstanding the forces. If the first wall is directly exposed to the plasma it may also be necessary to vacuum bake it in situ, requiring the cooling system also to function as a heating element.

The use of passive liners in front of the first wall have been suggested for several purposes. These include a method of reducing the radiation damage in the first structural wall through moderating the incident neutron spectra, as a means of smoothing the thermal cycle by reradiation energy during the dwell period, and as a scheme to protect the first wall from a plasma disruption. Since passive liners are radiatively cooled and face the plasma, graphite has been the usual materials choice. The beneficial effect on the temperature cycle experienced by the first wall has been analyzed, and it was shown that a carbon liner can reduce the variation of temperature differential across the first wall by a factor of four. The advantages of the liner concept include the ease of replacing the liner, if necessary, but must be weighed against the additional space required and the effect on plasma impurity levels.

Another approach exists to reduce the temperature cycles in the first wall. This is the regulation of the coolant flow rate in such a manner that there is a constant temperature during both the burn and off-time. This possibility depends strongly on the choice of design.

The use of low-Z coatings as a passive means of plasma impurity control is being widely studied, and a range of materials are being exposed to plasmas in existing tokamaks. Titanium diboride has been found to be particularly attractive from the point of view of its resistance to thermal cycling and ease of application. Of the available coating techniques, plasma spraying and chemical vapour deposition appear most promising for covering large surfaces. Techniques will need to be developed to allow in-situ coatings to be applied, so that the original coated surfaces can be periodically refurbished. Plasma discharge nitriding of titanium structures and plasma deposition of pellet injected materials are two possible solutions.

In the case of water coolant, it may be important to incorporate coatings on the interior of the coolant tubes for tritium containment.

8. DESIGN CONCEPTS AND DATA BASE ASSESSMENT FOR THE BLANKET

Within the INTOR study the blanket has been defined as a region in which tritium breeding occurs and/or heat is collected for the generation of electricity. It follows from this definition that the reference design of INTOR (see Section 5.1) does not contain a permanent blanket but only regions of cold shielding. On the other hand the alternative designs (Sections 5.2 to 5.4) contain blankets of increasing complexity. An outcome of this data base assessment should therefore be a technical judgement as to whether these alternatives are possible (as opposed to desirable) on the basis of our present knowledge. This data base assessment is also directly relevant to the design of the test modules to be inserted in INTOR, since it will be in these modules that tests are conducted of alternative methods for breeding tritium or extracting heat for electricity generation.

TABLE XI-1. BLANKET DESIGN FEATURES OF RECENT TOKAMAK STUDIE

Blanket design	Material		
	Structure	Coolant	Breeding
ORNL/TNS	Stainless steel	H_2O	None
GA/TNS	Inconel 625/ stainless steel	H_2O	None
PPL/SLPX	Stainless steel	H_2O	None
WISC/TETR	Stainless steel	H_2O	None
ANL/EPR	Stainless steel	H_2O (steam)	None
JAERI/PETF	Concrete/lead/ stainless steel	Boric H_2O	None
USSR/T-20	Stainless steel	Helium/lithium	Li/Li, A O_2
JAERI/JXFR	Stainless steel	Helium	Li_2O
JRC/FINTOR D	Stainless steel	Helium	Lithium
CULHAM/MK 2	Stainless steel	Helium	Lithium compound
JÜLICH/CPR	Stainless steel	Helium	Lithium/beryllium
MIT/HFCTR	TZM	$2LiF \cdot BeF_2$	Lithium
GA/DEMO	Stainless steel	Helium	Lithium
ORNL/DEMO	Stainless steel	Helium	Lithium
WISC/NUWMAK	Ti-6A/4 V	H_2O (boiling)	Li_{62}, Pb_{38} (EUT)

The data base for this area is very wide, consisting of many design studies of commercial reactors (not only tokamaks), experimental power reactors and TNS devices. In addition, specific designs have been presented by some delegations of possible INTOR devices and some of these have studied the blanket problems in detail. Some of the combinations of structural material, coolant, and breeder are illustrated in Table XI-1.

Early tokamak reactor studies proposed liquid lithium as both the breeding material and coolant for the blanket system. This was an obvious choice since liquid lithium provides a high potential for tritium breeding, and, at the same time, is a very good heat transfer fluid. Unfortunately, the problem of MHD induced forces caused by pumping the conducting fluid in the high magnetic field was later found to produce high pressures and pumping power requirements.

The trend in tokamak blanket designs developed over the past three years has been toward helium, gas or water cooling in combination with both liquid and solid compounds of lithium as the breeding material. The use of solid breeding materials such as lithium oxide (Li_2O), lithium aluminate ($LiAlO_2$) lithium orthosilicate (Li_4SIO_4) or metasilicate (Li_2SiO_3) have structural and safety advantages and are associated with low tritium inventories. Heat removal is generally more difficult, in most cases there is a lack of information on thier behaviour under irradiation, and breeding ratios are often marginal unless a neutron multiplier is used. More recently the use of lithium-lead alloys (e.g. Li_7Pb_2, which becomes liquid at 726°C) has been proposed and these appear to have attractive breeding properties. The use of helium coolant has been adopted in most reactor designs because of its chemical inertness and low contamination in the event of a leak. Water cooling has also been considered in the past few years, based on LWR fission reactor experience. The use of stainless steel is accepted in most designs because of the immense experience of fabrication existing in the fission reactor field. Nickel based alloys and titanium based alloys have also been considered, however, either because of their better corrosion or radiation damage resistance or lower long-lived activation level after irradiation. Further information on several of these topics is included in the chapters on materials and tritium.

Existing reactor design studies also cover a wide range of structural concepts. Most designs rely on a segmented arrangement to allow remote maintenance during reactor operation, with a cellular structure to sub-divide the blanket and allow for the poloidal variation of neutron flux and heating rate. Radially arrayed cells are most commonly used, but cylindrical cells with one edge facing the plasma have also been proposed. In all designs there is a conflict between the desire to minimize the quantity of structural material, both to improve breeding and reduce cost, and yet to maintain safety standards and reliability. It is also a feature of all fusion reactor blankets that there is a need for extensive ducting for the coolant in the region behind the blanket. The result is that whilst there appear to be solutions to all major design problems, the blanket region is a complex structure with many inter-relating constraints and it is not yet clear which are the best solutions or whether they will give high reliability and availability in service. The need to construct and test blanket mock-ups would probably delay the design and construction of an INTOR device which included an extensive permanent blanket.

Whilst many thermal and structural problems associated with the blanket have been studied in previous designs, there are some electrical problems which deserve further consideration. As with the first wall, it is necessary to ensure that electrical breakdown does not occur during start-up or a disruption. The need for electrical insulators in the blanket region of a fusion reactor has not been sufficiently clarified. The degree of sub-division, or inter-connection, of the blanket which is required to allow magnetic field penetration without distortion is unclear. Finally, the possible necessity for internal electrical conductors for equilibrium or stability control is at present unknown.

9. DATA BASE ASSESSMENT FOR FIRST WALL AND BLANKET LIFE

In view of the complex construction of the first wall and blanket regions, a long and reliable life will be essential for the success of INTOR. The analytical tools required to predict the life of such structures include those needed for the prediction of stress histories and the prediction of damage to the structures under specified stresses and temperatures. Both of these require a knowledge of materials behaviour under neutron irradiation and surface particle fluxes.

Stress analyses are available to calculate stresses under both static and dynamic loads and to predict elastic and inelastic response of structures. Application of these codes to determine stress histories, however, requires constitutive equations which model the material stress-strain behaviour and require the definition of applied loads. For the first wall and blanket structures of INTOR, uncertainties exist in both of these areas. In particular, constitutive equations which adequately represent material behaviour at high temperature are generally lacking, even without the complication of radiation damage effects. Radiation damage effects, such as irradiation creep, irradiation swelling, and time dependent strength changes, unless well defined, will cause uncertainty in resulting predicted stress-strain. Applied loads in an INTOR device will include thermal stresses from temperature cycles, coolant pressures, atmospheric pressure, dead weight, and magnetic forces. Uncertainties in these applied loads will result in uncertainties in the resulting stresses. One of the greatest uncertainties in the life of first wall and blanket structures may result from the occurrence of plasma disruptions. These loads will significantly affect the design and life of the structure. Other uncertainties result from imprecise knowledge of the neutron wall loading distribution in the torus, coolant flow variations, and magnetic loads resulting from interactions of the magnetic fields. In summary, although the analysis capabilities exist for accurate prediction of the first wall and blanket stresses, significant uncertainties exist because of the state of knowledge of material behaviour and applied loads. Radiation damage of the structural material will become a major concern. For long burn times therefore, it is important to assess the state of knowledge on radiation effects at about the time INTOR would be designed and built. The available material design property data base for INTOR will have to be developed primarily using existing fission test reactors. These reactors do not adequately simulate a fusion environment because of an inability to generate helium and displacement damage in anything but nickel containing alloys (i.e. 316 stainless steel). This means that the only structural materials for which there will be a substantial irradiation design data base are austenitic stainless steels and nickel base alloys and the data base for these alloys will be marginal until a correlation is established between the actual fusion spectrum and fission simulations. This correlation will be underway using RTNS and FMIT but will be incomplete because of the neutron fluence and test volume limitations.

Material fatigue test data show scatter in the data. Several theories exist for creep and fatigue damage accumulation, all of which have shortcomings. Better cumulative damage concepts for crack initiation need to be developed for the specific environments, materials, and service conditions of the INTOR reactor to provide a basis for adequate fatigue life predictions. Fracture mechanics is a well established analytical tool, but its application also requires definition of material properties such as fracture toughness and crack growth rates. Therefore, life predictions for the INTOR reactor first wall and blanket components, which are fatigue or fracture critical, will be subject to uncertainties unless the material properties are well defined.

In conclusion it appears that adequate computer codes are available but that materials data may be lacking, especially under neutron irradiation conditions appropriate to fusion. All calculations will be very design dependent, but the data base for assessing the life of the first wall or blanket components is probably adequate.

10. DESIGN CONCEPTS AND DATA BASE ASSESSMENT FOR THE SHIELD STRUCTURE

The shield is taken to include all cold (~100°C) structures surrounding the plasma which are not used for tritium breeding or electricity generation. In the reference design this covers the whole permanent structure of the torus apart from the first wall, limiters, beam dumps, etc. Its primary purpose is to attenuate the neutron flux from the reacting plasma to a level at which the superconducting magnets and other reactor components can operate. Since, however, it will be composed mainly of heavy materials such as steel it will also act as the main mechanical structure of the reactor. It may also incorporate the primary vacuum vessel.

Within the INTOR Workshop the questions of the composition of the shield, its minimum thickness and the effect of penetrations are dealt with by the Shielding and Personnel Access Group. It may simply be noted here that a typical composition is steel (50–70%) and borated water or boron carbide (50–30%) with an outer layer of lead. On the inboard side of the torus, where the thickness of the shield is critical, it is possible to substitute tantalum or tungsten for the steel and reduce the overall thickness by about 50 mm, but at considerably greater cost.

The wide range of existing reactor studies have all included conceptual solutions to the support structure within the shield. Few have pursued the question to any depth. However, since the shield operates at a low temperature, usually water cooled, the engineering problems associated with designing the support structure are somewhat relaxed. The details of the support from the first wall, blanket or test modules, and other internal components are also dependent on the specific designs selected for these items. Therefore, although the capability exists to design the support structure, the details must wait until other details of INTOR are more fully specified.

One major option in the shield design is the extent to which part of the structure is permanent. The outer region of the shield, which suffers little radiation damage, could form a fixed structure from which the first wall and inner shield is withdrawn for maintenance. If the vacuum boundary is at the shield, this fixed structure could include vacuum doors. The alternative is that the whole shield is removable, but this greatly increases the weight to be moved and implies that further structure is necessary elsewhere to support the shield.

A preferred location for the vacuum boundary is at the shield since this is the smallest volume with a cold surface to limit tritium diffusion. This boundary must pass in front of the shield, but the vacuum joints should be at the back for accessibility. If welded joints must be cut and re-made when a segment of shield is moved, space must be allowed in the design for the remotely operated equipment to accomplish this task. If segments are removeable, the gaps between them must also be arranged to prevent neutron streaming.

As with the design of the blanket, insufficient attention has been paid in previous designs of shield structures to electrical problems. In the Reference Design the shield may be required to act as a passive conducting shell to assist and maintain the vertical equilibrium of the plasma. It may also need to be sub-divided in some directions to allow the rapid penetration of magnetic fields. Its electrical resistance around the torus should be greater than 0.2 mΩ to limit induced currents.

Although the problem of penetrations does not fall within the task of this group, it should be remembered that large holes may be required for divertors, injector ports, pumping ports and the insertion of test modules.

11. DESIGN CONCEPTS AND DATA BASE ASSESSMENT FOR LIMITERS, BEAM DUMPS AND DIVERTOR TARGETS

These components have the common property that they are required to handle very high power densities and particle fluxes. Of all the components close to the plasma their design has probably been least extensively studied. Even in existing tokamaks the choice of materials for these components is more often the result of experiment than design.

11.1. Limiters

Under normal operating conditions the limiters provide the first contact between the hot plasma and solid material. Depending on the plasma conditions a fraction between about 10% and about 80% of the alpha particle power of the plasma could be dumped into the limiters. The limiters should be designed in such a way as to help spread the fluxes of charged particles which arrive from the plasma, over a maximum possible surface area. If INTOR is to be operated with a limiter, the total heat load might reach 100 MW over a pulse duration of 100 s, with a dwell time of 20 s. The corresponding particle flux would be of the order of 10^{23} particles/s, or 10^{25} particles/pulse. In addition to this "normal" load, plasma conditions have to be considered which cause even higher loads to the limiter:

(a) During the build-up phase of the plasma an additional heat load is expected due to the skin distribution of the current.
(b) During major plasma disruptions the thermal energy of the plasma might totally or partly be dumped into the limiter, in a time which could be 1/100 of the energy confinement time. This means the heat load would increase by a factor of 100.
(c) At low plasma densities, especially during the build-up phase, runaway electrons might hit the limiter and cause local melting.

The material selected for the limiter should have a high melting temperature, high thermal conductivity, good thermal shock properties, low sputtering rate, low outgassing and possibly low Z number. The data base of candidate refractory metals such as molybdenum or tungsten is fairly well known. Thermal shock properties, sputtering rates, and outgassing of low-Z materials such as B_4C, Al_2O_3, SiC, Si_3N_4, pyrolytic graphite, glassy carbon and BN have been studied. The prime criterion for the selection of low-Z materials seem to be the thermal shock properties. The critical thermal load, due to cracking, is of the order 20 MW/m^2 and increases in the order: Al_2O_3, B_4C, Si_3N_4, SiC, pyrolytic graphite. This would require a total limiter surface area in the order of 10 m^2 to take the heat load from the INTOR plasma. The thermal properties of candidate metals as well as low-Z material have been studied with respect to evaporation, melting and thermal stresses. As the best choice among the available low-Z materials, a composite material such as silicon carbide, covered with a graphite layer has been suggested. This would combine the good thermal properties of SiC with the low-Z number and low sputtering yield of graphite.

Several coating/substrate systems are currently being developed for fusion reactor limiters. Laboratory testing of the response of these coating/substrate systems to thermal shock, ion bombardment, arcing, chemical reactions with hydrogen plasma and hydrogen retention is currently being carried out. Promising materials from these screening tests will then be tested as limiters in plasma physics experiments.

At present, the coatings with the best response to the laboratory testing are CVD TiB_2, TiC and B on poco graphite and plasma sprayed TiB_2 and TiC on copper. These systems survived thermal shock testing of 50 cycles at a power density of 20 MW/m^2 without failure. The ion erosion rates of these materials are in the same range as those of stainless steel for 0.7 s pulses, and since they have a lower Z, there should be a less deleterious effect on the plasma. The same is true of the arc erosion rates with the exception of the boron coating. Since boron is an insulator, it did not arc in the laboratory experiments and will probably not arc in a tokamak environment. Hydrogen retention or chemical erosion has not been found to be a problem in any of these systems at this time. The interdiffusion of the TiB_2/graphite system has been studied at temperatures up to 2000°C and is not found to be a limiting factor. CVD coatings are also being developed for copper, which would allow a low-Z coating on an actively cooled limiter. The CVD coatings can also be applied to molybdenum and to certain types of stainless steels.

In summary, low-Z coatings offer the potential for substantial reduction in plasma losses due to eroded material. While the development of low-Z coatings for fusion applications is relatively new, the state of coating technology is such that relatively large components could be coated in the near future. A reactor designed with first wall components that meet these size restrictions would allow the option of adding a coating at a later date if it was determined to be necessary.

Further studies are needed to determine the effect of neutron irradiation on the coating.

The most severe condition for the limiter is a major plasma disruption in which the plasma thermal energy is transferred to the limiter surface due to thermal conduction. In this case about 100 MJ might have to be dumped within 20 ms over an area of a few square metres. These high loads cannot be handled with conventional means. A possible solution to this problem is a first wall protection device which is mechanically injected into the plasma region when the onset of a disruption is detected.

11.2. Beam dumps

According to the INTOR reference parameters, about 90% of the 75 MW of injected neutral beam power will be trapped by the plasma. Thus, a total beam power of 7.5 MW at 200 keV, or about 10^{20} particles per second will hit the walls opposite to the injection ports for 5 seconds. The power-density before injection is 25 MW/m^2. With an attenuation of a factor of 10 by the plasma and a further expansion of the beam within the plasma, a power density of 2.5 MW/m^2 at the wall can be assumed. At these locations, this heat flux requires special first wall design.

Recent surface studies in support of the TFTR beam dump development effort have shown that the use of molybdenum as a dump material is enhanced if the temperature is kept above 300° C. No blistering occured under these conditions for bombardment energies up to 120 keV while blistering was observed at lower temperatures. A similar study on graphite revealed a ridging of the graphite surface under bombardment though there was no serious surface erosion.

11.3. Divertor targets

The divertor target has to take about the same total energy and particle flux as a corresponding limiter. In the divertor case, however, the interaction between plasma and solid material takes place in a remote chamber, thus reducing the impact of plasma wall interaction on the main plasma. Furthermore, the contact area between plasma and solid material can be more easily controlled in the divertor case.

In general, two types of divertor targets can be chosen to fulfil the twofold task of removing the heat and particle flux.

(a) A reactive metal divertor target can simultaneously accommodate the incoming heat flux and pump the energetic particles by trapping. With this solution large divertor pumps can be avoided. On the other hand the target surface has to be reactivated or replaced periodically and a compromise between high trapping efficiency and good thermal properties for heat accommodation has to be found.

(b) The functions of particle trapping and heat accommodation are separated. In this design only the energy is dumped into the divertor, but the particles are reflected or re-emitted and have to be pumped with separate pumps. In this way the divertor target can be optimized for high heat transfer, with the drawback that large vacuum pumps have to be accommodated inside the divertor. An example of such a design is the Ta-10W target in TETR which can withstand 7 MW/m^2 on a steady state basis.

The trapping efficiency of reactive materials such as vanadium, niobium, tantalum, titanium, zirconium and lithium has been studied experimentally and can be calculated by the use of a model. For application in a reactor, only titanium, zirconium and lithium have been found to be appropriate. A zirconium target, for example, can trap a deuterium beam of 2×10^{20} ions/$m^2 \cdot s$ at a target temperature of 530 K with an efficiency of 90% for 291 seconds.

The high particle flux to the divertor target results in a high surface erosion due to sputtering. Assuming a sputtering yield of 2×10^{-3} and a target surface area of 1 m^2, the surface erosion would be 1 to 10 mm in 60 days. If a thin wall design was employed for good heat transfer, this erosion rate would be too high,

TABLE XI-2. R AND D NEEDS ASSESSMENT – FIRST WALL, BLANKET AND SHIELD

	Status*	
	Adequate	Inadequate
Development of methods to protect the first wall against plasma disruptions (e.g. liners or mechanical devices)		x
Determination of heat flux limits for limiters, divertor dumps and targets		x
Determination of heat flux limits under normal operation for first wall and liner materials, theoretical and experimental	x	
Clarification of the needs and operating conditions of electrically insulating parts in the first wall and shield region		x
Out of reactor tests of standard blanket, shield and first wall components and mockups (thermal hydraulics, pressure and thermal load response)	x	
Study of heat storage devices for smoothing changes in power output	x	

Adequate: Programmes are either planned or could be planned on existing or authorized facilities (with some upgrading if required) and using existing trained personnel.

Inadequate: Either the available and authorized manpower and facilities are inadequate or a substantial increase in effort is needed.

*The ratings are made on the world-wide fusion effort and not separately on the individual national programmes.

unless a moveable tape for replacement of the active getter surface was used. Thus there are three major criteria which determine the limit to the flux density at the target and the choice of the target surface area:

(1) heat dissipation,

(2) trapping efficiency,

(3) surface erosion.

From the viewpoint of heat dissipation, surface erosion and particle trapping, a target surface area of about 10 to 30 m^2 seems to be appropriate and could be achieved either with a poloidal divertor or with a modified bundle divertor.

It would yield a steady state heat load of 10 MW/m^2, and a trapping efficiency which would be high enough to pump the particles predominantly by direct trapping, but the surface erosion is still a problem. It should be kept in mind, however, that these considerations are valid only for the steady state load and transient loads well above these estimates can occur at particular plasma conditions. Moreover, for the case of a trapping target, means have to be found which allow periodic replacement or reactivation of the target.

12. R AND D REQUIREMENTS

Experience in the design, construction, and operation of fission reactors is relevant to many aspects of the first wall, blanket, and shield of INTOR and subsequent fusion reactors. The extensive research and development programmes associated with fission reactor development also support work being specifically undertaken for fusion reactors. Specific examples include neutronic codes and data assessments, materials testing, development of constructional techniques, quality assurance, etc. The state of the art in these areas is considered adequate for INTOR. Several items, however, have been selected as important areas in which existing work in fission or fusion laboratories is judged to be inadequate for the needs of INTOR, and are included in Table XI-2.

It should also be noted that many items relevant to this group will also be found in the chapters on materials, remote maintenance, and tritium.

13. CONCLUSIONS

1. In general, a wide range of design options have been explored in previous fusion reactor design studies, and wide experience relative to this area is available from fission reactor design and construction. An adequate data base exists for many components of the first wall, blanket and shield.

2. Having considered the possible objectives for INTOR, it appears that a preferred Reference Design would be a torus consisting of segments of cold shielding with 4 to 12 test modules in which different types of blanket construction for electricity generation and tritium breeding would be demonstrated.

3. In addition to the test modules, which will each cover a first wall area of 1 to 3 m^2, provision would be made for several test channels of 100 to 300 mm diameter for smaller scale tests. In many cases these channels would not penetrate the primary vacuum system, but in some cases they will allow plasma-surface interaction studies.

4. In addition to the Reference Design, three Alternative Designs have been considered which allow for more than 50% of the energy from fusion reactions to be used for electricity generation, or more than 50% of the tritium consumed to be recovered in a breeding blanket, or both of these objectives simultaneously.

5. The simple objective of greater electricity generation is not favoured.

The additional capital equipment would be expensive, and operating flexibility would be greatly reduced. A level of 10 MW_e is sufficient to demonstrate the principles and to test equipment.

6. The extended objective of net electricity generation is not favoured for the same reasons and because it places additional constraints on many components. Furthermore, it may be necessary to increase the size and gross power output in INTOR.

7. The objective of breeding more tritium from a cost standpoint has greater support. Breeding in the outboard regions will only allow breeding ratios of about 0.4 to 0.8 but with breeding in both the inboard and outboard regions a breeding ratio of about 1.0 may be possible, depending strongly on the divertor choice.

8. The extended objective of obtaining a breeding ratio of unity would overcome tritium production and transport problems but would increase the complexity of INTOR and require an inboard blanket, and hence increase the size of the torus. If this blanket was operated at a relatively low temperature including that of the breeding material, the technical problems would be reduced but the experience would not be relevant to future reactors.

9. The greatest step forward would be the operation in INTOR of a hot blanket generating both a substantial proportion of electricity and breeding at least 50% of the tritium consumed. Such a device would imply staged operation, however, since the best form of blanket would not be known initially. Such a device would be considerably more expensive.

10. A segmented structure for the first wall, blanket, and shield appears desirable from the point of view of remote maintenance. This question has been dealt with in Chapter XV. This approach would also allow the introduction of test segments and the possibility of staged operation.

11. The vacuum boundary is taken to be in the shield region, which is consistent with the conclusions of Chapter XVII. The joints between segments would be at the back of the shield to ease problems of access.

12. If all the alpha-particle energy is deposited on the first wall (as compared with limiters or divertor dumps) a separate first wall is desirable. This might be constructed from tubes, with a suitable coating or liners which cannot yet be specified. Thermal cycling of the first wall will be a serious limitation, and the temperature fluctuations must be reduced.

13. If all the alpha-particle energy is deposited on the limiters it is not certain that it will be possible to design these components to withstand the very high local fluxes which may occur. If the energy is mainly deposited on the divertor target, the same problem occurs unless it is possible to expand the exhaust plasma and increase the area of the divertor target.

14. The composition and necessary thickness of the shield are known. The form of structure is not yet clear, but in view of the high proportion of steel and low operating temperature, no serious design problems are foreseen.

15. It may be necessary to introduce electrically insulating gaps in the blanket or shield structure. The location and design of these insulators is unknown, and the possibility of suitable insulators withstanding the radiation damage levels in the front sections of shield is uncertain.

16. The magnetic fields required for plasma control will vary with frequencies up to 1 Hz, and must penetrate with little distortion. Little consideration has yet been given to the design of blankets and shields to fulfil these conditions.

17. Adequate calculational methods are available for assessing the life of the first wall, blanket, and shield under the mechanical stresses and radiation levels foreseen. The materials data is inadequate in some areas.

REFERENCES TO CHAPTER XI

[1] European Contributions to the 2nd meeting of the INTOR Workshop, Blanket, Shield and First Wall, Chapter II, EUR FU/XII 501/79/EDV 50, June 79 (2 Volumes).

[2] European Contributions to the 3rd meeting of the INTOR Workshop, Blanket, Shield and First Wall, EUR FU/XII 501/79/EDV 70, Oct. 79.

[3] Fusion Research Development Center, INTOR Workshop Report, Group 2, First Wall, Blanket and Shield, JAERI-8711 (1980).
Fusion Research Development Center, INTOR Workshop Report, Engineering Aspects of the JAERI Proposal for INTOR (I), JAERI-8503 (1979).

[4] Fusion Research Development Center, INTOR Workshop Report, Engineering Aspects of the JAERI Proposal for INTOR (II), JAERI-8518 (1979).

[5] USSR Report to INTOR, "Blanket, First Wall and Shield of INTOR" (see Appendix A.5.3 [8]).

[6] "Blanket, Shield, First Wall", Chapter in USA INTOR Report (see Appendix A.5.4).

[7] "Electricity Production", Chapter in USA INTOR Report (see Appendix A.5.4).

Chapter XII

TRITIUM

D. LEGER	– EC, Chairman
J. DARVAS	– EC, Expert
R. HIRAOKA	– Japan
B. KOLBASOV	– USSR
G. KULCINSKI	– USA
V. MARONI	– USA, Expert
G. SHATALOV	– USSR

1. INTRODUCTION

Tritium is one of the two fuel components in fusion reactors like INTOR that burn in a plasma containing a fifty-fifty mixture of deuterium and tritium. Tritium, the heaviest hydrogen isotope, is also radioactive and because of this last property, its handling is subject to strict safety rules. All the tritium handling takes place in a specific plant: the tritium facility.

The tritium facility itself is subdivided into two systems: the fuel recycle system and the bred tritium recovery system.

The different functions of the fuel recycle system are the following:

- recovery of the burned mixture of deuterium-tritium and impurities that are withdrawn from the plasma chamber
- processing of this mixture to obtain a pure hydrogen isotope stream
- isotopic separation of tritium, deuterium and protium
- tritium storage
- separation of fuel for reinjection into plasma chamber.

The bred tritium recovery system is less well defined. The function of this part of the plant is to recover tritium from the breeding material, but the process itself depends strongly on the lithium compound used for breeding. There exists a large variety of candidate lithium-containing materials: lithium itself, lithium oxide, lithium aluminate, lithium silicate, and lithium alloys. Following recovery from the blanket, the bred tritium is sent to a purification unit and then on for isotopic separation.

It is trivial to say that tritium has essentially the same chemical properties and about the same physical properties as hydrogen. One of the most important properties, well known with hydrogen, is its rapid permeation through most structural alloys. This effect is magnified exponentially when the material

TABLE XII-1. TRITIUM CONSUMPTION IN INTOR

Availability %	Time y	Tritium consumption kg
15	1.39	4
20	1.71	7
25	4	24
50	5	60
Total:	12	95

temperature increases. Consequently, the choice among different candidate metals for the reactor and the tritium facility building must take account of the problems associated with tritium permeation.

Because of its radioactive nature, the handling of tritium may require as many as three successive containments, in addition to a ventilation system and an emergency system to recover a major portion of the tritium in case of a large tritium release to the INTOR building.

Finally, the procurement and shipment of large quantities of tritium makes necessary an international agreement among the different participating countries, as well as new rules and the development of containers for tritium transportation. The question of whether an adequate supply of tritium could be made available to support the operation of a non-breeding INTOR needs to be thoroughly investigated.

2. INTOR TRITIUM CONSUMPTION AND INVENTORY

From the different assumptions about the burn-up, the availability, the burn time and the dwell time, the INTOR tritium consumption lies between 50 and 88 g/d.

According to the time schedule given in Table II-3 from the INTOR reference concept and the guiding parameters (peak power 620 MW, burn time 100 s, time between pulses 35 s, burn-up 5%), the INTOR tritium consumption is given in Table XII-1.

With these assumptions, the minimum tritium inventory could be 2.5 kg.

Individual contributions from each country start with different sets of parameters, and their results are compiled in Table XII-2.

TABLE XII-2. MAXIMUM TRITIUM CONSUMPTION AND INVENTORY

		USA	Japan	USSR	EC	Basic case
Assumptions	Power, MW	790	400		620	600 (760)
	Burn, s	100	200	100	100	100
	Dwell, s	35	50	35	50	35
	Availability, %	50	50	50	25	25–50
	Burn up, %	10	0.3–1.5[1]	5	5	5
	Feed, g/d avail. = 100%	880	4900[1]	1133	1220	1320 (1630)
	Burned tritium g/d	88	49	57	64	64 (78)
	Consumption, g/y	16 000	9000	10 000	5700	6000 (7500)
Active inventory	Vacuum pumps, g	37	2400[1]	1070	50	
	Days regeneration	0.042	0.5	1.	0.042	
	Fuel distillation and clean-up systems	47	460	1000	114	
	Fuel preparation	100	409	500	100	
	Sub-total	184	3269	2570	264	
Passive inventory	Structural, g	–	–	500	–	
	Cold storage, g	2650	1470	1700	1900	
	Day storage of full operation	30	30	30	30	
	Total inventory	2834	4739	4770	2169	

[1] Values are given for a burn-up of 1% (utilization of a divertor).

Two kinds of tritium inventory are considered. One is the active tritium inventory that takes into account the tritium stored in cryopanels, cryogenic distillation or other isotopic separation units and in the fuel preparation system. This inventory is very sensitive to any failure of the containment and the risk attached to this inventory is relatively high.

The other inventory takes into account the tritium storage for 30 days of full-operation and the tritium absorbed on machine structures. Tritium adsorbed in the machine structures could be considered also as a part of the active inventory. This second kind of inventory is less sensitive than the first to a failure of the containment and could be considered very safe. It is considered a passive inventory.

The different evaluations of active inventory lie between 180 and 3300 g. The main part of it is stored in cryopanels and depends strongly on the regenerating time foreseen for these cryopanels; 180 g correspond to one hour of regeneration and 2500 g to one day. It depends also on the burn-up, and the highest inventory in cryopanels corresponds to 1% of burn-up and 12 hours of regeneration time.

Looking at the different contributions compiled in Table XII-2 it seems that the total tritium inventory should not exceed 4 kg of tritium.

However, a great deal of attention must be given to the structural inventory. The only estimation of this special inventory is 500 g. A more realistic estimation requires a more accurate evaluation of the tritium system like the USA evaluation given in Table XII-3 and also a better knowledge of the permeation of tritium into materials.

3. FUEL RECYCLE SYSTEM

The fuel recycle system is divided into a fuel purification system and an isotope separation system. The fuel purification system separates all the gaseous chemical impurities, including helium, from the partly burned fuel mixture of hydrogen isotopes which has been pumped out of the torus. It also recovers all the tritium which is chemically combined with these impurities. The isotope separation system has three duties to perform. These are:

(1) to separate the purified fuel mixture of hydrogen isotopes into individual streams with different concentrations of D and T isotopes as required for fuel preparation and injection (not yet fully specified; e.g. if pellet fuel injection requires all the fuel to be split into D_2 and T_2, the TSTA system in Fig.XII-1 would require another distillation column and equilibrator to promote $2DT \rightarrow D_2 + T_2$),

(2) to separate the unwanted H isotope (mainly arising from inleakage) from the fuel mixture so that it can be discharged as a tritium free waste stream,

(3) to separate the unwanted H and T isotopes from the isotopically pure D_2 used in the neutral beam injector circuits. The T back-streaming into the beam lines would otherwise accumulate and lead to neutron production.

The fuel purification and isotope separation systems are discussed separately below.

TABLE XII-3. INTOR TRITIUM FACILITY PARAMETERS LIST

TRITIUM FACILITY AND VACUUM SYSTEM PARAMETERS FOR THE INTOR	
PLASMA AND BURN CYCLE	
*Burn time (s)	= 100.0
*Dwell time (s)	= 20.0
*Ramp time (s)	= 15.0
*Lag time (s)	= 0.0
Duty factor	= 0.796
*Plant availability factor	= 0.500
Plant factor	= 0.398
Energy per burn pulse (MJ)	= 8.49E 04
*Power per burn pulse (MW)	= 790.0
Cycle average power (MW)	= 629.1
*Ion density (ions/m^3)	= 1.40E 20
*Major radius (m)	= 5.00
*Minor radius (m)	= 1.50
*Plasma radius (m)	= 1.30
Plasma volume (m^3)	= 266.9
*Energy confinement time(s)	= 1.50
*Particle confinement time(s)	= 3.00
*Reflection coefficient	= 0.700
*Energy per fusion (pJ)	= 2.82
Fractional burnup	= 0.1204
TRITIUM AND VACUUM SYSTEMS COSTS ($)	
Tritium recovery	= 0.0
Piping	= 1.07E 06
Isotopic separation unit	= 1.12E 06
Gloveboxes and purifiers	= 3.55E 06
Misc. T facility costs	= 1.35E 06
Pellet fueling	= 3.53E 06
Total-fuel processing	= 1.06E 07
Cost – EDS	= 6.88E 06
Cost– Vac+ roughing system	= 1.87E 06
Cost – divertor vacuum system	= 2.62E 07
Cost – N. beam vacuum system	= 2.07E 07
Total–Tritium+vacuum systems	= 4.55E 07

* Input parameters.

TABLE XII-3 (cont.)

TRITIUM FACILITY AND VACUUM SYSTEM PARAMETERS FOR THE INTOR

TRITIUM AND FUEL PROCESSING PARAMETERS

Tritium initial load/burn (g)	= 0.0934
Number of burn cycles/day	= 640.0
Tritium gas load/hour (g)	= 2.4908
Tritium gas loading/day (g)	= 59.780
Tritium burnup per cycle (g)	= 0.1503
Tritium burnup per hour (g)	= 4.007
Tritium burnup per day (g)	= 96.18
Tritium exhaust per cycle (g)	= 1.10E 00
Tritium exhaust per hour (g)	= 2.93E 01
Tritium exhaust per day (g)	= 7.02E 02
Tritium fueling per cycle (g)	= 1.15E 00
Tritium fueling per hour (g)	= 3.08E 01
Tritium fueling per day (g)	= 7.39E 02
Tritium input per cycle (g)	= 1.25E 00
Tritium input per hour (g)	= 3.33E 01
Tritium input per day (g)	= 7.99E 02
Helium exhaust per day (g)	= 127.914
Protium exhaust per day (g)	= 3.965
Carbon exhaust per day (g)	= 23.792
Oxygen exhaust per day (g)	= 1.919
Nitrogen exhaust per day (g)	= 1.663
Argon exhaust per day (g)	= 15.861
Tritium bred per day (g)	= 0.0
Tritium decay per year (g)	= 175.82
Doubling inventory (g)	= 0.0
Doubling time (y)	= 0.0
Annual tritium consumption (g)	= 1.77E 04
Annual cost of tritium (s)	= 1.20E 08
*Breeding ratio	= 0.0
Tritium inventories (g) **	***************
Divertor	= 107.1
Vacuum pumps	= 29.3
Storage	= 2885.3
Surge tank	= 2.7
Liquefaction unit	= 12.0
Cryogenic distillation cascade	= 22.6
Fuel preparation	= 80.8
Breeder blanket	= 0.0
Tritium recovery system	= 0.0
Total tritium inventory	= 3139.7

* Input parameters.

TABLE XII-3 (cont.)

TRITIUM FACILITY AND VACUUM SYSTEM PARAMETERS FOR THE INTOR	
TORUS EVACUATION SYSTEM PARAMETERS	
Gas load-divertor ($Pa\cdot m^3/s$)	= 1.74E 01
*Divertor pump pressure (Pa)	= 1.00E-03
Divertor pump speed (m^3/s)	= 1.74E 04
Divertor pump speed-DT (m^3/s)	= 1.53E 04
Divertor pump speed-He (m^3/s)	= 4.20E 03
Cryopanel surface-Div. (m^2)	= 8.40E 01
Cost – He pumping Div. ($)	= 6.30E 06
Cost – DT pumping Div. ($)	= 9.20E 06
Cost – div. pump regeneration ($)	= 5.23E 06
Cost – div. hard seal metal valve($)	= 5.43E 06
Total cost div. vacuum system ($)	= 2.62E 07
Evacuation volume (m^3)	= 355.3
Surface area (m^2)	= 393.8
*Number of ducts	= 12.
*Duct diameter (m)	= 0.80
*Number of duct bends	= 0.
*Duct length (m)	= 6.00
*Rated speed/pump (m^3/s)	= 60.0
*Pump capacity ($Pa\cdot m^3$)	= 2.00E 04
*Pump loading factor	= 0.83
*Duct cond constant	= 0.860
*Outgas rate ($Pa\cdot m^3/s$)	= 0.0
Outgas speed (m^3/s)	= 0.0
Post-burn gas temperature (K)	= 673.20
Post-burn gas pressure (Pa)	= 5.47E-01
*Preburn pressure (Pa)	= 4.00E-04
Post-burn gas load ($Pa\cdot m^3$)	= 1.94E-02
Reqd tot pumping speed (m^3/s)	= 1.28E 02
Post-burn He pressure (Pa)	= 6.58E-02
Preburn He pressure (Pa)	= 1.00E-04
Required He speed (m^3/s)	= 1.15E-02
*Pump inlet temp. (K)	= 273.2
Required duct speed (m^3/s)	= 10.69
Duct conductance (m^3/s)	= 23.86
Effective duct speed (m^3/s)	= 17.07
Regeneration period (h)	= 38.470
Minimum duct diameter (m)	= 0.6504
Fixed regeneration time (h)	= 1.00
Reqd pump capacity ($Pa\cdot m^3$)	= 4.32E 02
Required pump speed (m^3/s)	=19.36

* Input parameters.

TABLE XII-3 (cont.)

TRITIUM FACILITY AND VACUUM SYSTEM PARAMETERS FOR THE INTOR

TORUS EVACUATION SYSTEM PARAMETERS (cont.)

Reqd. He pump speed (m^3/s)	= 24.26
Cryopanel surface area (m^2)	= 0.4852
Cost – He sorption pumping ($)	= 4.37E 05
Cost – DT condens. pumping ($)	= 7.70E 04
Cost – pump regeneration ($)	= 3.85E 04
Cost – hard seal metal valves ($)	= 1.32E 06
Total vacuum system cost ($)	= 1.87E 06

DEUTERIUM AND NEUTRAL BEAM SYSTEM

*Gas load/inj. – zone 1 ($Pa \cdot m^3/s$)	= 4.10E 00
*Beam pump pressure – zone 1 (Pa)	= 1.000E-02
*Gas load/inj. – zone 2 ($Pa \cdot m^3/s$)	= 4.100E 00
*Beam pump pressure – zone 2 (Pa)	= 1.000E-02
*Neutral beam power (MW)	= 75.00
*Neutral beam energy (keV)	= 175.00
*Neutral beam duration (s)	= 6.00
*Number of injectors	= 12.
T up beam ducts (g/m^2 per burn)	= 1.42E-04
Pump speed – zone 1 (m^3/s)	= 4.10E 02
Pump speed – zone 2 (m^3/s)	= 4.10E 02
Pump cost – zone 1 ($)	= 6.15E 05
Pump cost – zone 2 ($)	= 6.15E 05
Pump surface area – zone 1 (m^2)	= 1.78E 01
Pump surface area – zone 2 (m^2)	= 1.78E 01
Total cost beam vacuum sys. ($)	= 2.07E 07
D burned per day (g)	= 64.44
D injected per day (g)	= 34.15
D pumped/day-beams (g)	= 50.50
T pumped/day-beams (g)	= 1.0930
D fuelling/day (g)	= 463.40
D cold gas fill/day (g)	= 37.50

* Input parameters.

TABLE XII-3 (cont.)

TRITIUM FACILITY AND VACUUM SYSTEM PARAMETERS FOR THE INTOR	
EMERGENGY AIR DETRITIATION SYSTEM (EDS)	
Maximum conceivable release (g)	= 217.16
Initial T conc. ($\mu Ci/m^3$)	= 2.08E 07
*Cleanup time (h)	= 72.00
*Volume of reactor bldg. (m^3)	= 1.00E 05
*Decontamination factor	= 1.00E 06
*Cleanup efficiency	= 0.900
EDS flow rate (m^3/s)	= 5.547
EDS flow rate (% VBldg/min)	= 0.333
*T level stacked ($\mu Ci/m^3$)	= 50.000
Tritium release (Ci)	= 5.000
Vent rate (m^3/s)	= 1.67E 00
Tritium vented to env. (Ci)	= 6.26E 01
*Unit cost EDS ($)	= 1.06E 06

* Input parameters.

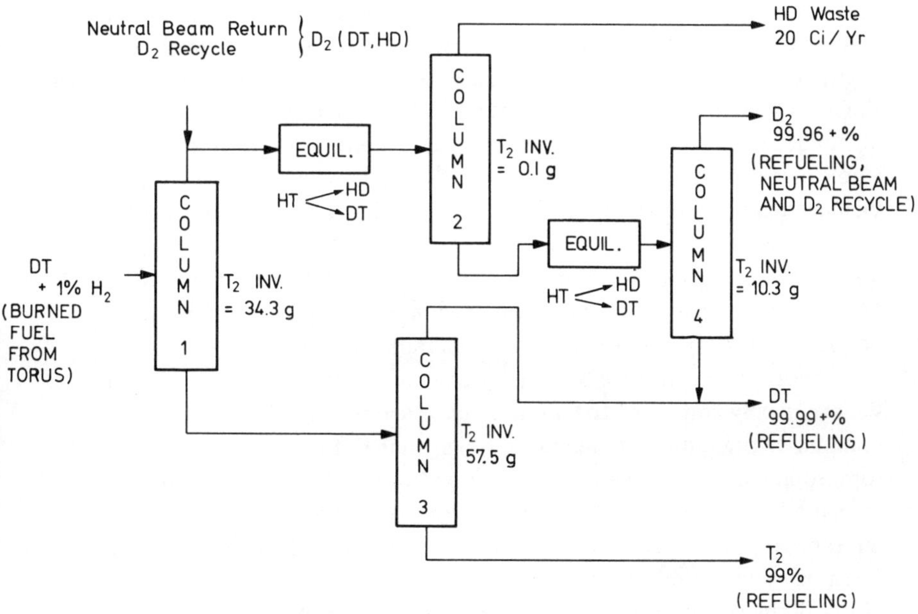

FIG.XII-1. Flow schematic of the cryogenic distillation isotope separation system.

3.1. Fuel purification

The exhaust fuel gas is expected to have roughly the following atomic composition:

D,T	~ 85–90%
H	~ 1–10%
^{4}He	≈ 5–10%
O, N and C	~ 1%
Ar	~ 0.001%

Oxygen, nitrogen and carbon are present mainly in the form of the chemical compounds water [DTO], ammonia [$N(D,T)_3$] and methane [CD_2T_2], respectively. After separation from the main stream, all these compounds must be chemically dissociated to separate and recover all the chemically combined tritium and discharge the impurity atoms, O, C and N to a tritium-free waste.

Various methods are known to effect the separation of impurities from hydrogen. One of the well-known ones is the use of palladium-silver alloy membranes, but in this method all the chemically combined tritium and also some molecular tritium passes out with the waste gas stream. Thus, although the main fuel stream is purified and the impurities are concentrated, the more difficult separations still remain to be done. In this kind of unit, the flow of gas to be purified is sent into a reactor equipped with thin membranes. The mechanism inside the reactor is the following: molecules of hydrogen at the upstream surface of the membrane dissociate into atoms, these atoms either react with the membrane material to form hydrides or are occluded in the interstices of the metal lattice. In either form the hydrogen atoms migrate through the membrane by bulk diffusion. At the other surface we find the inverse phenomena, i.e. recombination of the hydrogen atoms to form molecules. The whole mechanism is a physico-chemical phenomenon which in principle should yield a stream of pure hydrogen isotopes. On the other hand in order to have a high-enough flow rate, the membranes must be heated to temperatures of around 450–500°C. The combination of this relatively high temperature with a hydrogen pressure higher than 1 atm above the membranes, may cause a large tritium premeation through the walls of the membrane housing.

Another way to solve both the problems of purification and of recovering the chemically combined tritium is by chemical reaction with hot metals, like uranium or titanium, at temperatures high enough for the gaseous impurities to form solid oxides, nitrides and carbides and for solid hydrides not to be formed so that all the hydrogen isotopes remain in the gas phase. For the above gas mixtures and the degrees of separation required, both theoretical studies on thermodynamic equilibria and experimental studies on combined chemical and physical kinetics are needed. But one must strictly avoid any accidental introduction of air into a uranium bed to prevent a fire. With titanium, there is less of a problem in this regard, but the reactions of impurities with titanium are not as well-known as those with uranium. The helium and argon are not removed by hot metals and will need to be separated by physical processes. This hot metal technology involves the problems of diffusion of tritium through metal walls at high temperatures and the disposal of tritium contaminated solid wastes.

A third solution to the purification and tritium recovery problem, which requires only gaseous wastes with very low tritium content, is based mainly on cryogenic gas separation techniques and on electrolysis of DTO. It still requires developing and is most convenient for combining with cryogenic distillation for isotope separation and cryo-separation of the helium. All the impurities (except helium) are trapped, by freezing and adsorption down to 77 K in a conventional cryogenic hydrogen purification train. The chemically combined tritium is then recovered by oxidizing the trapped impurities to convert all the tritium into water (DTO) which is then cryo-separated and electrolysed to recover all the tritium. The impurity atoms are then all in the form of CO_2, O_2, N_2 and Ar which are non-toxic and, after cryo-separation from DTO, can be discharged as gaseous waste. Here is no need for containment of the main fuel stream at high temperatures and tritium partial pressures are low for the oxidation step.

Considering all possible options for purification, this last purification method may be the best choice for INTOR.

For the primary separation of the helium from the fuel stream, as alternatives to diffusion through hot Pd/Ag alloy there are two cryogenic methods which require developing. One is to separate the helium at the cryo-pump (assuming one is used) to evacuate the torus. This is done by adsorbing only the helium on a 4 K panel which is suitably separated from the panel on which the D-T solidifies. The helium panel is then regenerated while the D-T panel is kept cold. The second method is to condense the D-T at normal pressures down to 19 K in a falling liquid film condensor and pass the uncondensed helium stream through regenerateable adsorption beds at 77 K to remove all the remaining D-T from it. This last step will be required whatever method is used for the primary separation of the helium.

3.2. Hydrogen isotope separation

The burned D-T fuel consists of a mixture of six molecular species:

H_2, HD, HT, D_2, DT, and T_2.

For a given atom fraction of H, D and T the molecular composition will depend on whether the molecules are in thermodynamic equilibrium because this equilibrium (isotopic) can only be attained in hydrogen with a catalyst at about 200°C, above which temperature it does not change significantly. It is therefore assumed that the D-T fuel will be in high temperature molecular equilibrium (isotopic) when, for a 50/50 D-T mixture containing a small amount of H, it will have a molecular composition of:

DT $\cong 50\%$
D_2 $\cong 25\%$
T_2 $\cong 25\%$
HD = HT small
H_2 very small

Thus hydrogen (cryogenic) distillation, for certain separation duties, needs to be combined with chemical equilibration (200°C) of the molecular composition, for example (1) to promote $D_2 + HT \rightarrow HD + DT$ to split HT when it is necessary to extract more than half of the H from the fuel – as shown for TSTA in Fig.XII-1, and (2) to promote $2DT \rightarrow D_2 + T_2$ should it be necessary to split more than half of the D-T fuel into D_2 and T_2 streams.

In the water distillation process (which first requires oxidation of all the fuel) this molecular equilibration is rapid and takes place continuously in the distillation column itself. This is a distinct advantage over hydrogen distillation, depending on the exact separation duties required.

Hydrogen distillation is the process with which most experience with high tritium concentrations exists and it is also favoured because its high separation factor leads to a lower tritium inventory (see Fig.XII-1). Classical (isotope separation) theory predicts that, for a given separation duty and the known separation factors for hydrogen distillation and water distillation (at an optimum reduced pressure), the tritium inventory (liquid hold-up) is about two orders of magnitude higher for water distillation, if we ignore the advantages of continuous molecular equilibration in water. There is also the need to develop electrolytic cells for water distillation (for converting all product streams back to hydrogen) which have a sufficiently low liquid inventory and are made of materials which do not suffer any radiation damage from the high tritium concentrations.

Problems expected in hydrogen distillation are concerned mainly with controlling the operation of the several interlinked columns (see Fig.XII-1 for TSTA necessary to perform the various separation duties listed in paragraph 3 above).

Another method of separating hydrogen isotopes is the countercurrent mass diffusion process. The separation is carried out in very simple columns at a pressure of one atmosphere and a temperature of about 100°C. The main construction of the elements of the column consists of a grid cylinder and a sprayer (Fig.XII-2). A non-hydrogenous inert vapour is used as a carrier gas. This gas enters through the sprayer, crosses the grid cylinder and is condensed onto the walls of a cylindrical condenser, cooled with water on the outside. The vapour continuously presses the gas to be separated to the surface of the condenser

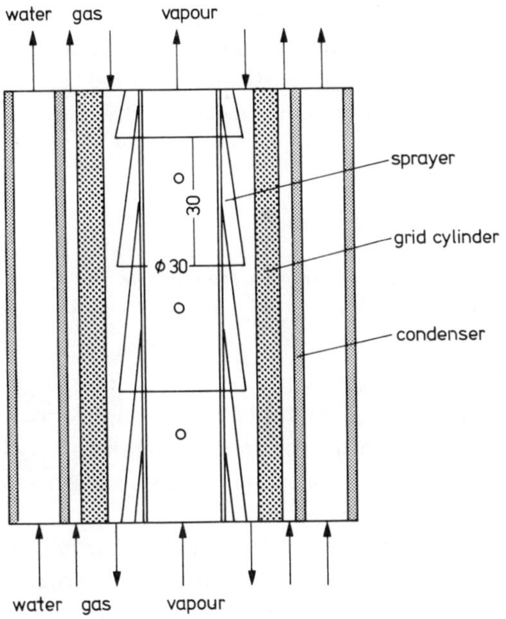

FIG.XII-2. Schematic diagram of a column for counter current mass diffusion hydrogen isotope separations.

so that the gas diffuses into the counterflow of the vapour within the grid cylinder, wherein the primary separation takes place. With this method the primary separation factor for the mixture H_2-D_2 is equal to 1.72 and for the mixture D_2-T_2 to 1.34. In the process of gas circulation along the column axis, a secondary counter-current separation occurs, however, the secondary separation factor is equal to the primary separation factor at the column height which is about equal to its diameter. The main advantage of this method lies in the fact that no preliminary purification is required. During the separation of the hydrogen isotopes, all the impurities are almost completely separated.

Other isotopic separation methods could be considered but they are not at the same level of development as the first three. Chromatography is a very interesting separation method in that it effects a high separation factor with a low height of transfer unit. Up to now chromatography has only used columns with a small diameter and in the case of INTOR we need columns with a diameter of at least 10 cm. New studies are necessary in this field. Another interesting method is the use of lasers. Lasers are a powerful method in the field of isotope separation but up to now the most important and most relevant data are classified. Thermal diffusion could be an attractive process for hydrogen isotopes where flow rates are small as in the final separation of tritium (and deuterium) from the waste protium stream. It would catalyse the formation of H_2 rather than HD waste and remove traces of tritium arising from degassing etc. while being able to operate at normal temperatures.

3.3. Storage

Obviously, there must be some storage unit in the fuel recycle system. A ageneral agreement exists for 30 days of burn up on hand which means that at least about 2 kg of tritium must be stored with a volume of about 7.5 m^3. Two ways are possible: storage of gas or of some hydride form.

Gaseous storage could be implemented, but in this case the gas must be kept at a pressure lower than atmospheric pressure to prevent any leaks of tritium out of the container. This container itself must be stored in an adequately ventilated secondary container.

Tritium storage as uranium or titanium hydride could also be used. This method involves heating the tritide to a temperature between 600–800°C to recover the tritium, but at this high temperature we find again the problem of tritium permeation through the container walls. The application of tritium storage on uranium beds has been extensively studied in the USA (See, for example, the Tritium Storage and Delivery System design for TFTR.)

3.4. Refueling

One of the purposes of the tritium system is to prepare the fuel for injection into the torus. Two modes of injection have been identified and retained for INTOR: gas puffing and pellet fueling. The first method seems to present no difficult problems; but the second case may entail some difficulties. If we want to obtain mixed pellets of deuterium and tritium, it seems practically impossible to prevent an isotopic separation between the two isotopes when the gas mixture is frozen. In this case, it seems necessary to exactly adjust the composition of the gas and this may be a very delicate operation. The same phenomena could also occur during the evaporation of the pellets and the combination of the two phenomena could add complexity to the fueling of the machine. Future studies are foreseen as essential in this area of technology.

3.5. Other technologies

Materials should be investigated for tritium/hydrogen embrittlement problems, tritium compatibility and radiation damage. The initial choice of materials, components, etc. should be made after considering all of the available reserve of information on tritium systems developed by existing, primarily defense-related programmes. In some cases, it will be necessary to scale up the dimensions, flow rates, etc. This should be done in a tritium facility such as TSTA.

It seems that there are no unsurmountable problems in the fuel recycle system, but, because of new separation requirements which arise with the use of tritium, there are a number of new problems which require research and development. The choice of the purification methods, isotopic separation process and storage systems must be done with the goal of minimizing the tritium inventory and the cost of the process. Some problems could arise in the manufacture of mixed pellets but nothing that cannot be solved with some research in this particular field.

3.6. Safety

Before final selections are made from the above processes and systems, detailed risk analyses are necessary in relation to protection of personnel and environment, particularly in connection with explosion risks (hydrogen) and relative radio-toxicity of the H_2 and H_2O based processes.

4. RECOVERY OF TRITIUM FROM THE BLANKET

The scope of this study is not to discuss the fraction of tritium that INTOR must breed, nor to discuss the need to breed tritium in test modules, large segments and/or outboard or inboard of the blanket. All these points are a function of the general philosophy and goals of INTOR. This part of the report attempts to define the choices among the different candidates for tritium breeding and the different processes involved in tritium extraction. More detailed properties of lithium compounds are compiled in Chapter XIII. Table XIII-8 of Chapter XIII summarizes the principal characteristics of ceramic lithium compounds.

4.1. Liquid lithium

Liquid lithium was one of the first materials suggested for breeding tritium. Methods for processing liquid lithium blankets have received a great deal of attention. The extraction of tritium, in this specific case, could be achieved by a continuous method which allows the separation to be carried out in small units directly coupled to a blanket module. The most promising methods that have been considered are: permeation of tritium through a "window" of highly permeable metal like niobium – or vanadium-base alloys, absorption of tritium by exothermic solid getters like yttrium and zirconium, extraction of tritium from liquid lithium into mixtures of liquid lithium halides (LiCl-LiF or LiF-LiCl-LiBr), and distillation. This last method has been shown to be thermodynamically possible because of the presence of the recently discovered lithium hydrides (Li_2D, LiD_2, Li_2D_2) in the saturated vapour over solutions of deuterium in liquid lithium; and experiments are under way to demonstrate enrichment of tritium by fractional distillation of solutions of tritium in lithium. With all these methods it is possible to maintain a tritium concentration in the lithium of less than 1 ppm. On the other hand, corrosion problems can occur with liquid lithium and these problems must be solved before using liquid lithium as the tritium breeder.

4.2. Lithium oxide (Li_2O)

Lithium oxide is the primary candidate material under study in several national programmes to breed tritium. The release rate of tritium from this oxide is very high. At a temperature or about 400–600°C, the mean residence time is so small that in-blanket tritium inventories of a few grams result. Nevertheless it is necessary to take care of the effects of neutron flux on the rate of decomposition of lithium hydroxide; some experiments are under way to study these effects, even though considerable changes in the tritium release rate have not been observed for $Li_2O(s)$ and LiOH(s) irradiated with neutrons to the fluences between 5×10^{15} and $9 \times 10^{17} n \cdot cm^{-2}$.

Lithium oxide in powder form is a very hygroscopic compound, and the manufacture of it in such a way that water vapour is not retained in the powder is just now beginning in the United States. Tritium is removed from the lithium hydroxide at a temperature of about 400–600°C in the water vapour form with the help of a helium stream that sweeps the powder material, probably with some impurities of an as yet unknown nature. If some light water is bonded with lithium oxide, as $Li_2O.(H_2O)_n$, this water will unfortunately lower the tritium concentration in the helium, i.e., there will be a large amount of tritiated water at a low tritium concentration. Also, the isotopic separation that must follow the electrolysis of this water will be more difficult. The other point against the utilization of lithium oxide as breeding material is the high temperature needed to remove the tritium from the blanket material. In this case, we have again the problem of tritium permeation through the surface of the helium loop containment structure and coolant tubes, and corrosion of the structure.

4.3. Lithium aluminate ($LiAlO_2$)

This lithium compound exhibits much the same behaviour as lithium oxide from the point of view of tritium extraction. Its rate of decomposition is also very high and the tritium concentration in the blanket could be kept as low as

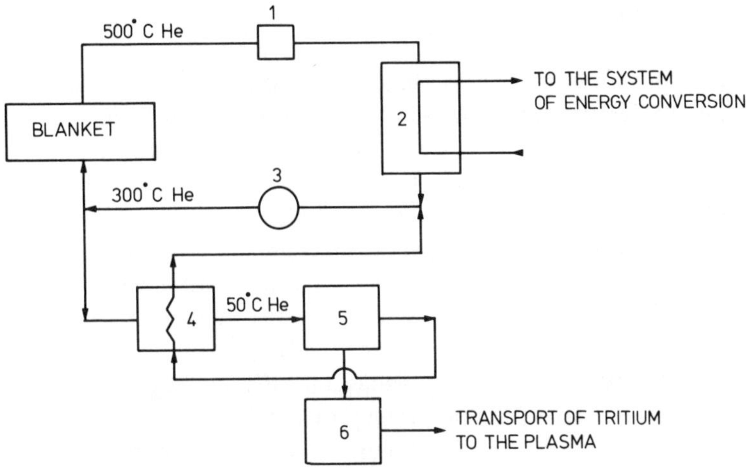

FIG.XII-3. Schematic diagram of the fuel cycle for the INTOR blanket.
1. *Catalytic oxidation* $T_2(H_2)$
2. *Primary heat exchanger*
3. *Compression*
4. *Heat exchanger*
5. *Absorption of water vapours*
6. *Tritium recovery.*

0.1 ppm, as is the case for the lithium oxide. Its affinity for bonding with water is lower than that of lithium oxide, but nevertheless it is necessary to pay great attention to this problem during the manufacture. The temperature of decomposition of lithium aluminate in its hydride form is much higher than that of lithium hydroxide, i.e., it is necessary to reach a temperature of about 700–800°C to obtain an appreciable rate of decomposition. In this case, lithium aluminate must be cladded inside a lower temperature wall to prevent a significant rate of tritium permeation through the blanket and helium loop structure.

4.4. Lithium orthosilicate (Li_4SiO_4) and lithium metasilicate (Li_2SiO_3)

The problem here is the same as for $LiAlO_2$. With these ceramic compounds, the rate of decomposition of the hydride form is very high, but in this case the tritium concentration in the blanket material could also be maintained below 1 to 0.1 ppm. Nevertheless, the temperature at which it is possible to remove the tritium from this type of compound is very high, at least 700°C, and with this temperature level the problem of tritium permeation through the structural walls may again be expected.

4.5. Tritium removal from ceramic materials

To remove tritium from the ceramic materials (Li_2O, $LiAlO_2$, Li_4SiO_4, Li_2SiO_3), it is necessary to force a helium stream through the blanket material. This helium stream can also be used as coolant. A schematic diagram of a typical flow plant for the blanket is given in Fig.XII-3 as an example. Helium, which circulates through the blanket at a temperature of about 500°C, leaves the lithium zone of the blanket and passes through the main primary heat exchanger, releasing its heat into the secondary circuit, connected with the energy conversion system. When helium leaves the heat exchanger, its temperature is about 300°C. The great amount of tritium removed from the blanket with helium is in the form of tritiated water vapour. In order to reduce the diffusion of tritium through the walls of the heat exchanger, all the tritium should be converted into the oxide. A catalytic oxidizer is provided for this purpose. A fraction of the helium stream is sent from the primary heat exchanger directly to the blanket. The other part passes through a secondary heat exchanger in which the water vapour is condensed at a temperature of about 50°C. The remaining fraction of tritiated water vapour is converted into tritium on a hot metal bed (uranium or titanium). Then, the hydrogen isotope mixture is adsorbed on a getter (titanium or zirconium). The recovered water is decomposed into hydrogen (tritium) and oxygen in an electrolytic cell. At the end, the removed stream of hydrogen (from the getter and from the electrolytic cell) is sent to the fuel recycle system for purification and isotopic separation. Experimental evidence is still required to establish all features of the tritium extraction scheme of the primary coolant.

Tritium release from ceramic materials is shown in Figs.XII-4 to XII-7.

4.6. Lithium alloys (LiAl, Li_7Pb_2)

Lithium alloys are among the first solid materials that were studied for tritium breeding. Nevertheless, greater uncertainty exists with respect to solid alloy blankets, although these materials do appear to approach liquid lithium and the ceramics, respectively, in terms of inventory and recovery characteristics. Finally, the temperature needed for tritium recovery is too close to the respective melting points of the alloys to yield a credible blanket design.

4.7. Summary

At this time, it seems difficult definitively to choose one form of lithium from among those mentioned above. The effects of some necessary neutron multipliers, as beryllium or lead, on the thermodynamic equilibria in the blanket itself are not known. In this field, there is a need for (1) some detailed studies

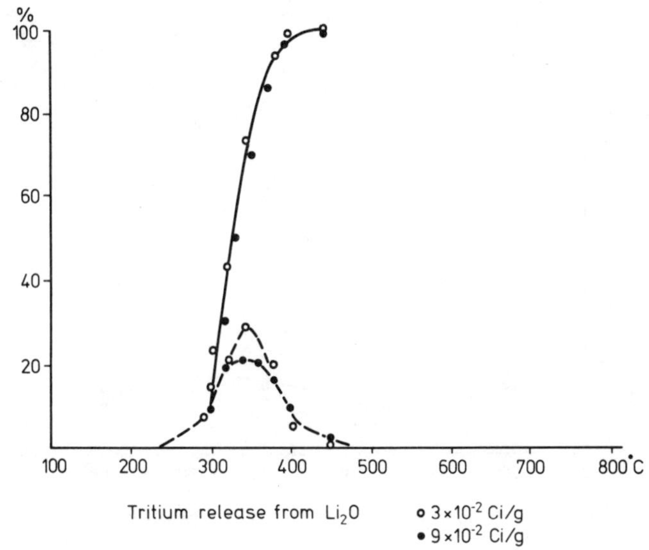

FIG.XII-4. Tritium release from Li_2O.

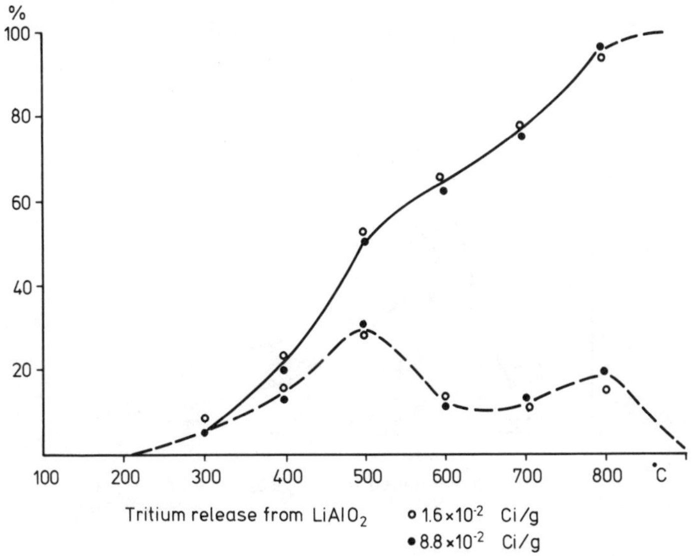

FIG.XII-5. Tritium release from $LiAlO_2$.

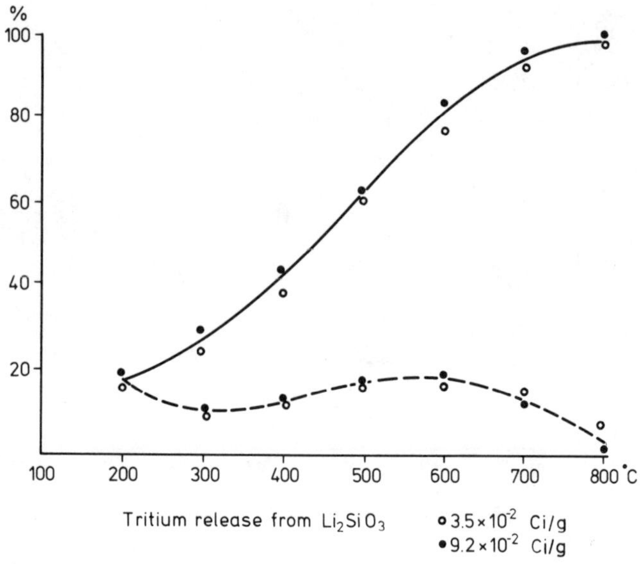

FIG.XII-6. Tritium release from Li_2SiO_3.

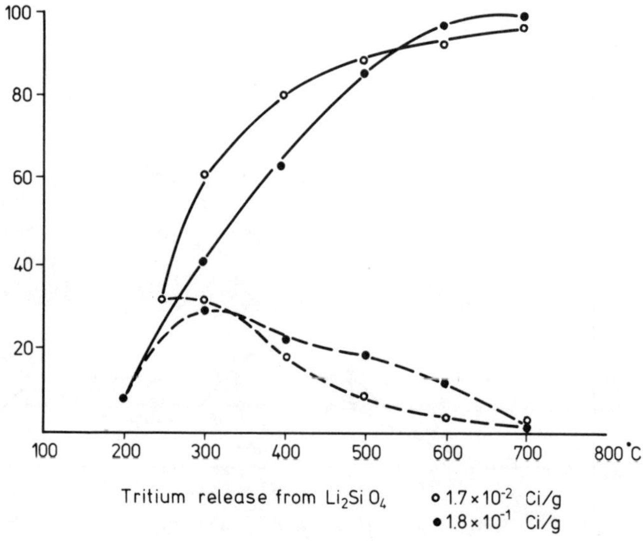

FIG.XII-7. Tritium release from Li_4SiO_4 separations.

on the chemistry of breeder materials and (2) irradiation experiments in thermal reactors to help identify the two or three lithium-bearing materials that should be retained for test module experiments on INTOR.

5. TRITIUM MATERIAL INTERACTION

The interaction of tritium with materials of construction is an important area of consideration in addressing tritium handling and containment for an INTOR-type reactor. The key questions relate to (1) dissolution, diffusion, and permeation of tritium into and through construction materials of all kinds; (2) effects of hydrogen isotopes on the integrity of materials; (3) consequences of the radiolysis accompanying beta decay of tritium; and (4) long-term performance and maintainability of hardware in a tritium environment. All of these phenomena could be influenced by temperature, impurities, neutron and gamma radiation, material imperfections, and mechanical stresses.

Tritium dissolution, diffusion and permeation into and through different materials are long-studied phenomena but, in the fusion field, these phenomena are understood only semi-quantitatively and there is an important lack of experimental data. Some results are given in Tables XII-4 toXII-6. In the specific case of a fusion device, it is important to also take into account the impact of the existence of fast ions and atoms. In the case of the first wall, these specific atomic entities could greatly increase the rate of permeation of tritium through the first wall. In this case, the first step of the whole diffusion mechanism (decomposition of a thermal molecule of hydrogen or tritium into two atoms, a step that is normally very slow in comparison with the others) effectively disappears. A research programme to measure the diffusion coefficient, the solubility, and the premeability of tritium through different kinds of materials under realistic conditions should be undertaken in the very near-term to allow appropriate choices between different candidate materials on the basis of minimum tritium outflow.

Embrittlement and other degrading phenomena associated with the presence of hydrogen in materials are also well known, but relatively few experimental measurements with tritium have been published up to now. Some of these embrittlement phenomena are likely to be enhanced by the beta-decay of tritium.

Another consequence of having tritium in or near materials is its corrosivity towards organic compounds, especially the compounds which contain halogens. In this particular case, even in the pesence of a reasonably low water vapour level, it has been shown possible to obtain a very aggresive mixture.

In summary, there is a general consensus of opinion that tritium permeation barriers will be needed in fusion reactor systems (particularly in elevated temperature environments) and that their development should be a high-priority item.

TABLE XII-4. PERMEATION CONSTANTS OF HYDROGEN ISOTOPES IN STAINLESS STEEL

Isotopes	Materials	Permeation constants (cc(STP)mm/cm^{2}h (atm)$^{1/2}$)	Temperature (K)	Pressure (atm)
H_2	SUS 316	1.15×10^3 exp(−15600/RT)	673–1073	1.3×10^{-2} –0.8
H_2	SUS 316	8.71×10^3 exp(−15130/RT)	673–1173	
H_2	SUS 316	1.53×10^3 exp(−16075/RT)	773–1173	
T_2	SUS 316	7.79×10^2 exp(−15900/RT)	773	1.3×10^{-13}
H_2	SUS 321	2.43×10^2 exp(−14197/RT)	423–1073	10^{-11}–10^{-14}
H_2	SUS 321	1.09×10^3 exp(−15700/RT)	659–1065	
H_2	SUS 310	8.46×10^2 exp(−17542±1399/RT)	425–873	0.001–1
H_2	SUS 304	2.26×10^3 exp(−17200/RT)	673–1073	1.3×10^{-2} –0.8
H_2	SUS 304	8.51×10^2 exp(−16100/RT)	773–1173	
H_2	SUS 304	6.70×10^2 exp(−15800/RT)	488–1003	
H_2	SUS 304	1.27×10^3 exp(−17000/RT)	823–1173	
H_2	SUS 304	1.16×10^3 exp(−17800/RT)	473–873	
H_2	SUS 304	8.68×10^2 exp(−16100/RT)	713–113	
T_2	SUS 304	5.81×10^1 exp(−16100/RT)	773	1.3×10^{-13}
H_2	SUS 304L	1.67×10^3 exp(−16100/RT)	500–713	
D_2	SUS 304L	5.80×10^2 exp(−15700/RT)	500–713	
T_2	SUS 304L	4.72×10^1 exp(−13750/RT)	373–473	
H_2	SUS 304	1.07×10^3 exp(−16130/RT)	869–1169	1
D_2	SUS 304	6.16×10^3 exp(−15400/RT)	650–1050	1.3×10^{-9} 1.3×10^{-6}

TABLE XII-5. HYDROGEN PERMEABILITY OF SOME REFRACTORY METALS

Materials	Permeation constants (cc(STP)mm/cm² h (atm)$^{1/2}$)	Temperature (C)	Pressure (atm)
Inconel 600	2.54×10^3 exp(−15800/RT)	800–1000	1
Inconel 600	3.05×10^3 exp(−16340/RT)	500–1100	1–5
Inconel 617	8.19×10^3 exp(−18860/RT)	600–1050	1–10
Inconel 625	1.57×10^3 exp(−14840/RT)	800–1000	1
Incoloy 800	2.73×10^3 exp(−16800/RT)	760–960	1
Incoloy 800	1.45×10^4 exp(−20570/RT)	600–1050	1–10
Incoloy 800	2.44×10^3 exp(−16500/RT)	800–1000	1
Incoloy 800	2.76×10^3 exp(−16800/RT)	500–1100	1–5
Incoloy 807	3.39×10^3 exp(−17480/RT)	800–1000	1
Hastelloy X	6.44×10^2 exp(−13910/RT)	400–600	0.5–1
Hastelloy X	2.29×10^2 exp(−16000/RT)	800–1000	1
Hastelloy X	1.90×10^3 exp(−15660/RT)	500–1100	1–5
Incoloy 800	6.24×10^2 exp(−15300/RT)	700–950	1–5
Incoloy 600	1.15×10^3 exp(−15300/RT)	700–950	1–5
Incoloy 800	1.46×10^3 exp(−15500/RT)	450–950	5–10
Incoloy 802	1.05×10^3 exp(−15200/RT)	450–950	5–10
Incoloy 807	1.30×10^3 exp(−15900/RT)	450–950	5–10
Incoloy 800H	2.14×10^3 exp(−15500/RT)	450–950	5–10
Incoloy 800	2.65×10^3 exp(−17700/RT)	600–950	5×10^{-4}–0.5
Incoloy 800	1.12×10^3 exp(−17700/RT)	649	0.001–1
Hastelloy N	2.97×10^3 exp(−18630/RT)	605	5×10^{-6}–0.04
Croloy	8.81×10^1 exp(−9350/RT)	400–500	$3 \times 10^{-6} \sim 1.7 \times 10^{-5}$
Incoloy 800	1.12×10^4 exp(19681/RT)	450–750	1.03

The tritium migration problem in heat transfer and energy conversion systems is well recognized, but the same problem in other reactor subsystems may be equally as serious and should be subjected to detailed analysis. There is need for a more expanded diffusivity, solubility and permeation data base encompassing austenitic steels, refractory-base alloys, and nickel-base alloys and including the effects of bulk and surface impurities, influence of alloying elements, and studies at low-driving pressures (i.e., $\leqslant 10^{-4}$ Pa) using tritium. Also the effects of neutron damage, energetic particle injection, and effects of helium bubbles formation (from transmutation and tritium decay) on tritium permeation through metals and on the integrity of permeation barriers should be studied. These effects are expected to be most pronounced in the vicinity of the first wall and in the blanket,

TABLE XII-6. PERMEATION CONSTANTS AND DIFFUSION COEFFICIENTS OF HYDROGEN ISOTOPES THROUGH Mo

Isotopes	Permeation constants (cc(STP)mm/cm^{2}h (atm)$^{1/2}$)	Temperature (K)	Pressure (atm)
H_2	$9.3 \times 10^2 \exp(-21.5\pm1/RT)$	1050–2400	1.3×10^{-9} –0.26
D_2	$8.27 \times 10^2 \exp(-19828/RT)$	523–730	1
D_2	$1.37 \times 10^2 \exp(-17400/RT)$	543–913	1.3×10^{-3} –2.6
Isotopes	**Diffusion coefficients (cm^2/s)**	**Temperature (K)**	**Pressure (atm)**
H_2	$5.9 \times 10^{-2} \exp(-14700/RT)$	1553–1973	1
H_2	$6.31 \times 10^{-4} \exp(-5900/RT)$	739–1123	1

their influence on materials performance in outlying regions as well should be taken into account.

Although the primary tritium containment systems for fusion reactors are expected to be constructed almost exclusively of metals, the use of organic materials cannot be avoided in many applications throughout the secondary containment systems (e.g., pump oils, seals, lubricants, gloves, gaskets, etc.). Radiation damage to these types of materials (caused primarily by the tritium that gets impregnated into them) is recognized as a serious problem. There is some experience in this area, but test data are available only for a limited variety of materials. Some incentive exists for developing coatings, paints, and other surface preparations that would resist tritium absorption. These might be particularly useful in guarding large surface areas (i.e., like the walls of the reactor hall itself) against massive tritium soaking following any sizable tritium release. Ceramics are known to be particularly suited as permeation barriers in and around secondary containment systems.

6. ANALYTICAL TECHNIQUES

Tritium analytical techniques can be logically divided into three main categories: methods used in inventory accounting, detection in liquids and breeding compounds, and ambient monitoring. This last technique is more properly located in the chapter on tritium safety.

6.1. Methods used in inventory accounting

6.1.1. Direct methods

Direct methods for hydrogen isotope analysis include standard procedures, such as mass spectrometry and gas chromatography. They also include infrared and magnetic resonance spectroscopy. There are a number of satisfactory methods for determining tritium quantities for inventory accounting purposes.

Mass spectrometry is well developed and continues to be improved as a hydrogen isotope analytical tool. The principal obstacle to effective mass spectroscopy is the near coincidence of the masses of ^4He, HT, and D_2 at m/e = 4 and of ^3He, HD, and T at m/e = 3. This results in a conflict between the selectivity (resolution) required to separate the various contributions at a given m/e value and the needs for sensitivity and/or stability required for high accuracy. By utilizing a combination of techniques, analyses can be routinely performed to accuracies of hundredths of a mole percent. This involves titanium sublimation pumping (to eliminate all hydrogen isotopes and permit unequivocal measurement of ^3He and ^4He) and uses different degrees of resolution (in which a rapid coarse scan provides relative abundances at m/e = 2, 3, 4, 5, and 6 and a high resolution scan separates HT from D_2 and HD from T).

Gas chromatography is well suited to determination of trace levels of impurities in hydrogen isotopes and can resolve impurities at levels below 100 ppm for each of them. Chromatography as an analytical tool to measure the isotopic hydrogen composition is also well known, and further development is not needed in this field.

Nuclear magnetic resonance is being used to measure properties of tritium and hydrogen in solids (e.g., metal hydrides), and it can be expected to be of some use in materials studies. Infrared and Raman spectroscopy may be of some value in hydrogen isotope analyses, but not much work has been done to develop their potential.

None of these methods (with the possible exception of nuclear magnetic resonance) measures tritium quantitatively. They measure concentration and, therefore, are only effective when used in conjunction with PVT or other measurements of total quantity. It should be noted that PVT measurements of tritium-containing gas mixtures are subject to errors arising from temperature gradients within the measuring volume. These gradients result from the heat generated by the tritium decay.

6.1.2. Inferential methods

These include methods of tritium detection which are not capable of directly demonstrating the presence of tritium or the relative isotopomeric concentrations but rather utilize the fact that tritium decays, releasing a beta particle and energy. The first of these is ionization counting. As mentioned earlier, ionization counters may be used up to and probably beyond 10%. Some development will be required to make accurate reliable instruments for this range available, as ionization counter development has largely focused on tritium concentrations below 1%. However, there seems to be no fundamental barrier to their use to 100%.

In addition to ionization chambers, solid scintillation photomultiplier tube combinations can be used to measure tritium at concentrations approaching 100%. A CaF_2(Eu) scintillator has reportedly been found to work over the range from 1 ppm to 100%. Solid scintillators must be protected from the presence of condensible forms of tritium such as T_2O or NT_3 which may adsorb on the detector surface, causing drift. With either ionization chambers or scintillators, PVT must also be used to obtain the tritium qualities.

The other inferential method that is widely used for tritium assay is calorimetry. As with ionization and scintillation counters, other radioactive sources can cause a false reading. On the other hand, a calorimetric measurement will yield a total tritium value from a single measurement as long as there are no other radioactive materials in the sample. Calorimetry has the further advantage of being capable of measuring tritium in solids, liquids, gases, or mixtures of the three, although measurement of tritium as gas is complicated by the large volume of the container. Currently available calorimeters cover the range above 0.1 watt.

6.2. Detection in liquids and breeding materials

6.2.1. Aqueous or organic materials

The data base for tritium detection in these liquids is quite well established. The principal means for detection of tritium in aqueous or organic materials is liquid scintillation counting. The principal requirement of the liquid being studied is that it be miscible with a scintillant. Tritium levels on the order of 10^{-9} Ci/litre are readily detectable.

6.2.2. Tritium in breeding compounds

In contrast to the situation in simple organic liquids or water, the data base for the detection of tritium in breeding materials (liquid lithium and solid compounds) is virtually nonexistent. Tritium in liquid metals or molten salts cannot be directly detected as can tritium in water, but must first be chemically separated. This can be done by almost any of the techniques being considered for large-scale recovery of tritium from breeder materials, for example by permeation through a membrane. The separated tritium may then be detected by an appropriate means, such as the current in an ion pump used to evacuate the space behind the membrane in the case of tritium gas permeating through a membrane. A tritium monitor has been available for use with liquid sodium, but because of the great affinity of lithium for hydrogen, a liquid lithium analogue has only recently been demonstrated. For breeding materials other than liquid lithium, such as LiAl, FLiBe (LiF-BeF_2), $LiAlO_2$, Li_2O, etc., there do not appear to be any satisfactory in-situ analytical procedures.

A major problem that must be addressed is how accurately tritium will need to be accounted for. Current accountability guidelines call for the reporting of tritium quantities to ~100 Ci(0.01 g). In a fusion reactor, where inventories will be measured in kilograms, this would require an accuracy of 10^{-3}% (0.01 g out of 10^3). This is far beyond current material accounting technology and is further complicated by the continuous creation, diffusion, burnup, and decay of tritium, combined with the fact that it will undoubtedly be present as a liquid, a gas, and a solid and as part of a number of solutions and chemical compounds.

7. TRITIUM MONITORING, CLEANUP AND DISPOSAL

7.1. Tritium monitoring

Tritium monitoring is taken here to refer to measurement of tritium concentration at levels that may be encountered in containment volumes (gloveboxes or rooms) or in the atmosphere. This covers the range from curies per cubic metre down to below 1 $\mu Ci/m^3$, which could also include levels in helium used as a reactor coolant. For air, nitrogen, or argon, which are the gases most commonly found in containment atmospheres, total real-time tritium activity measurement can be readily accomplished using available commercial ionization chamber counting equipment. Ranges covered are from 0.1×10^{-6} Ci/m^3 to 2000 Ci/m^3. Operation in a helium atmosphere is hampered by current leakage through the helium gas unless the counter is operated at low voltages (e.g., <1000 V).

At low levels and in the presence of either gaseous beta emitters or fields of gamma rays or neutrons, elaborate compensation (dual chambers) and other precautions (filters, shielding) may be required to avoid spurious results. By 1990, real-time tritium monitors that can distinguish tritium from other radioactive sources and HT (T_2, DT) from HTO (T_2O, DTO) will probably be readily available. Such instruments are currently being developed.

At levels below 0.1 $\mu Ci/m^3$, real-time monitors are not readily available. It seems probable that in 10 years improved electronic chamber fabrication and compensation techniques will permit a 10- to 100-fold increase in detection sensitivity. Integrating monitors are available which are sensitive to as little as 10^{-4} $\mu Ci/m^3$ in a 24-hour sample. Monitors of this type can also distinguish between HT and HTO.

7.2. Tritium containment

The most appropriate approach for the tritium containment problem is to use triple containment. It is the most convenient for the INTOR device.

In a fusion reactor, the primary containment must be taken to consist of everything within the fuel cycle: the injectors, gas feeds or pellet guns; the plasma chamber; the divertor, if one is used; the exhaust pumps; and the entire fuel purification and isotope separation system. It also includes the tritium breeding and recovery system, the blanket cooling system, and the storage and transportation systems.

Primary containment of tritium is accomplished through use of all metal components wherever possible. The preferred construction materials are the 300 series stainless steel (especially Type 316).

The ability for easy disassembly must be the guideline rule for most parts of the tritium system. Welding of the different parts of the loops must be avoided as far as possible to prevent large outgassing of tritium in the case of some change in this particular part of the loop.

Valves are all bellows sealed, and metal valve seats are preferred. Organic materials are generally avoided, although some materials, such as polymides for elastomer seals and polyphenyl ethers for pump fluids, are resistant to radiolytic attack. It is expected that the data base for primary containment will be substantially refined by 1990, but no major changes are likely to occur. Thorough cleaning of all interior surfaces of tritium-containing equipment will be necessary to prevent isotopic and/or chemical contamination. In general, established high vacuum system construction practice is a good foundation for primary tritium containment.

At higher temperatures, containment becomes much more difficult because of the rapid increase in tritium permeability through metals. For a nonbreeding INTOR, the necessary containment technology can be expected to be in hand, as the tritium levels in a nonbreeding blanket should be low. Large-scale breeding, a high-temperature breeding blanket, and a breeder coupled to the power extraction system will make the problem of primary containment progressively more difficult.

The purpose of secondary containment is to protect personnel in the event of either routine tritium loss through permeation, leakage, or a primary system failure, and to simplify tritium recovery in any instance. In many cases this is accomplished with gloveboxes, although there are a number of other kinds of secondary containment such as cryostatic vacuum jackets, jacketed tubing, and nested shipping containers.

Secondary containment volumes, in general, should be kept small to minimize the volume of containment atmosphere that must be processed in the event of a release from the primary volume. The containment atmosphere may be air, although it is usually argon or nitrogen in order to avoid the presence of oxygen and water vapour. Oxygen reacts with tritium to form T_2O, and water exchanges isotopically with tritium to form HTO. Because HTO is 200 times as toxic as HT and permeates more readily through organic materials (which are usually the weakest link in a secondary containment system), it is important to minimize the opportunity for HTO to form.

If the tritium is in the form of a metal tritide (particularly LiT or UT_3), there is an even more compelling reason for using a dry, oxygen-free secondary atmosphere. In the presence of either oxygen or water vapour, LiT will react rapidly to form Li_2O, and UT_3 will form UO_2; in either case, HTO or HT will be liberated. UT_3 will also react with nitrogen, hence only an argon atmosphere should be used near uranium tritide.

Secondary containment technology is rather well established. As long as the containment atmosphere is maintained at a low tritium level (less than 10^{-6} ppm) standard drybox practice is satisfactory for secondary containment construction.

Tertiary containment of tritium is much less well understood than are primary and secondary containment. This level of containment includes containment rooms and buildings. Because air detritiation is more closely related to tertiary containment systems than to primary or secondary systems, we will discuss them together here.

The two most important factors in this uncertainty are building size and soak-in. The first of these, building size, is important because the containment volume, V, the required cleanup time, T, and the flow rate through the cleanup system, F, are approximately related by $T = KV/F$, where K is a proportionality factor determined by parameters such as the size of the release, the degree of detritiation required, and many other less easily defined factors. K is of the order of 20; hence, if it is desired to clean a containment volume within a week after a spill, V/F should be on the order of 8 h. This requires a flow rate on the order of 7 m^3/s for a 2×10^5 m^3 building. If it is desired to clean the atmosphere in one day, the flow rate must be about 50 m^3/s, and the cost goes up accordingly.

It should be noted here that the degree of detritiation is only a secondary factor in the above case because the cleanup time (ideally) depends logarithmically upon the rates of initial to final concentration. Assuming a large release, say to a level of 10 Ci/m^3, it makes relatively little time difference whether one detritiates to 100, 10, or 1 $\mu Ci/m^3$.

Here the other factor, soak-in, enters. The more time atmosphere detritiation requires, the more tritium is absorbed by surfaces throughout the containment volume. This tritium will contaminate not only the absorbing materials, but also the containment atmosphere after the cleanup, as the tritium slowly desorbs from these materials over periods of weeks, months, or more. Hence, there is a tradeoff between the speed and expense of an atmospheric cleanup system that is difficult to resolve.

A factor that is related both to soak-in and to cleanup system flow rate is the nature of the tertiary containment structure. If a metal structure is used, even if it is only a thin shell, soak-in will be reduced and cleanup of a metal wall will be easier and more complete than in the case of a concrete structure. To the extent that the tertiary containment can be totally sealed, the environmental release of tritium may be greatly reduced, as well. This too is simpler with metal walls.

7.3. Atmosphere detritiation

In the case of secondary containment systems, there are two common methods used; continuous processing and detritiation as required. In either case, the processing method is almost always catalyzed oxidation (with O_2 supplied if necessary), followed by absorption of tritiated water vapour. One of the problems that must be solved in the catalyzed oxidation is the catalytic activity of the catalyst bed. Tritium must be present in the cover gas at a concentration as low as 10^{-6} ppm. At this level of concentration, the activity of the catalyst bed is not as good as in the case of higher concentrations. Some experiments show erratic results. An R and D effort is foreseen in this particular field to develop a more active catalyst. Detritiation by cryogenic absorption of tritium and water vapour could also be used with helium atmospheres.

Continuous processing is used in situations where the probability of tritium loss from the primary containment is high, as in a high-temperature system. In most cases, however, if the primary system is properly designed and operated, the secondary atmosphere can be kept clean by means of a slow purge of dry gas. The gas leaving the secondary containment can then be detritiated by a relatively small cleanup system. In the event of a tritium release into the secondary containment, a rapid purge can be used until the problem has been taken care of. A holding tank can be used to receive purge flow which is beyond the capacity of the cleanup system. This stored gas can then be cleaned at a later time. If there are organic vapours in the secondary atmosphere, high-temperature (>500°C) catalysts are required to break these compounds down and fully oxidize them.

7.4. Waste disposal

We can expect a substantial quantity of tritiated waste to be generated by INTOR. Two of these wastes will be tritiated water and tritiated oil. The former is a by-product of atmosphere detritiation systems, the latter will be principally vacuum pump oil and cleaning solvents. Water is immobilized by using it to form concrete and oil by absorbing it on vermiculite. In either case, the solidified waste is packaged in a nested series of watertight drums with vermiculite (for absorbing) and asphalt (for further sealing) in the annular spaces.

The remaining wastes will be solid waste (blanket test modules, parts of the tritium system, and any pieces that have been in contact for a more or less long time with tritium). Their level of activity, from the tritium point of view, is very low, but the real problem is an important outgassing of all these pieces. In any case these pieces must be stored in some plastic envelope. At the opening of this envelope, an important release of tritium would arise. Some momentary storage in a ventilated room is preferable. The definitive conditioning of these solid wastes must follow well-established procedures.

Because of its high value and limited availability, the materials management requirements are such that very accurate accountability records must be maintained. A semi-annual physical measurement of tritium inventories is required. Tritium must be accounted for to the closest 0.01 g. This presents some very serious problems to a fusion facility with a 2–3 kg inventory, much of which is tied up in cryogenic vacuum systems, structural material and storage.

The development of new requirements and new techniques for maintaining accountability data is a very real challenge that must be solved prior to using large quantities of tritium at a fusion facility.

8. TRITIUM PROCUREMENT, TRANSPORT AND SAFEGUARDS

The different estimates of the INTOR tritium inventory give a quantity of about 2 to 5 kg, and the INTOR tritium consumption rate is between a few kg/y at the start of DT operation and 10 to 14 kg/y at the end of its life. It is difficult to imagine that INTOR could itself produce all the tritium it will consume at least at the beginning of the experiment. In any case, a breeding ratio of one was not considered, which implies that it will be necessary to identify an external supply for the tritium.

Large quantities of tritium can only be made available by the neutron irradiation of ^{6}Li in fission-production reactors or with other neutron sources. Technology has been well developed for tritium production from both lithium-aluminium alloys and lithium aluminates. Tritium has been produced for thermonuclear weapons and commercial uses in production reactors by irradiation of lithium-aluminium alloys. Typical annual T_2 production rates of a 5 kg/1000 MW_{th} have already been achieved in tritium production reactors.

Some other external sources will be envisaged, such as heavy water reactors, (CANDU reactors are able to supply at least 20 kg of tritium up to 2000) reprocessing plants, etc., but none of these plants is actually equipped with a tritium recovery unit. Tritium is available to commercial users in quantities of about 50 000 to 100 000 Ci. But the needs of INTOR are largely out of these usual quantities. Under these conditions, it seems reasonable to think that the external tritium supply for INTOR will be the subject of some agreement between the INTOR participating countries.

Up to now, the largest, approved shipping container for tritium gas is a 50 l container approved for shipments of no more than 193 500 Ci (about 20 g) of tritium. A 3 kg inventory would therefore require shipment of 150 such containers and a tritium consumption of 12 kg/y is out of the scope. A bigger kind of container with higher pressure than the 1250 torr actually required will have to be developed and approved for the fusion programme. In some cases a solid tritium form could be developed in a parallel way. This task must be the basis of some international organization.

At the present time tritium is subject to rigorous material management requirements. Due to its limited availability and high costs (~\$7500/g), tritium is handled as a "sensitive" material, but is not considered a "strategic" material. This means there are no safeguard requirements for tritium as a nuclear material, but rather, the physical security of the material must be assured because it is a limited resource and very expensive.

9. CONCLUSIONS

The handling of tritium for INTOR is subject to varying degrees of difficulty depending on the portion of the tritium system in question. In the fuel recycle system the level of knowledge required for a design is reasonably at hand today. Some more extended research is needed on the isotopic separation methods, such as chromatography and lasers. The choice between the other methods is only with regard to minimizing the cost and the tritium inventory. Some experiments are also needed in the technology field and these experiments will be done in a tritium facility such as TSTA in the USA , in the TETL in Japan and the VTS in the USSR by the mid 1980's.

While considerably more research must be performed on the ultimate breeding materials for fusion, it seems possible now to choose materials for tritium test modules in INTOR; but the ability continuously to extract tritium from a breeder blanket module in a fusion reactor at the rate it is being produced and simultaneously to maintain a low in blanket tritium inventory remains to be demonstrated. Studies upon the thermodynamic equilibrium which could exist between lithium containing materials, neutron multipliers and the activation products in the blanket should be undertaken. Extraction methods and behaviour studies of breeding compounds under irradiation would and could be developed and performed in fission reactor facilities by the mid 1980's. It is also necessary to develop some analytical techniques to analyse the tritium concentration in solid breeding materials.

The containment of tritium within the fuel processing system, the blanket and associated processing systems, and the clean-up and waste treatment systems is an area of some concern. The data base concerning hydrogen/metal interactions is reasonably well in hand for INTOR. However, considerable work remains to be done on the tritium/metal interaction, on the development and proof-testing of a variety of tritium permeation, barriers, particularly for the heat transfer interfaces and for lining the wall of the primary containment materials to reduce the level of tritium dissolved in them. A definitive and successful demonstration of the adequacy of INTOR tritium containment practices is absolutely necessary. With appropriate modifications such demonstration could be done in TSTA, TETL or VTS.

The supply of tritium for INTOR (as much as 70–100 kg over its lifetime) appears not to present immediate difficulty in that the USA and Canada have affirmed their capability in this regard. The principal question is one of cost, with the lowest estimate of $0.4/Ci (Canada). However, such procurement requires international agreements for the transfer of tritium from one country to another and the format for these agreements should be established at an early stage.

TABLE XII-7. R AND D NEEDS ASSESSMENT – TRITIUM

	Status*	
	Adequate	Inadequate
Demonstrate performance, reliability and maintainability of integrated fuel handling system. Demonstrate performance of auxiliary safety (clean-up and back-up) systems and tritium monitoring instrumentation (TSTA, TETL, UTS)	x	
Study dynamics of enclosure clean-up operations following tritium spills	x	
Determine concentration and chemical composition of impurities exhausted from the plasma		x
Characterize tritium interaction with materials under realistic conditions		x
Develop effective barriers to tritium permeation		x
Develop and demonstrate removal of tritium from the coolant	x	
Resolve tritium shipping, accountability, and safeguard issues		x
Develop and demonstrate extraction of tritium from breeder materials	x	
Study tritium aspects of fuel fabrication and injection		x
Study the influence of a divertor on the tritium system		x

Adequate: Programmes are either planned or could be planned on existing or authorized facilities (with some upgrading if required) and using existing trained personnel.

Inadequate: Either the available and authorized manpower and facilities are inadequate or a substantial increase in effort is needed.

* The ratings are made on the world-wide fusion effort and not separately on the individual national programmes.

If current tritium shipping criteria were applied to INTOR (except in the USSR), hundreds of shipment per year would be necessary. A clear need exists to establish and implement criteria that will permit shipment of hundred-gram quantities within and across international borders. There appears to be no serious problem in this area insofar as all participating countries are in good agreement, however, a great deal of time will be required to design and license larger shipping casks.

Up to now, tritium is not regarded as "nuclear material" but the handling of large quantities of tritium should be subject to new safeguard requirements. The question of in hand accountability for tritium (i.e., to perhaps 0.01 of a gram in the USA) may become a problem area. This is another area of policy which must be examined in more detail.

Finally, the consolidation, handling, and storage of tritiated wastes present no serious problems for INTOR, in that a reasonably well established technology already exists. It would, however, be of great benefit for INTOR to adopt the philosophy of minimum tritium recycle (from waste steams) for economic reasons and to minimize tritium-active waste volumes.

10. R AND D NEEDS ASSESSMENT

The R and D needs assessment for tritium is summarized in Table XII-7.

REFERENCES TO CHAPTER XII

[1] "Tritium", Chapter in USA INTOR Report (see Appendix A.5.4).

[2] European Contributions to the 2nd meeting of the INTOR Workshop, Tritium, Chapter XIII, EUR FU/XII 501/79/EDV 50, June 79 (2 Volumes).

[3] USSR Report to INTOR, "Tritium in International Reactor Tokamak INTOR", (see Appendix A.5.3 [38]).

[4] Fusion Research Development Center, INTOR Workshop Report, Group 13, Tritium, JAERI-M 8512 (1979).

Chapter XIII

MATERIALS

G. KULCINSKI – USA, Chairman
P. REYNOLDS – EC
T. HIRAOKA – Japan
B. KOLBASOV – USSR

1. INTRODUCTION

The question of materials performance in INTOR is considered vital to whether or not we can build and operate such a tokamak in the early 1990's. The objective of this report is to state what data base is available today to make such a decision and what might be available by the 1984–5 time period.

At the present time, the member nations which have contributed to this report have all used somewhat different operating parameters for INTOR and these are listed in Table XIII-1. With such a wide range of parameters, it is understandable that different choices of materials might be made by the various countries. There were also variations in emphasis by the different countries and this, combined with the fact that the individual reports amounted to over 400 pages, means that we can only present the briefest of summaries here.

2. DATA BASE ASSESSMENT FOR UNIRRADIATED STRUCTURAL MATERIALS

2.1. Alloys considered

Several different alloys were considered for the structural components of INTOR These alloys fell into the general classes of:

Austenitic stainless steels
Ferritic stainless steels
Titanium alloys
Aluminium alloys

In addition, the Japanese considered the molybdenum alloy, TZM, for protection of the first wall.

Text continued on page 374.

TABLE XIII-1. SUMMARY OF IMPORTANT FEATURES OF INTOR THAT GOVERN THE CHOICE OF MATERIALS

	U.S.	JAPAN	U.S.S.R.	E.C.
Peak DT power	790 MW_t	670	500	600
Burn time	100 sec	200	100–200	>100
Dwell time	20 sec	35	20	30
Startup and shutdown time	15 sec	15	15	20
Maximum availability	50%	25	40	25
Pulses/lifetime	10^6	3×10^5	10^6	3×10^5
Impurity control	Improved divertor	Poloidal divertor	Improved divertor	Cold blanket bundle divertor
Maximum hydrogenic cavity pressure	10^{-3} Pa	–	–	–
TF coil S/C	NbTi, Nb_3Sn	Nb_3Sn	NbTi, Nb_3Sn	Nb_3Sn, NbTi
TF coil stabilizer	Cu	Cu	Cu	Cu
Maximum lifetime tolerable radiation	10^{10} rads	10^9	10^9	10^9
OH coil conductor	NbTi	NbTi	NbTi	NbTi
EF coil conductor	Cu and NbTi	NbTi/Cu	NbTi/Cu	NbTi/Cu
EF coil location	Inside and outside TFC	Outside	Inside/outside	Inside/outside
Vacuum vessel location	Behind shield, inside TFC	Inner sur. shield		NS
First wall	316 stainless steel	316 S.S.	Steel or Al alloy	SS, Al-alloys
Coolant	Pressurized H_2O	H_2O	H_2O	H_2O
Maximum structure temp.	350°C	~150	200–250 SS, 150–200 Al.	150
Av. neutron wall load	1.6 MW/m^2	1.3	~1.3	~1
Av. surface heat load (W/o div.)	0.4 MW/m^2	0.3	≤0.35	0.25–0.3
Peak neutron wall load	2 MW/m^2	1.6	NS	NS
Max. required FW lifetime	>7 $MW\text{-}yr/m^2$	NS	NS	NS

	U.S.	JAPAN	U.S.S.R.	E.C.
FW protection	Low-Z (coated?) armor plate	Prot. wall	Prot. limitor	Low-Z prot. coating
Electricity production	Module (outboard)	Segment (out/b)	Module (out/b or Ih/b)	segment (In/b/out/b)
Electricity producing blanket material	316 stainless steel	316 S.S.	S.S.	S.S. or Al alloy
Electricity producing blanket coolant	H_2O	He, H_2O	H_2O	He/H_2O
Electricity producing blanket thickness	0.3 m	0.5	0.3	1
Tritium producing blanket	Outboard only	Outboard only	Outboard	Inboard/outboard only
Tritium producing blanket material	316 stainless steel	316 S.S.	S.S.	S.S.
Tritium producing blanket coolant	He, H_2O	He, H_2O	He, H_2O	NS
Tritium breeding blanket breeder	Li_2O, Li	Li_2O	$LiAlO_2$, $LiSiO_4$, Li_2SiO_3	Li
Bulk shield modules	Inboard and outboard	Inboard and outboard	Inboard/outboard	NS
Bulk shield material	Stainless steel (100°C)	304 S.S. borated H_2O	S.S.	NS
Bulk shield coolant	H_2O	H_2O	H_2O, He	H_2O
Bulk shield thickness – inboard	1.1 m	1.2 heavy concrete	0.9–1.1	~0.9
outboard	>1.4 m	1.5	1.4	>1.0

NS = Not Stated

TABLE XIII-2. NOMINAL COMPOSITION OF MAIN ALLOYS CONSIDERED FOR INTOR.

	Alloy	Country	Fe	Cr	Ni	Mn	Mo	C	V	Ti	Al	Si	Other
Austenitic	316 SS	USA Japan EC	Bal	17	13	2	2.5	0.05	0.2			0.7	
	OX16N15M3B	USSR	Bal	15–17	14–16	0.8	2.5–3.0					0.6	0.25–0.5 M6
	1 X 18H12T (321 SS)	USSR	Bal	17–19	11–13	1–2		0.12		0.5		0.8	
	1 X 18H10T (304 SS)	USSR	Bal										
	0 X 18H9	USSR	Bal										
	ICLOO4BC	EC	Bal	18	7	7.5		0.004				1.0	0.2 N
	OCLO16BC	EC	Bal	18	7.5	8.0	1.5	0.004					0.02 N, 2 Cu
	ANCR	EC	Bal	10	0.8	17.5		0.3					
	14 Mn-2Cr-2Ni	Japan	Bal	1.8–2.2	1.8–2.2	13–14		0.55–0.65				0.5	
	18 Mn-5Cr	Japan	Bal	3.5–6	2	16–20		0.4–0.6	0.1–0.3			0.2–0.65	
	SS-Y4 S-170	Japan	Bal	24.5	13.1		0.84	0.031					
Ferritic	HT-9	USA	Bal	12			1	0.2	0.3				0.5 W
	9CR-1Mo	USA	Bal	9			1	0.08	0.15				0.5 W, 01 Nb
	0 X 13	USSR	Bal	11–13		0.6		0.08				0.6	
	1 X 13	USSR	Bal	12–14		0.6		0.09–0.15				0.6	
	X13M25QP	USSR	Bal	13	0.22	0.31	1.6	0.14				0.26	0.38 Mb

	Alloy	Country	Fe	Cr	Ni	Mn	Mo	C	V	Ti	Al	Si	Other
Aluminium Alloys	SAV-1	USSR	0.2		0.03						Bal	0.7–12	0.45–0.9 Mg 0.012 Cu 0.03 Zn
	AMG-2	USSR	0.4								Bal	0.2	1.8–2.8 Mg 0.1 Cu
Titanium Alloys	Ti-6Al-4V	USA, EC Japan							4	Bal	6		
	Ti-6242	USA					2			Bal	6		2 Sn, 4 Zr
	Ti-5Al-2Cr-1Fe	Japan	1	2						Bal	5		
	Ti-13V-11Cr-3Al	Japan		11					13	Bal	3		
	AT-6	USSR	0.25–0.4	0.5–0.9						Bal	5–7	5–7	
	AT-3	USSR	0.2–0.3	0.2–0.3						Bal	3–3.5	0.15–0.25	
	VT-6	USSR							4.5	Bal	6		
	VT-14	USSR					30		1.0	Bal	14.5		
	4201	USSR					33			Bal			

While most of the data base presented in this report centres around stainless steel, it is important to recognize the reasons for including the other alloys at this time. The ferritic steels have been considered mainly for their superior resistance to void formation during neutron irradiation, lower cost, and because they can handle higher heat loads than the austenitic steels.

The titanium alloys were also chosen for their better heat handling capabilities, resistance to void formation (at least up to a few dpa) and their much less long-lived induced radioactivity levels. The Al alloys were considered because of better heat load capacity, and low long lived radioactivity.

For various reasons, the alternative alloys did not measure up to the stainless steels and because of that fact we have only included brief references to the data base for these alloys. The reader is urged to consult the US report for more information on ferritic steels and titanium alloys, the USSR and EC reports for Al alloys, and the Japanese reports for Mo alloys. The composition of the typical alloys considered for INTOR vary from country to country and Table XIII-2 summarizes some of those alloy constituents.

2.2. Industrial base

In 1974, over 1 200 000 tonnes of stainless steel were shipped in the US. Of this amount, over 40 000 tonnes were of 316 SS sheet and strip. Figures for the other countries (USSR, Japan and EC) were unavailable but they also have large industrial bases for stainless steel production. It was determined that there was essentially no problem in any of the countries to produce the several hundred tonne quantities of 316 SS required by INTOR.

Chromium-containing alloy shipments in the US totalled over 3 400 000 tonnes in 1974. The Cr-Mo-V alloys comprised over 44 000 tonnes of that quantity. Presently, 9 Cr–1 Mo is a developmental alloy and only available in 25 to 50 kg heats. However, there are no perceived difficulties in expanding production to the multitonne level by 1985. No production figures on ferritic steels were given by the other countries but again, there seems to be no major problem in obtaining 100 tonne quantities for INTOR.

In 1977 the industrial production of Ti alloys in the US was over 20 000 tonnes, over half of which was of the Ti-6Al-4V alloy. The Ti industry in Japan also produces some 6000 tonnes per year. The USSR and EC teams also expressed the optimism that the needed quantities of Ti could be obtained.

The production of Al alloys in 1974 was approximately 400 000 tonnes in the US. None of the other countries anticipate any problems in providing Al alloys at the 100 tonne levels.

TABLE XIII-3. REPRESENTATIVE VALUES OF SELECTED THERMAL, PHYSICAL AND ELASTIC PROPERTIES, AND THE THERMAL STRESS FACTOR

Property (units)	Candidate alloy			
	316	HT-9	Ti-64	Al alloy**
Melting temperature (°C)	~1430	1427/1482	1668	660
Mass density (g/cm^3)*	7.81	7.65	4.3	2.7
Poission's ratio*	0.294	0.265	0.31	(0.3)
Young's modulus (GPa)*	167	155	85	60
Heat capacity (J/kg-K)*	567	725	705	(1000)
Mean thermal expansion ($°C^{-1} \times 10^{-6}$)*	17.7	11.3	10.0	27.0
Thermal conductivity (W/m-K)*	19.5	29	10.9	210
Electricity resistivity (10^{-6} Ω-cm)	97	90	186	5.6
Thermal stress factor, 10^{-6} GPa-m/W*	215	93	113	12

* at 400°C.

** at 200°C.

2.3. Physical properties

It is not possible to list all the physical properties of the alloys in Table XIII-2 and the reader is referred to the original reports for details. However, it is instructive to consider those properties that might influence the thermal stress induced into the first wall. Essentially all the **physical** property data for the alloys in Table XIII-2 are available in the open literature and there is no need for extensive programmes to provide more information. An indication of the relative value of the thermal stress which will develop in a given structural material can be estimated by the relationship.

$$\text{Thermal stress} = \pm \frac{1}{2} \frac{\alpha E}{K(1-\nu)} \left[\dot{Q}\Delta\ell + \frac{W_n}{2} \Delta\ell^2 \right] \quad (1)$$

where:
- α = Thermal expansion coefficient
- E = Young's modulus
- Q = Surface heat load
- W_n = Nuclear heating
- $\Delta\ell$ = Thickness of rigidly held first wall
- K = Thermal conductivity
- ν = Poisson's ratio

The physical properties for the main alloys considered for INTOR are listed in Table XIII-3.

Solving Eq. (1) for the value of $\Delta \ell$ at which the thermal stress just reaches the yield stress (see Section 2.6), we find the following values for the main INTOR materials ($\dot{Q}_{max} = 40\ W/cm^2$; $W_n^{max} = 15\ W/cm^3$):

Austenitic	(316 SS–20% CW) – 1.3 cm
Ferritic	(HT-9) – 2.2 cm
Titanium alloy	(Ti-64) – 2.4 cm
Al alloy	(6061) – 4.5 cm

Assuming a safety factor of 2 still allows the first wall thickness to be ⩾5 mm for austenitic steels and ⩾10 mm for ferritic and titanium alloys and ⩾20 mm for Al alloys. Consideration of other factors such as fatigue life, creep, and creep-rupture life could also lower the values by another factor of 50%.

2.4. Fabrication

The austenitic stainless steels possess excellent formability and fabricability characteristics. Some concern is expressed about the welds in the first wall and blanket region. First of all, existing welding conditions produce small amounts of ferrite in the weld zone and the effect of this magnetic phase needs to be examined. Secondly, the welded areas remove the beneficial effects of cold working and may make the heat affected zone more susceptible to radiation damage. Thirdly, the welds themselves are susceptible to neutron radiation and, therefore, may be the limiting feature in lifetime calculations. Therefore, considerable attention must be paid to removing the welds as far back in the blanket as possible.

The fabricability of the ferritic steels is also considered to be good but there is a major concern about the pre- and post-weld heat treatments required for the Cr-Mo-V alloys. It appears to be necessary to preheat the weld zone areas to 200–400°C and to provide final post-weld heat treatments at 600 to 750°C for one hour to avoid the formation of cracks in the heat affected zones. This may present some problems for the field fabrication of large components.

The Ti-6Al-4V alloys possess excellent formability characteristics as witnessed by their wide applications in the aeronautical industry. Welding is also straightforward in this system. The Ti-6242 alloys are also relatively easy to fabricate, but the welding of this alloy is somewhat unpredictable due to metallurgical phase instabilities. More work needs to be done in this system with respect to the irradiation stability of these phases.

The fabrication of Al alloys is no problem for INTOR.

2.5. Compatibility

2.5.1. Water

The corrosion rates of austenitic and ferritic steels in an aqueous environment depends on the purity (primarily the oxygen content) of the water, and the surface conditions of the steel. At 400°C, the corrosion rate of ferritic steels is generally about an order of magnitude higher than for the austenitic alloys (i.e., ~10 microns per year versus ~1 micron per year in 304 SS). This is not considered serious from the standpoint of the mechanical integrity of the structure, but the corrosion products could contribute to very high radioactivity levels in the auxiliary cooling equipment (heat exchangers, pumps, clean-up systems etc.) outside the reactor. Some concern is expressed by the USSR working group on the stress corrosion of cold worked 316 SS.

Aqueous corrosion data for titanium alloys are very limited for the conditions of interest. However, Ti is generally considered to be more resistant to high temperature water than Zr or Zircalloys which are currently being used up to 360°C in LWR's. Titanium alloys containing up to 1.75 wt% boron exhibit weight gains of about 1 g/m^2 after 3000 h exposure to 360°C water. Hydrogen, generated by the water-Ti reaction also seems to be more important than for Zr.

It is well known that high purity Al *disintegrates* in 150°C water after a very short time. Alloying improves the situation such that corrosion rates of 1 mil/year (6061-T6) to 8 mils/year (3003-H18) are obtained in 260°C water. However, some of these alloying elements (Fe, Cu, Mn, Cr, and Zn) remove the great advantage that Al may have with respect to long lived radioisotopes.

2.5.2. Helium as a fusion reactor coolant

Helium has functioned satisfactorily in gas-cooled fission reactors, establishing its reliability as a coolant and providing a substantial technology base relevant to fusion technology. However, several problem areas must be addressed to make a decision regarding the use of helium as a fusion reactor coolant, since the designs and requirements of gas-cooled fusion reactors may differ in some respects from the existing gas-cooled fission reactors. The principal compatibility problems arise not from helium, which is inert, but from impurities which frequently are difficult to restrict to concentrations which preclude corrosion reactions. A positive factor in helium utilization as a coolant is the fact that a substantial technology exists for helium purification and analysis, though it is not clear that existing procedures will fully meet the needs of fusion technology.

In contrast to the liquid-cooled fission systems, activity transport in gas-cooled systems has been minimal, mainly due to fission products. Thus, the

TABLE XIII-4. EQUILIBRIUM HYDROGEN CONCENTRATION IN CANDIDATE STRUCTURAL MATERIALS AS A FUNCTION OF PRESSURE AT 300°C

Hydrogen pressure Pa	Equilibrium hydrogen concentration (appm)[a]			
	Type 316 SS	Ferritic steel	Ti-6Al-4V[b]	Al
10^{-3}	<1	0.1	4 000	≪0.1
10^{-2}	<1	<0.1	10 000	≪0.1
10^{-1}	~1	~0.1	40 000	≪0.1

[a] Concentrations of deuterium and tritium are approximately the same as hydrogen at corresponding pressures.

[b] Kinetic considerations may not allow such high values to be achieved in normal operating times.

gas-cooled fusion system offers a major advantage from the standpoint of radiation fields. On the other hand, tritium inleakage to the gas coolant may impose the need for other activity control measures.

2.5.3. Hydrogen isotope effects and compatibility

The two major concerns with hydrogen compatibility involve effects on tritium inventory and the potential for hydrogen embrittlement, which occurs in some structural materials at relatively low hydrogen pressures. The equilibrium hydrogen concentrations in the candidate structural materials are given in Table XIII-4 as a function of pressure at 300°C. The concentrations of deuterium and tritium in atom fraction are approximately the same as those for hydrogen at corresponding pressures. The hydrogen-isotope concentrations in the austenitic and ferritic steels, and Al alloys are so low as to be of little consequence with respect to either tritium inventory or embrittlement. However, titanium alloys exhibit very high hydrogen solubilities even at relatively low hydrogen pressures. Substantial hydride embrittlement of titanium alloys occurs at low temperatures when the hydrogen concentration approaches 1.0 at-%. The hydrogen concentrations given in Table XIII-4 are the maximum possible and may not be attained in a real system because of kinetic considerations.

2.6. Mechanical properties

The unirradiated mechanical property data base for each of the four alloy systems considered for INTOR is fairly extensive and in general it should be

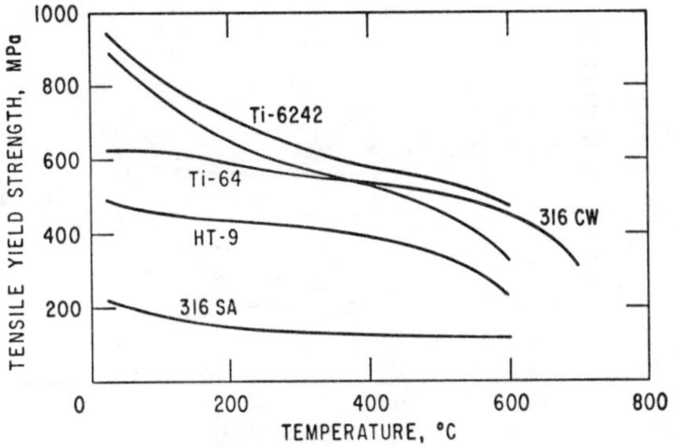

FIG.XIII-1. Tensile yield strength temperature for INTOR candidate alloys.

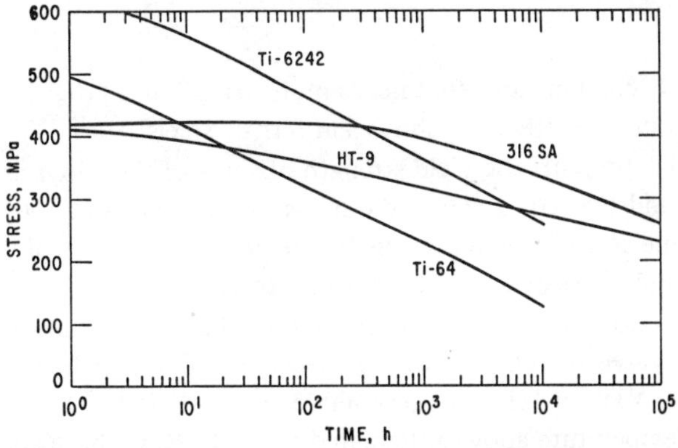

FIG.XIII-2. Stress rupture data at 500°C for INTOR candidate alloys.

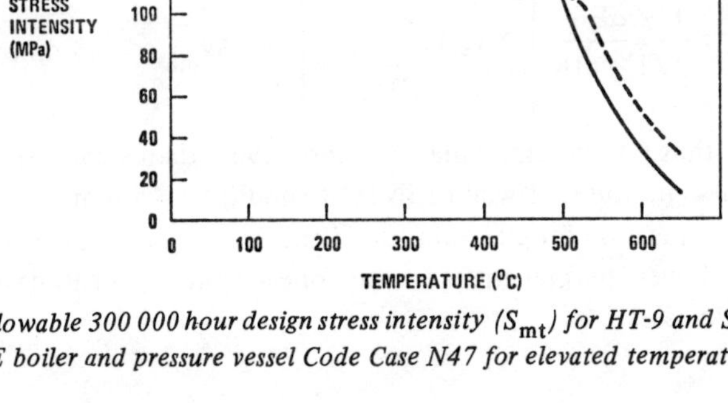

FIG.XIII-3. Allowable 300 000 hour design stress intensity (S_{mt}) for HT-9 and SA 316 SS based on ASME boiler and pressure vessel Code Case N47 for elevated temperature service.

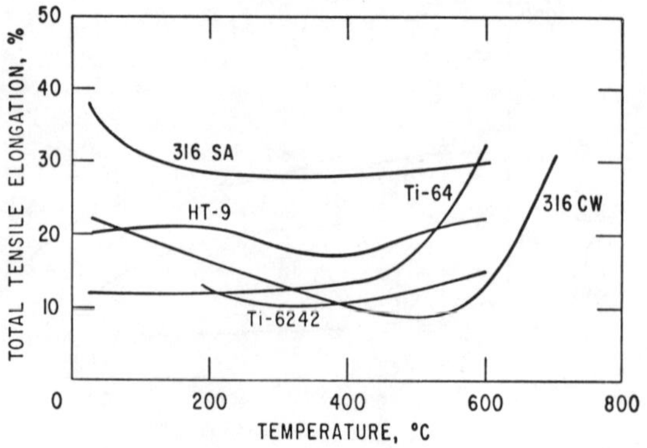

FIG.XIII-4. TOTAl tensile elongation versus test temperature for INTOR candidate alloys.

adequate by 1986. One area that needs more attention is fatigue data in the 300–400°C range for titanium alloys and ferritic steels.

Fig.XIII-1 presents the yield strength of some of the alloys as a function of temperature and Fig.XIII-2 shows the stress rupture values at 500°C. The stress to rupture time at 400°C will presumably be longer and values of greater than 200 MPa should be easily attainable for the steels.

The 300 000 h stress intensity allowed for the steels is shown in Fig.XIII-3. For the anticipated INTOR conditions, stress intensities less than 100 MPa for 316 SS or 200 MPa for HT-9 are acceptable. The data in Fig.XIII-4 show that for all the high temperature alloys anticipated for INTOR, we have adequate ductility for fabrication.

The allowable stresses and strains in 316 SS to achieve 10^5 to 10^7 cycles are presented in Figs XIII-5a and XIII-5b. It can be seen that if the peak-to-peak cyclical stresses are kept below 200 MPa (about 30 ksi), and total cyclical strains are kept below 0.5%, lifetime values of 10^6 cycles in INTOR should be achievable.

The limit of about 200 MPa cyclical stress on a 316 SS first wall also places a limit on the maximum thickness that can be used. If we use the form of Eq. 1 we find

$$\Delta\sigma_{\text{fatigue}}^{\text{limit}} = \frac{1}{2}\frac{\alpha E}{(1-\nu)K}\left[\dot{Q}\Delta\ell + \frac{W_{\text{n}}}{2}\Delta\ell^2\right] \text{ or } \Delta\ell_{max} < 4.3 \text{ mm}$$

Hence, we see that fatigue limits may be more severe than steady state thermally induced stresses and the first wall of INTOR should be ⩽4 mm if it is steel. Even though we have no fatigue data for Ti and ferritic steels in this range, because of the lower thermal stress factors, one would expect to have a higher allowable thicknesses.

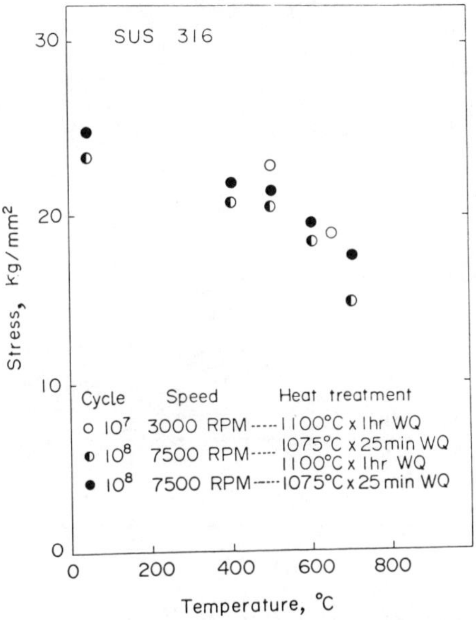

FIG.XIII-5a. Fatigue limit of SUS 316 in high cycle fatigue (rotating bending) (double swinging).

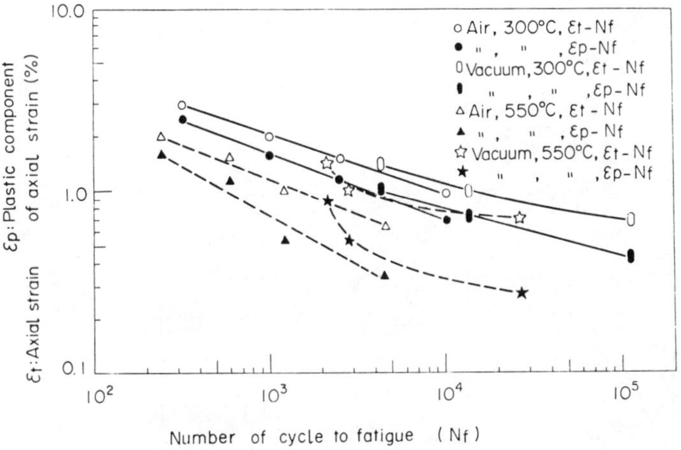

FIG.XIII-5b. Effect of vacuum environment and temperature on fatigue life of SUS 316 in high temperature fatigue (double swinging).

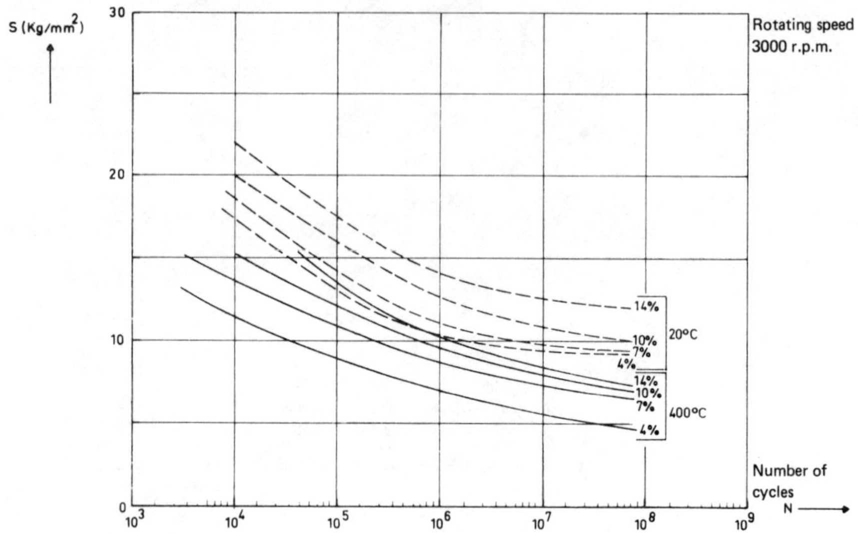

FIG.XIII-6a. Rotating bending tests at 20°C and 400°C on SAP ISML. Cumulative diagram (reference [1]).

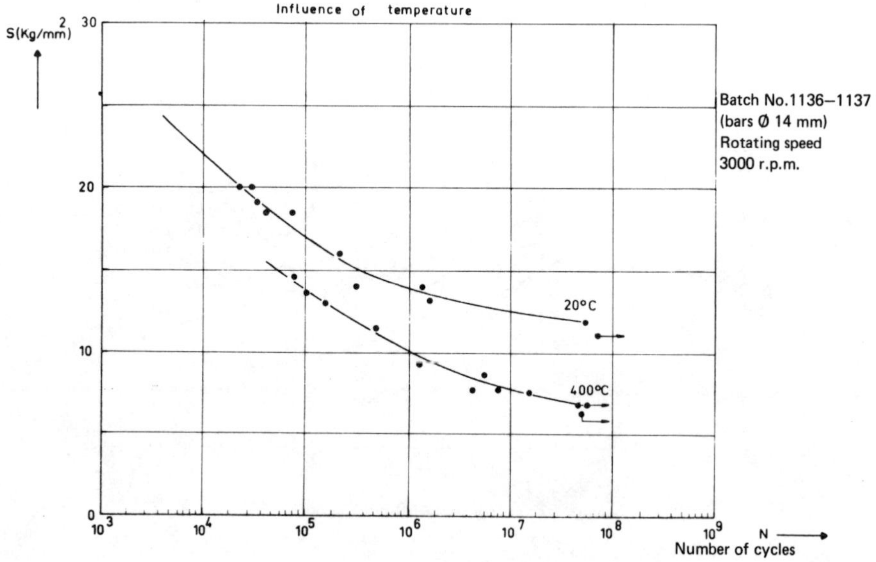

FIG.XIII-6b. Rotating bending tests at 20°C and 400°C on SAP ISML 14% (reference [1]).

Fatigue data on SAP at 20°C and 400°C is given in Fig.XIII-6 as a function of Al_2O_3 content. It can be seen from this data that allowable stresses are in the range of 100–150 MPa for 20°C operation or 50–100 MPa for 400°C operation in INTOR (10^6 cycles).

2.7. Stress corrosion

The combined effects of stress and high chlorine (≥600 ppm) or oxygen (about 1 ppm) can lead to a complex failure mechanism referred to as stress corrosion. Under appropriate conditions, this phenomenon can cause austenitic steels to crack at temperatures of interest to INTOR. If Ti alloys are stressed to more than about 200 MPa in the 300–400°C range, they too can be subjected to such cracking. One must also be concerned about the effects of stress on the precipitation of hydrides in Ti alloys below 300°C. However, it is not expected that these phenomena will be serious in INTOR if careful attention is paid to the water chemistry.

3. IRRADIATION EFFECTS ON STRUCTURAL MATERIALS

3.1. Anticipated maximum neutron damage in INTOR

It is well known that the major difference between fission and fusion neutron damage is the ratio of the transmutation products to the displacements. The damage produced by 1 MW/m^2 of neutron wall loading for a full year is given for various materials of interest in Table XIII-5. Also included in this table are the damage levels accumulated by the first wall after 10^6 shots which are 100 seconds long and produce an average neutron wall loading of 1.6 MW/m^2 (i.e., 10^8 seconds, about 3.2 full power years, or 6.7 $MW\text{-}y/m^2$ when a 30% poloidal peaking factor is included). These values can be scaled appropriately for more or fewer shots, different burn times, average wall loadings, peaking factors, etc.

The main point of Table XIII-5 is that all alloys considered for INTOR application will suffer maximum displacement damage equivalent to fast neutron fluences of 10^{23} cm^{-2} or greater, and they will contain 1000 to 2000 appm of helium. Before reviewing the present state of the art for radiation damage to some of the CTR materials, it is important to place into perspective the differences between data accumulated in a fast fission reactor (like EBR-II, DFR, Rapsodie, or BOR-60), a thermal fission reactor (like the High Flux Isotope Reactor (HFIR)) and in a magnetic fusion reactor (MFR) like INTOR. Fig.XIII-7 gives the rate of accumulation of helium and displacement damage in the three types of nuclear facilities. The high helium accumulation rate in HFIR is due to the

TABLE XIII-5. NEUTRON RADIATION DAMAGE TO POTENTIAL INTOR MATERIALS

Alloy	Per MWy/m^2			Maximum values at 6.7 MWy/m^2		
	dpa	appmH	appmHe	dpa	appmH	appmHe
Austenitic steel (316 SS)	11	532	147	74	3564	985
Ferritic steel (HT-9)	11	450	110	74	3015	737
Titanium alloy (Ti-64)	16	175	142	107	1173	951
Aluminium	14	296	316	94	1983	2117

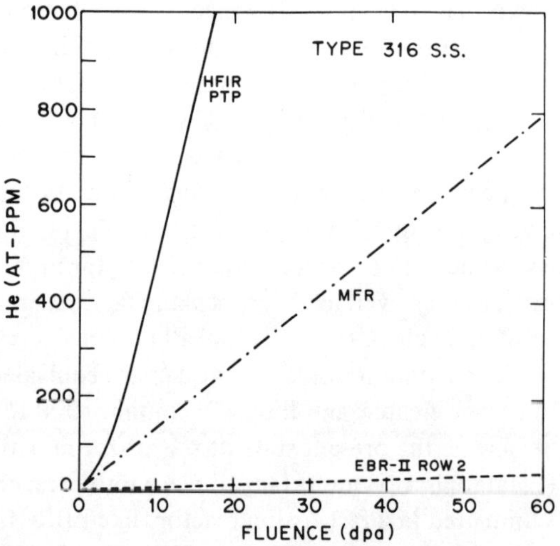

FIG.XIII-7. Helium and dpa production rates for 316 SS irradiated in MFR, HFIR and EBR-II. The lines end at: MFR – 5 $MW\text{-}y/m^2$; HFIR – 6 months; EBR-II – 1.5 years; both at 100% plant factor.

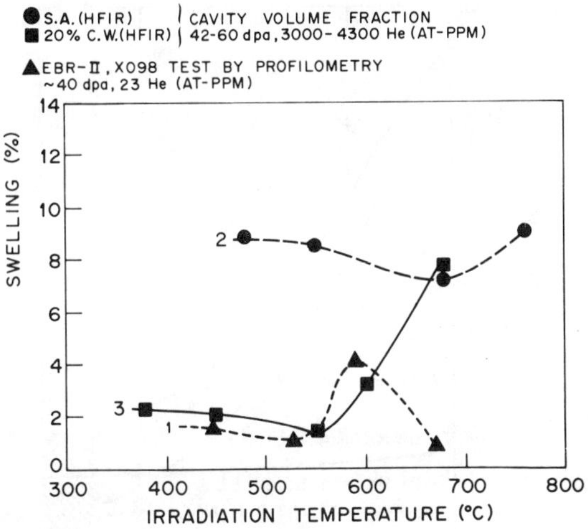

FIG.XIII-8. Swelling as a function of temperature for 316 SS with dpa and with dpa +He.
Curve 1 = EBR–II irradiation of annealed 316 SS.
Curve 2 = HFIR irradiation of ennealed 316 SS.
Curve 3 = HFIR irradiation of 20% CW 316 SS.

$^{58}Ni + n_t \rightarrow {}^{59}Ni + \gamma$ reaction followed by the $^{59}Ni + n_t \rightarrow {}^{56}Fe + \alpha$ reaction. This sequence of events is only important in Ni-containing materials and will not be applicable for Ti or Al alloys, or even ferritic steels with very low Ni content. Obviously such variations in transmutation reactions will have to be taken into account when attempting to arrive at a lifetime estimation.

3.2. Swelling and mechanical properties at fusion reactor helium levels

At helium concentrations up to a few thousands at ppm., information on irradiation response of stainless steels is available for swelling, uniaxial tensile and creep-rupture properties. Comparisons can be made to response in fast reactor irradiations to obtain an indication of the importance of large amounts of helium in the total irradiation damage picture.

The swelling behaviour of annealed 316 SS is very different for dpa only (i.e. fast reactor) and for dpa with large quantities of helium (mixed spectrum reactor) as shown in Fig.XIII-8. For irradiation in EBR-II to 40 dpa and 23 at. ppm He (Curve 2 of Fig.XIII-8), there is a maximum in swelling at ~590°C which is approximately one half of the melting temperature (0.5 T_m) of 316 SS. For temperatures above the peak swelling value the combination of reduced nucleation rates and reduced supersaturation levels of vacancies cause the swelling to be rapidly reduced.

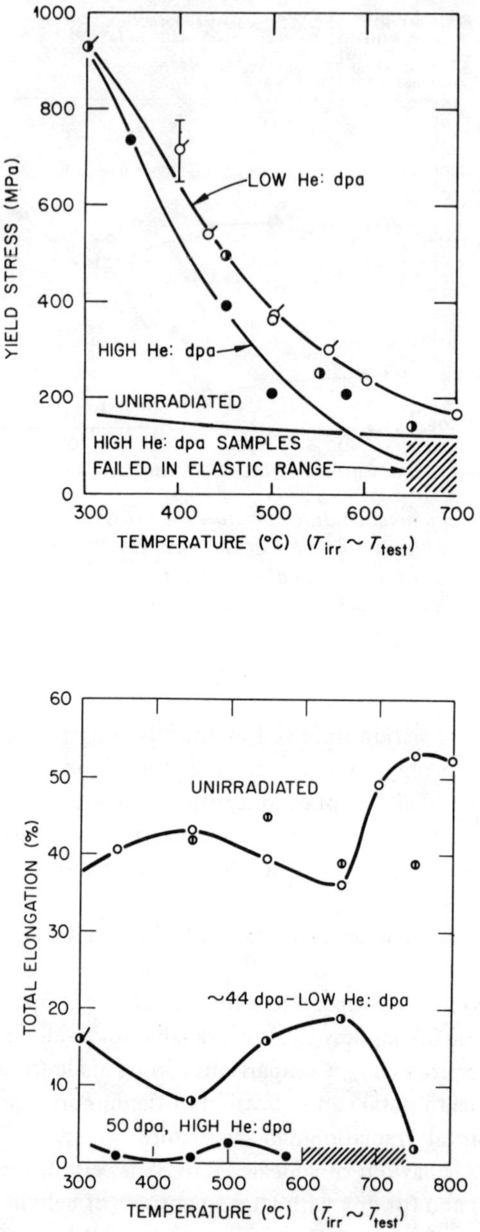

FIG.XIII-9. Yield stress and total elongation of annealed 316 stainless steel in the unirradiated condition and after irradiation in a fast reactor (low He : dpa) and a mixed spectrum reactor (high He : dpa ratio).

For annealed 316 SS irradiated in HFIR and containing from 3000–4000 at ppm He (Curve 2 of Fig.XIII-8), the swelling is higher than that for EBR-II irradiations at all temperatures. However, at $T > 0.5\ T_m$ for the HFIR-irradiated 316 SS, the helium atoms promote nucleation by agglomerating into helium bubbles. The helium atoms also enhance the growth of existing cavities so that the swelling does not decrease with increasing temperature as is observed after fast-reactor irradiations. This is the regime of gas-driven swelling.

Curve 3 of Fig.XIII-8 provides an example of microstructural control to alter irradiation response. For the 20% cold-worked 316 SS the swelling is very low for $T < 0.5\ T_m$ even with helium concentrations of 3000–4300 appm. The dislocations produced by the cold-work provide very effective nucleation sites for helium, and it remains trapped at the dislocations and results in much lower swelling than is observed for the annealed 316 SS. At $T > 0.5\ T_m$ the dislocation density is decreased by climb and recovery caused by thermal vacancies. The decrease in alternate sinks for helium with increasing temperature combined with gas-driven swelling results in rapid increases in swelling with temperature above $0.5\ T_m$. In summary, at equivalent dpa levels the presence of large amounts of helium increases the magnitude and changes the temperature dependence of swelling, but some reduction in swelling values can be obtained by cold working the steel. At INTOR conditions, we would not expect the swelling to exceed 4% at the end of life.

Fig.XIII-9 shows the effect of irradiation on the yield strength and total elongation for annealed 316 SS. The increase in yield strength, which results from displacement damage, tends to saturate at about 10 dpa. At low temperatures, near 300°C, the increase in yield strength is nearly the same for irradiations in which the He : dpa production is high or low. As the temperature increases, the yield strength for high He : dpa falls below that for low He : dpa. Above 650°C, the samples containing large amounts of helium fail in the elastic strain region of the stress-strain curve at stresses below the expected yield stress.

Total elongation of annealed 316 SS as a function of temperature, for tests in which the irradiation and test temperature are approximately equal, is also shown in Fig.XIII-9. At all temperatures in the range 300–700°C, the reduction in ductility is significantly larger at helium levels anticipated for fusion reactor systems than at levels produced in a fast reactor. Above 600°C the total plastic elongation drops to zero.

Fig.XIII-10 shows the effect of irradiation of the yield stress and total elongation of 20% cold-worked 316 SS. For irradiation to about 50 dpa and 4000 appm He, the yield strength is lower than the measured values for the low He : dpa fast reactor irradiations. Total elongation at failure is also reduced to much lower levels when the samples contain more helium.

A further important observation is that at 350°C the tensile fractures remain transgranular, (that is failure through the grains) whereas at high

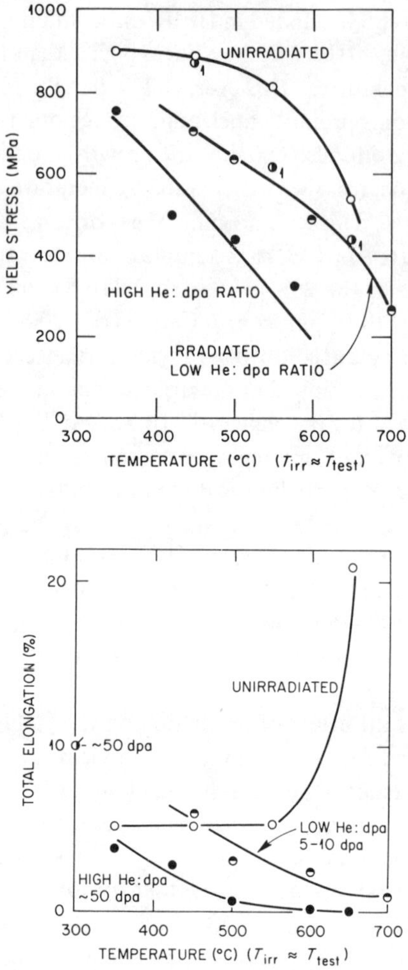

FIG.XIII-10. Tensile properties of 20% cold-worked stainless steel irradiated in neutron spectra which produce low He : dpa (0.3 appm He/dpa) and high He : dpa (100 appm He/dpa) ratios at 40–50 dpa.

temperatures they are clearly intergranular (that is failure along grain boundaries). This tendency for intergranular fracture is most likely responsible for the marked reduction in creep-rupture properties at high helium levels and temperatures as low as 500°C (Table XIII-6). The data in this table also illustrate why 316 SS operation should be limited to below 500°C in a high stress environment such as in fusion reactors.

TABLE XIII-6. CREEP RUPTURE PROPERTIES OF TYPE 316 STAINLESS STEEL AT 550°C AND 45 000 psi (310 MPa) AFTER IRRADIATION IN HFIR

Irradiation temperature (°C)	Fluence, >0.1 MeV (n/m^2)	Damage (dpa)	Damage: Helium (appm)	Creep results: Rupture life (h)	Creep results: Creep rate (%/h)	Creep results: Creep elongation (%)
			Annealed			
	0	0	0	2300	0.0014	17
535	3.8×10^{26}	52	1930	0.45	0.1	0.1
			20% cold-worked			
	0	0	0	3200[a]	0.0012	10
605	2.9×10^{25}	4.0	20	5120	NA[b]	[c]
575	5.0×10^{25}	6.8	48	66.4	0.0062	0.7
545	6.2×10^{26}	85	3190	0.06	0.068	1.1

[a] Extrapolation from tests at stresses in the range of 50 000 to 65 000 psi (345–448 MPa).
[b] Not available.
[c] Accurate value not available because of equipment failure. Elongation between 0.5 and 3.0%.

The conclusion with respect to INTOR is that in the 300 to 400°C range, the cold worked 316 SS is again clearly favoured over the annealed material. The yield strength of the cold worked alloy may be somewhat reduced during irradiation in INTOR but it should still be acceptable. The anticipated **total** elongation values after 300–400°C irradiation are in the 1–2% range. There is some concern for the actual **uniform** elongation values because they may indeed be below the desired 0.5% levels. This needs to be studied further.

3.3. Fatigue and fatigue crack growth (FCG)

Design studies have generally concluded that fatigue and fatigue crack growth are the mechanical properties which may control structural lifetime in reactors which have a cyclic burn. The importance of these properties stems from the cyclic heat load, and thus cyclic thermal stress, in the first-wall structure. Fatigue failures involve both the initiation and propagation of a fatigue crack. Fatigue crack growth is, in reality, the last stage in the fatigue life. The growth rate can be correlated with the stress intensity factor as calculated using linear-elastic

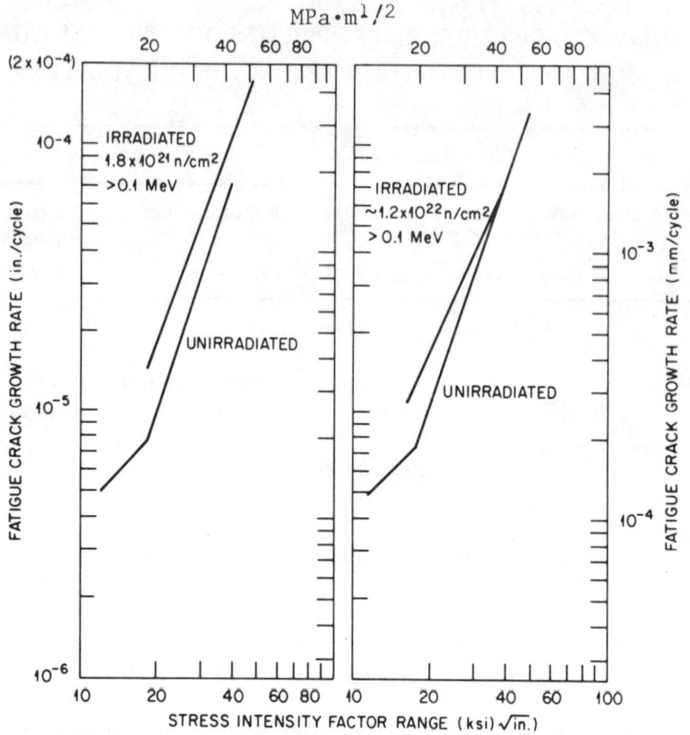

FIG.XIII-11. Effect of irradiation on fatigue crack growth in annealed 316 stainless steel when tested below $T_m/2$. Tensile fractures are transgranular for these conditions.

fracture mechanics (LEFM) techniques. Lifetime predictions based on FCG assume the preexistence of flaws of a given size. In stainless steels and nickel-base alloys, the mode of cracking and the crack propagation rates are sensitive to many variables. At low temperatures, in the regime where deformation is athermal, initiation and propagation are generally transgranular. As the temperature is increased and/or the strain rate is reduced, a transition to intergranular initiation and propagation occurs. The temperature at which this transition in fracture mode occurs is near $T_m/2$ for austenitic stainless steels. A similar transition in fracture mode occurs in tensile and creep fractures. Helium tends to reduce the temperature at which the transition from transgranular to intergranular fracture occurs.

For test conditions which produce transgranular fracture, the crack growth rate is not strongly dependent on microstructural features, including neutron irradiation to relatively low dpa and helium levels. Aging annealed 304 or 316 stainless steel at either 538 or 649°C for times to 6000 h reduced the crack growth rate at low ΔK levels (10 $MPa \cdot m^{-3/2}$) by approximately a factor of 2.

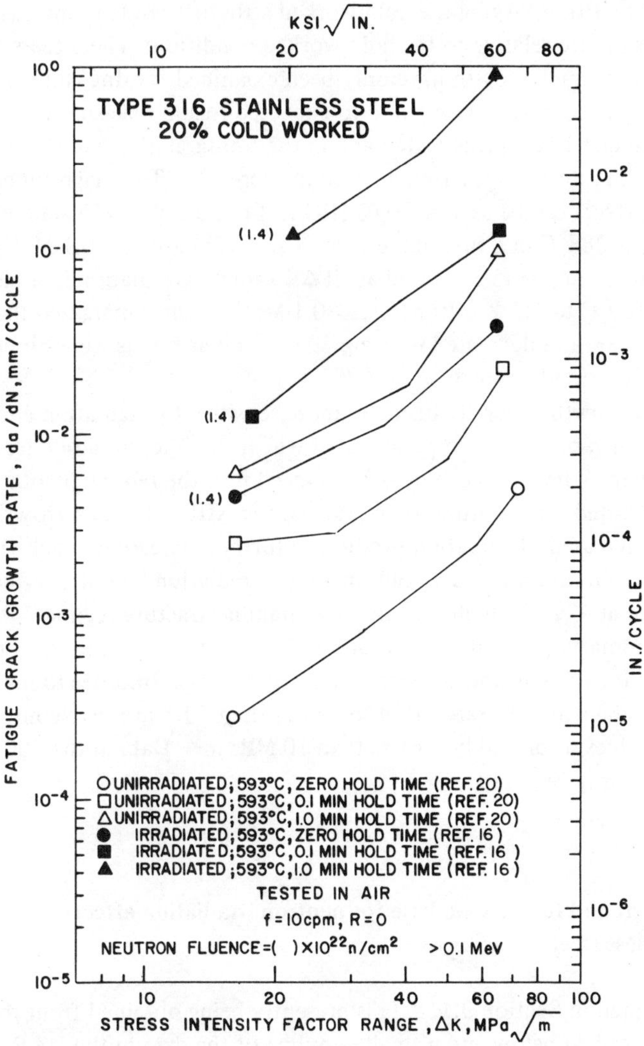

FIG.XIII-12. Effect of hold time in the tensile portion of the cycle in fatigue crack growth in 20% cold-worked 316 stainless steel in the unirradiated condition and after irradiation in EBR-II.

With increasing values of ΔK the effects of aging disappear Cold-working 304 or 316 stainless steel tends to reduce the crack growth rate relative to the annealed condition. Aging the cold-worked structure tends to increase the crack growth rate relative to the cold-worked condition. The effects of neutron irradiation on crack growth have only been examined for dpa and helium levels which are very low when compared to expected fusion reactor requirements. For the damage levels investigated and in the transgranular fracture regime the crack growth rates are either unchanged or increased. The magnitude of the observed effects can be seen in Fig.XIII-11. For annealed 316 stainless steel irradiated at 288°C in a thermal reactor (1.8×10^{25} n/m^2, >0.1 MeV), fatigue crack growth rates were increased at all ΔK values. Irradiation in a fast reactor at about 400°C to 1.2×10^{26} n/m^2, (>0.1 MeV) slightly increased the growth rate at low ΔK and decreased it at high ΔK. Similar trends were observed for annealed 304 stainless steel.

Crack growth generally becomes more sensitive to irradiation and to metallurgical variables at temperatures at which the fracture mode tends to be intergranular. When a hold time is introduced into the tensile portion of the cycle at elevated temperatures (e.g. 593°C, Fig.XIII-12), crack growth rates are generally increased. Irradiation produces a further increase in crack growth rate. In contrast, a hold time and a hold time plus irradiation had little effect on crack growth rate at 427°C which is in the transgranular fracture regime.

In summary, particular attention must be paid to the time structure of the burn cycle for INTOR and the stress intensity factors. In order to maintain safe operation, crack growth rates need to be less than 10^{-6} mm/cycle implying stress intensity values of probably far less than 10 MPa m$^{\frac{1}{2}}$. Data in the 300 to 400°C range is urgently required.

3.4. Present and future data base for neutron irradiation effects on austenitic stainless steels

As stated in Section 3.1, data is presently being obtained from the four sources described below and a small sampling of the data in the US Report is presented schematically in Figs XIII-13a and XIII-13b.

Fast Reactor Irradiation (EBR-II, FFTF)	Primary effects of displacement damage.
Mixed-Spectrum Reactor Irradiations (HFIR)	These irradiations will achieve goal helium contents in relatively short times. The He/dpa ratio will be higher than in a fusion system.

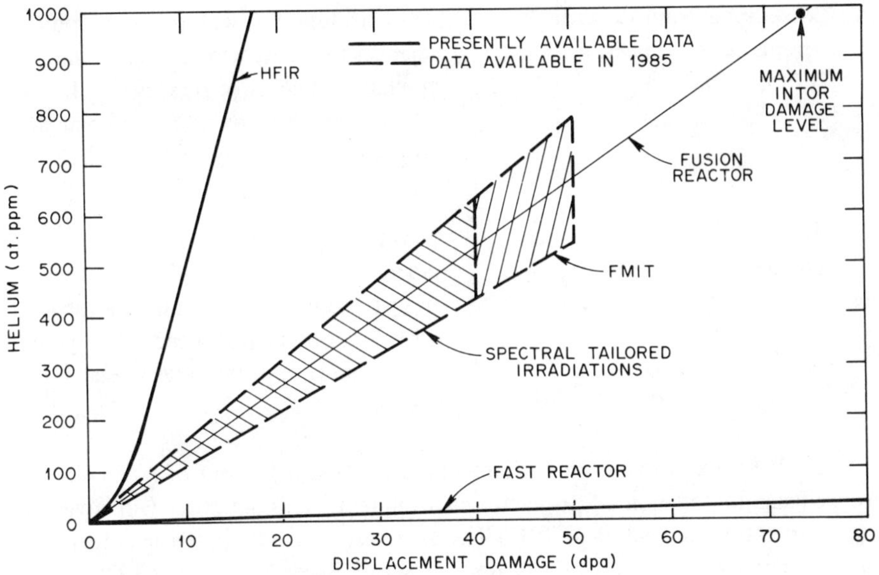

FIG.XIII-13a. Irradiation-induced swelling.

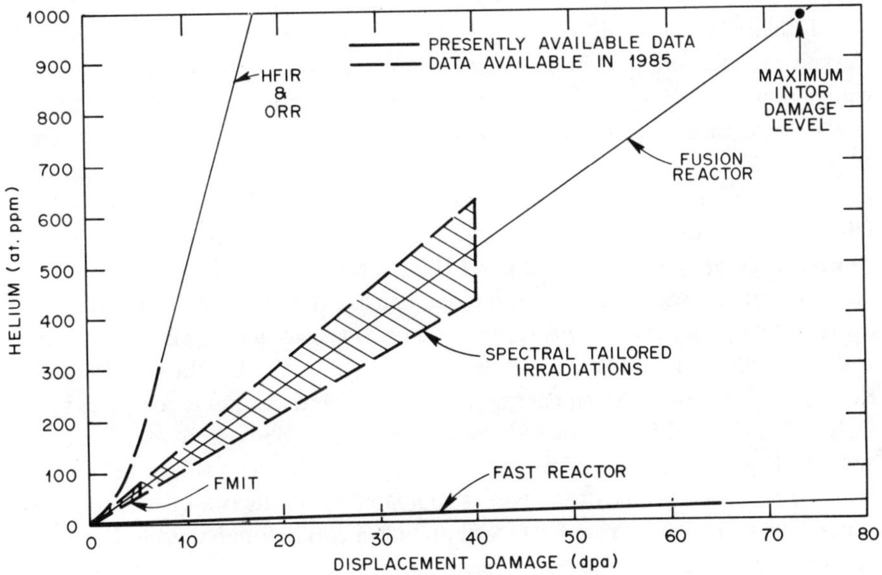

FIG.XIII-13b. Irradiation creep and stress effects on swelling.

Mixed-Spectrum Reactor Irradiations – Spectral Tailoring (ORR)	These irradiations will utilize periodic changes in the experiment surroundings to change the $\phi th/\phi s$ flux ratios. In this way the He/dpa ratio can be made to track more closely that of a fusion environment.
High Energy Neutron Sources (RTNS, FMIT)	This facility is key to obtaining a correlation between irradiation effects in fast and mixed spectrum reactors with that observed in a spectrum which more closely approximates a fusion spectrum.

By 1985, extensive information will exist on the behaviour of 20% cold-worked stainless steel in a fast-reactor environment. This data base will cover the temperature range 350–650°C. The data base on swelling, tensile, creep-rupture, and irradiation creep behaviour presently exceeds anticipated INTOR dpa levels. The data base on fatigue and fatigue crack growth will extend to about 10 dpa in 1985. There will, however, be limited fatigue and fatigue crack growth data to much higher dpa levels. In addition to the above described data base on key properties, extensive information on history effects (e.g., temperature change, stress change) is being developed under the breeder reactor materials programmes. This will be most valuable in reactor design and analysis.

Either in the near future, or by 1985, results of irradiations of 20% cold-worked 316 stainless steel in mixed spectrum reactors (without spectral tailoring) will be available for swelling, tensile and fatigue properties in INTOR goal helium levels. Information should also be available on irradiation creep and fatigue crack growth at 200–400 at. ppm He. Spectrum tailoring experiments will be initiated by about 1980 and by 1985 should provide swelling, irradiation creep, tensile fatigue and fatigue creep growth properties at roughly one half goal INTOR damage levels at approximately correct He/dpa ratios.

The Fusion Materials Irradiation Test Facility (FMIT) is scheduled to be operational by mid 1984. This facility will provide a volume of about 5–10 cm^3 within which the neutron flux is $\geqslant 10^{15}$ n/cm^2. Assuming that the necessary pre-experiment neutron flux and energy spectrum characterization is complete by early 1984, this facility can provide information on swelling, irradiation creep and tensile properties by the end of 1985 as shown. The data will be limited in terms of number and size of samples as compared to results from fission reactor experiments. The data from FMIT will provide a correlation between the effects of fission reactor and higher energy neutron spectra on some of the key properties.

3.5. Ferritic steels

The data base on the effects of irradiation on the properties of ferritic steels is presently derived from two major sources: Light Water Reactor Pressure Vessel Technology and Breeder Reactor Alloy Development programmes. The LWR industry has extensively investigated the properties of pressure vessel steels at relatively low irradiation temperatures (up to about 300°C) and low fluences (up to about 10^{25} n/m^2). The primary emphasis has been on the ductile-brittle transition temperature (DBTT) and fracture toughness. The Breeder Reactor Alloy Development programme is developing ferritic alloys for cladding and duct applications. Temperatures and fast neutron fluences of interest are in the range 300–600°C and up to 3×10^{27} n/m^2. Properties of interest in the Breeder Reactor programme include irradiation-induced swelling and creep, tensile and creep-rupture properties and shift in the DBTT. The programme is concentrating on alloys having 9–12% Cr. It is in this range that the best combination of high-temperature strength, phase stability and weldability is available. The commercial alloy, SANDVIK HT-9 (Fe-12 Cr-1 Mo-0.3 V), and a developmental alloy D-57 (composition not available) are being investigated. In addition to breeder reactor cladding and duct applications, ferritic steels may also have application as structural materials (containment vessels, piping IHX steam generators, etc.). A programme to develop 9 Cr-1 Mo steels for these applications is in progress. The elevated-temperature strength, toughness and phase stability of this composition is thought to be superior to HT-9. Since the available irradiation effects data base is derived from fast-reactor irradiations, the He : dpa ratio is much lower than that of a fusion reactor. Although the trends observed in these irradiations are very promising, it is not clear that they will be maintained in the fusion reactor neutron spectrum. The fast reactor data extends to about 50 dpa and these samples would contain only a few at. ppm He. At this damage level the void swelling in HT-9 is much less than observed in 20% cold-worked 316 stainless steel. Ferritic steels are not, however, inherently swelling resistant. Void formation has been observed in the developmental alloy in the Breeder Reactor Alloy Development programme after irradiation to about 50 dpa and in binary ferritic alloys as a result of nickel ion irradiation. In a fast reactor neutron spectrum, the ferritics exhibit lower irradiation creep rates than 20% cold-worked 316 stainless steel and equivalent or slightly higher elevated-temperature postirradiation tensile ductility. At present, there is no available information on the shift in DBTT at high damage levels. The present data base on swelling, as well as that anticipated in 1985, is summarized in Fig.XIII-14. Since some of the ferrites (e.g. HT-9) contain up to 1 wt% Ni, it will be possible to obtain a limited data base at helium contents in the neighbourhood of 100 at. ppm. This data base will not be adequate for reactor design needs. Optimistically, it will be possible to assess the potential of the ferritics from irradiation damage considerations in the 1985 time frame.

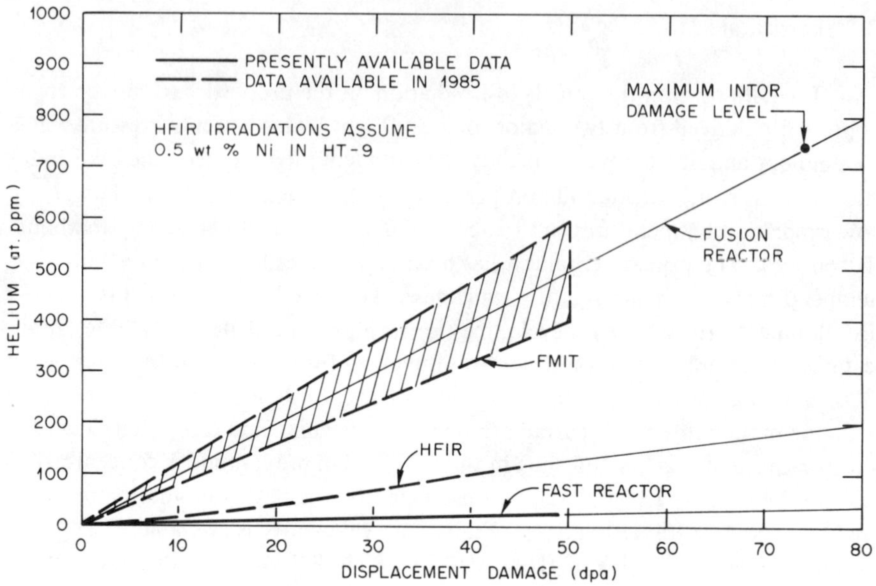

FIG.XIII-14. Irradiation-induced swelling. Summary of data base on effects of irradiation on key properties of ferritic steels. Presently available data is from fast reactor irradiations. By 1985, some information should be available from FMIT and from HFIR irradiation of ferritic steels containing small (~0.5–1.0 wt-%) amounts of nickel. Temperature coverage will be adequate for materials selection.

3.6. Titanium alloys

The presently available data base on the effects of irradiation on the physical and mechanical properties of titanium alloys is extremely limited. The maximum neutron fluence to which titanium or any titanium alloy has been irradiated is about 4×10^{21} n/cm^2 (>0.1 MeV). Void swelling has not been observed. However, the dpa levels and helium contents for these irradiations are much too low to make any projections as to behaviour in a fusion reactor environment. At these damage levels, nucleation and growth of dislocation loops occur. One interesting observation has been the formation of vacancy loops, a type of defect not normally observed in irradiated fcc or bcc alloys. The only data available on effects or irradiation on mechanical properties is for low fluence $\sim 1 \times 10^{21}$ n/cm^2 irradiations at ~70°C followed by tensile and fatigue testing at ~20°C. As expected, the tensile strength was increased and ductility reduced as a consequence of the strengthening produced by the dislocation structure. Fatigue properties were unaffected for these conditions. In summary, the available irradiation effects data are so limited in irradiation and test temperature coverage, properties examined, and damage (He and dpa) levels, that a realistic assessment of the properties of titanium alloys cannot be made at this time.

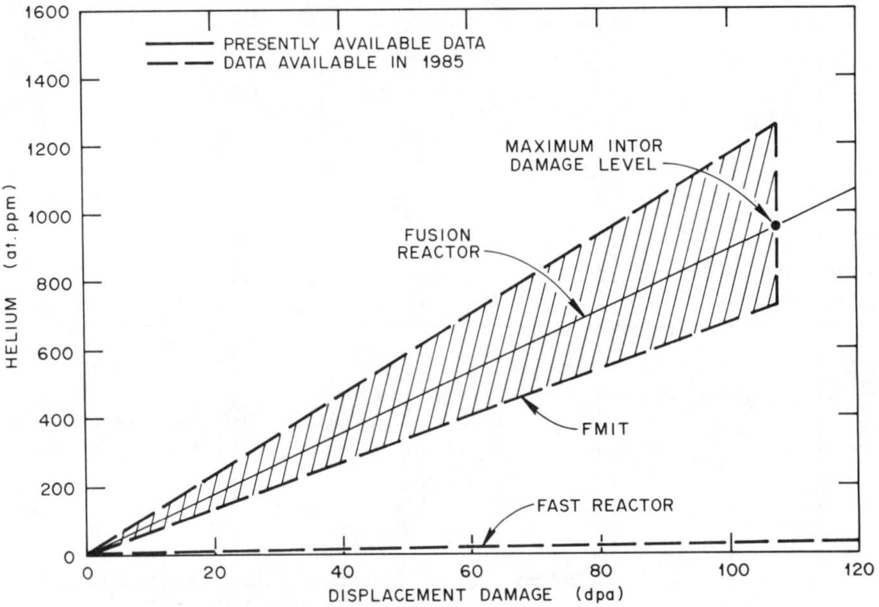

FIG.XIII-15. Irradiation-induced swelling. Summary of data base on titanium alloys expected to be available by 1985. The data will come from fast reactor and FMIT irradiations. Temperature coverage will be adequate. The total data base may allow a decision on potential use of titanium alloys, but will not be adequate for design purposes.

Irradiations of titanium alloys are presently under way in the EBR-II and ORR. The irradiation effects data base on swelling expected to be available in 1985 is summarized in Fig.XIII-15. For these alloys and fission reactor spectra, the (n, α) cross section is very low. Thus, in the 1985 time frame the majority of the data will relate to the effects of displacement damage on properties in the "absence" of helium. In this time frame, FMIT should provide data on swelling, irradiation creep and tensile properties at damage levels of importance to INTOR.

3.7. Aluminium alloys

The situation with respect to Al alloys is similar to that of the ferritic alloys, there is a significant amount of low (50°C) temperature data up to 10^{23} n/cm^2 (~140 dpa) but with very low helium contents. Because of the low melting point of Al, it has a very limited range of application, e.g. less than 200°C. Post irradiation annealing at temperatures above 100°C (after irradiation at 55°C) causes rapid degradation of ductility (Fig.XIII-16) such that by less than 1 MW-y/m^2 of displacement damage, the uniform elongation values fall well below the 1% value. When helium is added, the rate of embrittlement is likely to be accelerated.

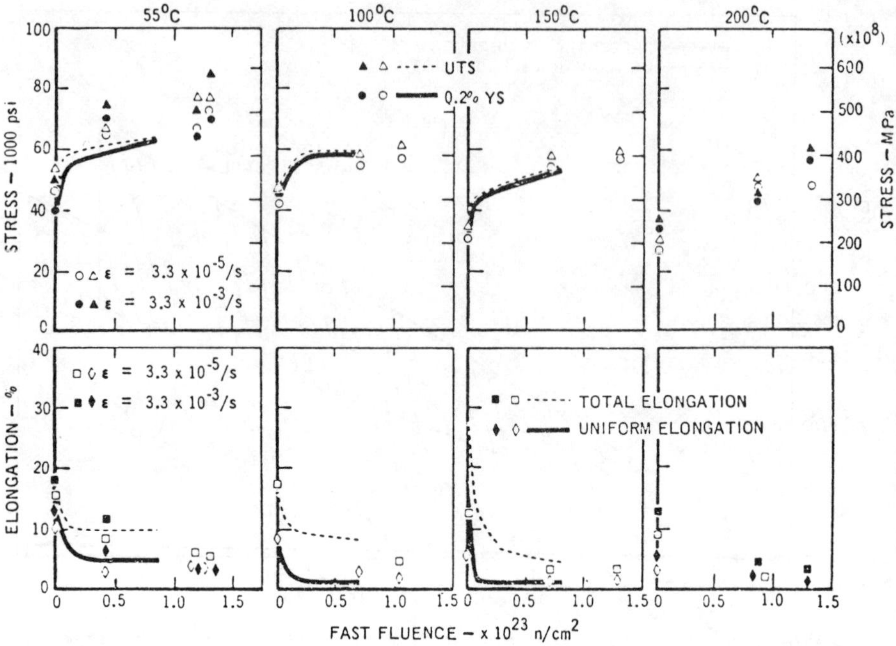

FIG.XIII-16. Tensile properties of irradiated 6061-T6 at two strain rates and four test temperature

TABLE XIII-7. SUMMARY OF DATA BASE FOR APPROPRIATE FUSION MATERIALS BY 1985 IN FISSION REACTOR NEUTRON SPECTRA

Property	316 SS in ORR	316 SS or ferritic steels in HFIR	FMIT		Fast reactors (a)	
			Steels	Ti	Steels	Ti
Swelling	3.6	>4	4.5	~7	>10	>8
Tensile prop.	3.6	>4	1.8	~1	>10	>8
Irradiation creep	3.6	–	0.5	<1	>10	>8
Post irradiation fatigue	3.6	>4	–	–	>10	>8
Post irradiation crack growth	3.6	–	–	–	>10	>8
In reactor crack growth	1.8	–	–	–	–	–
Fracture toughness	0.9	>4	–	–	>10	–

[a] Without appropriate helium content. Equivalent displacement damage only.

Such an effect would essentially prohibit bake-out temperatures above 100°C. A discussion of the irradiation effects to Al alloys can be found in EPRI-ER-328-SR, March 1977, and various papers by scientists at Oak Ridge Laboratory in the US.

Finally, the use of Sintered Aluminium Product (SAP), an alloy of Al + Al_2O_3, has been proposed by the European delegation and the reader is referred to their report for more details. Irradiation at 55°C, in which the appropriate helium was added for INTOR conditions by Li-6(n,α) reactions, reveals that after 0.3 MW-y/m^2 of displacement damage and 5 MW-y/m^2 of He production, the as irradiated ductility is well below 0.5%. The data also shows that if the samples are heated up to 300°C or more, uniform ductility values at failure fall to the 0.1% to 0.5% level. (See DOE/ET-0058/6, June 1979, p. 122). This is well below reasonable engineering design values and it is doubted that use could be made of the higher SAP temperature capabilities (i.e. above 200°C).

3.8. Summary of data base

Figs XIII-13 to XIII-15 are schematic to indicate the kinds of data that will be available for candidate INTOR alloys by 1985. Since some types of tests require much more reactor volume than others, the ability to obtain high fluence data on these properties is more limited than on those such as swelling, which require only a small test volume per sample. A summary of the expected data base is given in Table XIII-7. Except for tests in EBR-II/FFTF, all results will be for samples containing high helium contents.

4. DATA BASE FOR COOLANTS

4.1. Water as a fusion reactor coolant

The principal attraction of water as a fusion primary system coolant is the large technology base. Approximately 200 water-cooled fission reactors have come on line, and another 315 are planned or are under construction. For systems cooled by liquid water, nature has set the temperature limit at 371°C, the critical point of water. Existing water-cooled nuclear systems operate up to ~340°C (PWRs) and 10 to 15 MPa. Coolant purity control in pressurized water reactor primary circuits has generally been satisfactory. However, approaches to control activity transport are still being developed. Major impurity control problems have developed in steam generators of the U-tube design. Contaminants from condenser water inleakage appear to be a substantial source of the impurity inventory. In a fusion demonstration reactor, that source of impurity could be precluded by use of dry cooling towers.

Activity transport is a problem common to all liquid-cooled fission reactors. Iron-base, nickel-base, copper-base, and cobalt-base alloys release a fraction of their corrosion products to the coolant by solution or spallation. The corrosion products circulate through the core, adsorb on fuel element surfaces, undergo neutron activation, desorption, transport, and deposition on previously non-radioactive circuit components. The radioactive species are in both ionic and particulate forms. The ionic species are incorporated into the oxide films on the circuit surface, while the particles may develop relatively loose deposits overlying the oxides.

4.2. Helium as a fusion reactor coolant

Helium has functioned satisfactorily in gas-cooled fission reactors, establishing its reliability as a coolant and providing a substantial technology base relevant to fusion technology. However, several problem areas must be addressed to make a decision regarding the use of helium as a fusion reactor coolant, since the designs and requirements of gas-cooled fusion reactors may differ in some respects from the existing gas-cooled fission reactors. The principal compatibility problems arise not from helium, which is inert, but from impurities which frequently are difficult to control to concentrations which preclude corrosion reactions. A positive factor in helium utilization as a coolant is the fact that a substantial technology exists for helium purification and analysis, though it is not clear that existing procedures will fully meet the needs of fusion technology.

The problem with the use of helium coolant is that the relatively low heat transfer coefficient requires considerably higher structure temperatures adequately to remove that heat. For example,Al alloys would be completely ruled out with helium coolants.

In contrast to the liquid-cooled fission systems, activity transport in gas-cooled systems has been minimal, mainly due to fission products. Thus, the gas-cooled fusion system offers a major advantage from the standpoint of radiation field buildup. On the other hand, tritium inleakage to the gas coolant may impose a need for other activity control measures.

4.3. Summary

It is concluded that, with respect to corrosion, either coolant could be used with the prime candidate alloys at low temperatures in INTOR.

TABLE XIII-8. PHYSICAL PROPERTIES OF CANDIDATE BREEDING MATERIALS

Property	Li	Li_2O	Li_7Pb_2	$LiAlO_2$
Melting temperature (°C)	180	1427	726	1610
Boiling temperatue (°C)	1330	1527	–	–
Density, (g/cm^3)	0.484[a]	2.01	4.60	2.55
Molecular weight	6.94	29.88	463.36	65.91
Lithium atom density (g/cm^3)	0.484[a]	0.931	0.482	0.268
Thermal conductivity (W/m–K)[b]	50	1.7	20[c]	
Vapourpressure (Pa)	8	131	–	Low
(Temperature (°C))	600	955	–	<1000

[a] At 500°C.
[b] Approximate values at probable operating temperatures.
[c] Estimated.

5. DATA BASE FOR TRITIUM-BREEDING MATERIAL

5.1. Candidate breeding materials

Three types of materials have been considered for the tritium-breeding material in fusion reactors, viz., liquid metals, metallic compounds or alloys of lithium, and nonmetallic compounds. Leading candidate materials for INTOR are listed in Table XIII-8 with their important physical properties.

5.2. Physical properties of Li_2O

Because of the great interest in Li_2O we summarize briefly here some of the information which has been recently developed by the Japanese (see their original report for details). We will only include a few of the most important properties here and only to illustrate that considerable work is in progress to develop this material.

5.2.1. Preparation of sintered pellets

Sintered pellets of a wide density range (70 to 93%) have been made by conventional hot pressing techniques. Some very high density pellets (99% of theoretical) were also prepared by an isostatic hot pressing method at 1000 to 2000 atm, and 1000 to 1200°C.

5.2.2. *Vapour pressure*

The equilibrium vapour pressures over solid Li_2O have been measured in the 1300 to 1700 K range. From this information, the atomization energies of $LiO(g)$, $Li_2O(g)$, $Li_3O(g)$ and $Li_2O_2(g)$ were determined.

5.2.3. *Thermal properties*

Thermal diffusivity of sintered Li_2O pellets having a wide range of density (70.8 to 93.4% theoretical) was measured in the temperature range of 473–1173 K by the laser pulse method. Porosity dependence of the thermal conductivity for Li_2O was determined by using the Maxwell-Eucken formula. The thermal expansion coefficient was also measured in the same temperature range.

The heat capacity of Li_2O was measured in the temperature range 306–1073 K using an adiabatic scanning calorimeter, and any heat capacity anomalies such as phase transition were not observed. The heat capacity equation obtained by the least squares method was given by

$$C_p = 75.24 + (9.55 \times 10^{-3}\,T) + (25.05 \times 10^{5}\,T^{-2}) \quad (\mathrm{J/mol \cdot K})$$

5.3. Chemical compatibility

5.3.1. *Lithium compatibility*

Even though liquid Li is not likely to be used as a major component of INTOR it is likely to be included in some test modules. Compatibility with the structural or containment materials is a major concern for the viability of such modules. The two major compatibility problems relate to (a) deterioration of the mechanical integrity of the structural containment; and (b) mass transfer of corrosion products that leads to plugging of circulating systems or excessive deposition of radioactive material in unshielded regions. It is generally concluded that these effects will be important long before uniform wall thinning will lead to failure under conditions of interest. Lithium is known to be much more corrosive to stainless steel than is sodium. Corrosion rates varying from 0.3 to 3 μm/Ms at 600°C have been reported for lithium under a variety of conditions compared to a generally accepted 0.4 μm/Ms corrosion rate of Type 316 stainless steel at 600°C in high-velocity reactor-grade sodium. Although the data are limited, the corrosion rates are known to depend on velocity, purity, temperature, and temperature gradient in the lithium as well as microstructure and stress state of the steel. The corrosion rate of stainless steel in lithium at 500°C for conditions of interest is estimated to be about 0.3 μm/Ms (10 μm/y), or nearly

TABLE XIII-9. OPERATING LIMITATIONS OF STRUCTURAL MATERIALS IN LITHIUM

Structural alloy	Max. Temp. (°C)	Limiting mechanisms
Austenitic stainless steel	450	Mass transfer – plugging; radioactive mass transfer; stress corrosion
Ferritic steels	550	Mass transfer – plugging; radioactive mass transfer; stress corrosion
Titanium alloys	<500	Possible leaching of aluminium; mass transfer
Aluminium alloys		Not compatible

two orders of magnitude higher than that for high-velocity sodium. The important corrosion mechanisms and the proposed maximum operating temperatures in lithium for each of the candidate alloys are summarized in Table XIII-9.

The second major concern regarding the use of stainless steel in a lithium environment relates to the interaction with stress and the resultant effects on the mechanical integrity of the structure. This may be particularly important because of the cyclic plasma burn. Mechanisms of major concern involve (a) stress effects on the microstructure that may enhance intergranular penetration by lithium; and (b) effects of stress that may reduce the tendency to passivity with time, e.g., fracture of the ferrite layer typically observed after exposure.

The ferritic steels are generally more resistant to corrosion by lithium than are the austenitic steels. The rates are affected by alloy composition, and microstructural effects, e.g., carbide precipitates at grain boundaries, can also affect the compatibility. There is also concern about the effects of stress on the compatibility. Further work is required to assess the importance of stress effects and microstructure on the corrosion. Based on the limited available data, the temperature limits for ferritic steels in flowing lithium systems are taken as 500–550°C.

Very little data are available on the corrosion of titanium by lithium and essentially no data are available on the corrosion of commercial alloys. Under limited test conditions, titanium is fairly resistant to corrosion by high purity lithium. However, a strong driving force exists for transfer of nitrogen and carbon from lithium to titanium. There is also a major concern regarding the compatibility of lithium with commercial titanium alloys, most of which contain several percent of aluminium and tin. Both of these alloys are highly soluble in lithium and will probably be leached from the titanium, similar to the case of nickel leaching from stainless steel.

5.3.2. Compatibility with solid compounds

Although data are limited, compatibility of $LiAlO_2$ with candidate structural materials is not expected to be a significant problem since this compound is very stable. The compatibility of Li_2O with candidate structural materials has received little attention. Recent Japanese data indicate that lithium oxide is very corrosive to stainless steel at temperatures near 1000°C. This may be caused by impurities since lithium oxide is highly soluble in water and is also hygroscopic. Small amounts of moisture normally present could react to form LiOH which is known to be extremely corrosive. There are essentially no compatibility data known for Li_7Pb_2 with the candidate structural materials

5.3.3. Reactivity with water

The major consideration regarding a water-cooled breeding blanket relates to the fact that all of the candidate breeding materials, except $LiAlO_2$, are chemically reactive with H_2O. The degree of chemical reactivity of the breeding materials with water poses both a safety concern and a maintenance problem in the event a breach of the cladding should occur. The following hydrolysis reactions will occur on contact of water with the breeding materials:

$$Li + H_2O \rightarrow LiOH + 1/2\ H_2 \qquad \Delta H_{hyd}(298\ K) = -58.7\ kcal/mol$$

$$1/7\ Li_7Pb_2 + H_2O \rightarrow LiOH + 1/2\ H_2 + 2/7\ Pb \qquad \Delta H_{hyd}\ (298°K) = -47.9\ kcal/mol$$

$$1/2\ Li_2O + 1/2\ H_2O \rightarrow LiOH \qquad \Delta H_{hyd}\ (298\ K) = -15.4\ kcal/mol$$

All three breeding materials react to form LiOH which has a melting temperature of 470°C. Since liquid LiOH is extremely corrosive to nearly all structural materials, the integrity of the structure would be severely degraded in the event of a lithium water reaction to form LiOH.

Recent unpublished results by S. Nasu of Japan show that the reaction $LiOH(s) \rightarrow Li_2O(s) + H_2O(g)$ is virtually complete (>99%) above 310°C in a vacuum. The activation energies for the thermal decomposition range from 112 kJ/mol at 242 to 323°C to 82 kJ at 323 to 419°C.

The reaction of Li_2O pellets with air has also been recently studied,with the following general conclusions drawn:

(1) Li_2O pellets do not seem to react with air below 50% relative humidity at 25°C.
(2) Above 60% relative humidity at 25°C, considerable reaction takes place. For example, 1% of a Li_2O pellet reacted in 110 h.
(3) CO_2 seems to play an important role in the Li_2O–H_2O reaction.

5.3.4. Compatibility of Li_2O with Fe-Ni-Cr alloys

The reaction of type 316 stainless steel, Incoloy 800, Hastelloy X-R (a modified Hastelloy X), Inconel 600 and pure nickel with sintered Li_2O pellets was investigated in the temperature range of 1073–1373 K under a dynamic vacuum by using the reaction couple method. The reactions of the alloys with Li_2O pellets were studied by the weight loss method and by the reaction depth method. The reaction rates were measurable in the range of 1073–1173 K and quite appreciable at 1273 K, being greatest with Incoloy 800 and least with Hastelloy X. These reactions all displayed a similar pattern as follows. Among the primary alloy constituents, chromium was preferentially attacked by lithium and oxygen diffusing from the Li_2O into the alloys to form $LiCrO_2$. The $LiCrO_2$ phase grew into a reaction bond (subscale) of uniform thickness beneath the surface of each alloy. Preferential growth of $LiCrO_2$ along the grain boundary was observed only in the case of Inconel 600 at temperatures below 950°C. On the other hand, iron diffused toward the Li_2O pellets to form a volatile double oxide, Li_5FeO_4, above 1000°C. However, any reaction product associated with nickel was not detected and nickel metal was attacked very little by the Li_2O pellet in the whole range of reaction temperature.

5.3.5. Compatibility of Li_2O with molybdenum

The reaction of molybdenum and molybdenum-base alloy (TZM) was also investigated in the temperature range of 1073 to 1373 K under a dynamic vacuum. The reaction proceeded measurably at about 1223 K and appreciably above 1273 K. Metallographic and X-ray diffraction analyses showed that a reaction product scale was formed at the interface and the scale consisted mainly of Li_4MoO_5. TZM showed slightly higher reactivity with Li_2O compared to molybdenum.

In addition, the vapourization of Li_2O has been studied by molybdenum, tantalum and graphite Knudsen cells in the temperature range 1200–1500 K to measure the distribution of the partial pressures under these conditions.

5.3.6. Compatibility with commercial-grade helium

Compatibility of reactor-grade helium with the lithium compounds is a major consideration since most blanket designs propose direct contact of breeder and

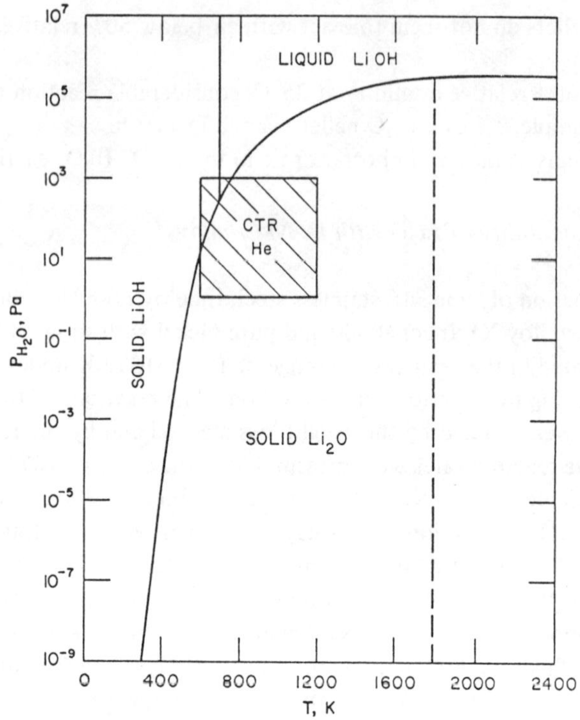

FIG.XIII-17. Phase equilibria for Li_2O-LiOH system.

coolant. Fig.XIII-17 shows the LiOH-Li_2O phase equilibria as a function of P_{H_2O} and temperature. The shaded area indicates the range of conditions anticipated in a fusion reactor environment. These results indicate that Li_2O is stable for expected values of $P_{H_2O} \leqslant 10$ Pa if the temperature is maintained above 600 K. A P_{H_2} above 100 Pa is required before the formation of liquid LiOH is predicted. However, the vapour pressure over Li_2O becomes substantial when the temperature approaches 1000°C.

Quantitative estimates of reactions with Li_7Pb_2 are more difficult because of lack of appropriate thermodynamic data. It appears that oxidation of the Li_7Pb_2 will occur if oxygen is added to the helium for purposes of tritium processing. The oxidation potential of Li_7Pb_2 probably does not differ substantially from the lithium and, as such, is extremely high. Therefore, the oxygen concentration of the helium must be maintained very low. In fact, it is very questionable whether the helium purity required to prevent oxidation of Li_7Pb_2 could be achieved in a high pressure helium coolant. Three consequences of this oxidation reaction are: (a) removal of oxygen from helium, which leads to higher tritium pressures; (b) effects resulting from the phase change of the

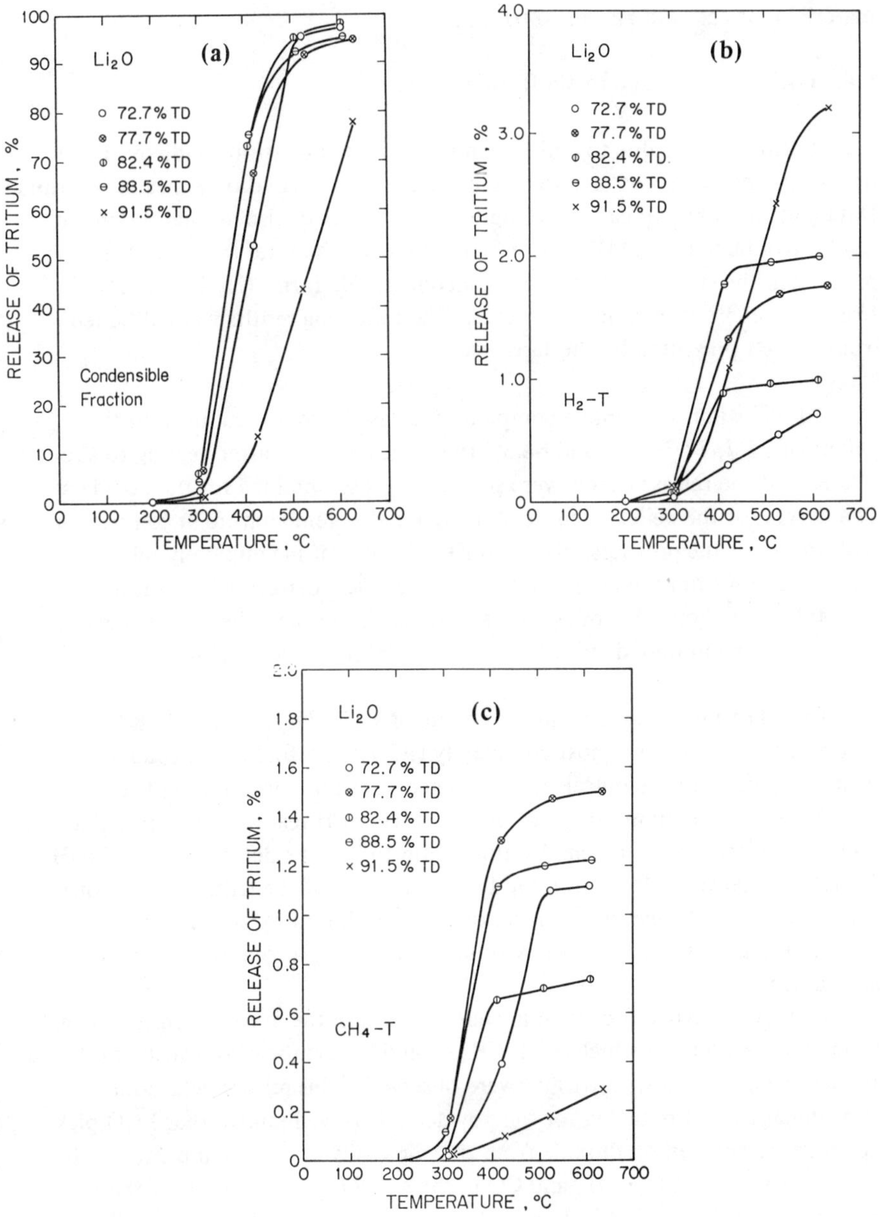

FIG.XIII-18. Total fractions of tritium release curves for the condensible form (a) and the non-condensible forms [H_2-T(b) and CH_4-T(c)] from neutron irradiated Li_2O pellets having various densities under vacuum.

breeding material (for example, excess lead and lowering of the melting temperature); and (c) possible formation of an oxide film on the breeding material that acts as a barrier to tritium.

5.4. Tritium release due to a diffusion process

Tritium release due to a diffusion process from neutron irradiated Li_2O pellets and powders was investigated using the post irradiation annealing method. Data corresponding to each working temperature of the isochronal treatment has been obtained. Fig.XIII-18 shows the curves of the total fraction of tritium released in the condensible and the noncondensible forms (H_2-T and CH_4-T) from the pellets with various densities. The following results were obtained from reports presented by the Japanese.

(i) Tritium was almost completely released under vacuum from the pellets of 72.7, 77.7, 81.5 and 88.5% theoretical density after heating to about 900 K, and the release curves were similar to those obtained for the powders. The curves for the pellets seem to shift to a higher temperature in comparison with those for the powders. By contrast, tritium was incompletely released under vacuum from the pellets of 91.5% theoretical density after heating up to 900 K, and about 19% of tritium still remained in the pellet. This suggests that pore size and pore distribution in the pellet may play a role in the tritium release.

(ii) Tritium released from the pellets of 72.7, 77.7, 81.4 and 88.5% theoretical density was almost completely (95.1 to 98.0%) in the condensible form. Although the chemical state of the condensible form has not been identified yet, if tritium produced by the $^6Li(n,\alpha)^3H$ reaction was present as LiOT in the specimens, tritiated water would follow the dehydration of LiOH. Tritiated water may also be formed by the reaction of the tritium with moisture. It is, however, not clear whether the tritium is oxidized by the reaction cited above, by the exchange reaction with moisture, or by a combination of the two mechanisms.

(iii) A few percent of tritium were released in the non-condensible form from all the pellets, in which H_2-T, CH_4-T, and C_2-hydrocarbons such as ethylene and acetylene tagged with tritium were observed. This release behaviour was also similar to that obtained for the powders. It is well known that Li_2O picks up moisture and carbon dioxide from the air easily at room temperature. In fact, mass signals of H_2, CH_4 and CO in the gas chromatograph show the presence of hydrogen and carbon in the pellets. In addition, proton backscattering experiments on Li_2O showed that carbon atoms were distributed over the surface of the pellet, and not homogeneously in the pellet. Since the pellets are covered with carbon (the chemical form of the carbon is not identified), the CH_4-T seemed

TABLE XIII-10. SUMMARY OF ANTICIPATED OPERATING LIMITS FOR INTOR MAGNETS

Component	Limits in fission neutron spectra equivalents, n/cm^2 [a,b]	
	Superconducting magnets	Normal magnets
Superinsulation	$>10^{22}$ (Al foil)	Not applicable
Electrical insulation	10^{18} to 10^{19} (organics)	10^{19} to 10^{21} (inorganics)
Stabilizer	1×10^{16} (Al) 10^{17} (Cu)	Not applicable
Conductor	3×10^{18} (Nb_3Sn) $>10^{20}$ (NbTi)	10^{21} (Cu) 10^{20} (Al)
Structural material (welds)	$>10^{20}$	$>10^{22}$

[a] Fission reactor spectra equivalent, 10^{22} n/cm^2 (fission) $\cong$ 14 dpa (Al) $\cong$ 7 dpa (Cu).
[b] Conversion factor of 10^9 n/cm^2 = 1 rad was used.

to be formed by the reaction of tritium with the carbon at the surface of the pellets.

(iv) The ratio of the total fraction of H_2-T released to that of CH_4-T for the pellet with 91.5% theoretical density was considerably larger than those for the pellets with lower densities. (e.g. at 600°C this ratio is 10 for 91.5% TD and 1.5 for 73% TD).

6. MAGNET MATERIALS

6.1. Introductory remarks and summary

The operating lifetime of magnets for INTOR will be determined by the most sensitive limits on the individual components of the magnets. A summary of the anticipated limits for both superconducting and normal magnets is given in Table XIII-10. For superconducting magnets, the Al stabilizers become a problem in the 1×10^{-5} dpa ($\sim 10^{16}$ n/cm^2) level while the Cu stabilizers are a problem above 5×10^{-5} dpa ($\sim 10^{17}$ n/cm^2, fission neutrons). Next, the organic electrical insulators start to degrade at about 10^{18} n/cm^2 ($\sim 10^9$ rads), but it is possible to anticipate now that epoxy insulators can be developed to withstand 10^{10} rads.

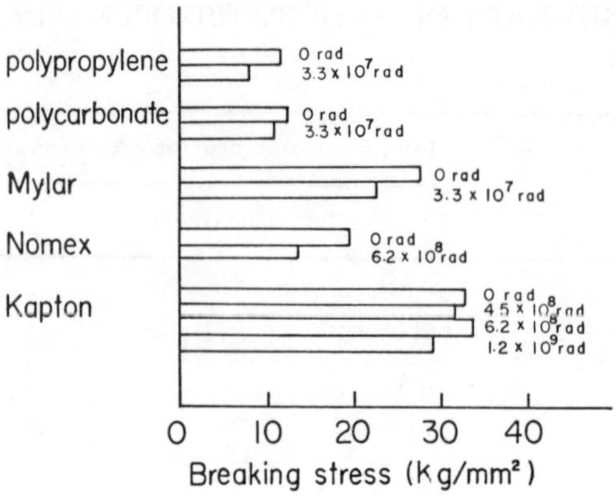

FIG.XIII-19. Relation between breaking stress and absorbed doses for several organic materials.

The Nb_3Sn conductors start to degrade rapidly at 3×10^{18} n/cm^2 but the NbTi superconductors could withstand more than 10^{20} n/cm^2 before their critical currents were reduced by a factor of 2. Structural components or pure Al foil superinsulation could stand 10^{20} to 10^{22} n/cm^2 exposures.

The limit on the normal coils appears to be the electrical insulators and these will probably have to be made of inorganic materials to withstand 10^{19} to 10^{21} n/cm^2. The conductors themselves probably will not present problems below the 10^{20} to 10^{21} range. Finally, the support structure should be able to withstand $>10^{22}$ n/cm^2. We will now examine the reasoning behind each of the limits in Table XIII-10.

6.2. Superinsulation

The current method of thermally insulating superconducting magnets is to use aluminized Mylar or Al foils alone in the vacuum space between the magnet casing and the liquid-helium cooled magnet. The mechanical and dimensional stability of Al at cryogenic temperatures is not in question (below $\sim 10^{22}$ n/cm^2) but the organic materials are notoriously susceptible to radiation. Irradiation of Mylar at 5 K and then mechanically testing it at 77 K showed a 20% loss in breaking stress after 3.3×10^7 rads ($\sim 3.3 \times 10^{16}$ n/cm^2) (Fig.XIII-19). Due to this high sensitivity to radiation, superconducting magnets in fusion devices will probably rely on Al foil alone and hence limits on the Mylar are not applicable for INTOR.

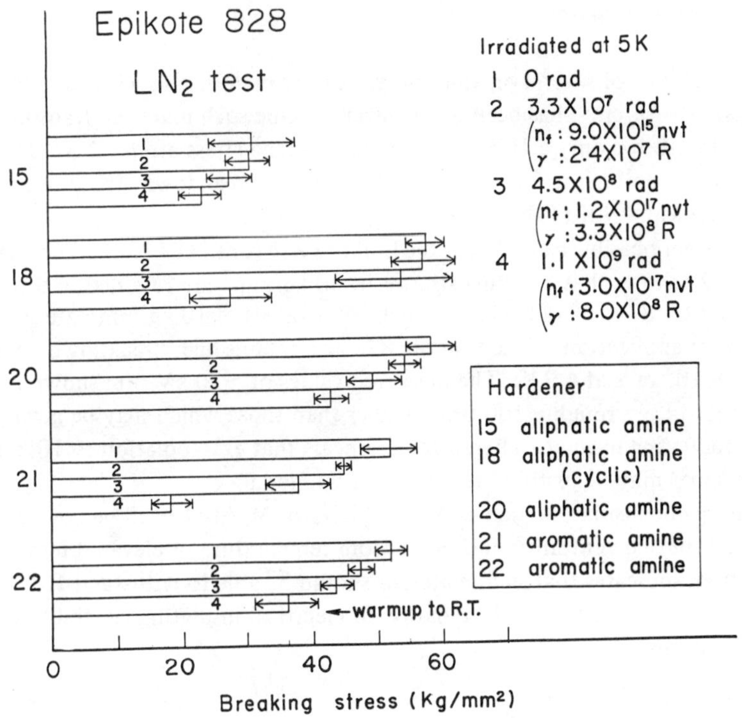

FIG.XIII-20. Relation between breaking stress and absorbed dose for Epoxy resins.

TABLE XIII-11. POST-IRRADIATION ELECTRICAL MEASUREMENTS AT ROOM TEMPERATURE

Irradiation dose 2×10^9 rads at 4.9 K

Specimen	2850 FT Epoxy	EPON 828 Epoxy	B-stage glass-cloth	Fiber-glass G-10CR	Nomex 410	Kapton F
Resistivity, ρ (10^{15} Ω-cm)						
Control	14	10	12	7	13	15
Irradiated	5	6	$\{5 \times 10^{10}\}$	7	14	19
Electrical breakdown, E (kV/mm)						
Control	33	24	31	22	38	70
Irradiated	32	23	11	22	39	57

6.3. Electrical insulators

The ability of several organic and glass fiber materials to withstand 2×10^9 rads (2×10^{18} n/cm^2) has been demonstrated. One such material, Kapton (Figure 19) shows only a 10% reduction in breaking stress after 1.2×10^9 rads and it can retain high resistivity and high electrical breakdown strengths after 2×10^9 rads at 4.9 K (Table XIII-10).

Another possible candidate is the Epikote 828 (or EPON 828) epoxy material. Fig.XIII-20 shows that breaking stresses of 40 kg/mm^2 are retained in this material after 8×10^8 rads. It is thought that values of 10 kg/mm^2 are sufficient for magnet applications. The Epikote 828 also retains high breakdown strength after 2×10^9 rads at 4.9 K. The measured values of ~20 kV/mm shown in Table XIII-11 are roughly 100 times higher than those which may be needed in a superconducting magnet. Therefore, it appears that extrapolation to 10^{10} rads (10^{19} n/cm^2) may be justified for the epoxy material.

Inorganic insulators such as Al_2O_3, Si_3N_4 or $MgAl_2O_4$ will be covered in Section 7, but current information on room temperature or elevated temperature irradiations indicates that such materials should be able to withstand 10^{19} to 10^{21} n/cm^2 without serious degradation of electrical insulating capabilities at the 0.2 kV/mm level.

6.4. Stabilizer materials

There are currently only two choices for this type of material, high purity Cu and Al. The main goal is to keep the resistance of the stabilizer below a level which will prohibit recovery of the coil should it momentarily go normal.

The resistivity of the stabilizer can be considered as the sum of 4 components: the residual resistivity ρ_0, the magnetoresistivity, ρ_m, the increase in resistivity due to deformation ρ_d, and the irradiation produced resistivity increase, ρ_{irr}.

$$\rho_{tot} = \rho_0 + \rho_m + \rho_d + \rho_{irr}$$

Even though one can obtain initial resistivity of Al much below that of Cu, the other three resistivity contributions tend to overwhelm ρ_0 so that initial resistivity is usually not a major consideration.

The magnetoresistivity increase for Cu is much higher than for Al. At 8T, the increase is:

40×10^{-9} ohm-cm for Cu
5×10^{-9} ohm-cm for Al

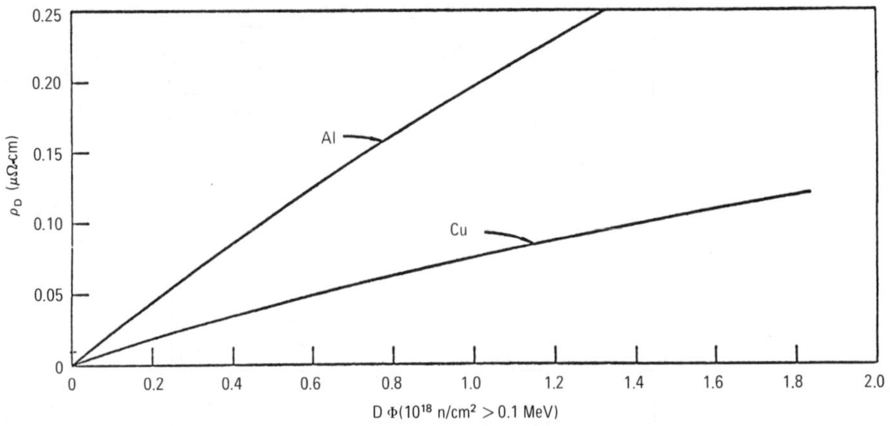

FIG.XIII-21. ρ_D, radiation-induced resistivity increase for Cu and Al vs. fast-neutron fluence (E > 0.1 MeV). The irradiation temperature was 4.9 K.

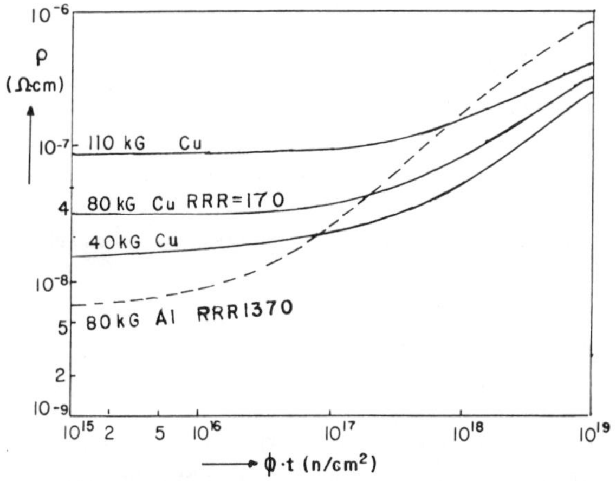

FIG.XIII-22. Schematic of behaviour of the resistivity of stabilizer material, Al or Cu, as a function of fission neutron damage and the interaction of that damage with the induced magnetoresistance.

On the other hand, the irradiation induced resistivity increase for Al is much greater than for Cu (Fig.XIII-21). Combining the three effects, one finds that at high magnetic fields and low damage levels, Al is clearly superior to Cu (Fig.XIII-22). Above $\sim 10^{17}$ n/cm^2 ($\sim 10^8$ rads) Cu becomes more favourable. As the magnetic field increases, this crossover point moves to slightly higher fluences.

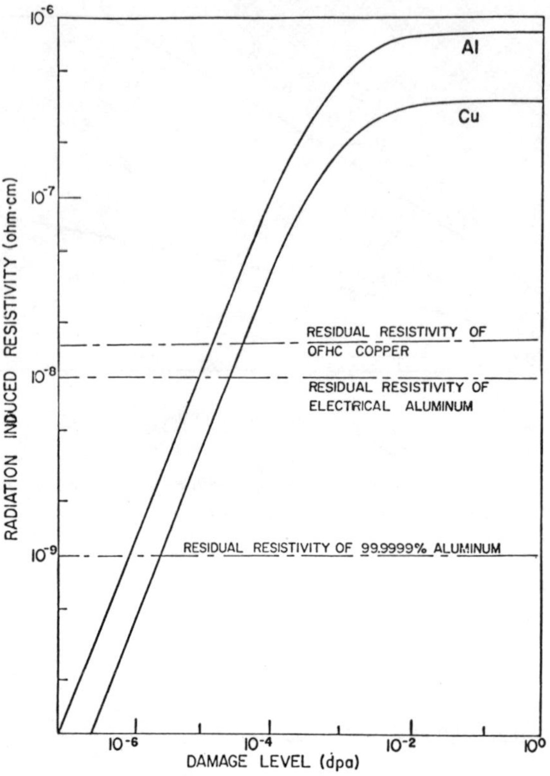

FIG.XIII-23. Radiation induced resistivity of copper and aluminium.

The damage level at which the resistivity of the stabilizer becomes of concern depends to a large degree on how much extra stabilizing material can be placed into the magnet without compromising the design of the magnet. This number is higher for Al than for Cu. For example,

$$\rho_0 + \rho_m + \rho_d \cong 5 \times 10^{-8} \text{ ohm-cm} \quad \text{for Cu}$$

$$\rho_0 + \rho_m + \rho_d \cong 1 \times 10^{-8} \text{ ohm-cm} \quad \text{for Al}$$

Because of the smaller amount of Al required to provide the stabilizing action, one can put proportionally more Al in the magnet to offset the effects of irradiation without unduly increasing the magnet size. It is estimated that one could afford to double the amount of Al in the magnet while increasing its

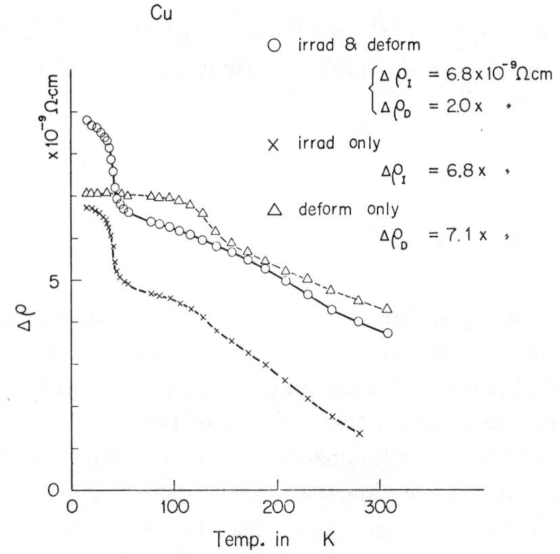

FIG.XIII-24a. The isochronal recovery curves for Cu specimens of three different treatments: 30 h irradiation at low temperature of as-annealed specimen (treatment 1) (x), as-annealed specimen is first irradiated for 30 h, then deformed by twisting at 4.2 K (treatment 3) (initial resistivity increase is normalized to that of treatment 1) (○), specimen deformed at 4.2 K without irradiation (treatment 2) (Δ).

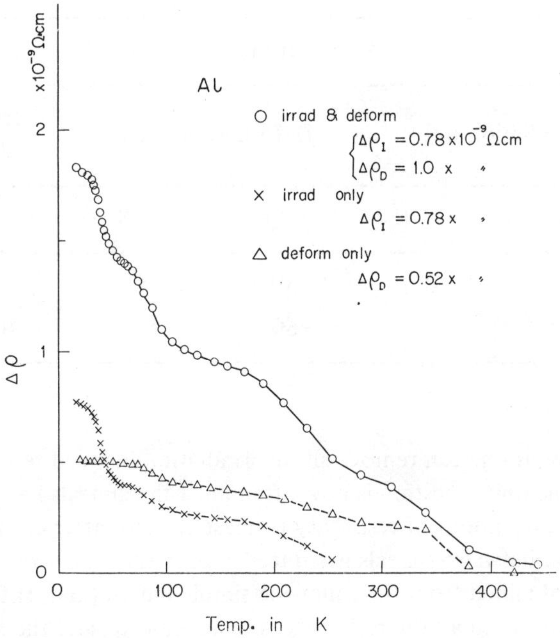

FIG.XIII-24b. The isochronal recovery curves for Al specimen three different treatments, treatment 1 (x), 2 (Δ) and treatment 3 (○).

thickness by less than 12%. The same increase in magnet thickness would result from only a 50% increase in Cu stabilizer. Hence, allowable irradiation induced resistivities would be

Al: $\sim 1 \times 10^{-8}$ ohm-cm

Cu: $\sim 2.5 \times 10^{-8}$ ohm-cm

Such values can be estimated from Fig.XIII-23 to be 1×10^{-5} dpa for Al and 5×10^{-5} dpa for Cu. These limits are roughly 1×10^{16} n/cm^2 for Al and 10^{17} n cm^{-2} for Cu (neutron fluences in fission spectra) and they should be interpreted to no more accuracy than a factor of two.

The fluence at which the stabilizers become no longer useful is not as serious as it may seem, because all, or almost all, of the damage can be annealed out by 300 K. However, recent results in Japan (Figs XIII-24a, b) show that if the specimens are deformed prior to irradiation (which may occur during the initial testing periods) then the amount of resistivity damage retained after room temperature annealing is appreciable. The residual resistivity remaining after room temperature (300 K) or hot water temperature (375 K) annealing is, from Figs XIII-24a, b, as follows:

% Resistivity remaining						
Metal	Irradiation only		Deformation only		Irradiation and deformation	
	300 K	375 K	300 K	375 K	300 K	375 K
Al	~0	~0	~50	~0	~20	~10
Cu	~20	<5	~60	~50	~40	~30

Obviously, if one can remove all the irradiation induced resistivity during an anneal, then the only question is how often must the anneals be? On the other hand, if a certain amount of resistivity increase remains after each anneal, then the time period between anneals is reduced. An idea of the seriousness of this effect can be obtained from the following simple illustration. If the time period between anneals when all the resistivity is removed is t_0, and the accumulation of the residual resistivity is linear, then the subsequent anneal periods for some residual resistivity are as follows:

	Period between anneal		
Annealing period	All resistivity removed in anneal	90% removed in anneal	70% removed in anneal
1	t_0	t_0	t_0
2	t_0	0.9 t_0	0.7 t_0
3	t_0	0.81 t_0	0.49 t_0
4	t_0	0.73 t_0	0.34 t_0
5	t_0	0.66 t_0	0.24 t_0
6	t_0	0.59 t_0	0.17 t_0
7	t_0	0.53 t_0	0.12 t_0
8	t_0	0.48 t_0	0.08 t_0
9	t_0	0.43 t_0	0.06 t_0
10	t_0	0.39 t_0	0.04 t_0

Such residual resistivity levels can obviously be a problem if the initial period between anneals is of the order of 1 year. The effect on availability, especially because the time for warming up and cooling the superconductors is of the order of one month, could be very serious.

6.5. Irradiation effects on superconductors

The effect of neutron irradiation is, in general,to lower the T_c, H_{c2}, and I_c of the superconductors. Information on the exact amounts is included in the various reports of the countries. An example of the magnitude of the effect is given in Fig.XIII-25. It is shown that the critical current of Nb_3Sn is severely degraded above 3×10^{18} n/cm^2 ($\sim 1.5 \times 10^{-3}$ dpa), while that of NbTi changes by less than a factor of two at 10^{20} n/cm^2 (5×10^{-2} dpa).

The combined effects of increased resistivity and reduced superconducting properties was studied by the Japanese(Fig.XIII-26). It was found that high-purity Cu is favoured over high-purity Al at all exposure levels above $\sim 10^{17}$ n/cm^2 (fission spectra) for both Nb_3Sn and NbTi superconductors. In general, it was found that the degradation of the stabilizer still outweighs the loss in superconducting properties.

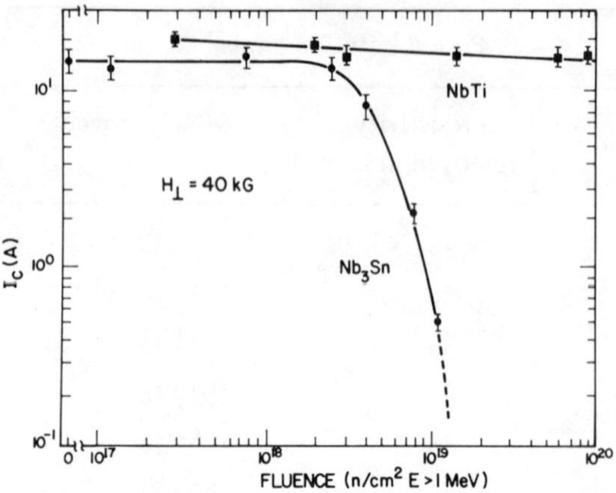

FIG.XIII-25. Effect of neutron irradiation on the critical current in NbTi and Nb_3Sn.

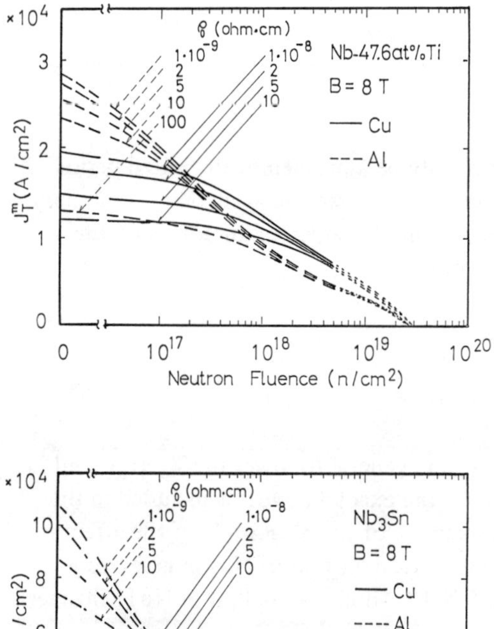

FIG.XIII-26a. Maximum transport current density J_T^m at 8 T is plotted as a function of fluence in case of Nb-46% Ti. The influence of initial residual resistivity of stabilizer is also shown.

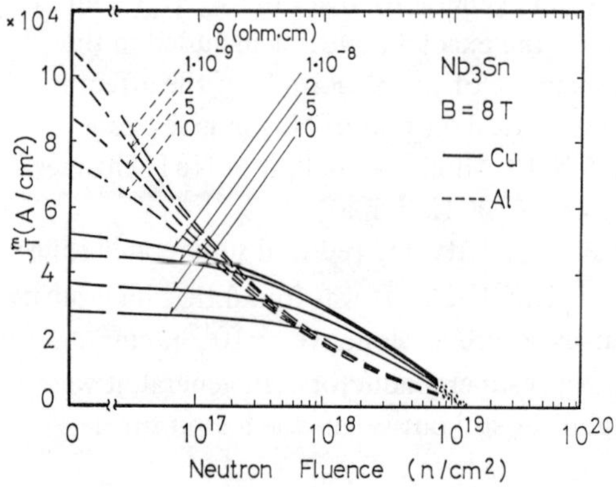

FIG.XIII-26b. Maximum transport current density at 8 T is plotted as a function of fluence in case of Nb_3Sn (19 multifilament conductor.

TABLE XIII-12. POTENTIAL DESIGN EXPOSURE OF CERAMICS IN INTOR

Possible application	Location	Fluence nvt (n/cm^2-E > 0.1 MeV)[a]	dpa	He (appm)	Proposed material
Low-Z first-wall liner	First wall	8.3×10^{22}	83	10 700	SiC
Insulators for RF systems	First wall or in blanket	$<8.3 \times 10^{22}$	<83	<5 500	$MgAl_2O_4$
Toroidal current breaker	Middle of blanket	8.3×10^{21}	8.3	550	$MgAl_2O_4$
RF coil insulation	Rear of blanket	8.3×10^{20}	.8	55	Si_3N_4
Neutral-beam injector insulators	Ports behind shield	8.3×10^{19}	.08	5.5	Al_2O_3

[a] EBR-II equivalent.

7. IRRADIATION EFFECTS ON CERAMICS

At this time, there is very little information as to what irradiation exposure the ceramics in INTOR might have to withstand. This could vary from the full exposure to ceramic coverings on the first wall to several orders of magnitude less at the back of the blanket or in the neutral beam injectors. A summary of possible applications, exposures and proposed materials is given in Table XIII-12. The reason that Al_2O_3 is not proposed for a greater role in INTOR is because of its dimensional changes under neutron irradiation at modest temperatures. A summary of recent information on swelling of Al_2O_3 irradiated at 400°C is given below.

Fission neutron exposure, n/cm^2 at 400°C	% Swelling
10^{20}	0.3%
10^{21}	1.5%
3×10^{21}	2 to 3%

TABLE XIII-13. LOS ALAMOS SWELLING DATA FOR IRRADIATION TO 2×10^{22} n/cm^2 AT 650°C AND 850°C

Swelling (%)		Material
Low swelling <0.5%		Single crystal $MgAl_2O$ (spinel)
		Single crystal $Y_3Al_5O_{12}$ (Yttrium aluminium garnet)
		Polycrystalline Si_2ON_2
		Polycrystalline Sialon
Moderate swelling 0.5–1.5%		Polycrystalline Si_3N_4
		Polycrystalline Y_2O_3
		Polycrystalline Y_2O_3/ZrO_2
High swelling	4%	Single crystal Al_2O_3
	10%	Polycrystalline BeO

TABLE XIII-14. THE ANTICIPATED JULY 1985 IRRADIATION DATA BASE ON ELECTRICAL AND STRUCTURAL PROPERTIES FOR TYPICAL INTOR CERAMICS

Material	Dpa	He (appm)	H (appm)	Irradiation source
$MgAl_2O_4$	80	Low	Low	Fast-flux fission reactor
	20	2 800	Low	Mixed-spectrum fission reactor[a]
	60	9 900	6 000	FMIT (TEM only)
Al_2O_3	80	Low	Low	Fast-flux fission reactor
	20	3 200	Low	Mixed-spectrum fission reactor[a]
	60	10 200	6 000	FMIT (TEM only)
Si_3N_4	80	Low	Low	Fast-flux fission reactor
	20	Low	4 300	Mixed-spectrum fission reactor[b]
	60	9 600	8 100	FMIT (TEM only)
SiC	80	Low	Low	Fast-flux fission reactor
	60	19 200	5 400	FMIT (TEM only)

[a] Utilizing ^{17}O doping and the $^{17}O(n,\alpha)^{14}C$ thermal-neutron reaction.

[b] Utilizing the $^{14}N(n,p)^{14}C$ thermal-neutron reaction and ^{15}N doping to reduce reaction rate.

There have been some recent experiments at LASL on some other insulating materials. These are listed in Table XIII-13. The data is separated into 3 categories: low swelling, moderate swelling and high swelling. It was concluded that tentatively, Al_2O_3, even if of high quality (high purity, small grain size, etc.) would **not** be expected to withstand more than 10^{21} n/cm^2 in a fusion environment.

Finally, the data base expected to be available from both fission and fusion sources by 1985, is summarized in Table XIII-14. Only the FMIT can correctly simulate the gas to displacement ratio of a fusion environment. The use of ^{17}O in Al_2O_3 or ^{15}N in Si_3N_4 can also result in a better simulation of a fusion environment in mixed spectrum fission reactors. By the use of all available techniques, there should be a reasonable data base on ceramics for use in INTOR by 1985.

8. IRRADIATION EFFECTS ON GRAPHITE

8.1. Introduction

Graphite is under consideration for use as a first wall liner, coating, or blanket; as a limiter; as wall armor; and perhaps as a beam stop. A large number of solid graphites offering a variety of properties are available in bulk or as coatings. Major variables are: grain and crystallite size, shape, and orientation; degree of crystallinity; density, and degree of anisotropy. Graphite cloth is also available. Here the nature of the precursor fibres, processing steps, and weave are important variables.

The most important factor governing lifetime of graphite in fusion reactor applications is its irradiation behaviour. The present data base assessment considers dimensional, thermal, chemical and mechanical behaviour of graphite, and concludes with comments on fabrication, advanced materials, and fusion neutron effects.

8.2. Assessment

8.2.1. Dimensional stability

Graphite typically shrinks and then expands (i.e. exhibits breakaway swelling) with increasing neutron fluence, as shown by Fig.XIII-27 and XIII-28. The detailed behaviour of many graphites as a function of particle size and orientation, temperature, and fluence has been evaluated and dimensional changes in graphite fibres after irradiation at 743 K have been reported. There is some disagreement as to dimensional changes at higher temperatures (~1300–1700 K); more work is needed here. No data are available above ~1700 K, a temperature range of importance for fusion graphites subjected to high heat loads.

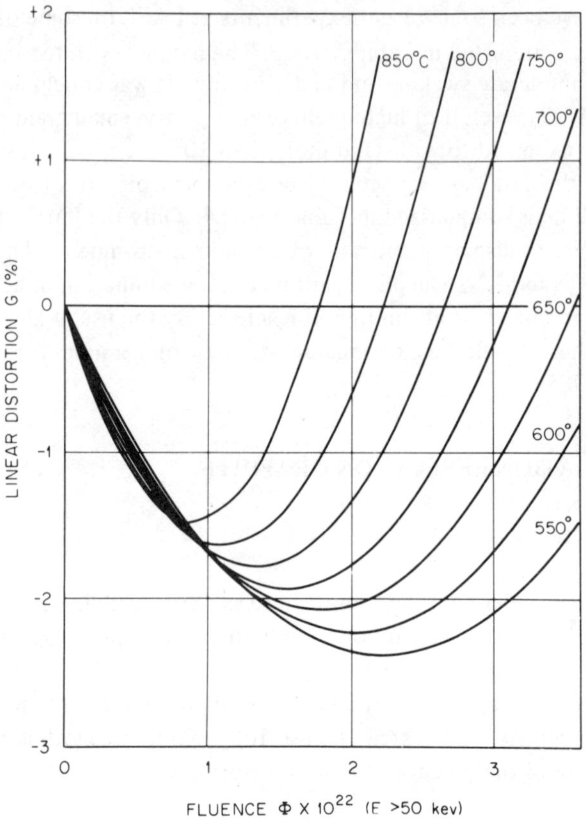

FIG.XIII-27. Standardized design curves for nuclear graphite demonstrate the gain in lifetime as the irradiation temperature is lowered.

A connection between breakaway swelling, microstructural changes, and plastic deformation has been established, but not quantified. Further studies in this area are required in order to allow fission neutron damage results to be related to a fusion neutron environment.

The data base for graphites of interest to fission reactors is expanding significantly and will continue to do so in the early 1980s. The value of this information to the fusion energy programme is circumscribed, however, by two considerations: the graphites being studied are probably not optimum for tokamak applications, and the neutron energy spectrum is that for fission rather than fusion.

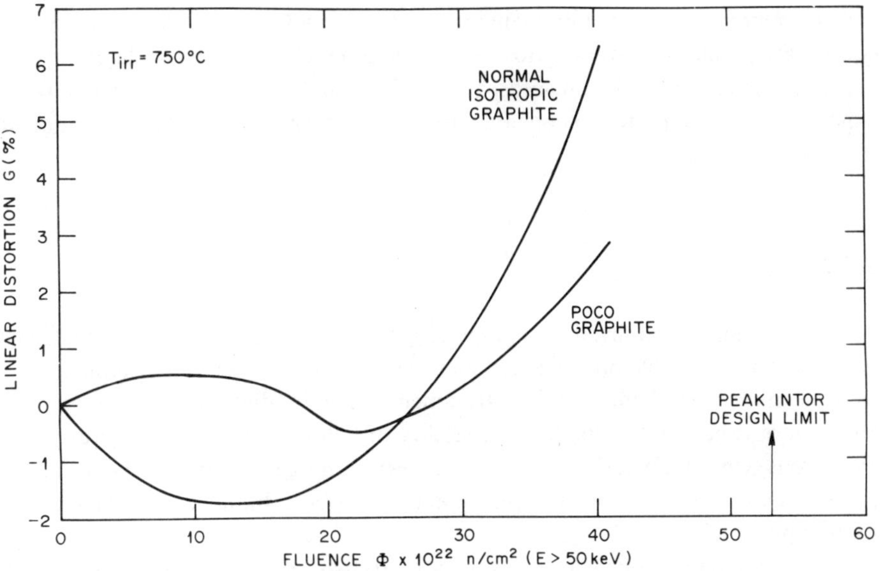

FIG.XIII-28. Nuclear graphites approach the lifetime required for the peak INTOR design fluence, by operating below 550°C, this limit is probably obtainable.

8.2.2. Thermal behaviour

Graphite crystals display strongly anisotropic thermal expansion coefficients, an effect which is reflected in polycrystalline bodies of preferred orientation. The data base for thermal expansion behaviour is good, both for unirradiated and fission neutron-irradiated materials. Work is needed to extend the data to higher temperatures (>1700 K), and to evaluate the apparent correlation between onset of thermal expansion changes and of breakaway swelling in the context of anticipated fusion neutron effects.

Thermal conductivity is decreased by irradiation-induced defects and by the onset of disintegration at the breakaway swelling point. Considerable information is available on this subject, but a systematic evaluation of thermal conductivity changes at high temperatures and fluences is needed. Since microstructural changes induced by fusion neutron irradiation may differ from those induced by fission neutrons, a reassessment will likely be required for fusion reactor applications.

Several of the anticipated applications of graphite in a tokamak device require high resistance to thermal shock and thermal cycling, a problem not seriously encountered in fission reactors. Several governments have established elaborate data bases in this area for graphites of interest in aerospace applications, and the microstructural requirements for thermal shock resistant graphites are qualitatively well understood. However, there is a dichotomy in the data: no neutron-damage data exist on aerospace graphites and little thermal shock data exist on nuclear graphites. There are currently no programmes planned to address this lack of a data base.

8.2.3. Mechanical behaviour

The strength of unirradiated graphite is fairly well understood, but insufficient information is available on irradiated material. There is evidence that either strengthening or weakening can occur, depending on irradiation temperature. Much work is needed to establish probability of failure under fusion reactor conditions (e.g., thermal shock, cyclic stresses). A data base can be established by testing in fission reactors, but fusion neutron effects should be evaluated as soon as possible.

The apparent Young's modulus is found to increase initially upon irradiation, perhaps as a result of dislocation pinning and microstructural changes. A decrease is noted at higher doses where structural disintegration begins. The sensitivity of modulus to low-dose irradiations has allowed its use as a criterion to assess relative damage effects from fission and fusion neutrons. Such low-dose comparative damage studies are needed to guide extrapolation of irradiation data from fission reactors to fusion reactor applications.

Helium will be generated in large quantities in a fusion neutron environment, and this could strongly affect mechanical properties if retained in the lattice. Diffusion of helium and its escape from some graphites has been studied. This parameter should be evaluated for all candidate materials, and the effect of concurrent helium generation on damage microstructure determined in order to assist fission-fusion damage correlation. However, it seems unlikely that helium in graphite will be a significant problem unless graphite is to be employed at cryogenic temperatures.

Graphite shows a greatly enhanced creep rate under fission neutron irradiation. A data base exists up to ~1500 K, but higher temperature measurements are needed. Creep-to-fracture is observed to be increased by irradiation, and in view of its importance to lifetime considerations this phenomenon should be thoroughly evaluated for candidate materials. More microstructural studies are needed to establish creep mechanisms, so that existing data can be considered in light of anticipated fusion effects.

The data base for mechanical properties is rapidly expanding for graphites of interest to the fission-energy programme. Again, however, these graphites may not be applicable to tokamak devices because of their poor thermo-mechanical properties. The question of fusion vs. fission energy spectrum also remains moot.

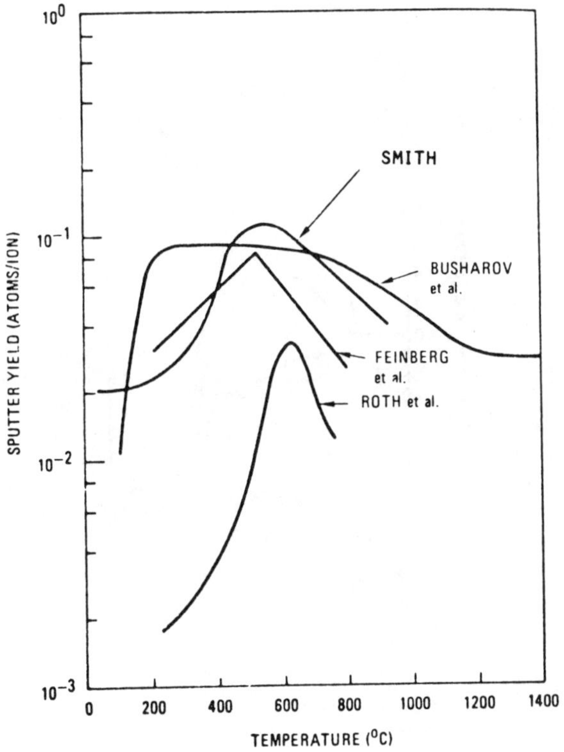

FIG.XIII-29. Temperature-dependent sputtering yield of carbon bombarded with hydrogen.

8.2.4. Further considerations

The wide variety of graphite forms as determined by raw materials and fabrication techniques should be recognized as both an advantage and impediment to establishing a data base for INTOR: an advantage in that graphite can be tailor-made to the tokamak requirements as necessary; an impediment in that a data base can only be established by careful selection of appropriate graphites from the myriad of commercial and experimental materials available. The two requirements of stability against radiation damage and resistance to thermal shock are reconcilable, but do represent somewhat opposite needs.

Neutron damage in graphite is primarily a displacement effect, and hence data at fission energy spectra should be translatable to fusion spectra. We have indicated throughout the above assessment that fusion spectrum damage rates must be established to remove the uncertainty in correlation before the mass of fission damage data can be employed with confidence.

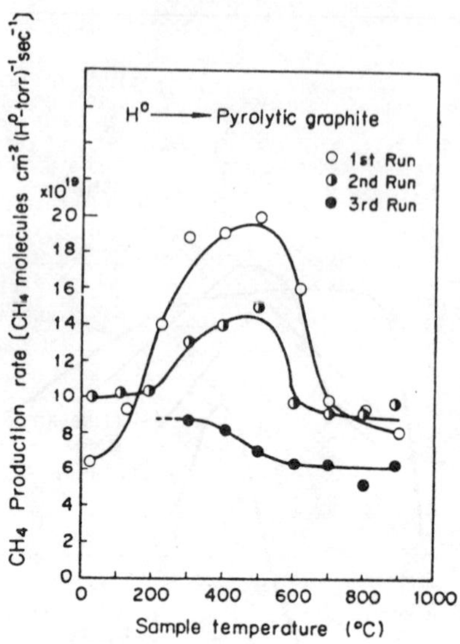

FIG.XIII-30. Production of methane during three successive cycles of heating pyrolytic graphite in a hydrogen atmosphere.

8.2.5. Chemical erosion effects

Chemical interaction between reactive plasma particles and first-wall materials can influence the erosion yields because of effects of compound formation on lattice displacement energies, on sputtering mechanisms, and on the nature of sputtered products. Although this so-called chemical sputtering occurs in a number of systems, it is most commonly associated with the formation of hydrocarbons upon bombardment of carbon or graphite by the hydrogenic plasma particles. Fig.XIII-29 shows some erosion or sputter yield data obtained from hydrogen bombardment of different kinds of graphite at different experimental conditions temperatures to ~1200°C. Although there are some discrepancies in the published data, the general trend is for an increase in the yield as the temperature is increased from room temperature to ~400°C and then a rather sharp drop in the yield above 800°C. The yields observed at 400–800°C are approximately a factor of ten greater than the minimum observed yields for physical sputtering alone.

Recent experiments by the Japanese have implied that a graphite surface can be made passive to chemical sputtering by hydrogen through a surface treatment. Pyrolytic graphite was heated three times between room temperature and 900°C in a neutral hydrogen atmosphere where the flux of hydrogen atoms was 10^{17} cm^2 s. The methane production rate was measured to peak at ~500°C on the first and second runs but the magnitude of the peak was greatly reduced (see Fig.XIII-30). The third time the 500°C peak was removed and the magnitude of the CH_4 was again reduced. Such results give hope that chemical sputtering will not be as serious as previously indicated.

It is unclear as to what extent the erosion yield of carbides is enhanced by chemical interactions with energetic hydrogenic plasma particles. It appears that some chemical interaction occurs during initial bombardment, which tends to leave the surface depleted of carbon. At present, the chemical sputter yields for stable carbides of interest are believed to vary from negligible to approximately three times the physical sputter yields for the compounds.

8.3. Conclusions

The data base for neutron damage in graphite for nuclear fission reactors is excellent and currently being extended. The thermomechanical data base for aerospace-type graphites is also excellent and growing. Unfortunately these two data banks do not overlap.

An orderly development of a graphite data base for INTOR by the mid-1980's requires the following steps:

1. The methane and acetylene reaction rates need to be established to determine temperature restrictions on the use of graphite.
2. Over the appropriate temperature range, fission damage data must be accumulated for the specific graphites appropriate to fusion neutron environments.
3. The ability to translate fission damage to fusion damage needs to be established to take advantage of the extensive data already available on radiation effects.

Sufficient information currently exists to establish the feasibility of graphite to withstand the expected INTOR neutron damage and thermal-mechanical requirements. The need is to establish data on a specific grade of graphite and develop this to the extent required for engineering design purposes.

9. DATA BASE FOR PLASMA-MATERIAL INTERACTIONS

9.1. Classification of materials and their functional use

Three types of materials have been proposed for first-wall components of near-term tokamak reactors. The structural materials considered for both present and near-term tokamaks include several transition-metal alloys, viz., austenitic stainless steels, ferritic steels, titanium alloys, and high-nickel alloys. Because of the relatively high-Z (atomic number) of these alloys, low-Z coatings on the first wall have been proposed to reduce the in-flux of high-Z impurities into the plasma. A major advantage of the coating is that it can be selected primarily on the basis of its surface properties. Low-Z materials most often considered include beryllium, boron, carbon, and compounds of these elements. Major internal components that may be subjected to high surface heat loads, e.g., limiters, liners, and beam dumps, require more refractory materials with high thermal shock resistance, high thermal conductivity, and large heat capacity. Leading candidates include graphite, ceramics, and refractory metals. Because of plasma purity requirements, coating of these materials may also be necessary. Aspects related to coating processes and properties are discussed in Section 10.

9.2. Neutron sputtering

Considerable effort has been expended during recent years on the investigation of 14 MeV neutron-sputter yields and large variations in the yields have been reported for various transition metals. A recent review of the literature suggests a value of $\sim 10^{-5}$ atoms/neutron for the most reliable sputter yield data. This value is in reasonable agreement with values predicted from atomic collision theory.

9.3. Physical sputtering – ions and neutrals

Considerable effort has been placed on physical sputtering in the last few years. This effort includes both experimental measurements as well as analytical and theoretical work. Important parameters considered for physical sputtering include:

- incident ion (or neutral)
- ion energy
- angle of incidence
- wall (target) material
- incident flux dependence
- incident dose dependence
- energy of sputtered specie
- angular distribution of sputtered specie
- wall temperature
- morphology of wall surface

The importance of these various parameters depends on the first-wall component of interest.

9.3.1. Experimental data base

The experimental data base for physical sputtering relevant to tokamak reactor conditions has improved substantially in recent years. The status of

TABLE XIII-15. MATERIALS FOR WHICH NORMAL INCIDENCE SPUTTER YIELDS FOR HYDROGEN, DEUTERIUM, AND HELIUM HAVE BEEN MEASURED

Elements			Compounds and alloys	
Be	Co	Ag	Al_2O_3	TaC
C	Ni	Ta	B_4C	Ta_2O_5
Al	Cu	W	BeO	TiC
Si	(Zr)	(Ir)	Ni alloys	WC
Ti	Nb	(Pt)	SAP	ZrC
(V)	Mo	Au	SiC	
(Cr)	(Pd)	(U)	SiO_2	
Fe			Stainless steels	

light-ion sputtering, i.e., hydrogen, deuterium, and helium, is in relatively good shape with respect to normal-incidence total yield as a function of energy and in fair shape with respect to the following parameters: angle of incidence, flux, dose, and temperature. Normal incidence yields have been measured for many elements and compounds as a function of energy and are moderately well known with respect to the following parameters: angle of incidence, flux, dose, and temperature. Materials for which some experimental data have been reported are listed in Table XIII-15 and the reader is referred to the review of data in the European Report to the October Workshop.

Fig.XIII-31 shows the experimentally measured total sputter yields for normally incident light ions on stainless steel. Only limited experimental data are available for plasma impurity ion ($Z < 2$) sputtering of relevant materials. Much of the higher-Z ion sputtering data has been obtained with inert gas ions. In general, the best experimental data for physical sputtering are no better than ± 20% and probably no worse than a factor of two. Experimental data on the effect of angle of incidence on the sputtering yields are much more limited. The angular effect gives a greater-than sec θ relationship ($\theta < 70°$) for most systems investigated (Fig.XIII-32). Fluence and flux dependencies, although not investigated in depth, tend to show rather modest effects. The major concern with experimental data involves residual impurity effects, particularly for reactive metals such as titanium and beryllium. It is generally conceded that some oxide exists on the surface of these materials during testing.

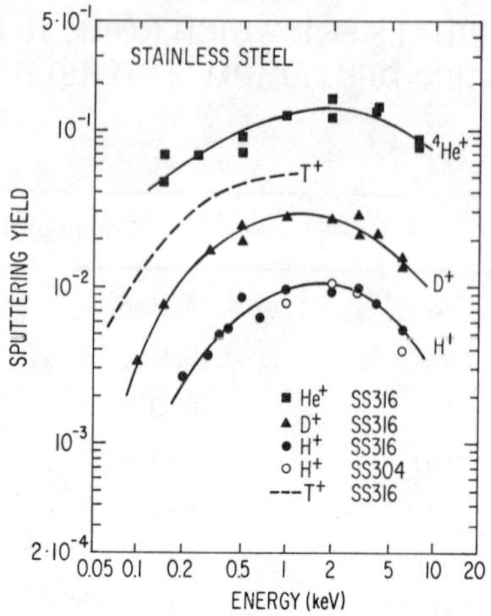

FIG.XIII-31. Sputtering yields of stainless steels (types 304 and 316) for irradiation by H^+, D^+ *and* $^4He^+$ *ions at normal incidence, as a function of the ion energy. The dashed line indicates estimated values for tritium.*

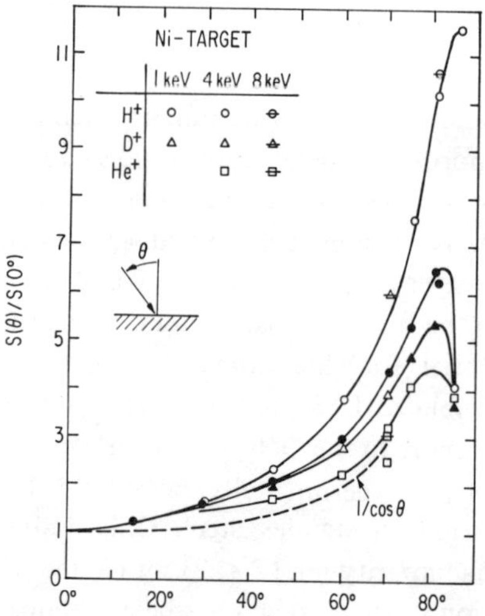

FIG.XIII-32. Angular dependence of physical sputtering yield for light ions incident on nickel.

A recent study by Guseva et al., has shown that the hydrogen sputtering yield of Fe-Cr-Ni alloys increases with increasing Ni concentration. Such variations with composition have not been systematically investigated thus far. Guseva et al. have also shown that when sputtering yields of a polished surface are compared to a 'sponge like' surface, the yields are reduced by a factor of 2–3.

9.4. Blistering phenomenon

Considerable information has been obtained in recent years on the "blistering" phenomenon produced by injecting high-energy light ions (primarily hydrogen and helium) into the surface regions of candidate fusion reactor first-wall materials. Several investigators have observed extensive blister formation and high erosion rates for metals bombarded with ~0.1 to 1000 keV helium under a variety of conditions. Blister size and skin thickness have been correlated with incident particle energy and penetration depth profile. The blister erosion rates are strongly temperature-dependent for some materials and material preparation history is also important. The results obtained from helium-injected nonmetals generally show a flaking or spalling rather than the typical blister formation. This phenomenon is generally attributed to low ductility which limits the plastic strain needed for classical blister formation. It has also been shown that blistering of sintered material is greatly suppressed compared to vacuum-cast materials. This reduction in blistering, which was observed for both aluminium and beryllium, is attributed primarily to modifications of the surface microstructure, viz., grain size and porosity. In summary, one can show from both theory and experiment that the surface erosion character changes with increasing temperature in the following way: blistering ($T \leqslant 0.1\ T_m$, T_m – the melting point in K) – flaking ($0.1 < T \leqslant K_1(E)T_m$) – blistering ($K_1(E)T_m < T < K_2(E)T_m$) – porous structure ($T > K_2(E)T_m$) without blistering or flaking. Both K_1 and K_2 are proportional to lnE, where E – ion energy, at $E < 100$ keV $K_1 \approx 0.4$; $K_2 \approx 0.5$.

Reductions in the erosion rates by as much as three orders of magnitude were observed at both room and elevated temperatures. This type of data indicates that blistering can be minimized by appropriate modifications to the surface structure. The plasma spray coating process, which permits wide variation in grain size and porosity of the coatings, has been proposed as a method of obtaining desirable properties.

The effective erosion rates in Ti, V, Nb, steel, and Al reach values of equal to or greater than 1 at energies greater than 40 keV for doses of $\leqslant 6 \times 10^{18}/cm^2$. For example, the erosion rate of 304 SS is 3 atoms per ion at 450°C, E = 100 keV and a dose of $3 \times 10^{18}/cm^2$.

With the above information in mind, a major question is, "how high is the flux of such high energy helium particles leaving the plasma? " Another major question is, "how many of those helium ions actually reach the first wall with such energies? " If the flux to the first wall is below $3 \times 10^{10}/cm^2 s$ then the effect will be negligible over the full lifetime of INTOR.

The erosion rate due to blistering is also dose dependent. The theory and experiment show that sizes of the blisters and exfolited area decrease from one generation to the sequential generation and there is a minimum blister size that exists. Hence only finite number of blister generations is possible. After the sputtering of the layer with a width equal to the blisters skin thickness, a porous structure is formed on the surface. Further investigation of synergistic effects of simultaneous He^+ and H^+ ion bombardment are required. The first results of simultaneous irradiation by D_2^+ and He^+ ion of stainless steel have shown significant enhancement of blister erosion rates.

9.5. Vaporization of wall materials

The major surface damage in present-day machines is normally at the limiters and to a smaller extent on the surface of the first wall. It appears that the damage is due mainly to runaway electron beams and to disruptive discharges and is therefore atypical of a stable, well-behaved plasma. Arcing has been observed in many tokamaks and can introduce impurities either as atomic particles or micron-size particles. The propensity for arcing appears to be dependent on surface condition and on the plasma MHD instability. A complete understanding of arcing is lacking at present but there are relevant laboratory and tokamak studies which would shed sufficient understanding so that arcing can be minimized. Vapour pressure curves of relevant materials are fairly well known but plasma control in the startup phase and during plasma disruptions leaves much to be desired and makes predictions of vapourization rates extremely difficult.

10. RELEVANT COATING TECHNOLOGY

The application of coatings, particularly low-Z materials, for fusion-reaction first-wall components has received considerable attention in the last few years. The properties of the materials used for the various components of a fusion reactor depend on the environment to which it is exposed, the performance desired, and the effect of the component or material on the performance of the reactor. Coatings (or claddings) on internal components, vacuum walls, neutral beam injectors, and divertors of fusion reactors may provide a desirable and possibly necessary engineering design flexibility because the coating can be chosen to perform only one function; to minimize plasma contamination. It does not have to be compromised by the need to be a structural material with the attendant criteria of capability for fabrication in thick sections, weldability, high strength, etc. In addition, because of the relatively small amount of material involved in a coating, possibilities are opened up for the use of exotic materials that would otherwise be considered too expensive or scarce. A coating system does impose the additional requirement of coating-substrate adhesion which must be maintained under normal and fault operating conditions. Successful adhesion of a coating to a surface may be more related to the factors which cause fracture

TABLE XIII-16. COATING/SUBSTRATE SYSTEMS UNDER DEVELOPMENT

Coating	Substrate	Application technique
TiB_2	POCO graphite	CVD
TiB_2	Mo	CVD
TiB_2	Cu	Plasma spraying
TiB_2	Stainless steel	Plasma spraying
TiC	POCO graphite	CVD
TiC	Copper	Plasma spraying
B_4C	POCO graphite	CVD
B	POCO graphite	CVD
Be	Cu	Plasma spraying
Be	Stainless steel	Plasma spraying
VBe_{12}	Cu	Plasma spraying
VBe_{12}	Stainless steel	Plasma spraying

than to chemical bonding. For this reason, the mechanical and thermal properties of the coating and substrate must be matched and such things as gas accumulation or eutectic formation at the interface must be avoided. Coatings (or claddings) on components could have thickness greater than 20 μm and may be formed by a variety of application techniques, e.g. chemical vapour deposition (CVD), physical vapour deposition, plasma-enhanced CVD, plasma spraying, electroplating, cementation processes, explosive cladding, glow discharge reaction, ion implantation and many other techniques.

In addition to the relatively thick coatings normally considered for components, it may be possible to form in situ coatings which are very thin, but renewable by a variety of techniques including: (1) deposition by plasma decomposition of a compound; (2) vapour deposition from a heated solid; (3) solute segregation; or (4) redeposition of eroded material as is now occurring in tokamak reactors.

Several coating/substrate systems have been studied and are under development for fusion reactor applications. These are summarized in Table XIII-16. At the present time, the materials of most interest are TiB_2, B, B_4C, Be and TiC on graphite or copper for limiter and beam dump applications. The titanium compounds may pose a problem if they decompose on erosion to elemental titanium which will absorb the hydrogenic species. Important factors being investigated include: adhesion, hydrogen retention, erosion, gas recycle, and mechanical properties.

Recent tests on copper coated with 0.16 mm thick TiC and TiB_2 have been reported by Sandia Albuquerque Laboratory in the US. It was found that there were no deleterious effects of 50 pulses with hydrogen at 2 kW/cm^2 for 0.7 seconds each. The energy of the ions was varied from 250 to 1000 eV. Similar tests with stainless steel substrates caused considerable alloying and substrate melting after 10 shots. Unsatisfactory results were also obtained with VBe_{12} on Be coatings on both Cu and stainless steel. It therefore appears that the TiBc/TiC combination with Cu is the most promising at this time.

Initial testing of coated limiters is planned for late 1979 in both ISX and PLT. Results from the ISX-coated limiter experiments should prove feasibility and lead to expanded use of coated limiters and other components in FY 1980 in ISX, PLT, PDX, and DOUBLET III.

11. OVERALL CONCLUSIONS AND OBSERVATIONS FOR MATERIALS DATA WITH RESPECT TO INTOR

In general, it is concluded that the data base either exists now, or that it will exist by the mid-1980's, to be able to make a reasonable judgement on the mechanical and electrical performance of materials in INTOR. Furthermore, there is reasonable optimism that the structural materials can be used for the lifetime of the device. More specific conclusions are given below.

(1) Of the 4 main classes of potential structural materials considered for this meeting (austenitic steels, ferritic steels, titanium alloys and aluminium alloys), it is clear that the austenitic system is the most favoured choice for INTOR at this time.

(2) The unirradiated data base on representative alloys, such as 316 SS, HT-9, Ti-6Al-4V, or Al 6000 series alloys is either sufficient now or could be obtained by the mid-1980's

(3) The higher thermal stresses induced in 316 SS compared to the Ti, Al and ferritic alloys will make first wall design more difficult for stainless steel. However, it should be possible to withstand heat fluxes of 50 to 100 W/cm^2 with water cooled tubes of 1.5–2.5 mm thickness.

(4) The post-weld annealing temperature of 600–750°C for the ferritic alloy may present some practical difficulties in the field fabrication of an INTOR first wall, blanket and shield.

(5) The higher solubility of Ti alloys for hydrogen isotopes could present an inventory problem for tritium and/or an embrittlement problem for the alloy below 300°C.

(6) Fabrication of Sintered Aluminium Product (SAP) appears to be very difficult and it is questionable that vacuum tight welds can be reliably obtained.

(7) In general, the tensile properties of both the ferritic and titanium alloys are superior to SA 316 SS at expected INTOR temperatures. However, cold-working the 316 SS reduces this difference.

(8) There is only one material, 316 SS, that has been irradiated under appropriate helium/dpa conditions expected in INTOR. The similar data base does not exist for any other alloy considered for INTOR. However, by 1985 it is expected that such information can be obtained on a few properties such as swelling and post-irradiation ductility in a high energy neutron spectrum.

(9) It appears that irradiation induced swelling of 20% cold worked 316 SS irradiated to ~6 MW-y/m^2 (~60 dpa and 900 appm He) at temperatures less than 400°C will be less than 3%.

(10) The most severe problem for the 20% CW 316 SS at 400°C is the irradiation induced embrittlement. Total elongation values may be as low as 2% and uniform elongation values even less at INTOR damage levels.

(11) Fatigue crack growth rates in 316 SS are slightly increased with irradiation and, depending on the stress intensity factor and the nature of the loading, survival after 10^6 cycles needs to be demonstrated under anticipated INTOR stress conditions.

(12) A sufficient data base on irradiated 316 SS could exist by the mid-1980's from which to make predictions of useful lifetime in INTOR.

(13) There is a sufficiently encouraging amount of data on irradiated ferritic alloys at the present time to anticipate good performance in INTOR. However, the data are generally obtained at low helium concentrations and it will be the mid-1980's before the effects of He on this alloy are clear. Major concern over the irradiation induced increase in the ductile to brittle temperature is expressed if operations are in the 300–400°C range.

(14) There are very little irradiation data on which to base a judgement of Ti alloys at this time. High fluence fission neutron data will be available in the INTOR range within the next few years, but it will be the mid-1980's before sufficient irradiation in high energy neutron spectra can be conducted to test the effects of helium.

(15) Even though there is considerable information about Al irradiated at low temperatures (50°C) in fission reactors, there is essentially no data above 100°C with appropriate amounts of He. Post irradiation heating to more than 100°C for Al alloys and 300°C for SAP greatly reduces the ductility. This brings into question the viability of surface cleaning procedures such as baking.

(16) There is presently adequate data base on coolants such as H_2O and He for INTOR. Corrosion rates of steel look like they are manageable although there will be significant amounts of radioactive Co, Fe, Mn, Cr, and Zn from steel in the water coolant to cause concern for maintenance.

(17) A considerable amount of information on the physical properties of Li_2O has been developed and pellets of 70 to 99% of theoretical density have been produced. Some physical data exists for $LiAlO_2$ and Li_4SiO_4 but more information is required for INTOR.

(18) Compatibility of Li_2O with steel and Mo alloys reveals that severe attack occurs above 800°C in steel and nickel base alloys and above 950°C with TZM. The design of tritium breeding modules must take such limits into account, even for other Li ceramic compounds.

(19) Studies of the release of tritium from neutron irradiated Li_2O reveal that 90% is released from pellets of 88.5% density or less at 500°C while some 50% remained in pellets of 91.5% density. Similar studies on $LiAlO_2$ and Li_4SiO_4 reveal that 90% of the tritium can be removed at 800°C and 700°C respectively. Effects of radiation on tritium release rates (resulting from sintering and defect pinning) have not been investigated at high fluences and high temperatures characteristic of INTOR or DEMO.

(20) There are major compatibility concerns regarding the use of lithium, Li_2O, Li_2AlO_2, Li_4SiO_4 and Li_7Pb_2 in a water-cooled system in the event of accidental leaks.

(21) There are limited data on the stability of Li_2O, $LiAlO_2$ and Li_4SiO_4 under conditions of interest for INTOR. Of particular concern is the potential for weight loss via LiOH formation, and the degree to which water can be excluded from the Li_2O inserted into the reactor.

(22) At the anticipated radiation levels to the INTOR TF magnet system, it appears that the radiation damage to the stabilizer is the limiting feature. The critical damage level is 1×10^{-5} dpa ($\sim 1 \times 10^{16}$ n/cm^2) for Al and 5×10^{-5} dpa ($\sim 10^{17}$ n/cm^2) for Cu.

(23) It has been shown that deformation of both Al and Cu prior to irradiation reduces the amount of resistivity increase annealed out at room temperature. Even at 100°C, as much as 10% of the resistivity increase in Al and 30% of that in copper may remain. This raised serious questions about the length of time between periodic annealing phases or would require increased shield thickness of 5 to 8 cm.

(24) Irradiation data up to 10^9 rads on the electrical insulators are presently available. The data indicate that useful lifetimes can be obtained in the 10^9 rad regime and perhaps 10^{10} rad limits are even reasonable. However, more testing at cryogenic temperatures under appropriate stresses and voltage gradients are desirable.

(25) The current information on neutron degradation of the superconductors indicates that Nb_3Sn is most sensitive with a damage threshold of 3×10^{18} n/cm^2 (~ 0.002 dpa). The NbTi conductors show much more resistance to radiation damage with a threshold for significant property degradation (a factor 2) above 10^{20} n/cm^2.

TABLE XIII-17. R AND D NEEDS ASSESSMENT – MATERIALS

	Status*	
	Adequate	Inadequate
Generation of irradiation effects data (e.g. mechanical properties) on nickel containing alloys for design purposes	x	
Environmental effects on fatigue of stainless steel at elevated temperatures	x	
Determine synergistic effects of stress/strain and irradiation on key properties of stabilizer materials	x	
Generation of irradiation effects data on superconductor and insulator materials as required for engineering design	x	
Development of fabrication techniques and data base on bulk irradiation effects for non-stainless steel structural materials and non-structural first-wall or liner materials (ceramics, graphite, insulators)		x
Determine chemical erosion characteristics of materials and coatings exposed to plasma at high temperature and high fluence conditions	x	
Develop retention/release characteristics of T and He on materials exposed to plasma (first-wall, liners, limiters, divertor collectors) under design conditions		x
Development of low-Z coatings for first-wall or liner components (fabrication, irradiation effects and thermal shock resistance)		x
Determine chemical and radiation stability of breeding materials under irradiation at elevated temperatures and chemical reactivity with structure and/or coolant	x	
Study the effect of gaseous transmutants on mechanical properties of structural materials which do not contain Ni		x

Adequate: Programmes are either planned or could be planned on existing or authorized facilities (with some upgrading if required) and using existing trained personnel.

Inadequate: Either the available and authorized manpower and facilities are inadequate or a substantial increase in effort is needed.

* The ratings are made on the world-wide fusion effort and not separately on the individual national programmes.

(26) There will be limited data on the electrical and mechanical properties of ceramics ($MgAlO_4$, Al_2O_3, Si_3N_4, and SiC) exposed to the INTOR front wall environment by 1985. Fast reactor data will suffice for ceramics for displacement damage conditions of interest in INTOR. However, gas production levels will be 1 to 3 orders of magnitude lower than at the first wall of INTOR. It is also important to investigate the scalability of small test specimen results to reactor size components in INTOR.

(27) Below 1250°C there are adequate data on fission reactor graphites for the design of INTOR. Present data indicate that operation at 550°C or lower would be required to achieve a full INTOR lifetime. Unfortunately, the grades studied are not the most thermal shock resistant and further tests are required of materials appropriate to the thermal conditions in INTOR.

(28) There are presently sufficient data on the sputtering of candidate wall materials with hydrogen isotopes and helium to predict wall erosion rates in INTOR (within a factor of two) once the particle spectra and angle of incidence are specified. Wall erosion by this mechanism does not appear to be serious in INTOR if an efficient divertor is used and the energetic particle flux is reduced to below $10^{15}/cm^2s$ or if the edge temperature of the plasma is $\leqslant 100$ eV.

(29) A great deal of progress has been made in the area of low-Z coatings for first walls, limiters and particle collection plates. However, as of now there is no clearly superior candidate that has been demonstrated to be compatible with both the plasma and cyclic operation of the structural components. TiB_2 and TiC hold great promise, but more tests are necessary for a final decision in INTOR.

12. R AND D NEEDS ASSESSMENT

The R and D needs assessment is summarized in Table XIII-17.

REFERENCES TO CHAPTER XIII

With a few exceptions, individual references have not been cited in this Chapter. The reader can consult the following reports for more details.

[1] European Contributions to the 2nd meeting of the INTOR Workshop, Materials, Chapter VI, EUR FU/XII 501/79/EDV 50, June 79 (2 Volumes).
European Contributions to the 3rd meeting of the INTOR Workshop, Materials, EUR FU/XII 501/79/EDV 70, Oct. 79.

[2] Fusion Research Development Center, INTOR Workshop Report, Group 6, Materials, JAERI-8510 (1979).

[3] USSR Report to INTOR, "Materials for INTOR" (see Appendix A.5.3 [18]).

[4] "Materials", Chapter in USA INTOR Report (see Appendix A.5.4).

Chapter XIV

SYSTEMS INTEGRATION AND SUPPORT STRUCTURE

G. CHURAKOV – USSR, Expert, Chairman
C. BAKER – USA, Expert
J. GILLELAND – USA
R. HANCOX – EC, Expert
B. KOLBASOV – USSR
P. REYNOLDS – EC
K. SAKO – Japan

1. INTRODUCTION

1.1. Contents

This report is a summary of the work on Systems Integration and Support Structure. The findings of the report are based upon an assessment of the existing data base and upon those additions to our experience anticipated for the early 1980's. The specific work scope and objectives of this group are delineated below.

1.2. Objectives

A primary objective is to evaluate the options identified by each team of experts from the point of view of the over-all system integration and to suggest those combinations or choices best meeting the programmatic and technical goals of INTOR. Other objectives include assessments of the structural features of major components, design philosophy, standards and procedures.

Activities required to meet these objectives include the following:

(1) Plan overall machine layout so as to optimize overall performance while considering all aspects of components' functions, machine availability, initial assembly, repair and maintenance.

(2) Make recommendation(s) about the location of the vacuum vessel (load bearing structure).

(3) Identify and quantify all significant system interactions (mechanical and electrical) and insure systems compatibility.

(4) Identify feasible support structure philosophies and their effects on component design.

(5) Suggest philosophies for applying design standards, establishing safety margins and address the required ability of components to withstand faults.
(6) Identify potential problems with the vacuum vessel, vacuum seals, divertor and/or limiters.
(7) Review Guiding Parameters for accuracy and consistency.

2. DESIGN OPTIMIZATION CONSIDERATIONS

2.1. Alternative machine topologies [1–6]

A prescription of the "optimum" machine layout is not possible at this phase of the INTOR assessment activity. Thorough tradeoff studies, constructed specifically for INTOR objectives and parameters, are required. Considerations of maintainability, reliability, and degree of experimental flexibility all have a profound impact on the topology of the machine. Nevertheless, certain key observations can be made at this point which can act as a guide for further work.

The first of these observations is that all machine components should be repairable or replaceable. INTOR will be the first facility of its type, utilizing or, indeed, testing new technologies and concepts generic to fusion. It is only reasonable, therefore, that plans for component replacement become an important design consideration from the outset (see also Chapter XV). This does not mean, however, that all systems must be capable of "fast replacement". It is acceptable that certain systems, judged to be very reliable, might take several months to replace; design of a component for ease of maintenance can, in some cases, cause the intrinsic reliability of that component to suffer. An example of a reliable component is the toroidal field coil. There exists, apparently, world-wide consensus that the toroidal coil can be designed and fabricated in such a way that failure during the life of the machine is very unlikely. The impact on machine design is significant. Since the topology of the machine need not necessarily incorporate rapid (i.e. a few weeks) coil replacement, simplifcations of a centre bore design may be possible which would have otherwise been excluded from consideration.

A second major observation is that the interlocking of components is undesirable. Every possible effort should be made to place the poloidal coils outside of the toroidal coils. Advantages of general machine access and maintainability outweigh the disadvantages associated with higher toroidal field coil loads and high poloidal system power requirements. Further, the toroidal coils should, and can, be made large enough that internal systems can be installed in modular form between the toroidal field coils. These internal components include the first wall, blanket, shield, and possibly the vacuum vessel. The size and shape

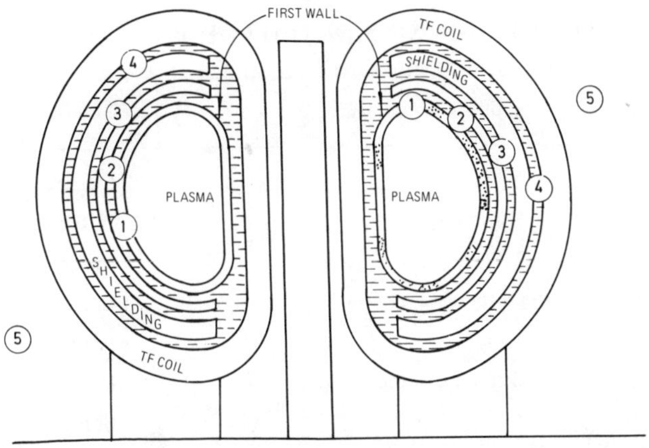

FIG.XIV-1. Characteristic machine arrangement showing regions possibly separated by vacuum walls.

of experimental blanket modules, which might be used for initial tritium breeding tests, electric power production, materials studies, and surface effect studies, have not been determined. From the systems integration viewpoint, it is clear that small experimental modules are preferred for quick experimental turn-around, good module reliability (simplicity) and high availability. It is also recognized that reactor relevance may require large modules (perhaps whole segments) for some types of testing.

Finally, a major reason for unlinking the components of the reactor is the strong favourable influence it has on component fabrication. The equipment manufacturing and testing processes should be organized to promote the maximum amount of work at the factory, with only fitting and jointing operations to be accomplished in situ. At least in this respect, sector approaches are advantageous in which already existing integrated industrial capability may be utilized.

2.2. Vacuum/atmosphere boundary [1–6]

In discussing a recommended location of the load-bearing vacuum wall, it is convenient to refer to Fig. XIV-1, which shows a typical machine arrangement in which concentric volumes are defined as follows:

(1) The torus around the plasma up to and including the first wall;

(2) The volume between the first wall and the interior wall of the (neutron absorption or tritium breeder) blanket;

(3) The volume between the outer wall of the blanket and the inner wall of the neutron shield;

(4) The volume between the outer wall of the shielding and the inner bore of the TF coils;

(5) The volume outside the TF coils and between the reactor cell wall.

TABLE XIV-1. LOCATIONS CONSIDERED FOR VACUUM WALLS

Case	Primary boundary	Load-bearing boundary
A	*	Outer edge, region 1
B	Outer edge, region 1	In region 2
C	*	In region 2
D	Outer edge, region 1	In region 3
E	*	In region 3
F	Outer edge, region 1	In region 4
G	*	In region 4
H	Outer edge, region 1	In region 5
I	*	In region 5
J	In region 2	In region 5
K	In region 3	In region 5

* Primary boundary coincident with load-bearing boundary.

Referring to the figure, reasonable choices for primary vacuum boundary and load-bearing vacuum boundary are shown in Table XIV-1.

In all cases where the primary and load-bearing boundaries are separate, additional intermediate vacuum boundaries may be inserted without invalidating any of the technical considerations set forth.

Although all possible configurations from Case A through K may not necessarily have to be studied, the engineering tradeoff will essentially involve the evolution of the above candidate approaches with respect to the following considerations:

(1) As the load-bearing boundary moves outward toward region 5, the pumping volume increases, the amount of equipment within the boundary increases and the size of the vacuum vessel structure (which has to take the full one atmosphere pressure differential) increases. These have negative impacts, leading to higher complexity and costs. Problems with auxiliary equipment located within that boundary, possible outgassing, requirements to seal all services, less reliable tritium containment and the need to operate in a vacuum environment place severe demands on remote maintenance capability.

(2) By the same token, as the load-bearing boundary moves outward, working room within the reactor interior is improved, the problem of opening and closing vacuum-tight joints or ports eases, since it is done in less constricted real estate, facilitating, in general, maintenance of components in regions 1 to 3.

(3) As the load-bearing boundary moves outward, the requirement for two vacuum boundaries (Cases B, D, F, H, J and K) becomes more urgent, since plasma contamination by outgassing of reactor components within the high vacuum boundary has to be avoided. In these cases, location of the primary boundary in region 1 is preferred, although regions 2 and 3 could be considered.

(4) If the load-bearing boundary is located within the neutron shield (region 2 or 3), mechanical degradation may require too frequent vessel replacement. In any case, intensive R and D would be required to develop a segment vacuum joint not dependent on in-situ welding.

Strong preferences have been voiced for cases B, F or H. The advantages of Case H (use of the building as the load-bearing vacuum boundary) is the simplification of the toroidal field coil design if no local cryostat is required. The disadvantages of Case H are greatly diminished personnel access, increased reliability requirements (for cooling systems, diagnostics, and other equipment which must now operate in a vacuum), and lower overall availability. Case F has been discussed intensively as an approach which balances many of the conflicting requirements. Among disadvantages of this choice are problems of single turn resistance of the vessel. Resolution of the issue awaits a thorough tradeoff study.

It should be noted that expanded cryostat designs (USSR and Japan) were presented in the meeting which ameliorate some of the toroidal coil design problems associated with the need for a local cryostat. In this design, the toroidal field coil and poloidal coils all share a common vacuum. This is to be distinguished from other previous reactor designs (TNS) which have a common vacuum for only the centre leg of the toroidal field coil and the centre solenoid of the ohmic heating coil. It also should not be confused with designs which use the reactor hall for the load-bearing vacuum surface.

2.3. Reference concepts

2.3.1. USA suggestions [5]

One embodiment of the proposed Guiding Parameters featuring individual dewars for TF and PF coil systems, location of the load-bearing vacuum boundary between the blanket modules and the bulk neutron shield (Case D, Table XIV-1), location of PF coils outside a twelve-coil TF cage and allowing (if required) a relatively weak PF coil system to provide fast-acting control at plasma initiation is depicted in Figures XIV-2 and 3.

This choice for location of the load-bearing vacuum wall takes advantage of the shielding properties of the blanket modules without significantly adding to the already existing need for mounting the modules in vacuum-tight enclosures capable of supporting low pressure differentials.

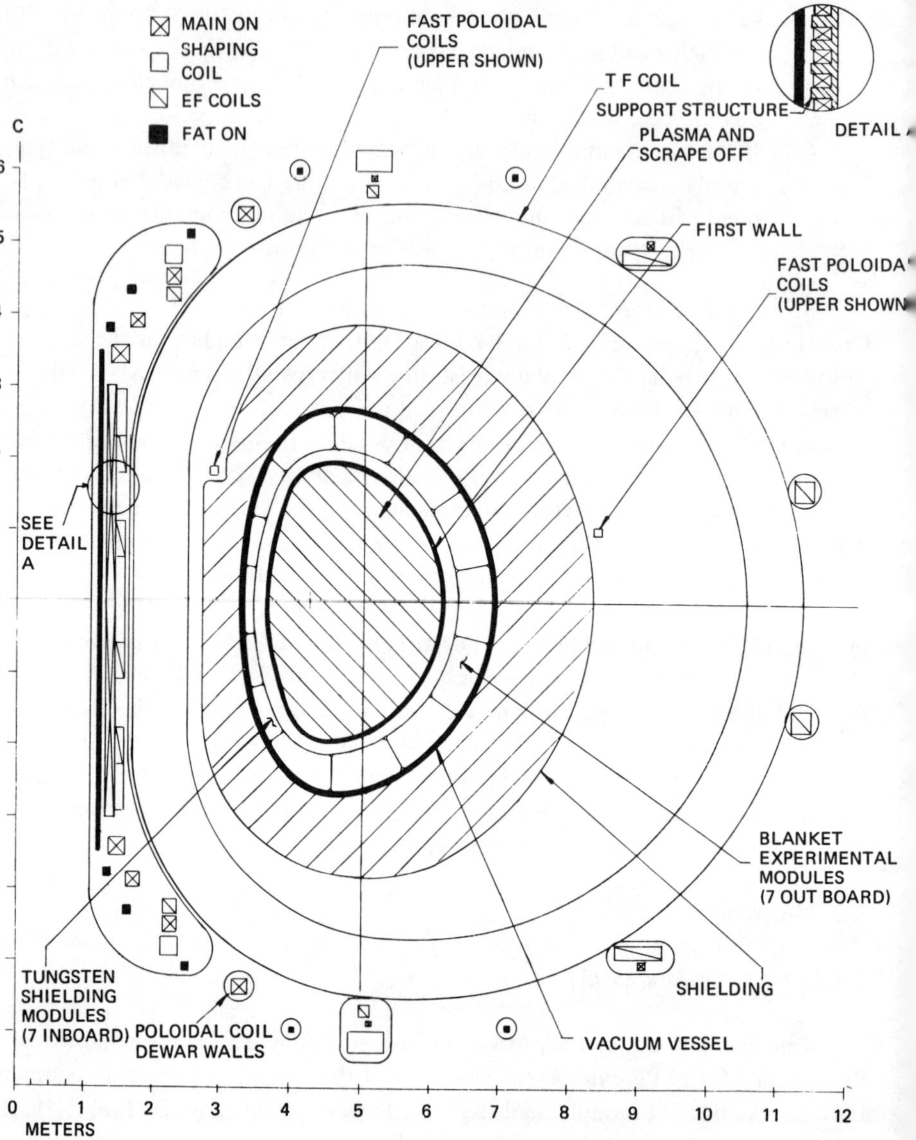

FIG.XIV-2. Elevation view of an embodiment of the reference concept.

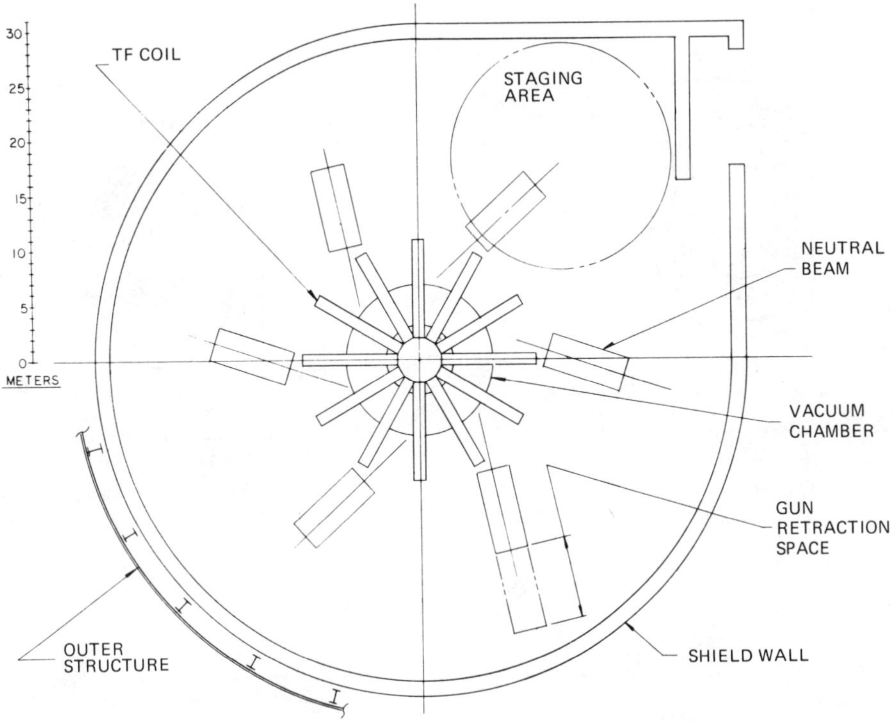

FIG.XIV-3. Plan view of reactor chamber for INTOR.

The TF ripple at the plasma boundary is 0.73% maximum deviation from average; the OH and EF coils jointly deliver 90 volt-seconds to the plasma with the fast PF contributing an additional 4 volt-seconds at 100 volts per turn.

The arrangement shown in the plan view provides the lay-down space needed during assembly and meets access requirements for operation and maintenance of the neutral beam injector systems.

2.3.2. European Community suggestions [1, 2]

Fig. XIV-4 shows two basic approaches to the assembly problem:

A – a fixed toroidal vacuum vessel with access doors;

B – a system in which the tokamak is enclosed in a large vacuum vessel.

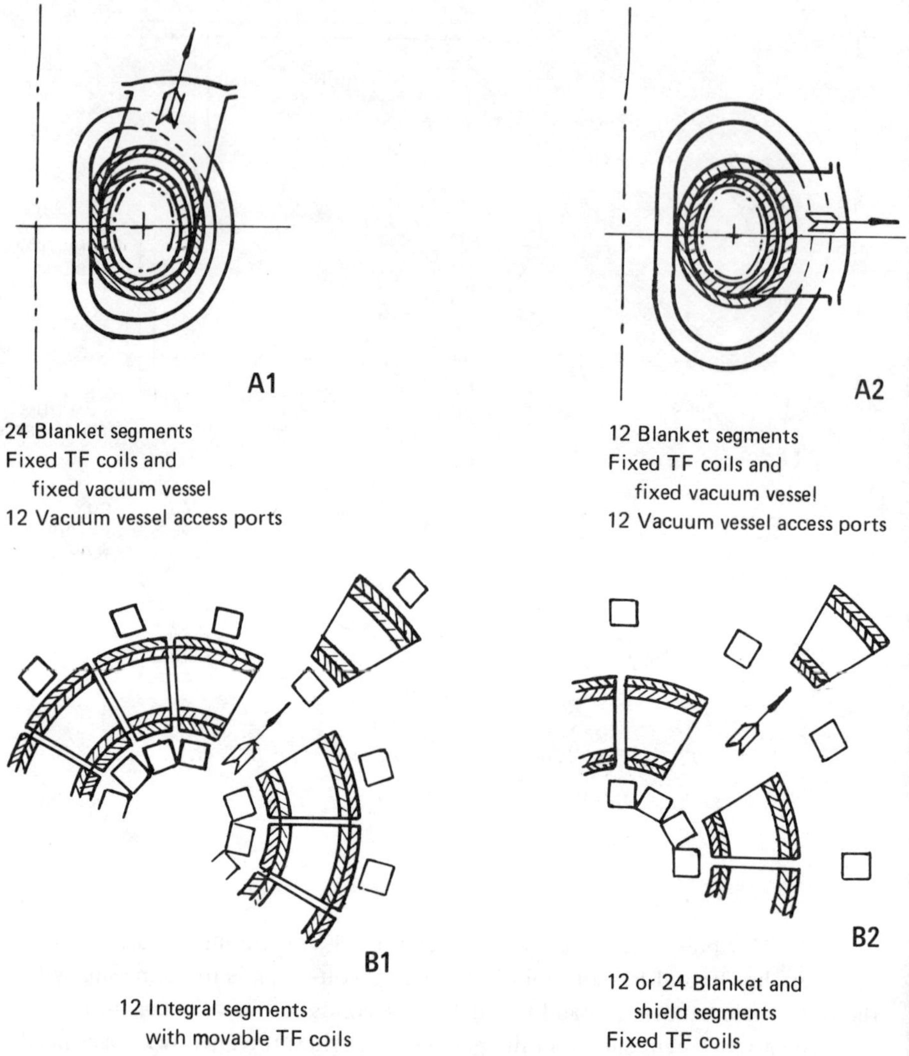

FIG.XIV-4. Possible assembly solutions for INTOR.

Each of the above is divided into two systems:

A_1 — corresponding to the latest Fintor design, has doors in the top of the torus permitting almost vertical removal of the blanket segments into a region clear of pumps and injectors;

A_2 — corresponding to the Culham Mk II tokamak reactor uses radial removal of blanket segments into a region expanding in size with radius;

B_1– corresponding to the Jülich design employs simultaneous removal of a TF coil and blanket segment;

B_2– enables a blanket segment to be removed without moving a TF coil.

In some of the above designs, more than one vacuum containment region may be required.

The relative advantages and disadvantages of these systems can be assessed fully only after detailed designs have been made and these have not yet been undertaken. Preliminary discussions can be found in the June and September European Community contributions to Chapters XI, XIV, XV and XVII.

2.3.3. Japanese suggestions [3, 4]

Overall reactor system designs including all of the main components and repair and maintenance system were performed. The reactor has a poloid divertor and consists of six modules, each of which has two toroidal coils. A dual purpose port for evacuation and neutral beam injection is placed on the outer side of a module. NBI's are installed on four of the six modules, and on the remaining two, blanket test sections are placed. Fig. XIV-5 shows the general reactor concept.

The concept for the repair and maintenance of blanket structure is one of the main design alternatives. The concept of withdrawing radially a module with two toroidal magnets (Concept A) and the concept of removing a part of the shielding in situ and taking out the blanket (Concept B) are both studied. In both concepts, repair of magnets is also feasible. In relation to repair and maintenance design, the vacuum boundary was investigated for magnets of both the belljar and separated types. The former is applied to the Concept A and the latter to Concept B, as the promising concepts. In both cases, the vacuum boundary of the plasma chamber is located on the inner side of the shielding structure, while the outer side of the shielding structure forms the vacuum boundary for the toroidal field magnets.

2.3.4. USSR suggestions [6]

General principles of the INTOR layout include the following:

(1) The structure adjacent to plasma/vacuum vessel and blanket/shield should be repairable or replaceable. Therefore, they should be of module type and the possibility of their remote maintenance has to be provided.

(2) Magnetic system components placed behind radiation shield (TF, OH, and EF coils) should be highly reliable and should be operable during the whole lifetime of INTOR. The period of commissioning and hydrogen plasma operation will be used to identify and eliminate possible misfunctions in these components.

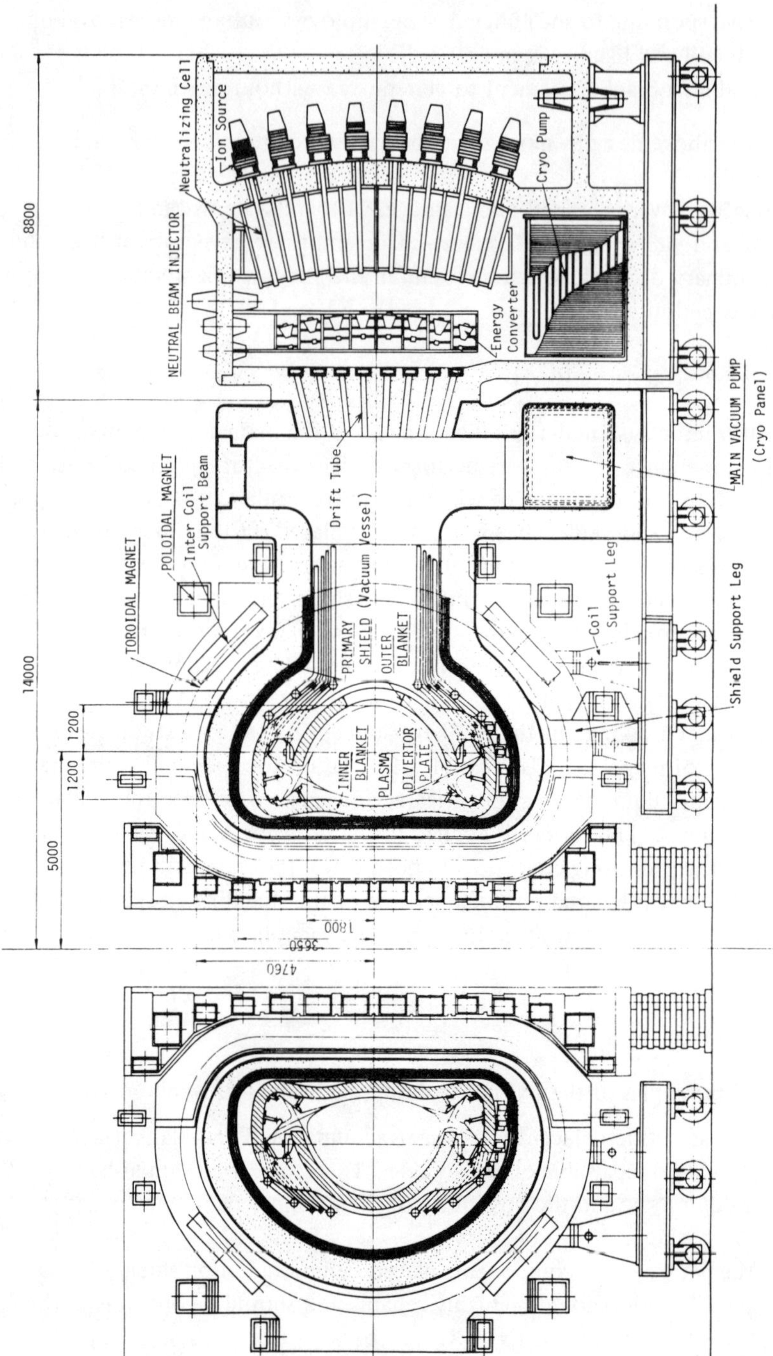

FIG.XIV-5. General reactor concept of INTOR-J.

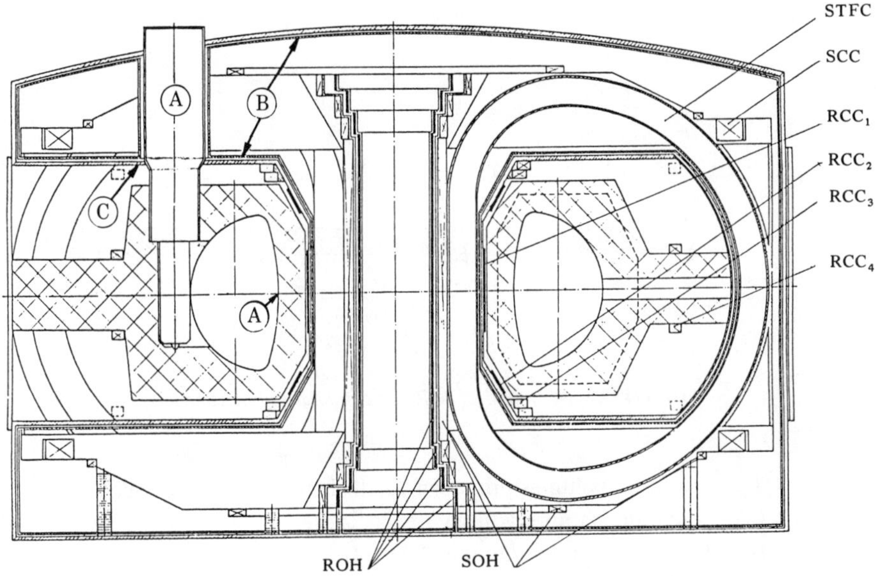

FIG.XIV-6. Diagram of construction of EMS of INTOR.

(3) Dismounted activated components should be repaired in hot cells. Components removed for repair should be replaced with spare ones which are already available on the site.

The cryostat vacuum volume is used as thermal insulation of the superconducting cryogenic magnetic coils. It should be separated from the discharge vacuum chamber and blanket/shield with an intermediate vacuum region. Such an arrangement provides a possibility of having short-time access to the blanket/shield and first wall for their repair without warming up TF, OH, and EF coils.

In accordance with these principles, the layout shown in Fig. XIV-6 is proposed.

(1) The TF system consists of 12 coils. Support structure withstanding interaction forces present themselves as arc spreaders between the power case of coils.

(2) The OH and EF coils are situated outside the TF coils. A small number of water-cooled copper control coils are placed inside the TF coils behind the cryostat wall.

(3) The cryostat for the complete magnetic system consists of a vacuum volume restricted by the toroidal vacuum chamber and the general outer shell. The shell has 12 lateral passages on the outer side of the forms between TF coils. The size of passages permits replacement of the segmented chamber and blanket sections without warming of the magnets.

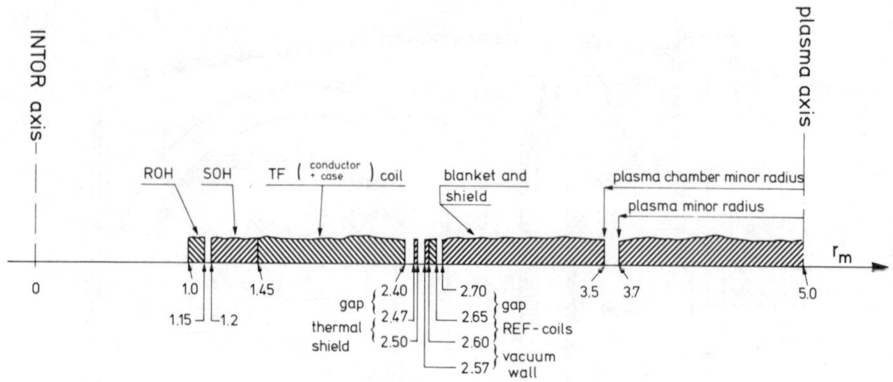

FIG.XIV-7. The main INTOR radial inboard dimensions.

Three vacuum volumes proposed for INTOR include the following:

(a) Vacuum chamber volume (~10^{-8} torr);

(b) Cryostat vacuum volume (10^{-5}–10^{-6} torr);

(c) Buffer volume between the above mentioned two to prevent possible tritium leakage (10^{1}–10^{-2} torr);

(4) The blanket consists of 12 segments welded together through the mounting;

(5) The blanket contains experimental test modules and channels in the outer side of the torus between TF coils.

3. SYSTEM INTERFACES AND INTERACTIONS [1–6]

3.1. The main INTOR sizes

The overall layout and structural design of the INTOR is determined by specific features of the tokamak device and its subassemblies. In Fig. XIV-7 (for example) the main INTOR radial inboard dimensions are shown schematically. Analysis of these main INTOR sizes may be expected to yield information regarding their effects on each other and on the parameters of the related systems. It will be subject for further optimization.

3.2. Specification of main INTOR systems and subsystems

INTOR will be a very complicated system which has many interactions and interfaces. The INTOR structural diagram is shown on Fig. XIV-8. This diagram was prepared for the Assembly and Remote Maintenance report in connection with the equipment reliability assessments. Fig. XIV-8 shows a division of the INTOR machine into systems and subsystems.

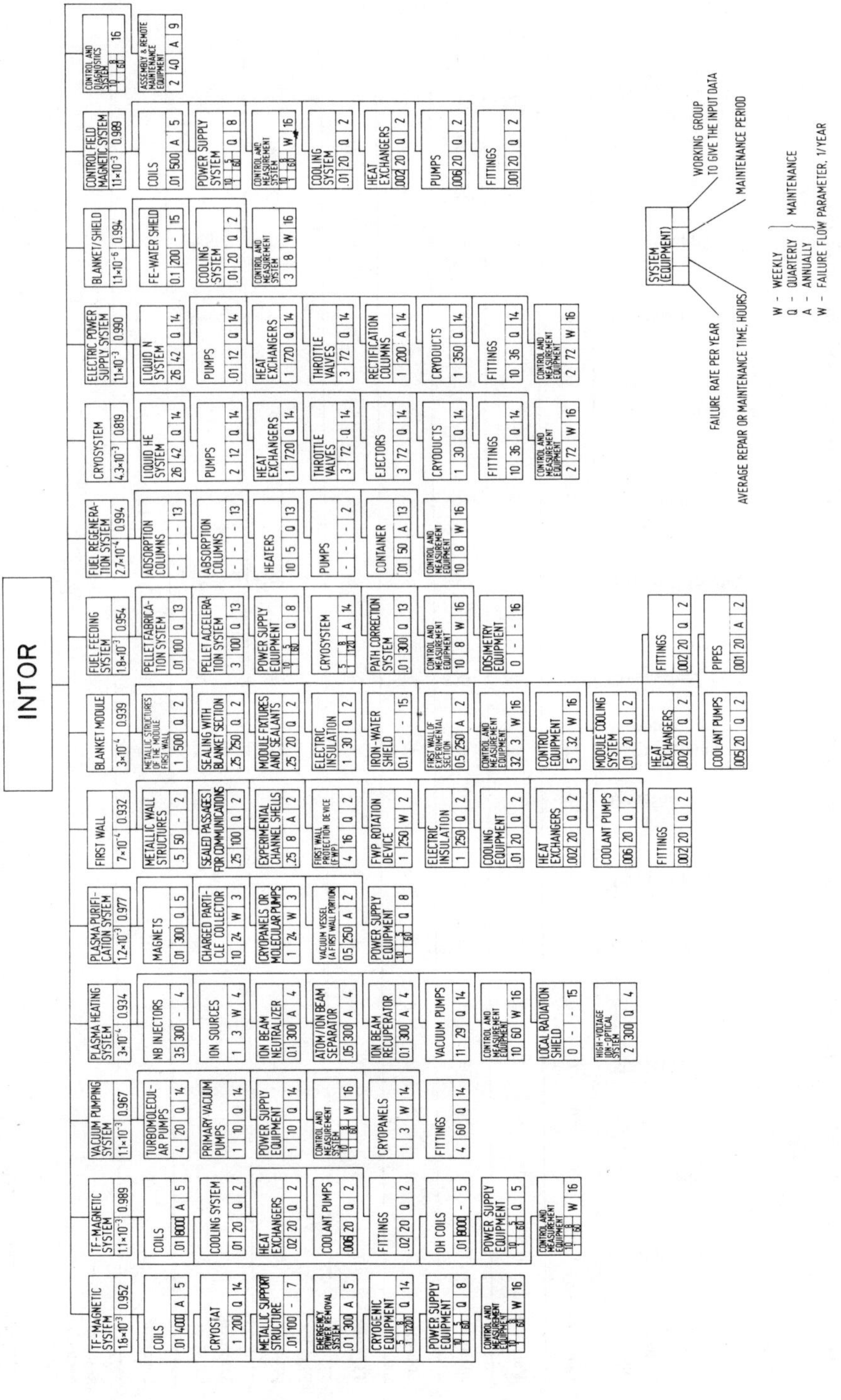

FIG.XIV-8. INTOR structural diagram.

TABLE XIV-2a. INTERFACE CHART (PART 1)

Relative impact on system listed in left-hand column caused by system identified at the top of each column.

Due to / Impact on	Central core	Vac. vessel	Blankets	Shielding	TF coils	External PF coils	Trapped PF coils	Divertor	Vacuum pump system	Cryosystem	Support structure	Remote maintenance	Plasma heating system	Diagnostics	Power supplies	Total
Central core	0	0	0	0	10	5	0	0	0	3	3	5	0	2	0	28
Vac. Vessel	0	0	10	7	7	4	7	3	6	0	5	8	5	6	2	70
Blankets	1	10	0	5	5	0	2	0	6	0	5	8	0	4	3	49
Shldg.	2	0	2	0	8	0	5	5	2	2	8	5	1	7	0	47
TF coils	7	3	0	2	0	0	4	5	4	8	10	10	5	1	1	60
Ext. PF coils	6	2	0	2	2	0	4	0	0	8	0	5	8	2	1	40
Trapped PF coils	1	2	0	8	10	2	0	5	2	0	5	10	3	3	4	55
Divertor	0	9	0	6	5	0	5	0	3	0	5	10	2	1	3	49

Scale: 0–10
No impact 0
Minor 1–3
Major 4–7
Extremely significant 8–10

TABLE XIV-2b. INTERFACE CHART (PART 2)

Relative impact on system listed in left-hand column caused by system identified at the top of each column.

Due to / Impact on	Central core	Vac. Vessel	Blankets	Shielding	TF coils	External PF coils	Trapped PF coils	Divertor	Vacuum pump sys.	Cryo. sys.	Support structure	Remote maintenance	Plasma Htg. System	Diagnostics	Power supplies	Total
Vac. Pump Sys.	0	5	6	3	3	0	0	8	0	0	3	6	3	3	0	40
Cryo Sys	10	0	0	0	10	7	4	3	4	0	7	2	3	3	1	54
Support structure	7	7	0	8	10	1	6	5	4	2	0	7	2	3	0	62
Remote maintenance	5	10	8	7	7	5	8	6	3	4	7	0	5	8	2	85
Plasma Htg. Sys	0	7	5	3	5	0	3	0	1	5	4	7	0	2	1	43
Diagnostics	2	10	7	7	2	2	2	2	5	3	1	6	1	0	0	50
Power supply	0	2	2	0	3	3	6	8	3	3	0	0	3	3	0	36
Total	41	67	40	58	87	29	56	50	43	38	63	89	41	48	18	768

Scale: 0–10
No impact 0
Minor 1–3
Major 4–7
Extremely significant 8–10

3.3. Interface chart [5]

The relative impacts on fifteen major systems caused by each of the remaining fourteen were estimated with the idea in mind of identifying or verifying those most likely to present extraordinary challenges during planning and design. The Interface Chart shown in Tables XIV-2a and XIV-2b shows that, even though the values assigned to each interface are somewhat arbitrary, the impact of remote maintenance is of overriding concern. Remote maintenance of TF coils can be expected to be involved in about a quarter of the interface problems. With the addition of the vacuum vessel, suport structure, vertical field (fast) coils and shielding, nearly two-thirds of the interface problems must be addressed.

The innate design difficulties peculiar to a single system (e.g. divertor), not easily shown on such a chart, may be strongly dependent on interface decisions.

4. STRUCTURE [1–6]

4.1. System loading interactions and responses

Tables XIV-3, 4 and 5 list many of the operating, testing and transient fault loadings which must be considered as part of the structural stress analysis. They show, by the use of short descriptive phrases, a qualitative assessment of the effect of each loading on the various structural elements.

The loadings defined in the tables are, in the main, caused by transients in either electrical currents, magnetic fields, temperatures or pressure. It is clear that if components were designed to be supported, or restrained directly by one another, structural/mechanical interactions would multiply to an overwhelming degree. It must, therefore, be an objective in design to decouple components from each other as much as possible. This was attempted in TFTR and led to some degree of success. Here follow a few examples from TFTR:

(1) TF coils interact only with each other via shear compression panels, the central structural core, and the substructure (machine bed).

(2) Outer PF coils have their own support structure which is connected directly to the substructure.

(3) The vacuum vessel is posted directly to the substructure.

In TFTR, there remains one interaction between components which has caused significant design problems, the inboard OH and EF coils are enmeshed in the central support core, resulting in a situation where two unrelated components each of which presents difficult design problems in its own right, exhibit operating stresses created only by virtue of their joint support mechanism.

For INTOR, it is important that the OH coils be decoupled from the central structural core even at the expense of efficiency in producing the required volt-seconds.

TABLE XIV-3. SYSTEM LOADING INTERACTIONS AND RESPONSES – OPERATING LOADS I

System → Loadings ↓	Central structural core	Ohmic heating coils	Vertical field coils	Toroidal field coils	Secondary vacuum & lateral support shell	Blanket		Shield
						First wall	Inner structure	
Operating loads								
Gravitational	Supports TF and OH coils, low stress	Negligible	Small effect	Small effect	Small effect	Small effect	Small effect	Important loading
Thermal due to cool down to cryogenic temp.	Moderate effect	Moderate effect	Moderate effect	Moderate effect	None	–	–	–
Thermal due to nuclear heating	Negligible	Negligible	Small effect	Small effect	–	Major effect	Moderate effect	Small effect
Distortional due to nuclear damage and helium generation	Negligible	Negligible	Governing effect on polymeric insulation	Governing effect on polymeric insulation	Negligible	Major effect	Moderate	Negligible
Steady state magnetic from TF coils	Governing axisymmetric radial comp.	–	–	Governing circumferential tension	Provides stabilizing stiffness for magneto-elastic effect	–	–	–
Steady state or slowly varying magnetic from OH coils	Small axi-symmetric radial tension	Governing circumferential tension in stabilizer	Negligible	Negligible	–	–	–	–

TABLE XIV-4. SYSTEM LOADING INTERACTIONS AND RESPONSES – OPERATING LOADS II

System → Loadings ↓	Central structural core	Ohmic heating coils	Vertical field coils	Toroidal field coils	Secondary vacuum & lateral support shell	Blanket First wall	Blanket Inner structure	Shield
Operating loads								
Rapidly varying magnetic from OH coils	Small pulsed axisymmetric radial tension	Varying tension in stabilizer	Some induced effects	Small lateral loading small eddy current heating of stabilizer	Negligible	Small induced effects	Small induced effects	Negligible
Steady state or slowly varying magnetic from VF coils	Negligible	Negligible	Governing circumferential tension or compression	Small lateral loading	Small stresses	Negligible	Negligible	Negligible
Rapidly varying magnetic from VF coils	Negligible	Very small pulsed stresses	Varying tension or compression	Small pulsed lateral loading. Eddy current heating	Small pulsed loads from TF coils and from induced currents.	Small induced currents	Negligible	Negligible
Self field magnetic during testing	Negligible	Small effect	Small effect for most coils; maximum for a few	Small bending stresses but possible max. shear stress	–	–	–	–
Coolant pressure	–	–	–	–	–	Major effect	Major effect	Major effect

TABLE XIV-5. SYSTEM LOADING INTERACTIONS AND RESPONSES – TRANSIENT LOADS

Systems → / Loadings ↓	Central structural core	Ohmic heating coils	Vertical field coils	Toroidal field coils	Secondary vacuum & lateral support shell	Blanket: First wall	Blanket: Inner structure	Shield
Transient fault loadings								
Response to an earthquake	Major effect	Vulnerable to central core response	Major effect on support structure	Vulnerable to central core response	Vulnerable to central core response	Moderate effect	Major effect	Major effect
Sudden failure of an OH coil	Small local change in radial pressure	Local heating in failing coil	Negligible	Small lateral loading; small eddy current heating of stabilizer	Small load changes	Plasma may quench	–	–
Sudden failure of an VC coil	Negligible	Negligible	Local heating in failing coil	Small lateral loading; small eddy current heating of stabilizer	Small load changes	Possible large thermal effects if plasma goes to the wall	Negligible	Negligible
Rapid plasma failure	Negligible	Large induced currents	Large induced currents	Small lateral loading; small eddy current heating of stabilizer	Small load changes	Induced currents and magnetic forces large thermal effects	Small effect	Negligible
Sudden failure of a single TV coil	Loss of axi-symmetry of radial compression; Local bending stresses	Negligible	Negligible	Local heating of failing coil; Bending induced in failing coil	Critical design loads produced in nearby shells	Plasma will quench	Negligible	Negligible

5. DESIGN STANDARDS

Many aspects of the INTOR structures do not readily lend themselves to application of existing structural codes. However, in many instances, the ASME Section III Nuclear Code Philosophy and Procedures are useful as guidelines and usually result in conservative design. In general, in the normal operating condition, the code prescribes that the design shall have the primary stresses bounded to the yield limit and a combination of primary and secondary stresses allowed to go beyond the yield point, however, not exceeding twice the yield strain (equivalent to 3 times the design stress-intensity value).

It is prudent not to permit critical structures of the tokamak to go beyond the yield point (except under fault conditions, where a multiplier of 1.2 is proposed) in any of the following circumstances:

(1) Dimensional stability is critical in disassembly and reassambly.

(2) A high-strength alloy (e.g. Ti-6Al-4V) for which yield strength is only slightly below its ultimate strength.

(3) Extremely thick-walled members (e.g. components of the central structural core).

Electrical design criteria consistent with normal industrial practice are appropriate for INTOR except for unusual environments (e.g. radiation, vacuum, unusual wave shape, etc.); in which case, sufficient long-term data will be required to establish desired confidence levels in experimentally-determined criteria.

6. TEST FACILITIES AND DIAGNOSTICS [2, 4, 5, 6]

The types of test facilities required for INTOR have been discussed in Chapter XI and consist of 'test channels' of 100–300 mm diameter, 'test modules' of 1–3 m^2 area, and possibly complete test segments. The number of test channels has not been discussed, but between 4 and 12 test modules have been suggested. The types of diagnostics required close to the plasma are discussed in Chapter VIII, but the number and space required have not yet been identified. It is clear, however, that both the test facilities and diagnostics equipment will interact with other components of INTOR and must be considered within a system integration study.

Some of the test facilities and diagnostics will need to be within the plasma vacuum region; and, therefore, will affect the design of the vacuum boundary and the operation of the vacuum system. In particular, a high degree of reliability will be required in these items so that the integrity of the vacuum system is maintained. These items will also represent possible leakage paths for tritium and may need to be enclosed within an additional tritium barrier.

The introduction of test equipment and diagnostics will also require penetrations through the shielding, and care will be necessary to maintain adequate shielding whether they are in operation or withdrawn. Additional shielding will be necessary around such penetrations, requiring additional space.

To allow access for the test facilities and diagnostics, they should occupy the areas of the torus free from torroidal and poloidal field coils, principally in the out-board region. The pressures for space in this region, however, for auxiliary heating, divertors, access to limiters, remote maintenance equipment, etc., will limit the number and accessibility of test modules. The directions from which they are introduced, vertically or radially, must also be consistent with the reactor maintenance scheme for the whole system.

7. CRITICAL ISSUES [2, 4, 5, 6]

There are four critical issues which must be resolved before a meaningful systems integration analysis can be performed. These are:

(1) the method of impurity control;
(2) the need for poloidal coils within the toroidal field coil bore;
(3) a final analysis of the possibility of a local cryostat for the toroidal coil; and
(4) a final resolution of the size of the experimental modules needed for electrical power production and tritium breeding.

All affect the basic nature of the systems interactions and, thus, machine topology.

The problem of impurity control is, of course, a key issue in the area of theoretical and experimental physics. It affects the vacuum vessel in the sense of complicating the design, fabrication, repairability and maintainability of the vessel structure if a divertor chamber and/or surface pumping capability needs to be incorporated. The coils associated with a divertor affect the magnetic coil system and the overall reactor accessibility, maintainability and repairability. The reader is referred to Chapter IX for a discussion of various divertor designs and to the Impurity Control Section for an extensive discussion of feasible options. Clearly, from the systems integration viewpoint, the omission of a divertor represents a significant reduction in the overall complexity of the machine. If it can be demonstrated convincingly, either theoretically or experimentally, that a concept, such as impurity flow reversal, which essentially does not affect any feature of the vacuum vessel is feasible, the issue is closed and the approach to be taken is defined. Because of the difficulties with implementing proposed alternates (divertors), every effort should be made to achieve the above-mentioned demonstration.

If a divertor is necessary, then a concerted attempt must be made to find simplified designs which will allow reasonable reliability and accessibility, even at the price of ultimate performance.

The only arrangement of poloidal coils that does not present extremely difficult design, assembly, reliability and down-time problems is the siting of all coils outside the TF coil cages. This can be accomplished for OH and most of the EF coil systems. However, the requirement for EF control response times less than 10 ms may require the fast EF coils (one or two pairs) to be located within the TF cage and perhaps even within the neutron shield.

If fast EF coils are placed between bulk shield and TF coils, maintenance considerations require these coils to be segmented, require a significant amount of working room in that region, and place significant requirements on the minimum TF coil-to-coil spacing. Most tentative machine arrangements that have been proposed do not provide sufficient space for these needs. Questions remain as to whether or not the requisite magnetic design is achievable, the response time is adequate, or the resultant power supply demands are manageable. The arrangement is feasible, provided these questions can be favourably resolved, even for superconducting coil segments.

The need for a consensus on the possibility or impossibility of a "local" TF coil cryostat is important because of the major impact on machine design and access. Were a local cryostat design not possible, then expanded cryostat or vacuum building designs would be necessary. These in turn affect machine access and availability.

Choice of experimental module size has, of course, a significant impact on machine topology and design, as discussed in Section 2.1.

8. R AND D REQUIREMENTS

No specific R and D needs have been identified in the area of Systems Integration, since the group is not concerned with the design of the specific components of INTOR but rather with their relative interactions. It is generally felt, however, that the group has only begun to touch upon these interactions, and that a more extensive and methodical study is required before the full implications of the various design choices become clear.

9. SUMMARY

(1) The system integration goals for the other groups were defined. Requirements concerning evaluation of specific features of main components, design philosophy, goals and procedures were addressed to other groups.

(2) Machine topology in connection with maintenance requirements, reliability and experimental component flexibility was considered. The main conclusions are:

(a) TF-coils should be designed and manufactured with such a degree of reliability which should avoid failures through the whole lifetime of the machine.

(b) OH and EC coils should be situated outside of the TF coils. The advantage of access and maintenance in this case essentially outweigh any disadvantages connected with the high overturning loads of TF coils and increased PF power supply requirements.

(c) Inboard reactor components (vacuum chamber and blanket) should be segmented. The dimensions of experimental modules should be sufficient for initial tritium and electricity production tests and for material studies. For specific material experiments, small test channels are preferred. Test segments may be required for the third stage of operation to provide adequate reliability.

(d) Manufacturing technology requirements strongly affect the structure design. The major part of the work should be done in fabrication shops and only assembly should be performed on the site. Such an approach is the best method to encourage the use of industrial capacities and technology.

(3) The experts discussed the recommendations concerning the location of the cryostat boundary and decided that the vacuum volumes of cryostat and plasma chamber region should be separated to provide the access to the blanket-shield components without warming up the TF magnets.

(4) A preliminary assessment of the interactions between various subsystems was carried out [5]. The various interactions were weighted according to the designers' judgement and experience. The most important critical systems in terms of their interactions with other systems were remote maintenance, the TF coils, the vacuum system and the support structures.

(5) The experts agreed to use common design codes and regulations.

(6) The participants reviewed different variants of proposed INTOR topologies and concluded that those contain significant amounts of identical engineering approaches:

- the number of TF coils is 12;
- OH coils are outside TF coils;
- EF coils are outside TF coils or a hybrid version is possible with the major part of the EF coils outside and the TF coils and only small EF coils inside;

– cryostat vacuum volume is separated from the discharge chamber volume;
– discharge chamber and shield are of modular type; modular mounting joints are outside the shield;
– there are identical approaches to the structural requirements from the viewpoint of machine remote maintenance.

However, there are some unsolved problems connected with INTOR topology. Such problems should be solved before the implementation of the detailed analysis of the system integration. Among them are:

– choice of divertor type;
– choice of the types, dimensions and allocation of test modules;
– choice of magnetic coil cryostat (general for all the coils, or separate for each coil);
– engineering decision for EF coil situated inside the TF coils.

All affect the basic nature of the systems interactions and, thus, machine topology.

(7) The group members agreed that the design development for INTOR could be made at this time and that no special R and D tasks are necessary in the field of INTOR subsystems interaction.

Investigations should be performed in accordance with technical requirements developed by other group specialists. Important layout decisions should be made as soon as possible to proceed with the design.

REFERENCES TO CHAPTER XIV

[1] European Contributions to the 2nd meeting of the INTOR Workshop, System Integration, Chapter VII, EUR FU/XII 501/79/EDV 50, June 79 (2 Volumes).

[2] European Contributions to the 3rd meeting of the INTOR Workshop, System Integration, EUR FU/XII 501/79/EDV 70, Oct. 79.

[3] Fusion Research Development Center, INTOR Workshop Report, Home Task Report of Group 7, JAERI-M 8711 (1980).

[4] Fusion Research Development Center, INTOR Workshop Report, Engineering Aspects of JAERI Proposal for INTOR (II), JAERI-M 8518 (1980).

[5] "System Integration and Support Structure", Chapter in USA INTOR Report (see Appendix A.5.4).

[6] USSR Report to INTOR, "INTOR System Integration" (see Appendix A.5.3 [21]).

Chapter XV

ASSEMBLY AND REMOTE MAINTENANCE

J. GILLELAND – US, Chairman
C. BAKER – US, Expert
R. HANCOX – EC, Expert
B. KOLBASOV – USSR
P. REYNOLDS – EC
K. SAKO – Japan

1. INTRODUCTION

Fusion test reactors present a significant challenge to assembly and maintenance disciplines. With the start of D-T operation in advanced machines, numerous operations for maintenance and repair must be performed remotely. Successful accomplishment of remote maintenance requires that the reactor system be specifically designed to operate in this remote mode. Consequently, remote maintenance must be a major design criterion starting with the basic reactor concept design.

In addition to designing to facilitate remote handling in order to achieve the desired availability, specific component reliability goals must be achieved. An assessment of reliability requirements must be performed and the result optimized. The resulting reliability goals for each component must become a design requirement for the component.

2. OBJECTIVES

The major responsibilities of Assembly and Remote Maintenance are to:

1. Formulate possible appraoches to remote handling and maintenance for INTOR consistent with the highest possible machine availability. Toward this end, compare alternative machine layouts and disassembly/assembly procedures and make recommendations as appropriate.

2. Assess the adequacy of current remote handling technology to fulfil the needs of INTOR and identify any areas needing further development.

3. Provide an assessment of the reliability and redundancy required of various components to attain the desired availability golas of INTOR. Compare the reliability requirements to the components existing reliability data base and, upon such comparison, comment on the practicality of the specified availability goal.

4. Specify general guidelines for assembly that are compatible with practical, realistic designs.

The remote handling and maintenance assessment for INTOR was based on experience to date on JET and TFTR and numerous fusion reactor designs. Considerable remote handling experience in the fission reactor, space and oceanographic fields will have to be drawn upon to supplement the fusion work. The result is a compilation of observations, guidelines, and recommendations for remote handling and maintenance on INTOR.

Preliminary failure frequency and availability analyses by the USA and USSR are reviewed in Section 4. The objectives of these analyses are as follows:

(a) Develop operating scenarios consistent with an availability of 25% and 50% respectively.
(b) Define component failure frequency constraints required to achieve availability goals.
(c) Identify components whose failure characteristics may be critical.
(d) Develop rationale for improving availability from 25% to 50%.
(e) Define possible accident conditions which may impact reliability requirements (see also Section 3.5).
(f) Develop general approach to design of non-standard equipment where there is a lack of reliability data.

Guidelines for machine initial assembly are also developed as a major area of group involvement. This assessment addresses assembly facility requirements as well as machine layouts and subassembly packages. A review of reasonably attainable assembly tolerances for initial setup of major systems is given.

An assessment of the research and development required for future machines is developed (see Section 6). This list is based on systems required specifically for remote maintenance as well as those programmes needed for attainment of high availability for INTOR. The general emphasis of this report is to provide an initial assessment of what needs to be done to meet the goals of INTOR from a maintenance and assembly point of view. Less emphasis was placed on a detailed review of the state-of-the-art of remote maintenance technology. In general, it is believed that a sufficient data base exists or can be readily developed to meet INTOR's requirements.

3. IMPLICATIONS OF REMOTE MAINTENANCE FOR INTOR

3.1. Remote maintenance as a requirement of machine design

The requirement for remote maintenance generates an entirely new approach to reactor design based on the need to assemble and disassemble components using remote and semi-remote devices. Successful accomplishment of remote maintenance requires that the reactor systems be specifically designed to operate in this remote mode. Therefore, remote maintenance must be a major design criterion starting with the machine concept design.

Initially, a basic approach to maintenance operations must be established. A plan for maintenance must be developed to establish a method of operation and design approach for maintainability starting with the concept design. This plan should consider all aspects of the approach to maintenance. It should define the method of interaction with reactor system designers, major systems requiring remote maintenance, major maintenance tasks and method or procedure development. The design of the reactor hall and support facilities should carefully reflect remote maintenance requirements.

This plan should provide a means of interaction with component design to assure maintainability. Each component designer must integrate the requirements for remote maintenance. A method of component review and approval involving maintenance systems must be established and followed. An administrative plan for review and an approval of component design must be defined to insure review and acceptance of all machine systems by the remote maintenance group.

3.2. Procedures for remote maintenance

In conjunction with successful component design for maintenance, the method for component maintenance must be developed. This method must initially take the form of a general sequence of events. These events are a logical series of steps required to disassemble, maintain, and reassemble major machine systems.

Each step of the maintenance sequence is a major step in the overall maintenance operation and must be further defined. Definition of these major steps must take the form of a detailed maintenance procedure.

The maintenance procedure provides several functions in defining the detailed maintenance approach. The procedure progresses, in detail, through every step of the maintenance operation. It defines requirements on the design to permit completion of each step, identifies the maintenance equipment required in the operation, and identifies the most efficient means of accomplishing the operation.

3.3. Maintenance methods

Maintenance operations can be subdivided into three basic modes, contact, semi-remote and fully remote. These are defined as follows:

(a) Contact maintenance – The direct use of hands or conventional hand tools for manipulating an object needing maintenance.

(b) Semi-remote maintenance – The use of extended reach devices such as long handled tools for manipulating an object that is physically separated from the operator by distance or a radiation attenuating barrier.

(c) Remote maintenance – Those operations involving the use of remote manipulator systems. Maintenance tasks performed without the aid of personnel within the area of operation.

The maintenance methods utilized are dependent on radiation exposure levels which, in turn, are dependent on reactor structure design.

The requirement of an eventual level of 50% availability on INTOR dictates a reactor design which makes maximum use of rapid maintenance methods. As noted in Chapter XVI, since the time to perform a task remotely may be considerably larger than by "hands-on" methods and, since INTOR is experimental in nature and one cannot predict all of its maintenance requirements, it is necessary to incorporate contact and semi-remote maintenance techniques into the design wherever feasible. It is recognised that capability should be developed to perform as many maintenance tasks as possible by remote maintenance should contact or semi-remote maintenance not be feasible under actual operating conditions.

It is generally conceded that all operations inboard of the bulk shielding are to be performed remotely. However, outboard of the shielding, numerous operations may be performed "hands-on". Typical operations to be considered for contact or semi-remote maintenance include:

Servicing of vacuum pumps and associated vacuum system components;
Assembly/disassembly of water coolant lines to the bulk shield (possible radioactive corrosion products will have to be considered);
Electrical and coolant connectors to diagnostics which penetrate the shield;
Neutral beam service and cryogenic lines;
Bus bar connections to PF and TF coil lead stems;
Assembly/disassembly of external support structures, frame members, etc.;
Miscellaneous facility repair such as servicing maintenance equipment and cranes, etc.

3.4. Mock-ups for remote maintenance

Involvement of remote maintenance personnel in initial design, considerations of remote maintenance requirements by each subsystem designer, and the development of remote maintenance sequences and procedures form the basis of successful maintenance system design.

Verification of the procedures is the next logical requirement for maintenance. A full scale mock-up of a portion of the machine must be constructed to verify component design and procedures for maintenance. First generation or simulated maintenance equipment must be tested and evaluated on this reactor mock-up and the results of these tests used to identify problem areas in component or servicing equipment design.

Early identification of problem areas by mock-up tests will result in design changes during the early machine design phase. Iterations in the component and tooling design can then be incorporated at an early phase and retested prior to component fabrication.

The usefulness of mock-ups continues even after the final design is fixed. These components then become the training arena for optimizing procedures and for training personnel. Time and motion studies using trained technicians are the first real-time assessments of maintenance scenarios and they become the basis for quantified availability estimates. In addition, the mock-up allows opportunities to "try-out" new procedures and equipment during the device life not originally considered in the final design. JET and TFTR are utilizing mockups in preparation for their remote operations.

3.5. Facilities for remote maintenance

The general facility which houses INTOR should contain as a minimum the following items for remote maintenance:

(a) A primary reactor hall equipped with adequate crane and manipulator system.

(b) Hot cell area for maintenance and repair of modules which have been removed from the reactor. This area must be shielded from the reactor hall to permit work in the cell during reactor operation.

(c) Remote maintenance control room – this area contains all viewing equipment, drive stations for manipulator, etc. This room should view the reactor hall through shielded windows.

(d) Access and decontamination/change areas for personnel performing contact or semi-remote operations.

(e) A mock-up area equipped with prototype cranes and manipulators suitable for performing all simulated operations.

(f) Sufficient storage and laydown areas within the reactor hall or adjacent hot cell area for temporary storage of reactor components during servicing.

(g) A restricted area for decontamination, storage, and repair of maintenance equipment.

An example of such a layout is shown in the appendix of the Japanese INTOR report on remote maintenance.

3.6. Major systems recommendations

As a basis for machine concept development, certain recommendations and requirements can be established as a baseline to facilitate remote handling. These recommendations are tentative in nature and will change to some extent during design iterations.

3.6.1. First wall (if separate from the blanket)

The first wall is an item which may require replacement on a periodic basis due to thermal, plasma, and neutron radiation damage or emergency replacement after an unexpected failure. Due to its location, removal and replacement impacts many other systems, e.g. the shield, vacuum vessel, and blanket. Therefore, considerable attention must be given to the first wall from a remote maintenance standpoint.

There are two approaches to first wall replacement which can provide changeout in reasonable time frames. Selection of a method must be based on detailed reactor design and maintenance studies.

3.6.1.1. Segmentation and removal of large blanket sectors. Utilizing this method for first wall repair or replacement requires removal of an entire sector of the vessel. A pie-shaped sector is generally 1/12 of the reactor torus. The bulk shielding is removed and the vacuum vessel parted (either by cutting and rewelding or by mechanical joints and seals). The sector, containing vacuum vessel, blanket, and first wall, is then removed as a unit for first wall changeout. Obviously, this is a major maintenance operation requiring performance of many complex tasks.

This method has several distinct advantages and disadvantages.

Advantages

A replacement sector unit can be inserted immediately after removal of a damaged wall unit. The damaged sector unit can then be repaired in a hot cell.

Blanket components are removed with the sector and the blanket replacement operations can then be accomplished during first wall replacement, if required.

Disadvantages

Sector removal involves disassembly of numerous reactor components and is a complex operation. Sectoring may require removal of vacuum vessel and blanket as well as first wall. Disassembly and reassembly of non-failed components will undoubtedly have a detrimental effect on reliability of these components.

Some blanket servicing operations must coincide with first wall replacement to make maintenance efficient.

First wall sector must be replaced on a rotating schedule. This schedule is based upon removal of all sectors prior to the end of available lifetime of the final first wall.

3.6.1.2. In-situ first wall repair and replacement. This method of first wall removal permits removal/replacement without a major disassembly of the machine. The wall is disassembled in pieces small enough to be removed through a vessel port. The advantages and disadvantages associated with this method include:

Advantages

Disassembly of the vacuum vessel is not required unless the first wall serves as the primary vacuum vessel.

The entire first wall can be replaced at one time, if required.

Blanket servicing is no longer a function of first wall maintenance.

Disadvantages

Complex equipment must be inserted through the penetrations of vacuum vessel, blanket and magnets and operated completely remotely.

All work must be completed within the critical path of the maintenance downtime. Any complications such as broken fasteners, etc., must also be repaired in situ and will add to the first wall service time.

Alternatively, a combination of the above methods might be utilized. Removal of a segment for servicing leaves the two adjacent sectors exposed. In situ removal of first wall and blanket in the adjacent sector can be accomplished prior to segment replacement.

3.6.2. Blanket maintenance

A modularized design which attaches to or sits behind the first wall and is removed by removing a minimum number of fasteners is recommended for the basic blanket of INTOR. In addition, there will be a number of experimental test modules. Connections to the module cooling lines must be assembled remotely. Development of these joints requires considerable development work. The blankets can be removed with the vessel sector or by in-situ uncoupling of coolant and removal through a port. Experimental test modules can be removed by vertical or horizontal access from outside of the blanket through the bulk shield. Final assessments must be based on detailed design studies.

3.6.3. Vacuum vessel

For clarification, the primary vacuum vessel is defined as the boundary which is subjected to a pressure differential of approximately one atmosphere. The secondary boundary is subjected to minimal pressure differentials of several inches of water, and serves primarily as tritium containment.

Possible candidate positions range from the first wall to the walls of the reactor hall. While actual location of the vacuum wall must be based on more complete tradeoff studies, the use of the first wall as a primary vacuum boundary seems to have a serious disadvantage because of the severe environmental conditions. Design of the first wall will be sufficiently complex without adding the requirement for high vacuum integrity. The first wall may, however, form a vacuum chamber of limited integrity to reduce leakage requirements on the primary vacuum vessel and to reduce the problem of outgassing of components located within the primary vacuum chamber.

Particular attention has been given to the use of the reactor building as a candidate for the primary vacuum vessel. From the standpoint of machine maintainability, the building vacuum vessel concept has the following advantages and disadvantages:

Advantages

The main toroid within the reactor no longer has high integrity vacuum welds which must be cut and rewelded remotely.

If the toroidal field coil must be made removable on a routine basis, then the use of the building as a cryostat is an advantage. It should be noted, however, that the general belief of the INTOR participants is that the toroidal field coil can be thought of as a quasi-permanent installation (i.e. it has high reliability).

A thinner toroidal vacuum chamber within the toroidal field coil would be allowed, thus potentially simplifying the insulating break problem.

The structural problems associated with the toroidal field coil design could be simplified.

Disadvantages

If the building is used to provide the vacuum to the toroidal coils, then personnel access to the reactor hall is severely limited. This results mainly from the approximate two month warm-up and cool-down time for the toroidal field coil. Thermal cycling of the TF coils is also undesirable.

Severe problems associated with piping and penetrations would be encountered. Several hundred megawatts of cooling would have to be piped through the vacuum. Radiation to the toroidal field coil could be a problem. The reactor hall would have to provide vacuum tight penetrations for this thermally exercised system and for electrical services. Doors for men and equipment are less of a problem.

If low vacuum cannot be guaranteed everywhere, then the chances for electrical arc damage are much greater.

Hundreds of additional components must be made to work reliably in a vacuum environment. This implies more vacuum seals.

Small leaks in hundreds of cooling connectors now become significant.

The general attitude of the INTOR participants is negative toward the vacuum building concept.

It appears that the most suitable location for the vessel is a position outboard (relative to the plasma axis) of the first wall and inboard of the TF coil outer radius. This location is consistent with the Chapter XIV recommendation, the actual location within these constraints must be based on components as well as maintenance requirements.

3.6.4. TF coils

The primary consideration for maintenance relative to the TF coils is to minimize the number of coils. This permits better access to first wall, blanket, and shield and, in the case of a segmented design, permits design of a more easily removed segment.

To decrease the downtime on INTOR, the TF coils should be designed with electrical and cryogenic connections located such that they can be assembled and maintained by rapid maintenance procedures. Although replacement of a TF coil may be an unlikely event, the coil must be designed to be removed by remote means. The machine design need not be predicated on "rapid replacement" scenarios, however.

3.6.5. PF coils

The design of the PF coil system should avoid the use of trapped PF coils within the bore of the TF coils. This location is highly undesirable from a maintenance standpoint. Removal of trapped PF coils requires that they be either lowered out of the way or disassembled in place. Removal by lowering, in order to remove a vessel segment, requires that the TF coils be sufficiently large to accommodate the nested coils in the lowered position. Disassembly of the coils in place requires the use of demountable joints in the coils with the associated problems of joint continuity and insulation. In either case, the trapped coils, if required, must be conventional copper coils.

External (to the TF coil) PF coils shall have their connections to bus and also the cryogenic connections external to the bulk shield to possibly permit hands-on or semi-remote maintenance and disassembly.

3.6.6. Shield

Shielding should be removable in modular sections by the station crane. All coolant connections should be designed to permit reasonably rapid maintenance and assembly. The use of borated water for inboard shielding is considered to be a practical approach from a maintenance standpoint. Provision to fill and drain this tank should be in a location which would permit semi-remote operation to drain and fill. In addition, the use of water filled shielding for the outboard shield should be investigated. This approach would minimize the weight of each module permitting the use of larger modules which could be removed and replaced faster.

3.6.7. Vacuum system

The vacuum system may offer the potential for the use of contact or semi-remote maintenance on its components. This is dependent on the incorporation of a device which restricts neutron streaming during the pulse as well as tritium levels in the pumping system. Roughing pump, turbomolecular pumps and associated equipment are normally located exterior to the bulk shielding. The location of this equipment relative to streaming paths should be given considerable attention and shadow-shielding provided where necessary. If fully remote maintenance is required, then provisions should be made for rapid pump and valve replacement.

3.7. Transport facilities

The problems involved in the transportation of the main reactor components and equipment may be grouped into two major categories:

- shipment of equipment from place of fabrication to the INTOR construction site; and
- organization of the transport traffic on the site of INTOR during the assembly and disassembly operations, both for noncontaminated and contaminated components.

An important feature of INTOR will be its large and heavy components which may be prefabricated and preassembled (e.g. neutral beam injector, divertor, blanket sections, etc.). This will influence the selection of the construction site, which should have necessary access to appropriate shipping services.

The operation of INTOR sets certain serious requirements for the intra-complex transport structure which will have to serve the transportation needs involved in the assembly-disassembly, repair and experimental device handling operation. The structure must handle the load traffic between the service rooms and reactor building.

The large-size components of INTOR must be so arranged as to allow shortest possible transport paths for units and elements having a high radioactivity level. For this reason, the hot cells are located immediately adjacent to the reactor room. It should also be noted that all the available vehicles must have provisions for both manual and remote control.

4. FAILURE FREQUENCY AND AVAILABILITY ASSESSMENTS

4.1. Introduction

This Section briefly reviews two preliminary failure frequency and reliability assessments carried out by the USA and USSR groups. These should be considered as an initial example of the more detailed reliability analyses required during later design phases of INTOR.

4.2. Summary of USA assessment

In order to assess the impact of 25% and 50% availability requirements, a definition of this availability as it relates to INTOR was assumed.

Based on the revised parameters, it is assumed that 50% availability is to be achieved only during a specific time period following physics and engineering testing. During this time, the machine configuration will remain unchanged. Major consumable items will be at a state in which replacement is not realistically expected during the availability trial period; example: first wall replacement is not anticipated. The availability "run" will be a period of testing dedicated specifically to achieving 50% availability with all other physics and engineering considerations being secondary during the period. The period of availability then includes all of that time in which the machine is available for operation. It does not include startup and shutdown, i.e., the proportion of the time for which a reacting plasma is maintained is the product of the availability times the duty cycle factor.

The analysis on which the following discussion is based interprets the overall availability goals in terms of scheduled and unscheduled downtime goal allocations. These allocations are made for major subsystems. The analysis was coordinated with the Radiation Shielding and Personnel Access Group (cf.Chapter XVI) for use in its analysis of the potential benefit of partially remote or contact-assisted remote techniques. This group was assigned the responsibility of determining the desirability of personnel access to the reactor hall after shutdown.

It is hoped that these allocations may be of help and guidance to the respective subsystem designers in planning the R and D that may be required in order to bring these subsystems to a reasonable level of maturity both from the standpoint of reliability but also for longevity of service and for accessibility in maintenance.

4.2.1. Approach to availability assessment

It should be pointed out that we are, in effect, asking the question, "What levels of reliability of each component do we think are required to achieve the overall reliability goals". In order to answer this question, it is necessary to make some assumptions as to the maintainability of INTOR. These assumptions deal with the location and accessibility of the reactor components and the modularity or replaceability of components.

The approach that we have employed is summarized as follows:

- Design assumptions;
- Operating assumptions;
- Identification of redundancy applications;
- Definition of scheduled activities:
  - Weekly;
  - Quarterly;
- Bottoms-up estimate of forced outages:
  - Failure frequencies;
  - Downtimes per outage;
- Estimation of manpower requirements;
- Identification of critical reliability components;
- Requirements to increase availability from 25% to 50%.

These activities are the basis of the analysis. It should be emphasized that the scheduled maintenance plan developed herein and the estimate of the unscheduled outages should be viewed as goals which, in their totality, are required in order to meet the availability that is desired.

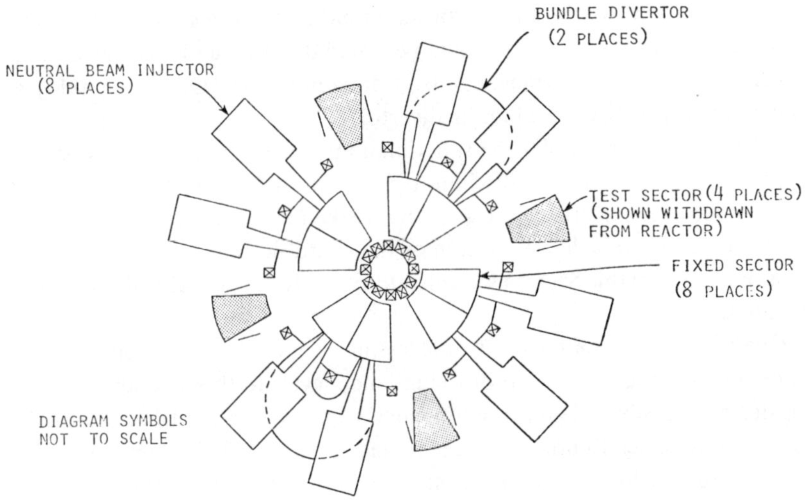

FIG.XV-1. Assumed machine configuration.

4.2.2. Design assumptions

It was necesary to define certain characteristics of the INTOR reactor concept in more detail than has been published officially at this time. Our basic assumptions are listed as follows:

Four modular segments as in Culham MK-II;
Twelve toroidal magnets;
Poloidal field coils outside of toroidal coils and rectractable;
Fixed limiters plus two bundle type divertors;
Eight neutral beam injectors accessible from overhead bridge crane;
Cryovacuum pumping chamber, neutral beams and divertors do not interefere with test sector removal;
Cryovacuum pumps accessible from overhead crane;
No cascading failures;
Shutdown and start-up in one day;
Remote maintenance.

Many of these features were determined to be desirable from the standpoint of maintainability in the study in [1]. They form the basis for the INTOR study. It should be noted that the above parameters are assumptions for this study and not a required configuration for INTOR.

It has been assumed that there are four modular test segments configured in a manner similar to the arrangement of Culham MK-II. The total number of sectors is 12 determined by the 12 toroidal field coils. The overall arrangement is shown in Fig.XV-1. Again, it should be noted that this layout is only an example of many possible INTOR arrangements. For example, INTOR may employ a poloidal divertor instead of a bundle divertor.

It has been assumed that the functional interfaces can be designed such that a failure in one component will not ripple through the rest of the subsystems. Also, in order to make effective use of the weekends for scheduled maintenance, it will be necessary to achieve the sum of the shutdown and startup times to not exceed one day. In this Section we will address only the use of fully remote techniques.

One of the important aspects of assessing each of the subsystems and their components was the assignment of redundancy levels to those components where it was deemed feasible. Typical redundancies are given in the USA Assembly and Remote Maintenance Report. It is the replacement of failed redundant components that comprise a major portion of the scheduled maintenance activities.

4.2.3. Operating assumptions

The basic schedule chosen is that the reactor is operated (except for forced outages) 9 weeks per quarter or 36 weeks per year. Four weeks a quarter it is shut down for scheduled maintenance activities. Out of each week, the weekends (2 days) are reserved for scheduled maintenance activities. If there were no unscheduled forced outages, the availability would be ~50%. To meet a 25% availability goal, we would be allowed 90 days of unscheduled outages.

4.2.4. Weekly maintenance activities

In order to assess whether it was reasonable to accomplish the required weekly preventative and other scheduled maintenance activities, typical operations for replacement of worn out items and replacement of failed redundant components were considered. It should be noted that there is available for scheduled maintenance activities in the reactor room only one day each week since shutdown and start-up require one day. For example, it is expected that while the divertor bombardment plates take a 3-shift 24-hour day to replace, they are accomplished in parallel with the neutral beam ion source replacement which is assumed to be replaced in one day. Because components such as ion sources require frequent replacement, techniques to accomplish such tasks within one day must be developed.

4.2.5. Quarterly maintenance activities

The quarterly activities, seventeen in total, were grouped in four units such that each unit could be accomplished in parallel and without intereference with the other three units. The times-to-accomplish were estimated for each unit and the critical path was identified (longest time). The critical path time was estimated to be 25 days, which fits within the 28 days allocated. The critical path activities are summarized below.

	Total time (days)	Net time (days)
Blanket		
Withdraw four sectors and re-coat walls	18	18
Limiters		
Replace 8 plates from 2 sectors	8	6
Experiment exchange		
Replace 3 first wall/blanket modules	8	0
		24
Maintenance equipment downtime (average days/outage)		0.4
Shutdown and start-up		1.0
		25.4

4.2.6. Bottoms-up estimate of forced outage frequencies and times

The failure frequencies were estimated for 34 types of failure events. These events were deemed critical in that they would shut down the reactor. These estimates are largely based on individual judgement and were reviewed as a group to assess whether their relative frequency of failure was credible. The downtimes to repair or replace each failed component were estimated based on the data contained in Ref.[1] for similar activities. The resulting data is summarized in Fig.XV-2 for the components whose failure resulted in 68% of the total downtime for an availability of 25%. The failure frequencies were adjusted from the original bottoms-up estimate to result in a level required to meet 25% total availability. An example of the meaning of failure frequencies is as follows:

A failure frequency of 0.641 per operating year for the experimental sector module means that there would be 6.4 failures every 10 operating years for every subassembly. This means that there would be 25.6 failures in the system over this operating period. For an availability of 25% there would be 6.4 failures in 10 years of elapsed time or one failure every 1.6 calendar years for the system.

COMPONENT (SUBASSEMBLY)	DOMINANT FAILURE MODE	NUMBER OF IDENTICAL COMPONENTS	FREQUENCY OF FORCED OUTAGES PER OPERATING YEAR[1]	DOWNTIME PER OCCURRENCE DAYS	DOWNTIME PER CALENDAR YEAR, DAYS
EXPERIMENT SECTOR	COOLANT LEAK STRUCTURAL WALL	4	0.641	17	10.2
FIXED SECTOR	FIRST WALL LEAK	8	0.063	68	8.6
EXPERIMENT MODULE	FIRST WALL LEAK	80	0.037	11	8.3
PRIMARY COOLANT LINE	VALVE SEAL LEAK	8	1.25	3	7.5
PRIMARY COOLANT LINE (EXP.)	VALVE SEAL LEAK	8	1.25	3	7.5
OHMIC HEATING COIL	CONDUCTOR SHORT	30	0.012	76	7.3
POLOIDAL[2] MAGNET	CONDUCTOR SHORT	10	0.012	180	5.7
DIVERTOR BOMBARDMENT PLATES	COOLANT LEAK	2	3.9	2	3.9
TF MAGNET POWER LEADS	BREAK	12	0.064	16	3.1
NEUTRAL BEAM POWER SUPPLY	POWER DISRUPTION	1	3.86	3	2.9
TOROIDAL MAGNET	DEWAR LEAK	12	0.0012	250	0.96

1 PER COMPONENT
2 SUPERCONDUCTING, OUTSIDE OF TOROIDAL BORE

FIG.XV-2. Major drives – unscheduled downtime. Availability goal = 25%.

4.2.7. Manpower requirements

In order to support the assessment of the benefits of partial contact maintenance activities established by the Radiation Shielding and Personnel Access Group, manpower requirements were established for total remote activities. These requirements are summarized below:

	Direct maintenance (manhours)	Work force (3-shifts)
Scheduled		
Quarterly scheduled	9746	
Quarterly failed unit repair	161	
	9955	44.4 (62)
Weekly scheduled (36 weeks)	240	
Weekly failed unit repair	35	17.2
Forced outages per calendar year	7568	
Total direct maintenance per calendar year	57288 (20.2 MY)	
Estimated work force (44.4 + 50%)		67 (93)

The work force of 44 people for the quarterly scheduled activities is based on a 56-hour work week. If a 40-hour week is assumed, the work force is increased to the numbers shown in parentheses. The quarterly work force is deemed sufficient for both the weekly activities and the unscheduled outages. The estimated work force is increased by 50% to allow for scheduling anomalies and productivity factors. No support personnel requirements are included in these estimates.

4.2.8. Requirements for 50% availability

To increase from 25% availability to 50% availability, there must be reductions in both scheduled and unscheduled maintenance times. We propose a scenario where the scheduled weekly maintenance is performed every other weekend. This results in an allowable unscheduled outage time of 35 days.

The reduction in allowable unscheduled downtime was then interpreted in an altered or reduced allowable failure frequency in the reactor components as shown in Fig.XV-3. The rather large improvement in failure frequency required for

COMPONENT (SUBASSEMBLY)	DOMINANT FAILURE MODE	NUMBER OF IDENTICAL COMPONENTS	FREQUENCY OF FORCED OUTAGES PER OPERATING YEAR [1]	DOWNTIME PER OCCURRENCE DAYS	DOWNTIME PER CALENDAR YEAR, DAYS
EXPERIMENT SECTOR	COOLANT LEAK STRUCTURAL WALL	4	0.120	17	4.2
FIXED SECTOR	FIRST WALL LEAK	8	0.012	68	3.3
EXPERIMENT MODULE	FIRST WALL LEAK	80	0.007	11	3.2
PRIMARY COOLANT LINE	VALVE SEAL LEAK	8	0.25	3	2.9
PRIMARY COOLANT LINE (EXP.)	VALVE SEAL LEAK	8	0.25	3	2.9
OHMIC HEATING COIL	CONDUCTOR SHORT	30	0.002	76	2.8
POLOIDAL [2] MAGNET	CONDUCTOR SHORT	10	0.002	180	2.2
DIVERTOR BOMBARDMENT PLATES	COOLANT LEAK	2	0.74	2	1.5
TF MAGNET POWER LEADS	BREAK	12	0.012	16	1.2
NEUTRAL BEAM POWER SUPPLY	POWER DISRUPTION	1	0.74	3	1.1
TOROIDAL MAGNET	DEWAR LEAK	12	0.0002	250	0.37

[1] PER COMPONENT
[2] SUPERCONDUCTING, OUTSIDE OF TOROIDAL BORE

FIG.XV-3. Major drives – unscheduled downtime. Availability goal = 50%.

50% availability is due to the fact that we reduced the scheduled outages by only 36 days out of 184.

4.2.9. Alternative operating scenario for 50% availability

As indicated earlier, an availability "run" will be attempted later in the machine life after the majority of physics testing has been completed. At this time, the machine configuration will be stabilized and it will be possible to run continuously for long periods. One mode of operation during this period has been suggested in Section 4.2.8. This mode utilizes bi-weekly shutdown periods. It is expected that the bi-weekly shutdown period is a more reasonable method of operation for INTOR in many phases of experimental operations. However, an alternate approach is a method which eliminates the bi-weekly shutdown and utilizes only the quarterly period. Due to the decreased downtime for scheduled maintenance, the time for unscheduled maintenance increases. This impacts the failure frequency assessment by permitting the frequency of forced outages and downtime per calendar year approximately to double, i.e. 71 days available for unscheduled downtown versus 25 days in the earlier scenario. This continuous mode of operation is more indicative of actual power reactor operation and such an operating scenario will prove invaluable in demonstrating reactor relevant operating characteristics.

It should be noted, however, that this scenario eliminates the maintenance tasks mentioned in Section 4.2.4. This heavily impacts the design of neutral beam sources and diverter plates. In addition, servicing of the failed redundant system is postponed until the quarterly period. This implies the use of additional redundant systems and significantly impacts maintenance scheduling.

4.3. Summary of USSR assessment

In order to study how the requirements of 25 and 50% availability could be met and to identify critical areas of the INTOR failure frequency, the reliability characteristics of INTOR systems and subsystems were analysed.

The availability A was calculated via the relationship

$$A = \frac{t}{t + t_{rm} + t_{rep}}$$

where t = total operation time of the machine,
t_{rm} = total routine maintenance shutdown time, and
t_{rep} = total accidental repair time,

the reliability analysis was carried out in accordance with the reliability calculation procedures used for such complex systems, e.g. for fission power plants, using "failure-tree" methods.

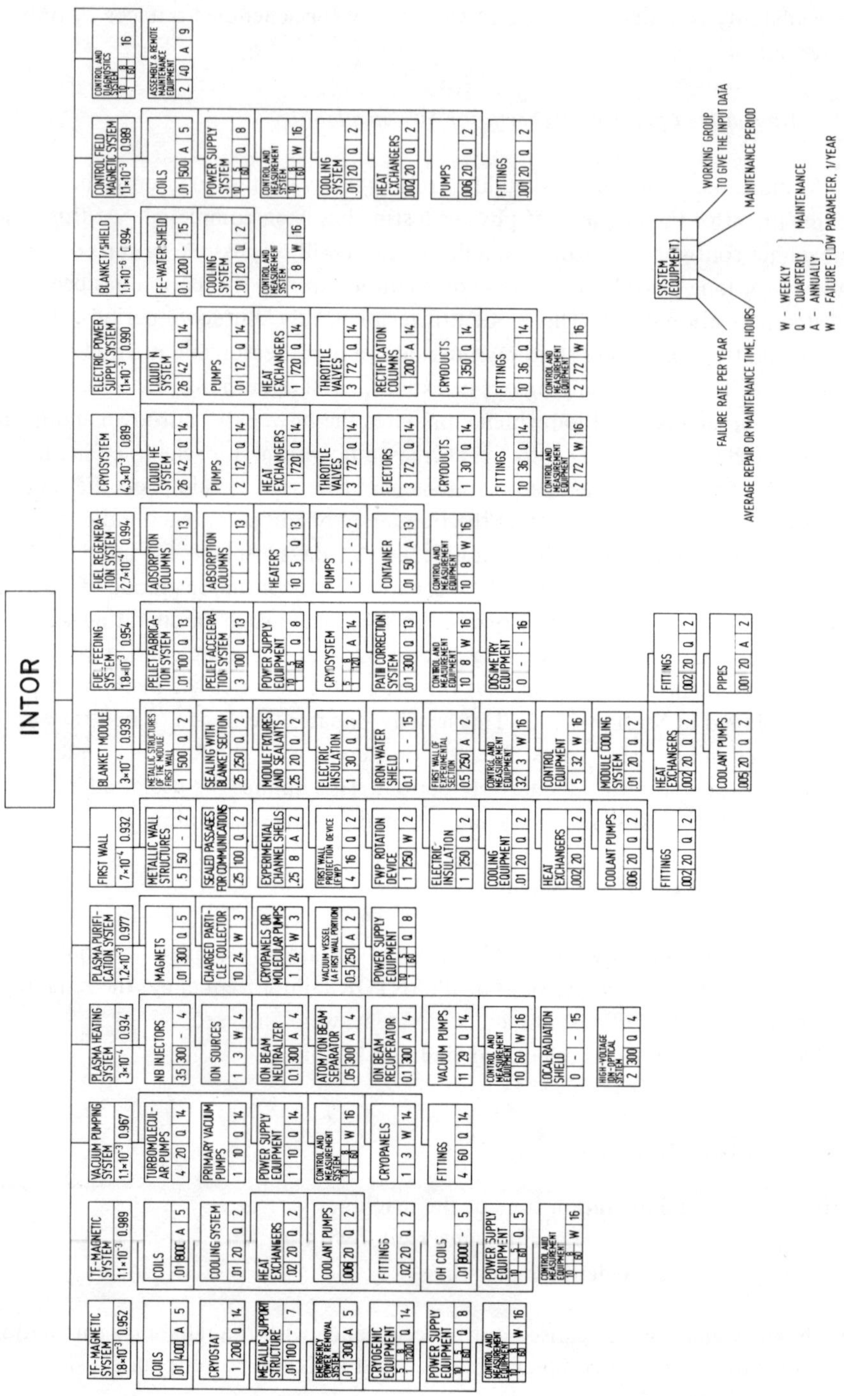

FIG.XV-4. USSR Block Diagram for Reliability Assessment.

For the purpose of calculation, the plant was divided into systems and subsystems (see Fig.XV-4). Each subsystem was regarded as an element in the calculations for which the reliability data could be determined. Some of the reliability data were gathered on similar equipment failures statistics at T-10, T-9, T-11, T-12 and T-7. The reliability data for cooling equipment were borrowed from similar statistical data for fission reactor equipment. For subsystems for which too little or no reliability data are available, the Delphi method was applied.

The acceptance of the exponential law of the failure versus time distribution used in the analysis, appears to be inadequate. For example, failure rates of the components exposed to high neutron fluency should increase with time. This will have to be accounted for in further studies for more adequate estimates of the machine availability, and for developing a schedule of maintenance operations.

It was assumed that INTOR failure occurs once any of its systems has failed. The calculations were carried out for a time spanning one calendar year. The details of the analysis are described in [10]. The analysis has shown that, depending on the routine maintenance schedule adopted, the machine availability will be 30–42% with the life to failure of about 90 days.

However, increases of lifetime of the most critical components (e.g. ion sources of the plasma neutral beam heating system) and improvement of the critical subsystems reliability as well as modification of routine maintenance procedures may provide a possibility of some availability increases.

At later design stages, it will be necessary to carry out a more detailed and comprehensive analysis of the system components reliability factors and determine the optimum level of redundancy with regard to both economic and reliability criteria.

4.4. Impact of accidents on maintenance systems

It is clear that the design of the remote maintenance system must consider potential off-normal or accident situations. During the conceptual design of INTOR, it will be necessary to evaluate various potential accidents such as a coolant leak into the vacuum chamber [10]. Such an analysis will impact the required reliability of each component as well as the overall design of the maintenance systems. The development of operational safety criteria will be required.

5. ASSEMBLY

5.1. Introduction

The assembly of INTOR is defined as the initial construction of the device prior to startup. The techniques used will be similar to some extent to those used for previous experimental fusion devices. It will also be required to check the remote maintenance techniques for all components during assembly. In this case, the remote method of assembly may be used but it will occur without a radiation environment.

It is expected that assembly requirements will not greatly impact the design except possibly in the area of tolerances and clearances. Since remote maintenance is a requirement for all reactor components, the design requirements for remote maintenance in most cases will be more stringent than those for assembly.

5.2. Objectives of machine assembly plan

An initial task for the assembly engineers should be the drafting of an assembly plan. This will specify the specific methods to be used to assemble INTOR. The objectives of the machine assembly plan are as follows:

(1) To assemble the machine to the configuration defined by the engineering documentation.
(2) To assemble the machine in the shortest possible time, consistent with quality and cost effectiveness.
(3) To minimize the cost of assembly operations.
(4) To prove the performance of the individual systems as they are assembled, and thus to minimize subsequent schedule delays caused by malfunctions.
(5) To proof test remote maintenance techniques to the greatest extent possible, consistent with the previously stated objectives.

5.3. Facility requirements

The assembly facility requirements are numerous, and contribute greatly to the successful completion of the assembly tasks. The following are major requirements for the facility:

Cranes of adequate hook height capacity and inching capability;
Building height adequate for largest subassembly over top of machine plus handling tool;
Temperature control;
Cleanliness control – dust-free floor surfaces;
Access from all sides and below;
Versatile work platforms;
Inventory storage and control;
Tool control and adequate supply;
High quality optical alignment tools;
Adequate staging and subassembly area;
Close but separate machine shop and welding facilities;

Truck/rail/barge access;
Quality assurance capability, including large surface plate for subassemblies;
Adequate lighting;
Available and functional safety equipment (fire extinguishers, safety harnesses, hard hats, breathing apparatus, eye protection,etc.);
Facility accessible to engineering personnel, either by proximity to engineering facility or by incorporation of on-site office space.

In addition to the above facility requirements, the tokamak mounting pad and adjacent walls must be designed in a manner to allow permanent, stable, and accessible fiducial marks to be placed on them. These marks would be used for establishing global reference points for positioning and alignment of the machine. Of particular interest are targets defining the vertical centreline, the midplane, and at least one azimuthal reference point. The structural elements supporting these fiducial marks typically are of reinforced concrete construction. It should be reinforced in a manner to make it act as a monolithic structure, with a minimum amount of distortion due to external loads or environmental effects.

5.4. Machine layouts

The machine layout should allow assembly tasks to be performed in parallel. A sector or modular design lends itself to this. If a number of sector modules of a particular system can be built and tested in another location, they can be assembled on the machine, while working in several locations at once.

If the sector approach is used, it is recommended that each system be pre-assembled and tested before placing it on the machine. An example of this would be a poloidal field coil located within the TF coil bore. The sectors of the PF coil should be assembled and tested for electrical integrity, cooling system leaks, and dimensional accuracy before installation. Once tested, this coil would be disassembled into sectors and reassembled inside the TF coil.

The concept of a bolted primary vacuum boundary within a secondary vacuum is particularly attractive. If the primary vacuum is a plasma chamber, designed in sectors, and jointed using mechanical seals, it can be pre-assembled, tested, and installed without cutting and welding. The inevitable small leakage at the mechanical joints can be tolerated as long as the secondary vacuum integrity is maintained. The secondary vacuum boundary should be close to the TF coils, where access is adequate for leak checking and repair.

5.5. Placement tolerances and design features

Tight placement tolerances can be held providing adequate means are established for adjusting and measuring the placement. The following conditions should exist for holding tight placement tolerances:

(a) Fiducial marks shall be on the component and the relationship of these fiducial marks to the critical features of the component shall be known. (For a coil the fiducial marks should relate to the magnetic centre of the coil).

(b) Fiducial marks or other global reference axes shall be available on the machine or the facility structure.

(c) Means of adjustment shall be built into the design (liquid shims, adjustment bolts, etc.). The range and precision of the adjustment shall be consistent with component tolerances, clearances, and installation tolerances.

(d) Adequate clearances shall be designed into the machine to allow realistic tolerances of components and still provide clearance for assembly purposes. A worst case radial clearance of at least 20 mm between large axisymmetric omponents should be maintained to allow assembly, with additional clearance provided if possible.

(e) Temperatures of all machine components and the facility surrounding the machine should be kept constant during assembly operations. If this is not accomplished, thermal deflections will greatly increase the minimum placement tolerance which can be held.

If the above conditions are met, and if good optical tooling is available, a placement tolerance of ± 3 mm or less should be achievable for major components. It should be noted that this is a possible limit and not a normal design value. In general, placement tolerances of at least a few centimetres are desirable if excessive components costs and assembly times are to be avoided.

6. MAJOR RESEARCH AND DEVELOPMENT REQUIREMENTS

A summary of the R and D requirements for assembly and remote maintenance is given in Table XV-1.

6.1. Development of remote handling systems

In general, there is a solid base of remote handling technology and experience. However, some further development will be required with respect to such areas as overall systems control and specialized manipulator positioning techniques.

6.2. INTOR reactor mock-ups

It is considered vitally important to construct accurate mock-ups of the INTOR device to develop and verify remote maintenance procedures. The general technology exists to accomplish this task.

TABLE XV-I. R AND D NEEDS ASSESSMENT – ASSEMBLY AND REMOTE MAINTENANCE

	Status*	
	Adequate	Inadequate
General remote handling apparatus	x	
Develop reactor mockup	x	
Develop remote welding and cutting systems	x[1]	
Develop viewing and inspection systems for remote welding		x
Methods for reliability analysis and verification of components		x
PF coil joints if required for internal coils		x

[1] Further development for specialized conditions in fusion devices is required.

Adequate: Programmes are either planned or could be planned on existing or authorized facilities (with some upgrading if required) and using existing trained personnel.

Inadequate: Either the available and authorized manpower and facilities are inadequate or a substantial increase in effort is needed.

* The ratings are made on the world-wide fusion effort and not separately on the individual national programmes.

6.3. Welding with manipulator systems

To date most remote welding has been accomplished using automatic welding equipment specifically designed for the welding application. A system of welding using a torch held and guided by an in-vessel manipulator system would prove invaluable in repairing vacuum leaks in vessel walls and bellows. Welding via this method has been attempted only on a limited basis. Welding on vessel, blanket or first wall structures implies metallurgical considerations relative to welding of materials irradiated in the tokamak neutron environment. Also, as part of this programme, the development of fixtures to assist in performing certain general welds (butt, corner filet, ID filet, etc.) should be addressed.

6.4. Viewing systems for remote welding

A major obstacle hindering development of remote welding is the ability to view the weld metal as it is being fused at the torch tip. The intensity of the flame cone is such that all other images are not visible due to the severe light contrast. Numerous systems of glass filters and metallic shields have been tested and found to be inadequate. Both colour and black-and-white television system have been used. Development of a viewing system capable of satisfactorily viewing the welding process is essential for the successful development of manipulator positioned welding systems.

6.5. Component reliability

In order to achieve 50% availability on INTOR, stringent component reliability goals must be achieved. All components will require a very careful reliability assessment during the design of INTOR. While failure rates of certain machine components can be assessed today, many components have little or no reliability data base. There is a need to collect and analyze reliability data from existing fusion devices and other similar devices. Subsystem development and testing must be accomplished to assure that the component failure rate is acceptable based on the failure frequency assessment. The implications of this requirement are considerable and impact the R and D assessments of all components.

6.6. PF coil joints

Demountable connectors for trapped PF coils must be developed if such coils are required in INTOR. PF coils which are trapped within the TF coil bore must be removed in sections. This requirement dictates a PF coil design which incorporates several mechanical joints. A joint design which fulfils the conductor, insulator and coolant channel requirements of the PF coils must be developed. The joint design must be able to be assembled/disassembled utilizing totally remote means. Therefore, as part of the development programme, the tooling to fabricate the joint remotely should be developed.

7. SUMMARY

The on-line availability of the INTOR reactor system is based on minimized down-time. Down-time is a function of the ease with which the system can be maintained and returned to service. The complexity of fusion devices poses a significant challenge to reactor designers as well as remote-handling and maintenance disciplines. It is essential that remote handling considerations be incorporated into the reactor design from the start of the concept design stage. Maintainability is completely dependent upon system design. Incorporation of remote features in the design phase is essential for producing a reactor system capable of performing at a high level of availability.

The methods and technology to maintain a device such as INTOR exist today, provided that the basic design of the machine is based on disassembly/assembly by remote means. The experience being developed during the design and construction of TFTR and JET will be useful during the conceptual design of INTOR. Nevertheless, these techniques and design methods will have to be greatly extended to meet INTOR's objectives.

The requirement for 50% availability presents a significant problem for INTOR. Achievement of 50% availability impacts not only remote handling, but it seriously impacts all systems. Development of machine subsystems that are sufficiently reliable to permit operation under the proposed scenario must be accomplished. Additionally, considerable attention must be given to the remote systems to assure that they are sufficient in number and capable to perform the required procedures in a realistic time frame. Preliminary failure frequency and availability assessments by the USA and USSR indicate that availabilities in the range of 25% to 50% should be achievable.

In summary, the key to 50% availability on INTOR is a design compatible with remote handling methods and development of subsystems sufficiently reliable to perform within the operating scenario. In general, it is believed that a proper data base can be established in the next few years to meet these objectives.

REFERENCES TO CHAPTER XV

This Chapter is generally based on the following references. Specific reference in the text is made in only a few cases.

[1] Fuller, George, M., et al., Developing Maintainability for Tokamak Fusion Power Systems, DOE Report C100-4184-6 (November 1978). Volume I – Executive Summary: Volume II – Study Results; Volume III – Appendices.

[2] Spampinato, P.T.,'Remote Maintenance Considerations for near term Power Reactors, ORNL (1977).

[3] Young, N. E.,A Basic Approach to Design for Remote Maintenance on Fusion Reactor, EP-D-054, 5C1-2, TFTR Internal Report.

[4] European Contributions to the 2nd meeting of the INTOR Workshop, Assembly and Remote Maintenance, Chapter IX, EUR FU/XII 501/79/EDV 50, June 79 (2 Volumes).

[5] European Contributions to the 3rd meeting of the INTOR Workshop, Design Options and Alternatives, EUR FU/XII 501/79/EDV 60, Oct. 79.

[6] European Contributions to the 3rd meeting of the INTOR Workshop, Assembly and Remote Maintenance, EUR FU/XII 501/79/EDV 70, Oct. 79.

[7] Fusion Research Development Center, INTOR Workshop Report, Home Task Report of Group 9, JAERI-M 8711 (1980).

[8] Fusion Research Development Center, INTOR Workshop Report, Engineering Aspects of JAERI Proposal for INTOR (II), JAERI-M 8518 (1980).

[9] "Assembly and Remote Maintenance", Chapter in USA INTOR Report (see Appendix A.5.4).

[10] USSR Reports to INTOR, "INTOR Assembly and Remote Maintenance" and "INTOR Assembly and Remote Service" (see Appendix A.5.3 [25, 26]).

Chapter XVI

RADIATION SHIELDING AND PERSONNEL ACCESS

K. SAKO – Japan, Chairman
C. BAKER – USA , Expert
J. GILLELAND – USA
R. HANCOX – EC , Expert
D. LEGER – EC
G. SHATALOV – USSR

1. INTRODUCTION

Radiation shielding and personnel accessibility to parts of the reactor are closely related. Shielding is essential for the operation of the reactor and to protect workers and the public outside the reactor hall. Since personnel access to parts of the reactor inside the reactor hall is considered desirable it will be necessary to use additional shielding. The area of activity of this Group includes all aspects of shielding, but particular emphasis has been placed on evaluating the additional shielding required for personnel access and the advantages of such access compared to fully remote operation and maintenance of the reactor.

The discussion of these topics may be approached in several ways. General arguments can be constructed, based on the typical features of tokamak reactors or even from experience with plasma physics experiments or fission reactors, which lead to conclusions as to the most desirable modes of operation or parameters for INTOR. These conclusions lead to design objectives. On the other hand, only by undertaking detailed studies of a specific design can it be demonstrated whether these objectives are obtainable. Both approaches have been adopted, although by different groups, and are found to be complementary. They have not led to any major differences of opinion.

The work of this Group interacts with several others, and it is necessary to define the limits of discussion. Consideration of shielding is also included in Chapter XI (first wall, blanket and shield), but it appears better to include all discussion of the composition and dimensions of shields here, leaving that group to deal with the mechanical structure. The level of shielding required for the reliable operation of reactor equipment will be taken from other specialist groups (see Chapter VI, ion sources; Chapter IX, magnets; and Chapter XIII, materials). Personnel access for maintenance is a partial alternative to remote maintenance, which is the subject of Chapter XV and, for convenience, details of maintenance operations, their frequency and duration will be based on the more extensive consideration of that group. There will also be conclusions from this study which are of importance to other groups, for example the exclusion of the reactor hall as a possible location of the vacuum boundary if personnel access for maintenance is required.

2. PURPOSES OF SHIELDING

Discussion of the shielding system can conveniently be broken into three parts:

1) The primary shield required to allow reliable operation of all reactor components within the reactor hall, such as magnets, ion sources, diagnostics.
2) The additional primary shield required if personnel are to be allowed access to the reactor hall for maintenance operations.
3) The secondary, or biological shield, forming the reactor hall to protect the operating personnel and public.

The radiation levels which can be tolerated by the reactor components will be determined by radiation damage, especially to the insulators and the stabilizing material for the SC magnet, by nuclear heating, or by the build-up of reaction products and activity. The composition of primary shielding to achieve the required levels will depend on the space available, and may be chosen for lowest cost elsewhere. The major difficulty in fulfilling the primary shielding requirements is usually the presence of holes or penetrations, as for example around divertors and NBI ports.

The desire for personnel access after shut down greatly extends the required primary shielding. The evaluation of the shielding required is more complex because it is necessary to calculate radiation levels after shut-down of the reactor due to the build-up of activity, rather than the instantaneous level of radiation during operation. Estimates of the radioactive inventories must extend to relatively minor items such as coolant circuits, including corrosion products, or airborne dust particles, at a future design stage.

The biological shield is much less complicated. Consideration needs to be given, however, to access for large reactor components during the life of the reactor and to the possible need for removing the reactor hall after de-commissioning of the reactor. Detailed consideration is necessary in the near future.

3. THE IMPORTANCE OF PERSONNEL ACCESS

Three methods of maintenance are possible: fully remote, partially remote, and contact maintenance. Fully remote maintenance operations are those which must be accomplished without the benefit of any human assistance within the immediate area of operation, and are performed by manipulators using special tools and fixtures. Such operations generally require much more time than the same tasks undertaken directly by a human operator. Nevertheless, it is recognized that provision must be made for fully remote maintenance of many components in a tokamak reactor, especially in the region of the first wall, blanket and shield where radiation levels are high (see Chapter XV).

Contact, or hands-on, maintenance requires the presence of a human operator, and includes many conventional techniques such as the use of hand-held and guided tools. Operations for which such personnel access is most desirable include inspection, damage assessment, and the repair of unforeseen damage for which special tools or procedures are not available. A wide range of partially remote operations also exists in which human access close to the work area will allow a considerable reduction in the time taken to complete a task. Contact or partially remote maintenance might be possible in many areas outside the magnetic field windings and around auxiliary equipment in the reactor hall.

An important issue, therefore, which has to be resolved in developing a complete maintenance plan for INTOR, is whether there is a net benefit from designing the reactor for human access into the reactor hall within a short time after shutdown to allow contact or partially remote maintenance. The advantage from personnel access is primarily a reduction in the total time required for maintenance, and hence an improved reactor availability. The principal penalty is the increased cost of shielding. It is concluded from this preliminary analysis that there may be a net benefit from designing INTOR for some personnel access. This preliminary analysis is only representative of the more detailed studies that are required for INTOR.

The time saved by personnel access has been estimated on the basis of a possible maintenance and repair scenario developed in conjunction with the assembly and remote maintenance group. The plan for all-remote maintenance includes scheduled and unscheduled operations, taking into account an expected failure frequency and probable redundancy provisions. Preventive maintenance was foreseen on a quarterly and weekly basis, with operations in parallel where possible. These maintenance times were then revised on the assumption that suitable operations could be undertaken by contact or partially remote means, the factor by which the time was reduced being dependent on the type of operation. As a result, it is concluded that for a particular example [2] personnel access may reduce the maintenance time by 120 days per year, corresponding to an increase in availability from 28 to 38%.

The economic penalty of personnel access is the capital cost of the additional shielding required around neutral-beam injectors, vacuum pumps, and other equipment located at the outer end of penetrations. Assuming that neutral beams are used for heating the plasma to ignition but that no divertor is used, the cost of the estimated additional shielding required is ~ \$90 M. A substantial saving in the cost of the additional shield is realizable if neutral beam injection is replaced with RF heating.

There is no established methodology at this time for a cost-benefit analysis of the advantages of personnel access in an experimental device such as INTOR, but a rough measure is the change in the ratio of capital cost to availability. For the example given above, this ratio is reduced by 17%, indicating an advantage for partially-remote maintenance with personnel access compared to an all-remote maintenance plan.

4. ASSESSMENT OF PRIMARY SHIELD

The primary shield of INTOR may perform the following functions:

a) Decrease irradiation of the magnets to a level ensuring trouble-free operation through the reactor's lifetime.
b) Reduce radiation energy deposition in the superconducting coils to the value of several kilowatts, which is comparable with thermal energy leakage into these coils.
c) Protect components in the outer part of the reactor from activation, to provide restricted personnel access.

The first two above-mentioned items put restrictions on the choice of inboard shield, the last one relates to the outer shield. The exact determination of the shield parameters and materials should be done as a part of extensive trade-offs for the detailed design. In this report, we can deal only with a preliminary estimation of shield thickness and location.

4.1. Allowable dose at magnet

The allowable radiation dose on the materials of SC magnets is summarized in Table XVI-1. The dose depends on the particular design of the coils and on the permissible variation in resistivity and critical current values for this design. Data in Table XVI-1 represent averaged parameters which may guide in the choice of shield components and thickness.

As shown in Table XVI-1, the limit on the radiation dose to the superconducting magnet is set by the stabilizer. For copper, the limit is set tentatively at 5×10^{-5} dpa which corresponds to a doubling of the original resistivity. This value

TABLE XVI-1. REPRESENTATIVE PERMISSIBLE DOSE ON SC MAGNET MATERIALS

Material	Main variable parameter	Representative Permissible irradiation (>0.1 MeW)
NbTi	Critical current	$(0.5-3)\times 10^{19}\ n/cm^2$
Nb_3Sn	Critical current	$(1-3)\times 10^{18}\ n/cm^2$
Epoxy	Electrical breakdown resistivity	$(0.5-1)\times 10^{18}\ n/cm^2$
Ceramics	Electrical breakdown resistivity	$10^{20}\ n/cm^2$
Al	Resistivity	$0.2\times 10^{17}\ n/cm^2$ (1×10^{-5} dpa)
Cu	Resistivity	$1\times 10^{17}\ n/cm^2$ (5×10^{-5} dpa)

corresponds to fluences of about $1\times 10^{17}\ n/cm^2$ (>0.1 MeV), though it changes a little with neutron spectra. However, because this limiting value is affected by a specific design, further investigations are necessary. In some cases a higher value is acceptable. Frequent annealing of the TF magnets must be avoided. There are difficult trade offs to be made between annealing periods (if any) and thickness at the inboard shield. The difference between annealing and non-annealing in the thickness of the shield is about 7 to 15 cm.

The maximum energy release in the SC magnets will, in this case, be on the level of $10^{-4}\ W/cm^3$, which is allowable.

4.2. Composition and thickness of inboard shield

A number of studies of the inboard shield composition and thickness have been carried out by the INTOR study teams. In general, these studies indicate solid shield thicknesses of 80 to 90 cm. The total distance from the first wall to the superconductor of the toroidal field coil must provide space for the shield coolant lines, assembly clearances, vacuum boundary, magnet cryostat and magnet thermal insulation. The overall total distance is thus in the range of 1.1 to 1.2 m.

The composition of the shield usually includes a heavy material and a light, neutron absorbing material. Candidate heavy materials include stainless steel, tungsten and lead. Light, neutron absorbing material candidates include water, borated water and boron carbide.

Preliminary trade off studies between the use of stainless steel and tungsten indicate that the use of tungsten may reduce the thickness by about 5 cm. However, the added expense and fabrication problems of tungsten, plus uncertainties in nuclear data, do not at this time indicate sufficient advantages for tungsten compared to stainless steel. It should also be noted that concrete partially filled with a suitable heavy metal, is a viable shield candidate material.

Studies indicate that it may be desirable periodically to anneal the copper stabilizer material in the TF magnets (by warming the magnets up to room temperature), and thus reducing the shielding requirement. Because the annealing process will take several weeks, it may be more desirable to increase the shield thickness and perhaps eliminate the need for magnet anneals. An order of magnitude reduction in the resulting radiation flux at the magnet is obtained for about every additional 15 cm of shielding. Further detailed trade-off studies are required on this point.

4.3. Composition and thickness of outboard shield

The purpose of the outboard shield is not only protection of the SC magnets, but also to decrease the induced radioactivity to the level permitting personnel access to the outer part of the machine. In the absence of penetrations, a properly designed shield, with a thickness of about 1.5 m, is capable of reducing the biological dose to the level of a few mrem/h within 1 to 3 days after shutdown. The outer shield may consist of an SS/H_2O mixture with borated water or borated stainless steel. Alternatives are heavy concrete, lead mortar and, possibly, lead layers instead of SS. The radioactivity decreases slowly versus time, which can be illustrated by an example for a 140 cm thick SS/H_2O shield as follows:

time after shutdown [days]	1	3	10	30	100	300
dose rate [mrem/h]	~15	10	7	5	3	2

The primary contribution to the dose received by a person in the reactor building comes from the outer layer (~10–20 cm) of the shield, as well as from the magnets and the structure. With careful selection of materials of the outer shield, the biological dose rate in the reactor building can be reduced to a sufficiently low level to permit access to the outer portions of the reactor.

4.4. Radiation streaming and additional shield requirements

The presence of many major penetrations (e.g. neutral beams, vacuum ducts, divertor, diagnostic holes, etc.) dramatically reduces the effectiveness of the bulk shield as they cause substantail radiation streaming to (a) many reactor components (e.g. magnets); (b) functional equipment located at the outer end of the penetrations (e.g. the beam injector chamber and ion source); and (c) the reactor building hall external to the TF coils. This compounds the difficulties of shielding and causes concern regarding two areas: (1) protection of components; and (2) protection of personnel, particularly during maintenance operations.

The only viable shielding option is a local exterior shield that surrounds all penetration ducts and functional end equipment (i.e. equipment located at the outer end of the penetration such as a vacuum pump, beam injector chamber or a detector). This shield can be designed to assure protection of all reactor equipment (e.g. TF magnets) except for the functional end equipment where many uncertainties remain. These uncertainties are discussed in the next subsection (4.5).

The thickness of the shield surrounding main penetrations can be in the range between 30 and 70 cm, depending on the size of the penetrations.

The magnitude of the attenuation that the special penetration shield has to provide is larger if personnel access to the reactor building after shutdown is required. Multidimensional transport calculations and nuclear analysis show that the largest and most costly requirement for personnel access is the additional shield around the functional end equipment. The largest portion of this shield is that required for the beam injectors because of their large size and the correspondingly large surface area that has to be covered. The shield reduces the dose only in the regions exterior to the shield. The dose inside the end equipment is, of course, not reduced by the exterior shield. Calculations show that in some places (for example in the area of NBI ion sources) the dose rate will be of the order of 10^3 to 10^4 mrem/h. Personnel access to this area will be restricted.

The most serious problem with penetrations and radiation streaming is the lack of any effective method to reduce the radiation level in the interior of the functional end equipment; particularly for the straight-through ducts such as those for neutral beam heating. This problem is discussed for key components below.

4.5. Radiation protection of other reactor components

This Section briefly reviews radiation protection considerations for neutral beams, diagnostics, vacuum pumps and divertors. Considerations for superconducting magnets were discussed in previous Sections of this report.

Neutral beams

The large size and straight-through penetrations for neutral beams result in high radiation fields inside the beam injectors. When the beam injection is off, shielding the interior of the beam injector is feasible although costly. One of the attractive methods appears to be the use of a movable shield shutter located inside the beam injector close to the beam duct opening. A movable shield plug that completely closes the beam duct is judged to be difficult.

Diagnostics

The radiation field expected at the outer end (detector location) of the diagnostic holes in INTOR was estimated. Two sizes were considered, one with duct diameter, d, of 15 cm and the other with d = 8 cm. Denoting the distance from the first wall to the detector as ℓ and assuming no bends are made in the duct path, the neutron flux at the end of the duct due to radiation streaming through the diagnostic hole (normalized to a neutron wall load of 1 MW/m^2) was estimated as follows:

$\ell = 2$ m,	d = 15 cm	Flux = $6 \times 10^{14}\ m^{-2}\cdot s^{-1}$
$\ell = 2$ m,	d = 8 cm	Flux = $5 \times 10^{13}\ m^{-2}\cdot s^{-1}$
$\ell = 5$ m,	d = 15 cm	Flux = $1 \times 10^{11}\ m^{-2}\cdot s^{-1}$
$\ell = 5$ m,	d = 8 cm	Flux = $3 \times 10^{9}\ m^{-2}\cdot s^{-1}$

Sharp bends in the duct path can significantly reduce the radiation flux if the same cross-section area can be kept unchanged. These fluxes present serious problems caused by instantaneous dose rate effects as well as the cumulative dose (fluence) effects.

The fluxes at the diagnostic detectors quoted above are only due to radiation streaming through the diagnostic holes. The background radiation flux at the diagnostic detectors due to radiation streaming from other major penetration equipment such as beam injectors and vacuum pumps can be even higher than the direct fluxes if this major penetration equipment is not shielded. Extensive shielding around all the diagnostic detectors may be necessary.

Vacuum pumps

Previous designs that located the vacuum boundary at the first wall required very large size penetrations with openings at the first wall. This leads to difficult problems arising from radiation streaming to the vacuum pumps. These problems are similar to those discussed earlier for neutral beams but their severity can be significantly reduced by introducing effective bends in the vacuum duct path.

Divertors

The most serious problem presented by radiation streaming through the divertor holes appears to be the radiation protection of the adjacent superconducting toroidal-field magnets. The heat load due to radiation streaming on the collector plates is small compared to the plasma particle heat load.

5. ASSESSMENT OF SHIELDING FOR PERSONNEL ACCESS

5.1. Radiation limits and regulation

At the present state of the INTOR device, it seems difficult to give precise radiation dose-rate limits taking into account the possibility of hands-on maintenance of the machine. For this a more detailed design is necessary, which is actually outside the scope of the present effort.

Nevertheless, one must, of course, respect the present international regulations on personnel exposure. This means, in particular, that working areas subject to ionizing radiation and personnel who have access into such areas will be classified and controlled. A classification of working conditions is given in the ICRP recommendation. In particular the annual dose equivalent limit of 5 rem/y for the whole body exposure cannot be exceeded for the workers. Within this limit exists the other limitation of 3 rem for 3 months; this last limit is not imposed by law, but by public acceptance reasons. Radiation workers should be divided into two categories: Class A, in which the individual might receive more than 3/10 of the relevant dose equivalent limit, and Class B, in which it is unlikely that an individual will receive more than 1/10 of the relevant dose equivalent limit. For workers not in these two categories, it is unlikely that any individual will receive more than 1/10 of the relevant dose equivalent limit.

A reasonable guiding criterion for the design could be the choice of the man rem/y exposure. In a light water reactor of about 1000 MW_e, the total occupational exposure is in the range of 500 man-rem/y. For an experimental machine like INTOR, with a power of about 500 MW_t, 200-man-rem/y has been chosen tentatively as a guiding criterion for the shielding calculations. Nevertheless, it may be possible to take into account the actual tendency to reduce the limit of the radiation exposure. Further study is required to establish a reasonable value.

5.2. Limitations on the physical location and time duration for the presence of a maintenance worker

This item is very important for the design of the shielding. However, detailed discussions, specifically aimed at INTOR, were not possible because of the absence of a sufficient design.

5.3. Maintenance procedures for reactor components, including time schedule

This item, one of the most important tasks of the Assembly and Remote Maintenance Group, is necessary for the evaluation in Section 5.2. Sample procedures for devices similar to INTOR have been suggested, but of course no detailed plan really exists for INTOR, since it has not yet been designed.

5.4. Source of radiation exposure during maintenance work after shutdown

Calculations of machine activation have been performed. Since the exact composition and geometry of INTOR is not known, exact calculations are not possible. Nevertheless, several of the participants have concluded that, in general, personnel access to the reactor hall is possible after one to three days for credible shield designs. This assessment recognizes the need for additional shielding near penetrations. Further work in this area is needed.

6. OTHER IMPLICATIONS OF PERSONNEL ACCESS

6.1. Impact of tritium on personnel access

It is possible that contamination of the reactor hall could negatively affect personnel access. If T_2 concentrations in air exceed 100 MPC, then cleanup is necessary before personnel access is allowed, even clothed with suits. The need to wear protective suits would also affect worker efficiency, but not to an extent which would significantly affect the outcome of the previous cost benefit analysis (see Section 3).

6.2. Impact of air activation (^{41}A) on personnel access

Account must be taken of the air activation in the reactor hall mainly in regard to argon activation. However, a preliminary estimate of this activation level indicates a value of 10^{-7} $\mu Ci/cm^3$. This low activation of air is compatible with personnel access to the reactor hall.

7. ASSESSMENT OF BIOLOGICAL SHIELD

The biological shield of INTOR must provide a normal environment outside the reactor building. In the main part, the biological shield coincides with the walls, ceiling and floor of the reactor building. It may be made of ordinary concrete with a shield thickness of ~ 2 m. No special heat removal from the biological shield is necessary except possibly at hot spots.

If sufficient shielding is provided around the equipment located at the end of the penetrations, as necessary for personnel access, then the design of the biological shield should pose no special difficulties beyond those familiar in fission reactors. However, if personnel access into the reactor hall is not required and no additional shielding is incorporated around the penetration end equipment, then the design of the biological shield will be very complicated. In this case, it is possible that the local heating rate at some locations on the interior of the biological shield would be so high as to require provisions for heat removal.

8. NUCLEAR DATA AND CODES FOR SHIELDING DESIGN

The basic calculation methods for the shield design have been developed in connection with fission reactors. The problem of the neutron and gamma-ray transport calculations can be solved with sufficient accuracy, using discrete ordinate or Monte-Carlo methods, in one or two dimensional geometries. However, the most difficult problem in this area is the solution of deep attenuation problems in complex three dimensional geometries.

Most 1–D or 2–D calculations use the discrete ordinate method. The most important advantage of this method is that it provides an accurate deterministic solution everywhere in the problem, even in the case of deep penetration problems with anisotropic scattering. However, the most serious disadvantage of this method is that it requires a computing effort which increases exponentially with the number of independent variables. Also, for computation involving complex large-size geometries and high energy neutrons with strong anisotropic scattering in fusion reactors, the present discrete ordinate codes are troublesome in terms of man-effort, computer time and memory storage.

The complex geometry of a tokamak requires that the greatest part of the shielding calculations to be carried out in 3–D. This means that the Monte-Carlo method will play a major role in the shielding design. The Monte-Carlo method has the advantage that the dependence of the computing time on the number of independent variables is much weaker than for the discrete ordinate method. The principal disadvantage of this method is known to be caused by its stochastic nature. It is necessary to handle a large number of particle histories in order to get a good accuracy; this method also requires long computing times. Nevertheless, present Monte-Carlo codes have been employed successfully by several authors to quantify the major effects of radiation streaming in fusion reactors. Detailed INTOR design requires some improvement in the state-of-the-art.

There is a need for improving Monte-Carlo variance reduction techniques that are tailored to fusion reactor shield applications. A need can also be identified for improving Monte-Carlo codes to facilitate input specifications for geometric representation of specialized fusion geometries, e.g., representing tori in tokamaks and particular types of major penetrations. This can save considerable man effort in fusion shield calculations.

The reactor shield will have to be divided into several geometrical and perhaps energy regions with the appropriate method of solution (S_n or Monte-Carlo) varying from one region to another. Although a code is now available to couple S_n and Monte-Carlo calculations, a new technique must be developed for volume coupling of these calculations on more routine bases. The difficulties anticipated in INTOR shield calculations also suggest that sensitivity analysis and integral experiments will have to be used often as a necessary part of the shield design effort. Because the reactor performance, power and cost are very sensitive to the shield design, optimization codes will play an important role.

Calculation of the activation at the composite structures is presently subject to large uncertainties. High energy DT neutrons may produce a large number of nuclear reactions and initiate decay chains where isotopes are produced for which no experimental information exists. Also, for natural isotopes, the high energy cross-section is, in general, measured only at 14 MeV. Nuclear models are applied for their extrapolation. Many multigroup cross-section sets for transport calculations have been produced by several laboratories. In this domain, an effort for the definition of standards is desirable. The acceptance of a common set of nuclear data would simplify and make more meaningful the comparison of different results. There is also the need for free access to all codes and data libraries useful in this domain.

Because of the difficulty in obtaining a suitable neutron source similar to that produced by the plasma in a tokamak, primary reliance for neutronics design will have to be based on calculation. In addition, because of the many reactions which can occur, as well as the wide spectrum of neutron energies in the structure of a tokamak, primary reliance for determining activity levels will have to be based on calculation also. Nevertheless, some key experiments must be performed to gain confidence in these calculations. The neutronics experiments should have as their objective the verification of the calculational techniques used in treating the various penetrations present in the first wall, principally the injectors, the divertor, and the various diagnostics. These can be done with point sources of presently available strength and should be carefully thought out to model the types of geometries likely to be encountered in INTOR. Additional experiments should be performed in support of the development of calculational techniques for other streaming situations likely to be a problem; for example the streaming between the finite segments of the blanket and shield.

In view of the large size of INTOR and the lack of suitable neutron sources, the use of mockup as opposed to the benchmark experiments mentioned above will, unfortunately, be seriously limited. These crucial experiments will have to be carefully chosen to limit them to areas of particular concern where an increase in the confidence of the calculated predictions is justified with respect to complexity and expense of the experiment. Particular examples which may be considered are the shielding characteristics of the blanket and shield, especially on the inner side for magnet protection and on the outside where penetrations are problematic to component and personnel protection.

There is a need for some important differential nuclear data measurements and evaluation for many materials that are likely to be used in INTOR. The specific needs have been detailed previously as a part of the nuclear data assessment carried out in December 1978 by the IAEA. Action on satisfying these needs must be made a part of the INTOR development effort.

9. CONCLUSIONS

There was general agreement on the following:

A. Importance and feasibility of personnel access

i) Personnel access for maintenance will be necessary for INTOR, since not all possible remote maintenance operations can be foreseen. A preliminary estimate presented by the USA indicated that personnel access for some maintenance operations will increase the overall availability of INTOR, and that this advantage more than compensates for the required additional cost of penetration shielding.

ii) Personnel access may be feasible for most of the reactor components outside the primary shield, but personnel access for those components at the end of penetrations will be limited.

B. Assessment of primary shield

i) The inboard shield should be designed to protect the superconducting toroidal magnetic system from radiation damage. The shield thickness is determined by radiation damage of the stabilizer of SC magnet. Frequent annealing of the stabilizer is difficult and there is the trade-off between frequency of the annealing and additional shield thickness.

For the shield design without stabilizer annealing, a value of the neutron fluence (>0.1 MeV) of 1×10^{17} n/cm^2 (5×10^{-5} dpa) for copper is tentatively recommended. However, because this limiting value is affected by a specific design, further investigations are necessary.

ii) Composition and thickness of inboard shield.

To minimize the inboard shield thickness is indispensable for realizing a reasonably-sized reactor. A number of studies have been carried out by the INTOR study teams. In general, these studies indicate the solid shield thickness of 80 ~ 90 cm and 1.1 ~ 1.2 m total distance from the first wall to toroidal coil conductor to allow gaps for structural arrangement and machine access for repair.

The shield must be manufacturable as well as have sufficient radiation shielding capability. At present, the most likely shield materials are stainless steel, water, borated water, boron carbide, concrete and lead. It is not clear whether the use of heavy metals (e.g. tungsten) provides sufficient advantage for the inboard shield to outweigh their additional cost.

iii) Composition and thickness of outboard shield

The outboard shield should be designed to reduce radiation levels to sufficiently low levels after a reasonable time after shutdown to permit maintenance operations. It should be possible to obtain ~ 2.5 mrem/h within a day or so after shutdown which would provide free access. The required outer bulk shield thickness is estimated to be about 1.5 m, based on calculations without penetrations.

iv) Radiation streaming and additional shield

The presence of many major penetrations (e.g. neutral beams, vacuum ducts, divertor, diagnostic holes, etc.) dramatically reduces the effectiveness of the bulk shield as they cause substantial radiation streaming. Radiation streaming is highly dependent on the size of penetration. In general, additional shielding of 30 to 70 cm will be required around all the penetrations to provide permissible dose rates in the main part of the reactor building. However, the dose rate in a straight line at the end of but inside the NBI system one day after shutdown is estimated to be 10^3 to 10^4 mrem/h.

C. Radiation limits and regulation

At the present state of the INTOR device, it seems difficult to give precise radiation dose-rate limits taking into account the possibility of hands-on maintenance of the machine.

TABLE XVI-2. R AND D NEEDS ASSESSMENT – RADIATION SHIELDING AND PERSONNEL ACCESS

	Status*	
	Adequate	Inadequate
Integral experiments for bulk shield	x	
Integral experiments for penetrations		x
General shield design methods	x	
Multi-dimensional transport techniques for streaming		x
Nuclear data for shielding	x[1]	

[1] Additional data required on induced activation levels.

Adequate: Programmes are either planned or could be planned on existing or authorized facilities (with some upgrading if required) and using existing trained personnel.

Inadequate: Either the available and authorized manpower and facilities are inadequate or a substantial increase in effort is needed.

* The ratings are made on the world-wide fusion effort and not separately on the individual national programmes.

A reasonable guiding criterion for the design could be the choice of the man rem/y exposure. Further study is required to establish a reasonable value.

D. Other implications of personnel access

It is likely that contamination of the reactor hall by tritium could negatively affect personnel access. Further investigations are necessary.

E. Assessment of biological shield

The biological shield of INTOR must provide a normal environment outside the reactor building. In the main part, the biological shield coincides with the walls, ceiling and floor of the reactor building. It may be made of ordinary concrete with a shield thickness of about 2 m. No special heat removal from the biological shield is necessary provided that adequate shielding of penetration is provided. The design of the shield is conventional for fission reactors and at present, no special difficulties are expected.

F. Nuclear data and codes for shielding design

A modest R and D programme for the nuclear shield design is necessary in the areas of calculational methods, nuclear data, and integral experiments. There is a need for improving Monte-Carlo codes to facilitate treatment of specialized complex tokamak geometries. There is also a need for improvement in the nuclear data base which includes specific differential data measurements and evaluation. For induced activation data, theoretical model calculations are needed. Integral and mockup experiments to validate methods and data represent a crucial part of the nuclear design effort for INTOR.

10. R AND D NEEDS ASSESSMENT

A summary of research and development needs is given in Table XVI-2.

REFERENCES TO CHAPTER XVI

This Chapter is generally based on the following references. Specific reference in the text is made in only a few cases.

[1] European Contributions to the 2nd meeting of the INTOR Workshop, Shielding and Personnel Access, Chapter X, EUR FU/XII 501/79/EDV 50, June 79 (2 Volumes).

[2] "Radiation Shielding and Personnel Access", Chapter in USA INTOR Report (see Appendix A.5.4).

[3] Fusion Research Development Center, INTOR Workshop Report, Home Task Report of Group 10, JAERI-M 8711 (1980).

[4] USSR Report to INTOR, "INTOR Radiation Shielding and Personnel Access", (see Appendix A.5.3 [21]).

[5] Fusion Research Development Center, INTOR Workshop Report, Engineering Aspects of JAERI Proposal for INTOR (I), JAERI-M 8503 (1979).

[6] European Contributions to the 3rd meeting of the INTOR Workshop, Radiation Shielding and Personnel Access, EUR FU/XII 501/79/EDV 70, Oct. 79.

Chapter XVII

VACUUM

V.I. PISTUNOVICH – USSR, Chairman
T. HIRAOKA – Japan
G.L. KULCINSKI – USA
D. LEGER – EC

1. INTRODUCTION

As a high-duty cycle tokamak with significant power production, INTOR will require advanced vacuum equipment which meets all of the criteria of commercial power reactors. Only deletion of the long pulse and high duty cycle requirements would simplify the vacuum system significantly. This report makes no attempt to design a vacuum to meet the needs of INTOR. It only provides a discussion of the range of conditions proposed for INTOR operation, the effects of these on the vacuum systems, alternative pumping, techniques which may be used, and the current state of data and experience relevant to INTOR.

The term "vacuum system" involves one or a number of sealed reactor vessels in which one should maintain a pressure lower than that of the atmosphere, facilities to provide the required pumps, gas inlet systems, etc., piping and communication means, as well as vacuum gauges, leak detectors, and vacuum control means. The thermonuclear reactor's vacuum system in general consists of:

vacuum system (VS) of the reactor (toroidal chamber),
neutral beam injector and RF heating VS,
superconducting magnet cryostat VS,
tritium and other technological facilities VS.

The production of a reliable vacuum vessel forming, to some extent, the first wall of the reactor, as well as providing a highly effective means of deuterium, tritium, helium and impurity gas pumping, appears to be a key question in designing the reactor vacuum system.

The vacuum system should be capable of removing D-T, helium and other impurities from a toroidal chamber, from the NBI, and from the RF heating systems at a rate equivalent to the particle loss rate, and maintaining acceptable pressures in the torus.

The pumps must also be capable of evacuating the torus to base pressures in a short time. All pumped gases must be recovered by the fuel recycle system within the allowable release limits for tritium to the environment and with the introduction of no impurity to the fuel system which would complicate the purification and fuel recycle. The pumping system should retain as low an inventory of tritium as possible, and regeneration of cyclic pumps must be rapid enough not seriously to reduce the duty cycle or availability of the INTOR experiment.

2. VACUUM BOUNDARY LOCATION

The functions of a D-T containing vacuum boundary appear to be most important features of the thermonuclear reactor. One can consider three positions for a vacuum boundary:

it can coincide with the first wall;
it can be placed at the back of the shield, inside the toroidal coils (TF); and
it can be placed at the building wall.

2.1. Vacuum boundary at the first wall

This would correspond to the design adopted for most experimental tokamaks up to the present – a toroidal vacuum vessel inside the toroidal magnetic field coils.

The use of the first wall as the vacuum wall means that all blankets and shielding are outside the vacuum enclosure. By locating the blanket outside the vacuum enclosure, the blanket outgassing surface area is eliminated and only the outgassing from the first wall surface remains. Furthermore, without the need to contain the blanket, the volume and surface area of the vacuum vessel will be smaller, and thus a low base pressure would be more easily obtainable, although some restriction of pumping speed would be caused by the length of the pumping duct.

If magnetic coils were required near the plasma for its positional control, these would be outside the vacuum and their insulation and maintenance problems caused by intense radiation would be eased. The location of the vessel inside the blanket results in a smaller vessel which, with all other parameters equal, will have a greater electrical resistance in the toroidal direction.

There would be a smaller tritium inventory inside the vacuum vessel than in the other cases.

Disadvantages would be:

the torus would need to be thin to give a high resistance, and would probably be a combination of rigid sections and bellows sections;

an insulating vacuum-tight break in the major circumference of the torus would be very difficult due to the high working temperature and intense neutron flux;

severe operational conditions for the joints would result;

such a vessel (without a protective liner) would probably not survive the lifetime of a commercial reactor because there will be considerable sputtering of the first wall by particles coming from the plasma, and blistering by helium would probably cause damage;

repair would be difficult due to the complication of the shielding panels, the double wall of the vessel, cooling pipes and the surrounding blanket and coils, the many penetrations for neutral beam injectors and pumping ducts and the intense radioactivity of the vessel and blanket;

the thinness and high temperature of the vessel would facilitate the permeation of tritium from the plasma region to outside the vessel.

2.2. Vacuum boundary at the back of the shield, inside the TF coils

Location of the first wall behind the shielding means that the first wall, blanket and shield are within the vacuum envelope. The immediate implications are an increase in vessel volume (and hence surface area) to contain these items, and an additional increase in outgassing surface area which results from the first wall/blanket and shielding. Furthermore, some shielding designs call for the use of concrete, which would present several outgassing problems under ultra-high vacuum conditions. To be compatible with the vacuum environment, the shielding elements would have to be encased in a vacuum-compatible material, probably stainless steel.

With this arrangement, the neutron and gamma radiation level on the vacuum boundary would be considerably reduced because of the requirement to protect the superconducting toroidal magnetic coils, and the temperature could be maintained below 100°C during operation. Large pumping ports could be used. Disassembly and assembly would be easier.

Disadvantages would be:

insulation and vacuum problems with the coils would have to be overcome if some of the control magnetic field coils have to be placed inside the vessel to overcome the magnetic field penetration problem and also to bring them sufficiently close to the plasma; and

the surface area within the torus would be greatly increased (by a factor 10 or more) by the presence of the blanket cells and shield.

2.3. Vacuum boundary at the building wall

Two cases are possible:

(1) Good gas conductance between the plasma region and the large evacuated building, and

(2) Poor gas conductance between the plasma region and the large building vessel.

In the first case, the advantages would be the following:

thermal insulation of the superconducting coils could be provided by the vacuum in the large building;

pumping of the divertor and neutral injectors would be eased;

easier maintenance (easily assembled seals);

reduced tritium permeation through the building wall.

The main disadvantages would be:

large volume to be evacuated, the base pressure would have to be 10^{-7}–10^{-8} torr;

all components and the large building would be contaminated with tritium, giving a much increased tritium inventory on the surfaces;

all components having high outgassing rates would have to be encapsulated to ultra-high vacuum standards and encapsulation of the pulsed poloidal magnetic field coils would need to be thin to avoid excessive induced currents;

long times would be required for maintenance and remote handling equipment for handling heavy loads under vacuum would need to be developed.

In the second case, the advantages would be:

vacuum requirements for the large containment vessel and contents would be relaxed, making baking unnecessary and leading to an acceptable pumping speed;

the toroidal vacuum vessel could be thin-walled because only magnetic forces and not atmospheric pressure would have to be supported;

there would be reduced tritium contamination of the components in the evacuated building, compared to the previous case without a torus.

The main disadvantages would be:

separate pumping systems would have to be provided for the building, torus, divertor and neutral injectors;

the presence of the torus would make maintenance and modification somewhat more difficult than in the previous case with no torus;

some points (the 3rd and 4th) raised against the single large evacuated building would apply also to this design.

2.4. Data base for large vacuum systems of today

One possible problem associated with cryopumped systems will be panel degassing. Techniques must be available that will allow continuous operation for long periods of time.

The space simulation programme provides broad, successful experience in fast pump down and low base pressures in very large, >100 m^3, vacuum systems. Examples are:

(1) The Lockheed Missiles and Space Co., Sunnyvale, California: The "High Vacuum Orbital Simulator" HIVOS –150 m^3 based 1×10^{-8} torr (1961, AVS Trans, page 1260). Three 1000 m^3 space chambers at G.E. Space Technology Center-Valley Forge: P_b about 10^{-9} torr (1962, AVS, page 282).

(2) Then there is the 7000 m^3 Mark I Accelerator vessel built at Livermore in 1950 as part of the MTA (Materials Testing Accelerator) programme. It operated routinely at ~ 10^{-6} torr with a large amount of surface associated with RF antennas, magnets and drift tubes.

(3) In 1958, the mirror machine Ogra was built in the USSR. The 110 m^3 vacuum chamber pumped with diffusion (oil and mercury) pumps and liquid nitrogen cooled cryopumps. The base pressure was 10^{-10} torr and pressure was increased with an operating high energy injection beam to 10^{-6} to 10^{-7} torr.

(4) More recent is the Mirror Fusion Test Facility (MFTF) now under construction. The 2000 m^3 vacuum chamber will be pumped with 1000 m^2 of LHe cooled cryopanels. The base pressure will be 10^{-9} torr and with an operating gas load of 2200 torr·litre/s will have a plasma chamber pressure of about 3×10^{-6} torr.

(5) The Space Environmental Simulation Testing Facility was built at Tsukuba Space Center of NASDA in 1975 [10]. The 1350 m^3 vacuum vessel is pumped with 200 m^3/s and 3000 m^3/s cryopumps (20 K), two 50 m^3 s Ti-sublimation pumps and three 0.8 m^3 s sputter-ion pumps. The ultimate pressure of about 10^{-7} Pa can be obtained within 15 hours from atmospheric pressure.

TABLE XVII-1. POSITION OF WELDED JOINTS IN THE VACUUM BOUNDARY FOR INTOR

	USA	Japan	EC	USSR	Approximate MW·y/m^2	T°C
Inside the blanket	x	x	x	x	2–5	200–300
Outside the blanket, inside the shield	xx	xx	xx	xx	0.02	100
Outside the shield	xxx	xxx	xxx	xxx	0.0002	100
Inside the blanket and vacuum building	x	x	x	x	2–5	200–300, RT
Outside the shield and vacuum building	xx	x	xx	xx	0.0002	100, RT
Only vacuum building	xx	x	x	x	–	RT

xxx preferred
xx possible
x not preferred
RT – room temperature

Almost all of the systems were or will be pumped with a combination of diffusion (oil and mercury) pumps and cryopumps. The Mark I was pumped with 50 cm diffusion pumps only. All these systems exhausted to atmosphere.

2.5. Conclusions

All three positions of the vacuum boundary have many "for" and "against" points. Different participants of the INTOR Workshop prefer different cases of the vacuum boundary location and their combinations.

Much design work, taking into account the requirements of all the INTOR groups, will have to be done before a choice can be made. However, this choice must be made before a detailed design can proceed since the design of many of the major components is drastically affected by the position of the vacuum boundary.

It is generally concluded that the position of the **welded** joints in the vacuum boundary for INTOR is the critical feature. The different possibilities for the position of the welded joints can be seen in Table XVII-1. All participants of the INTOR Workshop prefer positioning the welded joints in the vacuum boundary outside of the shield.

3. VACUUM SYSTEM (VS) OF THE REACTOR (TOROIDAL CHAMBER)

3.1. Effects of pumping different gases

The most important new vacuum requirements which arise when a tokamak begins to use D-T fuel are the need to recover **all** of the pumped gases and the requirement that no significant quantity of the gas (containing tritium) should escape to the environment. If the tokamak is to be a high-duty cycle power producing device, it is also necessary to recover the pumped gases quickly and achieve acceptably low inventories of tritium. Because the pumped gases must be recycled as fuel for the reactors, it also is desirable to avoid introduction of impurities to the gases. Fuel purification systems will remove impurities before the fuel is recycled, but larger quantities of impurities can complicate the process. The importance of an impurity is related to the difficulty of removing it, and hydrocarbons should be especially avoided.

The consequence of these restrictions is to eliminate the use of pumps with oil lubricated bearings, oil diffusion pumps, and slow or nonregenerating getter pumps. The oil-containing pumps (mechanical and diffusion) will contaminate the recovered gas and produce highly tritium-contaminated pump oils which are likely to release tritium to the environment and expose personnel to unacceptable quantities of tritium. Titanium gettering will also not be acceptable because quantitative regeneration (if possible) will require unacceptably long times and excessively high temperatures.

The remaining pumping techniques which should be considered for INTOR include cryopumps (condensation and sorption), turbomolecular pumps with bearings which do not expose the pumped fuel to oil, some getter pumps such as Zr-Al getters, and perhaps mercury diffusion pumps (see Table XVII-2). There is considerable experience with cryocondensation pumps in neutral beam injectors for present fusion experiments, and turbomolecular pumps are to be used in the high vacuum system of the tokamak. These pumping techniques are reasonably well understood, but new configurations for cyropumps and larger sizes of turbomolecular pumps may be needed for INTOR.

Getter pumps are not effective for pumping helium and should be considered only as hydrogen (or D-T) pumps. Getter pumps currently available commercially are not wholly suitable for many applications needed in INTOR, but perhaps redesign of the getter configurations would improve their utility.

TABLE XVII-2. AVAILABLE PUMPS FOR REACTOR PUMPING

Type	Capability or status	Remarks
Cryocondensation	Speeds of up to 10 ℓ/s·cm^2 (DT)	Neutron and activation heat shielding will not permit exposure to plasma
Cryosorption	Speeds of up to 3 ℓ/s·cm^2 (He)	Neutron and activation heat shielding will not permit exposure to plasma Lower refrigeration costs than condensation
Turbomolecular	Currently available in sizes to 10^4 ℓ/s (D-T, He) Capability 2 × 10^4 ℓ/s (D-T, He)	Appendage pumps must be located outside the magnetic field
Getters (Zr-Al)	Speeds of 5 ℓ/s·cm^2 (hydrogen)	New configurations needed to use with helium pumps. Slow regeneration rates will hamper use (especially in neutral beams)
Mercury diffusion	Not available in larger sizes (any gases)	Appendage pumps – Traps are needed to prevent contamination of plasma
Rotary blower	About 10^3 ℓ/s (D-T, helium)	Appendage pumps
Oil-free mechanical	About 10^2 ℓ/s (any gases)	Appendage pumps

Another difficulty with getter pumps, and perhaps their most serious difficulty, is the relatively long time required for regeneration. This can be a serious problem for INTOR or any high-duty cycle machine (especially for use in neutral beams). Additional getter pump development is needed to minimize regeneration time.

Pumping during the burn is the principal problem and uncertainty in the vacuum system. Adaption of vacuum seals and gauges for remote mainenance and replacement may also require significant research and development.

3.2. Methods of cleaning the first wall

The impurity problem in INTOR can be divided into two areas: impurities which are adsorbed on the wall (e.g., C, O, N) and impurities which are contained in the bulk material of the wall. The adsorbed species can be removed by discharge cleaning. Methods of chamber cleaning developed on tokamaks allow a sharp reduction in oxygen content of the plasma, and simultaneously decrease the amount of impurity on the metal wall surface.

Now there exist several methods of cleaning the first wall (see Table XVII-3). A comparative study of the various cleaning methods has shown that the most effective procedure for removing light impurities is a hydrogen discharge with low electron temperature. For example, in T-11 [1] a low temperature discharge in hydrogen (P_{H2} about 3×10^{-4} torr, $I_{pl} \sim 2$ kA, $B_o \sim 200$ G) uses a period of 40–50 h at the temperature of the molybdenum wall (about 400°C) in order to remove carbon and oxygen. After reduction of water vapour pressure to 3×10^{-6} torr the preparation of the chamber is concluded by a discharge in helium (2–4 h) to remove hydrogen. The surface of the chamber is then subjected to a glow discharge in krypton for 1–3 h or the surface is coated with titanium.

Baking the walls can reduce the surface concentration of absorbed and occluded molecules by thermal desorption. The outgassing rates of various wall materials have been measured at both room temperature and 500°C (see Japanese data on Table XVII-4).

The outgassing rate of unbaked stainless steel, Inconel, and molybdem is normally in the range of $10^{-4} - 10^{-7}$ $Pa \cdot m^3/s \cdot m^2$, depending on the pretreatments and the time for which they have been maintained under vacuum. The main desorbed species is water vapour, with lesser quantities of carbon monoxide and hydrogen. After baking for 100–200 hours at 500° C, the outgassing rate decreases to less than 10^{-8} $Pa \cdot m^3/s \cdot m^2$ at room temperature.

The vacuum walls of INTOR probably will have to be baked in a vacuum for several hours at a temperature of around 250°C, if the first wall will be constructed from SS. Baking at high temperatures does not seem very practical.

In the case when the first wall is constructed for aluminium alloys, the temperature of baking probably will be about 150–200°C. Three aluminium alloys (5456, 6061, 5083) have recently been evaluated for use in ultrahigh vacuum systems [9]. For baked surfaces (200° C), the lowest outgassing rates measured were less than 1×10^{-15} $torr \cdot litre/s \cdot cm^2$. In the design stage the question of low outgassing Al alloys will have to be investigated.

TABLE XVII-3. METHODS OF CLEANING TOKAMAK VACUUM VESSELS

Device	ORMAK	JFT 2	DIVA	TOSCA	CLEO (Stellarator)	DITE	ST
Contributor	R.J. Colchin	T. Tazima	T. Tazima	D. Robinson	D. Lees	J. Burt and G.W. Reid	E. Meserve
Conditions during discharge cleaning							
Gas used in discharge	H_2 and O_2	H_2	H_2	H_2	Neon	H_2 [b]	H_2, He
Plasma current (kA)	small	100	10	10	7	20–30	14
d.c., a.c. or pulsed	Pulsed and a.c.	Pulsed	Pulsed	Pulsed	Pulsed	Pulsed	Pulsed
Pulse length (ms)	0.5	100	30	1–2	10	20–30	5
B_T (kG)	≈1.0	5–7	5–6	3.0	2	1.5–1.8	8
Repetition rate	a.c. 120 Hz	1 shot per 60 s	1 shot per 60 s	1 shot per 30 s	1 shot per 10 s	1 shot per 10 s	1 shot per
How were these conditions optimized?	Discharge light maximized	CH_4 pressure maximized	H_2O pressure maximized	Not fully optimized	Maximum pressure rise during cleaning	Impurity release during cleaning maximized	Intensity c C-III and C impurity li maximized
Time or number of pulses required for cleaning							
After major opening of torus	3 days	~1000 shots	200 shots	1000 shots	~1000 shots	~1000 shots	2–3 weeks
After brief opening of torus		~500 shots			~200 shots	~1000 shots	10 hours
Daily	Overnight		10 shots	(Only 50 shots if pre-baked to 170°C)	Not required, torus gettered	~300 shots[c]	
Conditions during tokamak pulses							
Base pressure (best conditions) (torr)	8×10^{-8}	5×10^{-8}	1.4×10^{-8}	1.0×10^{-7}	5×10^{-8} [a]	2×10^{-8} [d]	1×10^{-8}
Maximum base pressure at which 'reasonable' tokamak discharges are obtained	1.5×10^{-7}	$>3 \times 10^{-7}$	1.5×10^{-6}	5.0×10^{-7}	1×10^{-7}	1×10^{-7}	1×10^{-5} a
Base pressure between tokamak discharges	1×10^{-7}	few $\times 10^{-7}$	$\sim 1.0 \times 10^{-7}$	3.0×10^{-7}	$5–8 \times 10^{-8}$	5×10^{-6}	Continuou
Composition of residual gas	Mainly H_2 and H_2O	H, N_2, H_2O CH_4, CO_2	H_2O, N_2, CO, O_2	CH_4, CO (H_2O only if torus is gettered)	Mainly N_2 due to leak. Masses 15 and 16 after cleaning	H_2, H_2O, CO, N_2	H_2, H_2O, N_2 (leak)

[a] Torus gettered.
[b] Argon used until mid-1976, gettered after April 1977.
[c] Not required when torus is gettered.
[d] 5×10^{-9} gettered.
[e] High T_e (2 keV) acceptable, τ_E (10 ms) possible in air or neon, $Z_{eff} \approx 8$.

4. VACUUM REQUIREMENTS FOR PLASMA HEATING EQUIPMENT

4.1. Neutral beam injection

The use of neutral beam injection for plasma heating can be expected to result in an increased pumping load from the plasma chamber. In addition to the hot neutral gas injected into the plasma, some "cold" gas streaming into the plasma chamber will occur. The magnitude of this cold gas streaming depends on the speed of the beam line pump nearest to the plasma. For TFTR, the total cold gas streaming into the plasma is limited to 1 torr litre per second. This "cold" gas could represent an impurity to the plasma.

TABLE XVII-3 (cont.)

Device	ATC	PLT	TFR	T-10	PULSATOR	ISX	ALCATOR
Contributor	K. Bol	K. Bol	P. Ginot	E.A. Moslennikov	O. Kluber	R.J. Colchin	A. Gondhalekar
Conditions during discharge cleaning							
Gas used in discharge	H_2	H_2	H_2	H_2, D_2	H_2	H_2	H_2 or D_2
Plasma current (kA)	16	5–10	50	150–200	30	Small	2–2.5
d.c., a.c. or pulsed	Pulsed	Pulsed	Pulsed	Pulsed	Pulsed	Pulsed, a.c.	a.c. 2 kHz
Pulse length (ms)	2–3	40	20	10–20	1	0.5	7
B_T (kG)	2.5	1.0	3.0	5.0	4	0.3	1.0
Repetition rate	1 shot per 2 s	1 shot per s	1 shot per 5 s	1 shot per 20 s	1 shot per 4 s	Burst of 24 pulses every 2 s	2–4 Hz
How were these conditions optimized?	Maximum power and rate set by power supplies and by temperature rise of torus	Partial pressure of H_2O maximized	Best heating of limiter	Stability of plasma in working discharges	Not optimized, but cleanliness determined by stability of high-density discharges	H_2 background pressure maximized	B_T and pressure turned to give best breakdown and optimum H_2O and CH_4 production
Time or number of pulses required for cleaning After major opening of torus	2 days	1 day					
After brief opening of torus							
Daily		2 hours	1–2 hours (1000 shots)	8–10 hours	300 shots		
Conditions during tokamak pulses							
Base pressure (best conditions) (torr)	$3\text{–}5 \times 10^{-7}$	$1\text{–}2 \times 10^{-8}$	7×10^{-10}	5×10^{-8}	3×10^{-8}	8×10^{-8}	4×10^{-10} 5×10^{-8}
Maximum base pressure at which 'reasonable' tokamak discharges are obtained	About twice the above pressure	$10^{-7}\text{–}10^{-6}$	5×10^{-8}	1×10^{-7}	1×10^{-7}		
Base pressure between tokamak discharges		$\geqslant 10^{-7}$ (H_2)	few $\times 10^{-7}$	$<1 \times 10^{-7}$	Continuous	Continuous	$6\text{–}9 \times 10^{-5}$
Composition of residual gas	Mostly H_2O + H_2	H_2, CH_4, H_2O, CO, N_2	Masses 2, 18, 28, 44	Masses 2, 18, 28, 44	H_2 O_2 and H_2O		H_2 or D_2 traces of CH_4, H_2O and N_2
Special comments		'Taylor' method					

The amount of cold deuterium passing from the injector into the plasma is determined by at least three factors:

(1) diffusion flow of gas through the drift tube (beamline);
(2) directed (unpumped) gas flow from the ion sources;
(3) flow of gas formed in the drift tube (PLT effect).

Diffusion gas flow is determined by conductivity of the drift tube and the pressure drop between injector and tokamak. Conductivity of the beamline will be about 3×10^5 ℓ/s. Gas pressure in the injector will be about 10^{-5} torr at the inlet to the drift tube and about 10^{-6} torr at the outlet. Consequently, the diffusion flow of cold gas will be about 3 torr·ℓ/s or about 10% of the flow of fast neutrals.

TABLE XVII-4. OUTGASSING RATES FROM SS-304L, YUS-170, INCONEL 625, HASTELLOY-X MOLYBDENUM, PYROLYTIC GRAPHITE AND SiC-COATED GRAPHITE (Pa · ℓ/s · cm²)**

		Before bake-out Time from starting pumping (hour)				During bake-out Time after the temperature reaches 500°C (hour)				After bake-out Time after the heater is turned off (hour)			
		1	5	10	20	1	5	10	50	1	5	10	20
Stainless steel SS-304L	Total	2.9E-6*4	5.3E-7	2.7E-7	1.3E-7	3.5E-5	7.6E-6	4.0E-6	2.3E-7	9.0E-9	2.5E-9	1.3E-9	6.7E-10
	m/e = 2 (H_2)	1.9E-7	5.6E-8	2.9E-8	1.7E-8	2.5E-5	6.5E-6	2.7E-6	1.9E-7	8.5E-9	2.0E-9	9.3E-10	–
	18 (H_2O)	1.5E-6	2.4E-7	1.3E-7	7.8E-8	3.9E-7	–	–	–	1.7E-10	–	–	–
	28 (CO)	1.5E-7	3.6E-8	1.5E-8	1.1E-8	8.2E-6	7.3E-7	2.3E-7	3.9E-8	1.5E-10	3.4E-11	–	–
Stainless* steel YUS-170	Total	5.2E-7	3.1E-8	1.3E-8	4.1E-9	3.7E-7	1.1E-7	5.7E-8	1.7E-8	3.4E-9*2	(4.0E-10)	(2.0E-10)	(8.8E-11)
	m/e = 2 (H_2)	8.4E-8	3.2E-9	1.6E-9	3.8E-10	5.7E-8	2.5E-8	2.3E-8	4.3E-9	3.1E-9*2	–	–	–
	18 (H_2O)	1.7E-8	1.3E-8	2.3E-9	2.0E-9	1.0E-8	1.3E-9	–	–	–	–	–	–
	28 (CO)	2.7E-8	2.9E-9	1.3E-9	5.6E-10	2.3E-7	4.9E-8	2.7E-8	4.9E-9	1.3E-11*2	–	–	–
Inconel-625	Total	9.3E-6	1.3E-6	6.4E-7	2.6E-7	–	–	5.2E-6	2.4E-7	3.3E-10	1.4E-10	1.1E-10	7.3E-11
	m/e = 2 (H_2)	9.0E-7	1.7E-7	1.2E-7	9.0E-8	–	–	4.1E-6	5.7E-8	1.7E-10	6.7E-11	6.9E-11	5.3E-11
	18 (H_2O)	2.1E-6	4.5E-7	2.2E-7	1.6E-7	–	–	–	–	–	–	–	–
	28 (CO)	7.7E-7	2.1E-7	1.3E-7	1.0E-7	–	–	8.3E-7	1.7E-7	6.0E-12	4.1E-12	3.4E-12	2.9E-12
Hastelloy-X	Total	1.7E-6	1.3E-7	6.7E-8	3.3E-8	–	1.6E-6	9.6E-7	2.0E-7	(3.7E-10)	(3.6E-11)	–	–
	m/e = 2 (H_2)	2.8E-7	1.9E-8	1.0E-8	5.3E-9	–	2.0E-7	6.9E-8	2.4E-8	(2.3E-10)	(1.1E-11)	–	–
	16 (CH_4)	5.1E-8	5.1E-9	2.8E-9	1.3E-9	–	–	–	–	(1.7E-11)	–	–	–
	18 (H_2O)	7.3E-7	5.3E-8	2.7E-8	1.1E-8	–	3.6E-8	–	–	–	–	–	–
	28 (CO)	8.0E-8	6.7E-9	3.7E-9	2.1E-9	–	1.2E-6	6.0E-7	1.3E-7	(1.1E-11)	(6.7E-12)	–	–
Molybdenum	Total	1.1E-5	1.6E-7	3.2E-8	4.7E-9	2.5E-6	6.3E-7	4.0E-7	1.6E-7	1.7E-10	9.3E-11	6.9E-11	5.1E-11
	m/e = 2 (H_2)	1.1E-6	1.3E-8	4.0E-9	5.3E-10	6.2E-7	1.6E-7	8.0E-8	2.0E-8	9.3E-11	4.4E-11	2.7E-11	1.5E-11
	18 (H_2O)	4.0E-6	7.6E-8	1.3E-8	1.9E-9	5.2E-7	1.3E-7	8.0E-8	1.7E-8	1.7E-12	9.3E-13	8.0E-13	8.0E-13
	28 (CO)	8.0E-7	1.2E-8	3.1E-9	3.3E-10	7.7E-7	1.7E-7	1.3E-7	7.3E-8	3.2E-11*3	2.5E-11*3	2.7E-11*3	2.7E-11*3
	44 (CO_2)	3.6E-7	2.9E-9	6.7E-10	–	–	–	–	–	–	1.6E-12	1.1E-12	7.7E-13
Pyrolytic graphite	Total	1.1E-6	1.3E-7	5.3E-8	2.0E-8	1.1E-5	4.7E-6	2.6E-6	1.0E-6	(1.2E-9)	(1.6E-10)	(1.3E-10)	(1.3E-10)
	m/e = 2 (H_2)	2.4E-7	1.7E-8	7.3E-9	3.7E-9	1.5E-6	1.8E-7	1.0E-7	4.0E-8	(8.6E-10)	(1.3E-10)	(1.0E-10)	(8.0E-11)
	16 (CH_4)	–	–	–	–	–	–	–	–	(6.0E-11)	(4.7E-12)	(2.7E-12)	(2.0E-12)
	18 (H_2O)	4.8E-7	6.4E-8	2.5E-8	9.3E-9	6.9E-7	6.7E-8	–	–	–	–	–	–
	28 (CO)	1.0E-7	1.1E-8	4.7E-9	1.9E-9	7.3E-6	3.3E-6	2.0E-6	7.3E-7	(3.3E-11)	(8.0E-12)	(5.6E-12)	(2.7E-12)
SiC coated onto graphite	Total	1.5E-5*1	3.6E-6	1.3E-6	4.8E-7	2.4E-5	5.2E-6	3.5E-6	–	–	–	–	–
	m/e = 2 (H_2)	4.8E-6*1	8.0E-7	4.0E-7	1.5E-7	1.5E-5	4.8E-6	1.9E-6	–	–	–	–	–
	16 (CH_4)	4.5E-7*1	6.7E-8	–	–	–	–	–	–	–	–	–	–
	18 (H_2O)	2.7E-6*1	4.0E-7	1.6E-7	3.7E-8	–	–	–	–	–	–	–	–
	28 (CO)	2.3E-6*1	8.0E-7	2.7E-7	1.3E-7	2.8E-6	5.3E-7	1.5E-7	–	–	–	–	–

*1: The outgassing rates at 2 hours from starting pumping. *2: The outgassing rates at 0.35 hours after the heater is turned off. *3: Nitrogen is a dominant gas in this stage. *4: The expression 2.9E-6 shows 2.9×10^{-6}.

* Chemical composition of SS-YUS-170 in Table XVII-4 is as follows: nickel 13.1, chrome 24.5, molybdenum 0.84, carbon 0.031, and iron balance.

** 1 Pa·ℓ/s·cm² ≅ 8.10^{-3} torr·ℓ/s·cm².

TABLE XVII-5. PUMPING SYSTEM PARAMETERS OF NBI

	EC	USSR	Japan	USA
Overall speed of pumping, ℓ/s	2.3×10^7	1.8×10^7	2×10^7	$\sim 2 \times 10^7$
Total area of cryopanel surface, m^2	300	250	250	216
Injection power, MW	75	75	75	75
Number of ports	6	4	4	4

The directed gas flow in the injector can be about 2–3% of the neutrals flow, depending on the geometry of the injector beam line.

Estimation of the blocking effects (PLT) shows that when the cross-section of the drift tube is about the same as the beam cross-section, the beam is completely "stopped" in the tube after 3 seconds. Therefore, direct pumping from the tube is required. There are two possible methods of pumping:

(1) by built-in systems (cryogenic, getter); and

(2) by outside systems (turbomolecular pumps).

In both cases, space is required above and below the drift tube in order to arrange the pumps and additional systems servicing these pumps.

In order to ensure effective transportation of fast atoms along the beamline, the cold gas is evacuated by means of a differential pump system. The vacuum chamber of the injector is divided into three parts: the first part included the ion sources and neutralizers; the second, the beam dumps; and the third, the drift tube region.

The gas is pumped by cryopumps cooled to the temperature of liquid helium. The pump rate for deuterium is about 7 ℓ/cm^2·s [2].

Pressure in the chamber parts can be 8×10^{-5}, 5×10^{-5} and 1×10^{-5} torr respectively, the cryopanel areas being about 20, 12 and 18 m^2, or 50 m^2 in total (USSR case, see Table XVII-5). The cryopanels are mounted on the side wall of the vacuum chamber. Japan is developing cryopumps for NBI of JT-60 [7]. Each pump has a speed of 1.3×10^6 ℓ/s. The required pump area is about 15 m^2.

During the injection of the neutral beam, energetic tritium particles will diffuse back into the beam and be pumped by the beamline cryopanels. Preliminary computer generated calculations for TFTR indicate a tritium back diffusion rate of approximately 2 Ci per beam line for a 0.5 s neutral beam pulse with a shutter valve between the injector and the torus capable of closing within 100 ms.

For the INTOR beamlines, cryopumping could provide sufficient pumping speed and acceptable regeneration intervals, but build-up of tritium inventory may require unacceptably frequent regeneration and cryopanel heating by neutrons may be significant. Zirconium-aluminium getter pumping should be investigated, as it would not release the tritium in the event of a vacuum leak and is not sensitive to small temperature increases.

4.2. RF heating

The use of RF heating with no gas injected into the plasma chamber could be expected to alleviate the aforementioned problems of gas load, impurities, tritium back diffusion associated with neutral beam injectors, and heat loads on neutral beam vacuum pumps.

5. SUPERCONDUCTING MAGNET CRYOSTAT VACUUM SYSTEM

The use of superconductive materials, cooled down to 4.2K for TF coils and slow poloidal coils (part of OH and EFC), as well as utilization of cryogenics (T = 20–77 K) for fast coils requires high-vacuum insulation. A pressure of $10^{-4}-10^{-5}$ torr is required to reduce heat fluxes into the magnets.

In order to create a reliable structure for the magnet system when the basic poloidal coils (OHC and EFC) are superconducting and are located outside of the superconductive TFC, it is reasonable to include the whole magnetic system within the general vacuum volume. In this case, all the mechanical joints between the TFC and the PFC could be kept at the one temperature level. This allows:

(1) reduction of high heat fluxes into structures which are kept at cryogenic temperatures through support units (such heat fluxes can exist if separate cryostats are used); and
(2) enlargement of the space for load bearing members leading to the vacuum chamber and blanket in the most highly stressed places.

Even with the above arrangement, there are highly stressed parts of the vacuum chamber and blanket inside the vacuum volume of the cryostat in this case. Failure of vacuum tightness is possible in such structures. Furthermore, short access times to these structures, for maintenance and repair, have to be provided. These circumstances can lead to problems and require separation of the heat-insulating cryostat vacuum volume from the vacuum chamber structure and blanket if we want to:

(1) avoid warming up all of the magnet system;
(2) increase construction reliability; and
(3) eliminate possible tritium leakage in the intermediate vacuum volume.

One possibility is to create an outer toroidal vacuum vessel having outboard pumping ports. The main vacuum chamber and blanket are placed inside this vessel. The gap between vessel and main chamber may be made vacuum-tight to exclude tritium leakage through the walls of the vessel.

When designing the cryostat vacuum volume, the following peculiarities must be taken into account:

(1) all surfaces facing the vacuum must have a low absorption coefficient which can be provided by polishing and special coatings (such as aluminium foils);
(2) the pumping system must have a margin to provide a vacuum in the presence of small leakages which corresponds to a pumping rate of approximately 10^5 ℓ/s;
(3) the cryostat has to have special valves to release gases in case of cryogenic liquid penetration into the heat insulated vacuum volume.

EXAMPLE OF INTOR CRYOSTAT EXTRAPOLATED FROM T-15 [8]

Case 1, Vacuum building

Overall vacuum volume	~3000 m^3
Overall surface facing the vacuum	
with temperature 300 K	~1500 m^2
with temperature 80 K	~3000 m^2
with temperature 4.2 K	~2500 m^2
Pressure in cryostat	10^{-4}–10^{-5} torr
Required pumping rate	~10^5 ℓ/s

Case 2, Cryostat with Separated Vacuum Heat Insulation

Overall vacuum volume	~ 500 m^3
Overall surface facing the vacuum	
with temperature 300 K	~1500 m^2
with temperature 80 K	~3000 m^2
with temperature 4.2 K	~1500 m^2
Pressure in cryostat	10^{-4}–10^{-5} torr
Required pumping rate	~2×10^4 ℓ/s

6. TRITIUM AND OTHER TECHNOLOGICAL FACILITIES VACUUM SYSTEM

The deuterium-tritium mixture is easily removed from the torus after burns with some cryoadsorption or cyrocondensation pumps. Such pumps work at a temperature equal to or below 4.2 K. In some cases, it is also possible to use turbomolecular pumps with bearings which do not expose the pumped fuel to oil.

It is necessary to recover all of the pumped gases and without allowing a significant quantity of the gas to escape to the environment. It is also necessary to recover the pumped gas quickly, because they must be recycled as fuel for the reactor, which also makes it desirable to avoid the introduction of impurities to the gases. This last point eliminates the use of pumps with oil-lubricated bearings, oil diffusion pumps and slow or nonregenerating getter pumps. The oil-containing pumps will contaminate the recovered gas and produce highly tritium contaminated pump oils which are likely to release tritium to the environment and expose personnel to unacceptable quantities of tritium. It is also needed to avoid utilization of solid organic compounds in joints, seals, etc., these compounds being corroded by the tritium action.

The problem with the use of cryopanels is the duration of the regeneration time. Important quantities of tritium could be stored in this kind of pump and this point has a strong impact on the tritium inventory. Some compromise must be found between the rate of regeneration of the pump (a quicker rate of heating could involve intolerable stress) and the tritium inventory. The different values of the regenerating time lie between one hour and one day. These values imply a tritium inventory in the cryopanels of 43 g for the first time to 3 kg for the last.

7. PUMPING OF HELIUM

Helium is in some respects the most difficult gas to pump of all those involved in fusion reactor atmospheres. The difficulty of pumping helium relative to other gases depends strongly upon the pumping method used. The most important differences are with cryocondensation pumps (operated at 4.2 K) and with gettering pumps using titanium or Zr-Al alloy. These latter pumps will simply not pump helium. The most promising of the remaining pumping techniques include cryosorption (using solid sorbents such as molecular sieves or frozen sorbents such as argon), turbomolecular pumps, mercury diffusion and ion implantation. Each of these techniques has limitations, and some have uncertainties concerning their usefulness for fusion applications.

In INTOR, helium particles will be born at the rate of about 2×10^{20}/s. As shown in Table XVII-6, the number of vacuum pumps and the areas of surface

TABLE XVII-6. SIMPLIFIED VIEW OF HELIUM PUMPING

Helium pressure in chamber, torr	Area of surface pumps, m^2 (speed of pumps – 3 ℓ/s·cm^2)	Number of appendage pumps (speed of pumps – 5 × 10^4 · ℓ/s)
10^{-7}	2000	1100
10^{-6}	200	110
10^{-5}	20	11
10^{-4}	2	1

pumps required to remove 2 × 10^{20} helium atoms from the toroidal chamber will be unreasonable, unless the allowable helium pressure can be greater than 10^{-5} torr.

Practically, this means that there is no method of pumping the INTOR chamber without a divertor or other special method of exhausting helium from the plasma. There must be a large area of surface pumps to pump helium from the toroidal chamber.

Since helium makes up only a small fraction of the gases to be pumped during a tokamak pulse, the required helium pumping speed needs to be established. In the divertor case, it would be desirable to pump impurities at a faster rate than hydrogen isotopes; it may be even better to pump impurities alone, if this were possible. Unfortunately, there is probably no way to pump helium at a higher rate than the hydrogen speed, and cryosorption pump speeds for helium will be significantly lower than hydrogen speeds particularly because the 4.2 K sorption panels are surrounded by hydrogen pumping panels. As is shown in Table XVII-2, helium can be pumped by several types of pumps. Mechanical pumps can be used for effective pumping of helium from the divertor chamber or other volume when the pressure is greater than 10^{-4} torr.

The principal limitation of mechanical pumps is the necessity for operating in appendage positions, that is, attached to the divertor or plasma chamber by ducts. This limits the capacity of a single unit to approximately 50 000 l/s (the approximate speed corresponding to a 1-m diameter duct with appropriate length to penetrate the blanket and shielding in INTOR). However, the largest units known to be commercially available are even smaller; these are the approximately 3500 ℓ/s units used in TFTR.

Cryosorption pumps are simple pumps that are relatively inexpensive and can be easily scaled to almost any size desired. Two types of cryosorption pumps have been studied in detail. The first uses a solid sorbent such as a molecular sieve, and the other uses frozen argon as the sorbent. Large cryo (condensation) pumps are currently used in neutral beam injector lines. Cryosorption pumps for helium must be protected from large doses of hydrogen isotopes, and this probably means that the helium pumping surface will be surrounded by two chevrons, one at liquid nitrogen temperature and the other (the hydrogen pump) at liquid helium temperature (about 4.2 K).

The principal shortcoming of cryopumps is their sensitivity to heat loads. This can minimize one of their principal advantages, the ability to operate as wall pumps. It is unlikely that a divertor chamber can be shielded adequately from neutrons to permit the use of cryopumps on the chamber walls. Recent tests at ORNL have demonstrated that cryopumps can handle hot ions (10 keV hydrogen ions striking the chevrons of a cryocondensation pump), but neutron heating presents different problems. Much of the ion or molecular energy can be sorbed on room temperature surfaces, as on liquid nitrogen-cooled chevrons. However, neutron heating will occur essentially equally in adjacent warm walls, liquid nitrogen chevrons and liquid helium-cooled sorption panels. Before concluding that cryopumps cannot be used along significant portions of the walls of divertor chambers, neutron transport calculations need to be made with realistic terms and divertor chamber geometries for INTOR, but the possibilities do not look good at this time.

Diffusion pumps have been used in several large vacuum systems and can be considered for fusion applications. However, all of the large diffusion pumps known to the authors are oil diffusion pumps which are not acceptable for D-T operations because of isotopic exchange with the oil (contamination of both the oil and the D-T fuel). Mercury diffusion pumps are acceptable from this standpoint. Mercury diffusion pumps were used briefly on PLT at Princeton and at the Kurchatov Institute on the open magnetic system. Two potential limitations are the lack of availability of large units and the potential for contamination of the plasma with a high-Z impurity (mercury). Diffusion pumps are by nature only appendage pumps. This places on them the same limitations just discussed for turbomolecular and appendage cryosorption pumps.

The development of vacuum pumps for helium should be considered the major vacuum problem for INTOR.

8. RADIATION EFFECTS ON VACUUM SYSTEMS

8.1. Cryopanels

(1) Nuclear heating

The major radiation effect on the liquid helium cooled cryopanels would be an additional heat load to the refrigerator due to neutron and gamma ray heating (nuclear heating). If the cyropanel is assumed to be operating at 4.2 K and if the heat capacity of copper at that temperature is 0.1 J/kg K, the nuclear heating should be suppressed to less than 1×10^3 W/m^3 to keep the small temperature rise of 1 K/s. This corresponds to a heat load of 10 W/m^2 for a cryopanel thickness of 1 cm. While this is a significant refrigeration load, it is well below the critical film boiling flux for liquid helium at one atmosphere. It could, however, mean high refrigeration costs for large vacuum systems.

To reduce energy deposition by nuclear heating, cryopanels must be placed in well-shielded locations without sacrifice of conductance to reach the panel.

(2) Cyclotron radiation

As plasma temperature is in the range of 1 keV, it is necessary to take cyclotron radiation into account. Wavelengths will be in the submillimetre to millimetre range. The surface reflectivity of wall material of the ports to these wavelengths will be large, and absorption on the inner surface of ports cannot be expected. Therefore, a large energy deposition on the cryopanel surface by cyclotron radiation will be possible.

(3) Desorption by radiation

The use of cryopanels as vacuum pumps can lead to conditions where large amounts of gas condense on the surface of the vacuum. It is possible for radiation from a plasma to impinge on the condensed gas layer, leading to desorption or sputtering (radiation-induced outgassing). However, further investigation is required on the mechanism of radiation-induced outgassing.

(4) Radiation effect on liquid nitrogen

In the case where liquid nitrogen is used as the coolant, some of the oxygen which condensed in the liquid nitrogen is converted to ozone due to ionization by radiation. This ozone may lead to a hazardous condition, as known from the field of reactor and accelerator experiments. Therefore, the purity of liquid nitrogen used as coolant must be as high as possible.

8.2. Seals

Vacuum seals in high radiation areas will be metallic and subject to heating, swelling and reduced ductility, as is the case for most metals. However, it does not appear to be a significant problem, as there is considerable fission reactor experience in the area of metal seals. The use of organic seals for whatever reason must be in areas of low flux, be readily accessible for maintenance, and have little exposure to tritium.

8.3. Valves

Valves, being nothing more than mechanized seals, will be all metal (probably double gates, differentially pumped for reliability) in high radiation exposure areas. The problems associated with lubrication of the mechanism may be of concern, but here again, there is considerable fission reactor experience.

8.4. Other materials

The greatest concern would be for electrical insulating materials, particularly high voltage insulators.

8.5. Effects of tritium

The effects of tritium on the above considerations are related primarily to inventory, and to its incompatibility with the organic components. The inventory will probably be much larger with a divertor than with wall refluxing or gas blanketing.

9. REMOTE LEAK DETECTION AND REPAIR

9.1. Detection of leaks

Some effort involving remote detection of vacuum leaks is under way at JET, JT-60 and TFTR. However, there is not much experience so far in this field. The procedure for detection is dependent on the tokamak machine structure and the location of the first wall/blanket with respect to the vacuum boundary.

In general, leak detection involves either (1) the substitution of a tracer gas for the actual leaking gas from outside the vacuum, using a detector sensitive to that tracer gas to analyze the pumped gases and thereby locate the leak, or (2) the detection of a local pressure increase within the vacuum vessel, caused by streaming from a leak. Limited access to the outer shell of a tokamak vacuum vessel after assembly usually precludes the first method for all of the vessel except for access ports and the possible determination of the general location of a leak, e.g., a vessel segment, a cooling loop, etc. Accurate location of such a leak, to which access by a man is impossible would have to be done by the second method. The accuracy and sensitivity of detection increases with decreasing distance between sensor and leak. Obviously, the sensor must be moved around the vacuum environment by some dexterous manipulator, designed and constructed such that it can perform in the high radiation environment without contributing contamination to the interior of the vacuum vessel.

It is open to question whether it would be possible to use helium as the tracer gas for an INTOR-type reactor in which helium will be produced by the fusion reactions and will be embedded in the torus wall. The technique may also be unusable where high pressure helium may be used as a coolant in the blanket and shield as well as liquid helium as the refrigerant for the superconducting coils.

The major development items for these methods are as follows:

(1) development of high resolution gas analysis system. This problem occurs, for example, during helium leak checking of tokamak systems previously used for studies involving deuterium;
(2) development of the movable mechanisms for leak sensors in three dimensional vacuum space. The mechanisms are remotely controlled, automatically and continuously;
(3) radiation affecting leak sensor and measurement systems.

9.2. Repair of leaks

While the conceptual designs for many tokamak power reactors include remote replacement of vessel wall and blanket sections, JET and TFTR are the only presently funded radioactive tokamak machines that will be operational in 1982. The design of the vacuum vessels and remote-handling equipment includes the capability of remotely removing and replacing a vacuum vessel segment (JET 1/8 and TFTR 1/10 of the vacuum vessel). This capability consists of a specially designed vessel joint, remotely operated cutter and welder, and handling fixtures for the vessel segment plus toroidal coils (JET, 4 coils; TFTR, 2 coils). The total time to remove and replace a vessel segment is estimated to be between six and nine months.

The segment replacement capability is intended to be used only in the event of irreparable failure of a segment; demonstration of the capability is not planned during the life of the systems. The segment replacement capability will be demonstrated in TFTR on a full-scale mockup of the Tokamak system consisting of approximately 2 1/2 segments of the vessel that can be evacuated to high vacuum, including toroidal and poloidal coils, vessel access ports, and a vacuum pumping system. Other components permitting replacement are diagnostics installed on flange ports that are sealed by mechanical seals.

Present TFTR design includes no capability for in-situ repair of leaks. Should such capability be included, it is thought that it would consist of cutters, grinders, and welding torches that are positioned within the vessel by a dexterous manipulator. Funding considerations have precluded incorporation of this in-situ repair capability in the TFTR design. Following successful preliminary tests, JET is planning the in-situ repair of leaks.

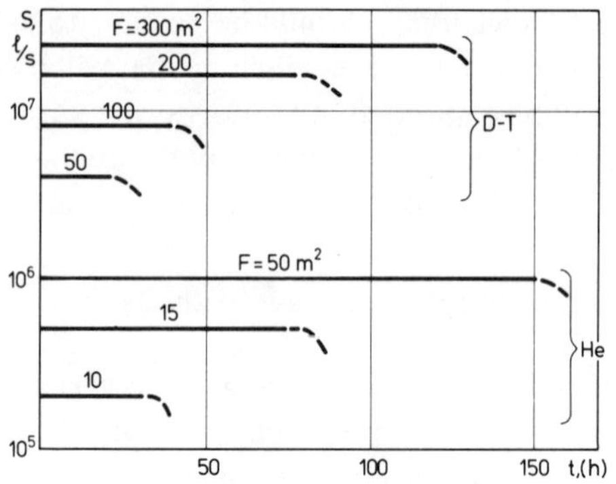

FIG.XVII-1. High vacuum cryopump D-T and helium pumping speed versus TR continuous operation time at various cryopanel surface areas.

It should be remembered that the JET and TFTR designs include no blankets. With the exception of areas that include a remotely removable thermal shield against unattenuated neutral-beam strikes, the vessel wall is the plasma first wall (JET has a double wall). The design of dexterous manipulators to accomplish in-situ repairs is simpler in a system in which the first wall is the vacuum vessel wall because one does not have to operate in the space between the blanket and the vessel wall.

10. CRYOPUMP REGENERATION

Regeneration is needed for the condensation pumps used for D-T pumping and the cryosorption pumps used for helium pumping in case of their use over long periods of time. The regeneration frequency should be determined by

(1) the allowed thickness of condensed D-T mixtures at which the sticking coefficient is reduced, owing to the temperature gradient;

(2) the allowed helium concentration in a layer of sorbent at which the sticking coefficient is decreased;

(3) the allowed amount of tritium deposited on cryopanels and limited, in particular, by the heat load due to tritium decay.

The USSR estimate of the maximum operating lifetime of the cryopanels is shown in Fig. XVII-1. In this estimate, the reflection coefficient for helium equals 0.9 and for D-T 0.1 (upper estimation). A specific pumping speed is 2–3 $\ell/s\cdot cm^2$ for helium and 7–10 $\ell/s\cdot cm^2$ for D-T, with an allowed gas load on the cryopanels for 1 $\ell\cdot torr/cm^2$ for helium [3, 4] and 30 $\ell\cdot torr/cm^2$ for D-T. The estimate is made only for helium pumped out on cryosorption panels. Due to this circumstance

TABLE XVII-7. CRYOPUMP REGENERATION SEQUENCE

Operation	Approximate time required
Closing of large valves	1 to 5 minutes
Removal of liquid He	1 to 2 minutes
Release of condensed and sorbed D-T and He	1 to 3 minutes
Pumpout of gases	5 to 15 minutes
Cooldown of cryopanels	1 to 10 minutes
Reopening of large valves	1 to 5 minutes
Approximate total time	10 to 40 minutes

one should place cryopanels in a pumping unit in such a way as to perform most hydrogen isotope pumping on condensation panels and helium pumping on panels with cryosorbent [5, 6]. Due to the fact that the D-T sticking coefficient at condensation is less than 1, the complete helium and hydrogen isotope separation is impossible even in this case, so the data in Fig. XVII-1 concerning the helium pumping time should be understood as an upper estimate. Because the cryopanel operating lifetime is limited by the amount of tritium produced and because this may not exceed 2–3 days, one gets, as Fig. XVII-1 shows, simultaneously operating condensation panel surface areas of 100 m^2 or more and cryosorption panel surface areas of 50 m^2 or more (with the simultaneous absorption on these panels of D-T and helium).

There are actually two times of importance for cryogenic regeneration periods. The first is related to the safety considerations, such that if an air leak were to occur, the resultant oxygen-hydrogen mixture would pose an explosive hazard. This establishes the longest time between regeneration periods. Normally, this period is determined by the time required to accumulate sufficient hydrogen isotopes such that if the leak were to occur in a closed volume, the partial pressure of the desorbed hydrogen would exceed 13 torr. Another limit would be that the DT not exceed 30 torr$\cdot\ell/cm^2$.

On the other side of this issue is the desire to minimize the tritium inventory on the cryopanels. This would dictate that the shortest possible period between regenerations be achieved. For INTOR, it is probably the shorter period which is of most importance.

Frequent regeneration periods will have an economic impact on INTOR from the increased cryogenic costs although that has not been quantified at this time. Regeneration periods of one day are certainly possible and values of

5 to 12 hours are currently the state of the art for PLT and TFTR. Recycle times as short as 1 hour could also be considered for INTOR and thereby reduce the T_2 inventory in the pumps to as little as 50–100 grams, depending on the number of spare pumps to be used. An analysis of the minimum pump regeneration time is given in the US report. It was concluded that the shortest regeneration time ranges from 10 to 40 minutes (Table XVII-7).

The economic impact of rapid recycling of INTOR cryopumps should be examined in more detail in the upcoming study period.

Regeneration time of the vacuum system may be restricted not only by technical limitations but also by safety considerations. An assessment of regeneratic time shows that the hydrogen accumulation to a value of about 2 torr·ℓ/cm^2 will limit the regeneration time.

11. THE EFFECTS OF A CATASTROPHIC FIRST WALL/VACUUM WALL FAILURE ON THE INTOR VACUUM SYSTEM

To the first order (without going through a detailed fault-free analysis), collapse of the vacuum wall or even a rapid depressurization accident could have the following far-reaching detrimental effects on the vacuum system:

missile generation;
containment room contamination with tritium and lithium;
blanket rupture;
liquid helium panel overpressure;
lithium interactions with water and concrete;
oxygen-hydrogen detonation;
contamination of the plasma chamber with pump lubricants from turbo-mechanical pumps;
damage to the pumps from cooling water.

12. LEAKAGE INTO THE VACUUM SYSTEM

12.1. Maximum permitted leak

The maximum leak that would not affect INTOR operation depends on the size of the pumping system and the methods employed for controlling impurities. For the purpose of this discussion, it is assumed that the leak is air coming through the plasma vacuum enclosure. Plasma physics experiments employing divertors and other impurity control mechanisms are presently under construction and should be fully operational in the 1980's. Data from these machines on the ability of controlling impurities entering the plasma will enable determination of the maximum tolerable leak rate that does not affect operation.

Damage to the vacuum system by large leaks could be caused, at a minimum, in the following ways:

accumulation of gas on cryopanels such that upon release of these gases during regeneration, pressures greater than atmospheric would result;

accumulation of air on cryopanels such that an explosive mixture of hydrogen and oxygen would result during regeneration or warmup;

damage to turbomolecular pumps due to a sudden inrush of air while pumps are at high rotational speed;

overpressurization of shutter mechanisms due to massive leakage on one side of a closed shutter;

freezing of water or coolant lines or changes in component alignments caused by cooldown due to excessive heat transfer to cryogenic surfaces;

rupture of cryopanels due to excessive heat transfer causing rapid boiling and expansion of cryogenic gases;

burnout of hot filaments on gauges.

From the above examples, it is evident that the damage attributable to a large leak is a function of its location as well as its magnitude. From the information presently available on the INTOR parameters, it is impossible to specify a quantitative value of a leak that would not damage the vacuum system.

12.2. Probability of leaks

Data on occurrences of large leaks in vacuum systems are not known to be available in a form that would permit a determination of the probability of occurrence of a certain size leak. This information could probably be obtained only by exhaustive review of the operating loss at plasma physics facilities, space simulation centres, and accelerator facilities. Based on discussions with other workers in the vacuum field and past experience, the following general statements regarding leakage can be made:

welded, heavy wall stainless steel vacuum vessels, once having been certified leak-tight during fabrication, generally do not develop leaks during operation;

thin-walled components, such as bellows and membranes, have developed leaks while in service;

catastrophic failures of glass ports have been observed in several systems and are usually caused by improper installation;

cryopanels can exhibit “cold leaks” that leak only at cryogenic temperatures and cannot be found at room temperature.

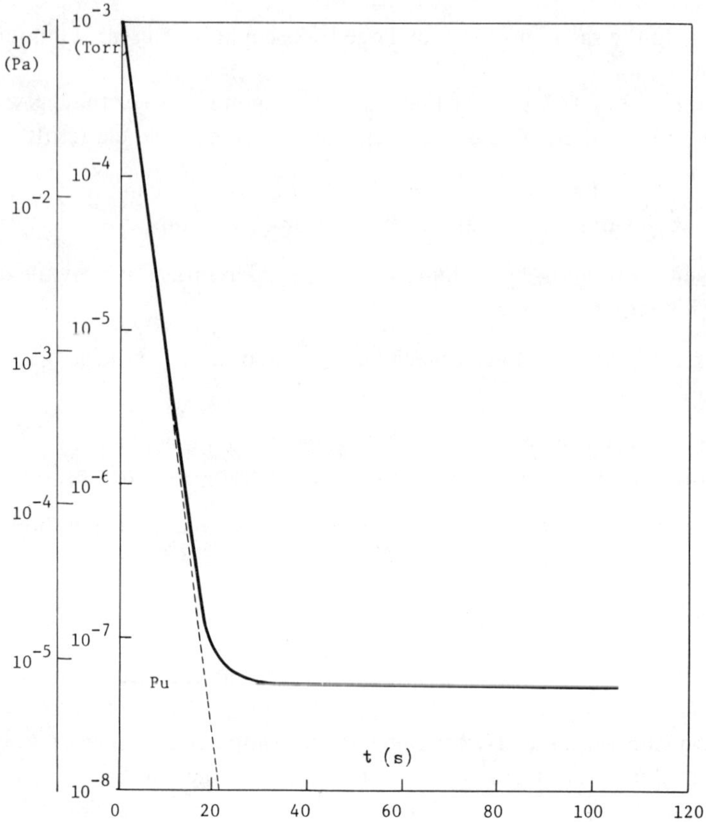

FIG.XVII-2. A predicted pumpdown of the torus vacuum vessel following a D-T plasma pulse.

13. MINIMIZING CHAMBER EVACUATION TIME

Minimizing chamber evacuation time would depend on the design of the INTOR vacuum system. Evacuating time will increase as volume and surface area increase due to enclosing the blanket and shield.

A Japanese estimation of a predicted pumpdown of the torus vacuum vessel of INTOR following a D-T plasma pulse is shown in Fig. XVII-2. The following parameters are used:

average ion density	$\sim 1.2 \times 10^{20}\ m^{-3}$
plasma volume	$\sim 200\ m^3$
vacuum vessel volume, V	$\sim 1000\ m^3$
ultimate pressure in the system	$\sim 5 \times 10^{-8}$ torr
effective pumping speed, S	$\sim 500\ m^3/s$ for D_2 and T_2
V/S	~ 2 s for D_2 and T_2

Thus, the dwell time of 30 seconds is long enough to pump out the free gases in the vacuum vessel.

In the actual machines, however, the transient behaviour of the residual gases are not simply related to the pumping time constants, and there is evidence that the surface adsorption and diffusion into the metal affect the evacuation time for the vacuum vessel.

14. SUMMARY

With the exception of one area, the data base for vacuum systems in INTOR is now available, or it could exist in 1982 to design and construct INTOR. The one area that could present a problem is that of removing helium from the divertor and/or reaction chamber during the burn. The solution to this problem must necessarily wait for the final impurity control method. It is felt that once such a method is determined, an intensive programme (with substantial financial and manpower support) can be developed to design and test such high-speed helium pumps by 1982. Without such an effort, the removal of helium ash during the burn could represent a major obstacle for INTOR.

Proposed particle pumping ranges can be reached by present mechanical equipment only if the total pressure at the pumped chamber (torus or divertor) is greater than 10^{-4} torr. Helium cannot be removed unless the helium pressures are greater than 10^{-5} torr. Unless higher pressures can be tolerated, an excessive number of pumps will be required which occupy an area equivalent to significant fractions of the outboard torus area. Novel pumping techniques such as ion implantation have been proposed, but **these are untested ideas.**

The vauum system for INTOR should be capable of removing D-T, helium, and other impurities from the scrape-off region surrounding the plasma or from a divertor chamber (poloidal or bundle) at a rate equivalent to the particle loss rate and maintain acceptable pressures in the torus and divertor chambers. Pumps must also be capable of pumping down the torus to base pressures in a short time. All pumped gases will have to be recovered by the fuel recycle system with acceptably low release of tritium to the environment and with the introduction of no impurities to the fuel stream, which would complicate purification and fuel recycle. The pumping system must maintain acceptably low inventories of tritium and regeneration of cyclic pumps must be rapid enough not seriously to reduce the duty cycle or availability of the INTOR experiment.

Much design work, taking into account the requirements of all the INTOR groups, will have to be done before a choice of vacuum boundary location can be made. However, the choice of vacuum boundary location is essential before detailed design can proceed and, therefore, a high priority on this topic is warranted.

TABLE XVII-8. R AND D NEEDS ASSESSMENT – VACUUM

	Status*	
	Adequate	Inadequate
Development of high speed helium pumps		x
Development and testing of large dry turbomolecular pumps		x
Test of the performance of cryopanels in the presence of high neutron and gamma fluxes		x
Development of baking, degassing, and discharge cleaning methods for stainless steel surfaces inside the vacuum vessel	x	
Development of baking, degassing and discharge cleaning methods for Al alloys surfaces inside the vacuum vessel		x
Evaluation of radiation heating in cryopumps for NBI and development of alternative pumping techniques	x	
Remote leak detection		x
Testing of vacuum seals suitable for remote disassembly and maintenance	x	
Design and fabrication of large cryosorption pumps and confirmation of their performance		x
Development of vacuum components with large cross section ($\gtrsim 1\ m^2$)		x

Adequate: Programmes are either planned or could be planned on existing or authorized facilities (with some upgrading if required) and using existing trained personnel.

Inadequate: Either the available and authorized manpower and facilities are inadequate or a substantial increase in effort is needed.

* The ratings are made on the world-wide fusion effort and not separately on the individual national programmes.

It is concluded that positioning the welded joints in the vacuum boundary outside the shield and inside the toroidal magnetic field is preferred.

There is no problem in pumping the torus before and after the working pulse, but difficult problems do arise in the pumping of gases from the NBI systems.

All vacuum seals, measuring instruments, and pumps will have to be capable of remote maintenance (or remote replacement). Seals and valves must be bakeable all-metal units. Surfaces will have to be baked in vacuum for several hours at temperatures of at least 250°C (SS first wall), and/or cleaned by a low-temperature discharge in hydrogen. The possibility of using aluminium alloys inside the vacuum boundary will have to be studied, particularly from the standpoint of baking and cleaning.

No organic elastomers should be used in the vacuum boundaries because of their sensitivity to bakeout temperature, cleaning, radiation and tritium.

Heat generation rates on vacuum pumps in NBI are likely to be high enough that refrigeration costs will be unacceptable if conventional cryocondensation pumps operating at temperatures up to 30 K and getter pumps (Zr-Al) are used. The latter pumps may, however, require relatively long regeneration times which could affect the INTOR duty cycle.

15. R AND D NEEDS ASSESSMENT

The R and D needs assessment is summarized in Table XVII-8.

REFERENCES TO CHAPTER XVII

It should be noted that this summary is based on the individual reports of the participating countries. These reports are listed above the numbered references.

European Contributions to the 2nd meeting of the INTOR Workshop, Vacuum, Chapter XIV, EUR FU/XII 501/EDV 50, June 79 (2 Volumes).

European Contributions to the 3rd meeting of the INTOR Workshop, Vacuum, EUR/FU/XII 501/79/EDV 70, Oct. 79.

Fusion Research Development Center, INTOR Workshop Report, Group 14, Vacuum, JAERI-M 8513 (1979).

USSR Report to INTOR, "INTOR Vacuum System" (see Appendix A.5.3 [41]).

USSR Report to INTOR, "INTOR Vacuum System" (see Appendix A.5.3 [42]).

"Vacuum", Chapter in USA INTOR Report (see Appendix A.5.4).

[1] Vlasenkov, V.S., et al., "Plasma heating and stability in T-11 with neutral injection", Proceedings of the 7th International Conference on Plasma Physics and Controlled Nuclear Fusion Research, Nuclear Fusion Suppl. v.1, p.211 (1978).

[2] Coupland, J.R., Proceedings of the 8th Symposium on Engineering Problems of Fusion Research, Knoxville, Tenn. 1977 (1977) p.1796.

[3] Yuferov, V.B., Condensation and Adsorption Pumping, KhFTI 72-9, Kharkov, USSR (1971).

[4] Halama, H.J., Aggus, S.R., J. Vac. Sci. Technol. **11** 1 (1974) 333.

[5] Karasev, B.G., et al., "Vacuum and Tritium Systems of Demonstration Tokamak Reactors", Preprint A-0318, Leningrad, USSR, NIIEFA (1976).

[6] Watson, T.S., Fisher, P.W., ' Cryosorption vacuum pumping under fusion reactor conditions", Proceedings of the 7th Vacuum Congress, Vienna, v.1 (1977) p. 359.

[7] Matsuda, S., Shirakata, H., JAERI-7773 (1978).

[8] Soviet Contributions to the 2nd meeting of the INTOR Workshop, "The Electromagnetic System", ed. N.A. Monsoon, June 1979.

[9] Bechtel Corporation, Selection of Chamber Materials for Ultrahigh Vacuum Service, Technical Report SSD-92-1, 1963.

[10] Nishikawa, S., Kanamori, Y., Sakakibara, S., J. Vac. Soc. Japan **19** (1976).

Chapter XVIII

DIAGNOSTICS, DATA ACQUISITION AND CONTROL

T. HIRAOKA – Japan, Chairman
R. HANCOX – EC, Expert
J. GILLELAND – USA
B. KOLBASOV – USSR
P. REYNOLDS – EC

1. INTRODUCTION

Although an analysis of the Diagnostics, Data Acquisition and Control System seems to be necessarily speculative at this early stage of the INTOR development, it is useful to carry out a preliminary assessment because of the new features of INTOR compared to existing devices. The diagnostics, data acquisition and control system will be reactor-relevant since there is no essential difference in the operation scheme between INTOR and future tokamak reactors. It may have little resemblance to those of large tokamaks now under construction.

In the following section, the impacts of the general features of INTOR are discussed. Then, its control and data processing system and instrumentation are qualitatively reviewed. Furthermore, new features provided by long D-T burning, accessibility to the tokamak torus and the physical environment of instruments are briefly discussed. Safety systems and reliability are also discussed. Finally, the discussions are summarized with conclusions and some recommendations.

No essential differences were found in participants' observations about this field of INTOR. However, it is particularly dependent upon input from the other groups, since this group has the overall control system responsibility. The present report has been produced without such input. It is the result of general discussions to identify agreements and disagreements which came up through discussions of the INTOR Workshop and to develop the integral outline of INTOR's diagnostics, data acquisition and control system. Since this is not a detailed technical review paper, the references are limited to four reports by the participants of the group.

It is essential that this group should consist of an extremely strong unit within the INTOR structure. Instrumentation and the data system have to integrate and interface with every other system.

Information and inputs from all other groups will be required to make discussions more quantitative or definite.

2. IMPACT OF GENERAL FEATURES OF INTOR

INTOR has as an eminent feature a pulse duration of more than one order of magnitude larger than that of large tokamaks now under construction. This long pulse length and high duty factor make INTOR essentially a steady-state experiment, which resembles more an industrial plant than today's fusion experiments. For example, at INTOR there will be a continuous data flow, as at a plant. Diagnostics must function for reactor-relevant operation, as well as for plasma physics. INTOR will need considerably more real-time feedback control than fusion experiments hitherto. For this control, and for the safety of the machine, it is likely that some complicated data (until now usually only regarded as physics data) will be necessary. This data must be easily managed by INTOR operators.

Another INTOR requirement is high availability and high reliability, while maintaining a D-T burning plasma at self ignition or very close to it. These features may also force INTOR's diagnostics, data processing and control characteristics to be similar in many respects to a power plant. Automatic maintenance, failure detection or repair will be required to achieve high availability.

In order that the control and diagnostics complex of the INTOR facility should function successfully it is necessary to:

- acquire and process large quantities of data,
- measure and control a large number of parameters,
- optimize the control procedure,
- archive and analyse the information stored throughout the installation lifetime,
- ensure that all the control operations are safe and reliable.

All these requirements can be met by creating a computer-based automatic control system.

Frequent stoppage of the machine to repair or to replace a component or a part may not be expected at INTOR to achieve high availability. A systematic approach should be based on safety and reliability studies. The technology to give a high availability exists in principle, but one has to make sure that sufficiently radiation-insensitive equipment is made available.

Since the tokamak machine will be activated by neutron irradiation, the instrumentation must have sufficient radiation resistance and redundancy to maintain high availability. A careful cost/benefit analysis must be carried out. This balanced redundancy is essential and reliability analysis on each component must be performed.

All the system equipment should, preferably, be standardized. The required unification and standardization may be attained through a modular structure approach allowing an easy and flexible modification of the functions of individual system elements. The modular structure is probably the only means to ensure

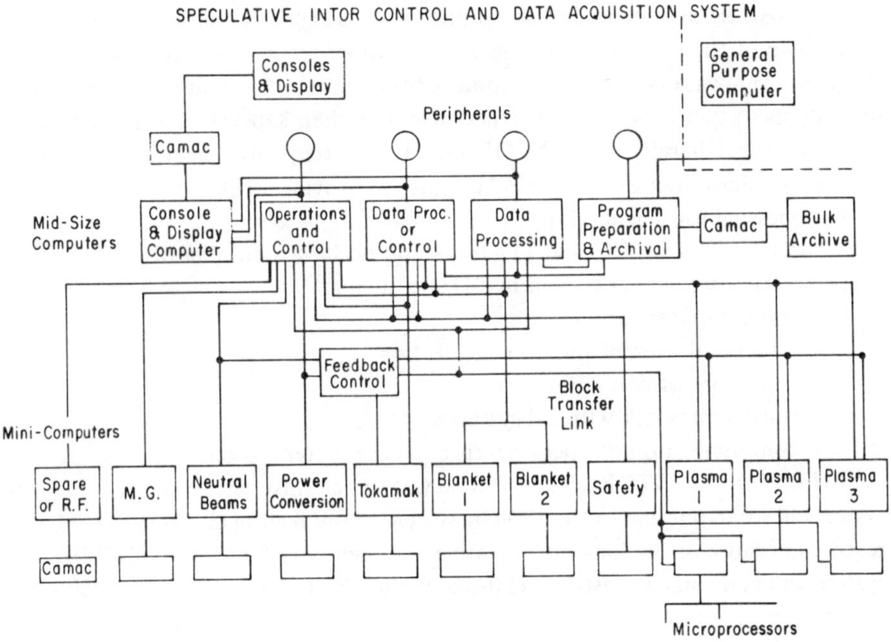

FIG.XVIII-1. Speculative INTOR control and data acquisition system.

a high reliability and maintainability of the system. The reliability index may be brought to the desired level by using back-up apparatus and automatic failure detection.

At the building stage of the machine, some international quality control or assurance procedure must be established. Although it may be quite difficult to achieve compromises because of the different technical backgrounds of each participant, country or community, INTOR must set the standards with an international review committee to prepare common conditions for manufacturers' bids.

3. CONTROL AND DATA PROCESSING SYSTEM

3.1. General configuration

The system will have a hierarchical structure with two levels, the upper, central level or the complex level, and the lower, subsystem level, as shown in Fig.XVIII-1.

In the architecture shown, there are three top level computers to allow for an element of redundancy for the control computer during the tritium burn phase when processing of large quantities of physics data is considered less likely. The breakdown of the lower level into many minicomputers is advisable because of the likely wide variety and distribution of the organizations building the various subsystems so that they can gain experience with their subsystem at their home-base before installation at the INTOR site. It may frequently be necessary for additional microprocessors in CAMAC crates to provide local control functions or local processing for diagnostics.

The JET, TFTR, T-15 and JT-60 systems do have a number of common properties that we expect to find in INTOR, such as:

- full computerization,
- integrated control and data acquisition,
- a network of computers,
- starlike network with subsystems,
- communication of operators through computerized consoles.

However, the INTOR system will be very much different from today's tokamak systems in some points. In the INTOR system, more feed-back control will be required because of the long burning time and high duty factor. The INTOR system will add some industrial plant-like feature by introduction of heat generating elements in the blankets.

Ideally, the control and data processing system is centralized. However, the size and complexity of INTOR makes likely a considerable involvement of industry in different countries, and it is essential that usable subsystem computers can be made available to these manufacturers at a sufficiently early time. A degree of partial de-centralization should be developed.

The INTOR facility will require a large general purpose computing capability for engineering design, physics modelling, off-time analysis and administrative processing.

3.2. Data processing

To some extent it is premature to examine the data-taking capacity. Since INTOR offers very long pulses (>100 s) with rather a short dwell time (20–35 s), real time operation within this period is necessary. An almost continuous data flow must be treated.

A total amount of data per discharge greater than for the large tokamak experiments now under construction seems unavoidable. It may be beyond the order of 10^7 words/pulse.

It is clear that the quantity of processed data will not simply scale as the length of the pulse, because, in the latter case, the physicists will be primarily interested in the start-up and decay periods of the plasma with only occasional sample periods in the relatively long steady burn period. Considerable use will be made of recycling memories in the diagnostic CAMAC racks with sufficient analysis

TABLE XVIII-1. PRELIMINARY INTOR DATA ASSESSMENT

(a)	Number of slow (100 Hz to 0.5 Hz) channels		10 000
(b)	Number of fast (~ 5 kHz) channels		10 000
(c)	Number of fast (~ 100 kHz) channels		1 000
(d)	Number of real-time channels included in (a)		1 000
(e)	Number of real-time channels included in (b) and (c)		100
(f)	Number of timing channels		1 000
(g)	CPU* capability for:	real-time comfort displays ~ 1% of all fast and slow channel data ~ real-time data	
(h)	Archival techniques:	local memories (many with rewrite capability) disks for individual specialist users bulk storage medium ($\lesssim$ 20 M words/pulse)	
(i)	Typical word size:	12 bits	

* The general purpose computer requirements are not included.

only for simple comfort displays of waveforms. Detailed analysis will only be triggered by unusual plasma events. On this basis, the data computers can function at an equivalent pulse fashion at all times; for the long pulses they can take data from storage and process it after the start-up phase, at occasional times during the pulse and during the plasma current decay. In the case of the current decay, a rather short processing time may be available, either restricting the data quantity or establishing rather fast data transfer and data processing specifications.

The US delegate presented Table XVIII-1, which shows some very approximate numbers of channels and sampling rates where rather simple extrapolations from PLT operation and TFTR studies have been made. Some numbers are enhanced by factors two or three to try to accommodate more tokamak component monitoring because of the more aggressive environment. It has been assumed that while some plasma diagnostics will be producing less data than in TFTR, others such as the neutron diagnostics will take considerably more. Digital control and monitoring channels and analog control and monitoring channels are not differentiated.

In assessing memory requirements, some consideration of likely operational requirements is necessary. The assumption that all the fast channels in Table XVIII-1 are sampling at full rate for 100 s leads to absurdly large archival requirements even if humans could assimilate the data. If one considers a more likely need for a fast-wave sampling plasma diagnostic, which is the driver in accumulating data, the requirement might be for 100 channels sampling 5 periods of 10 ms each at 10^5 Hz rate during the 100 s flat top, or 0.5 M words of permanently stored data.

Such assessments lead to total requirements for the bulk storage unit of about 20 million words per pulse.

The participants agreed that the data processing time during start-up and shut-down will be less than 10 ms, and that more definite requirements should come from the physics groups, especially from the Control Group. Thus, the maximum rate of data processing may be determined by the start-up and shut-down phases. For the steady burning phase, it could be modest depending on the need for feedback information to sustain burn and to detect precursors of disruption. Some data for decision making for the successive discharges will be processed during discharges and the dwell time.

3.3. Feedback control

At INTOR, substantially more feedback will be required since it will have a long burning period. The physical procedure for burn control at ignition or close to it must be found. A data processing time of about 100 ms will be sufficient for the normal steady state operation. To obtain confidence, some preliminary learning at INTOR itself will be essential. However, it will be necessary to carry out numerous simulation calculations during the design of INTOR. The major and essential physical plasma parameters for control should be identified by the physics groups. Analysis techniques with multi-input and multi-output have to be developed

The feedback control with data processing time less than 10 ms may be applied also to the start-up, heating, and shut-down phases, although a rather slow start-up time (about 5 s) and a secondary heating time of 5 to 10 s are recommended for INTOR. One of the important roles of control of INTOR will be the prevention of disruption which will greatly affect on the machine configuration. The precursor to disruption should be definitely detected and some actuator against disruption must respond quickly. The issue of disruption of INTOR plasma was intensively discussed in the physics groups, especially in the Stability Control Group (Chapter 4.1.4). Even if we find a slower indication for INTOR than present-day tokamaks, we should keep test sampling. We expect that a reliable procedure will be established by the valuable experience gained from large tokamaks scheduled to come into operation in the next few years.

3.4. Some remarks

Generally speaking, the hardware of the control and data processing systems exists at the present time for INTOR because of the flexibility of current technology and the expected advances in the state of art by the time INTOR is built. However, an analysis method for feedback control and control-related software should be developed.

TABLE XVIII-2. MAJOR PLASMA DIAGNOSTICS FOR INTOR CONTROL

Definitely required

Position measurement – Sine/cosine coils, saddle coil

Total plasma current – Rogowski loop

Average density – Interferometer (1 mm or FIR)

Background gas – Residual gas analyzer

Neutron flux – Time dependent measurement – burn

Radiated power – Bolometry, surface temperatures

(Divertor fluxes – Monitors in the divertor)

Probably required

Voltage measurement – Voltage loops

Inner magnetic surface location – X-ray intensity profile

Plasma density distribution – Multichannel interferometer

(CH_3OH or CO_2) or $2\omega_{ce}$, ω_{ce} cyclotron emission

Impurity influx – Spectroscopic monitors

Fluctuation precursors – X-ray wave and/or ω_{ce}, neutrons, B_θ loops

High-β control – Current density distribution – magnetic ($\beta_\theta + \frac{\ell i}{2}$) or new diagnostic

Helium density – Charge-exchange measurement, fluorescence spectroscopy

Triton density – Charge-exchange measurement

Since all computers should have international interchangeability with a common high-level structural language, INTOR should define specifications and commonability should be part of these specifications.

Since INTOR will have many components similar to an industrial plant, as mentioned in Section 2, information from fission reactors and industrial plants will be veıy helpful.

4. INSTRUMENTATION

The word "instrumentation" is used to express the integral observation system of INTOR including diagnostics, measurement, and monitoring.

TABLE XVIII-3. CONTROL SCHEME TO PROCEED SEQUENCE

Phase (duration)		Control purpose	Feedback control time const. (reasoning) *	Actuators: Magnetic fields	Actuators: Fuelling	Actuators: Heating	Diagnostics (sensors)
Start-up	Ionization and current initiation (~ 100 ms)	Loop voltage		Transformer	Filling pressure		One-turn loop
		Plasma current	10 ms (Joule loss)	Transformer			Rogowski coil
		Minor radius		Divertor			X-ray intensity profile
		Ionization				Pulsive	Multichannel interferometer
	Current rise and Ohmic heating (2 ~ 4 s)	Plasma current	100 ms ~ 1 s (Joule loss)	Transformer			Rogowski coil
		Current profile		Divertor		Pulsive	FIR polarimeter
		Elongation		Shaping field			X-ray intensity profile
		Electron temp.	100 ms ~ 1 s (τ_p)	Transformer			Cyclotron radiation
		Density			Puffing		Multichannel interferometer
	Additional heating (~ 5 s)	Plasma current	100 ms ~ 1 s	Transformer			Rogowski coil
		Ion and electron temp.	100 ms ~ 1 s (τ_E)				Doppler broadening (?)
		Density	100 ms ~ 1 s (τ_p)		Puffing or pellets	Continuous	Multichannel interferometer
		Density profile					
Burn (100 ~ 200 s)		Plasma current	~ 1 s (Joule loss)	Transformer			Rogowski coil
		Density	~ 1 s (τ_p)	Divertor	Puffing		Multichannel interferometer
		Helium accumulation					Charge exchange measurement Fluorescence spectroscopy Residual gas analyser Triton density measurement
		addition to above					
	Subignition	burn control	~ 1 s (burn confinement time)			Pulsive	Neutron measurement Doppler broadening (?) Multichannel interferometer Cyclotron radiation Bolometer
	Marginally ignited	Compression-decomp.		Vertical field			
		Reactivity			Puffing		
	Ignited state	Field ripple		Toroidal field			
		β limit		Shaping field	Puffing	Pulsive	
Shut-down (5 s)	Burn termination	Ripple enhancement		Toroidal field			
		Reactivity lowering			Puffing		
		Impurity injection			Impurity pellets		
		Decompression		Vertical field			
	Cooling and current run-down	Plasma current		Transformer			Same with the case of current rise and Ohmic heating phase
		Cooling			Impurity pellets		

* Feedback is unnecessary unless specified.

4.1. Plasma diagnostics

At INTOR, plasma "diagnostics" may be merged with "instrumentation". Diagnostics will be divided into two groups. One of them is the diagnostics of essential plasma parameters for control and the other is for developing or understanding of the various physical phenomena. However, the latter will be forcibly limited by the restricted availability of the space for instrumentation and of the time for physics experiments. Detailed plasma physics studies will be performed on other tokamak research devices.

All participants agreed that the number of plasma diagnostics will be restricted to the essential minimum. When choosing the number of diagnostics, it is necessary to take account of the conditions under which they operate, the complexity of processing the data obtained from the given technique, and the possibilities for installation of redundant parts as well as for replacement of separate components without disturbing the complex operation of INTOR.

Most INTOR plasma parameters are not very different from those at large tokamaks now under construction, except for the long D-T pulse duration. Applicability of certain diagnostic techniques to INTOR will be well defined at these tokamak experiments. However, diagnostics for control must be reliable and simple. Some current techniques will simply not work under the full burn neutron flux. We must find "rugged" diagnostics. A detailed assessment of alternative detectors and special analysis techniques themselves were not discussed at the Workshop. Therefore, this Section will not describe techniques but will investigate the plasma properties and related parameters to be measured and try to give a preliminary assessment of the techniques.

Table XVIII-2 shows the preliminary assessment of diagnostics for INTOR plasma control. Time sequential control schemes and instability control were assessed as shown in Table XVIII-3 and XVIII-4 respectively. Diagnostics for control will be required with a time resolution in the range of 0.1 to 10 ms and a spatial resolution of around 5 cm. Joint efforts between physicists and diagnostics experts should be initiated for definite identification of simple, reliable and effective diagnostics of INTOR plasma control and the process of control.

For D-T burning control, neutron measurements may occupy a more important position among current diagnostics. A well collimated neutron detector will give the line-integration of fusion reactions. It will be supplemented by some neutral particle scattering. Neutron spectroscopy could be used to measure the ion temperature.

There is a significant data base concerning neutron measurements in the fission field. Information on fast neutron spectroscopy is available from fast reactor physics experiments and some nulcear experimental physics.

Since INTOR is the first long burning D-T machine, the diagnostics will face new situations. Some of the participants recommend the staged advance of diagnostics from the hydrogen plasma to D : 50% – T : 50% plasma by the stepwise

TABLE XVIII-4. INSTABILITY CONTROL

Control purposes	Control measures	Feedback control		Actuators	Diagnostics (sensors)
		Time const.	Reasonings		
MHD inst. and disruption	m = 2/n = 1 mode supression	1 ms	MHD growth rate	Helical windings	X-ray, ω_{ce}, neutrons, magnetic probes, etc.
	Feedforward by disruption precursor	10 ms ?	Time delay between precursor and disruption	?	”
	Avoidance of dangerous regime				
	Impurity	10 ms ?		Divertor	Spectroscopy
	Density lowering			Puffing speed	Multichannel interferometer
	Speed of density increase				
	Speed of I_p increase			Transformer	Rogowski coil
	Current profile	10 ms ?		Pulsive heating, puffing	Multichannel FIR polarimeter
Equilibrium control	Horizontal vertical and shape control	0.5 ~ 1 s	Time constant of eddy current in the blanket	V. H and shaping fields	X-ray intensity profile magnetic probes

increase of tritium content. At the initial stage by a hydrogen plasma, initial problems in the diagnostics will be eliminated and some diagnostics will be applied to give a more solid foundation for the calibration of the plasma for operation in the following stages. However, the staged approach to establish the diagnostics system for D-T operation should be accomplished within the rather short period of a few years.

4.2. Engineering instrumentation

INTOR will have many monitoring requirements similar to fission reactors in the blankets and other engineering subsystems; e.g., temperature and fluid flow monitors and local neutron flux measurements, tritium levels, stress concentrations, etc.

Since one of the major purposes of INTOR is to provide testing of blanket concepts for which considerable design analyses will have been made, it is clear that rather extensive diagnosis of the blankets should be provided. The collection of this data will replace much of the plasma diagnostic information in significance once successful burn operation is achieved.

The neutron measurements will obviously complement those plasma diagnostics that determine the source strength, but are necessary for comparison with model calculations. The other diagnostics will be monitoring the physical characteristics of the blankets. The initial experimental blankets may have a larger quantity of each diagnostic type, but even the "production-run" INTOR blankets will have to be equipped with a fairly extensive array.

Other than conventional monitoring, some unconventional instrumentation will be needed depending on experimental purposes.

Table XVIII-5 shows a preliminary listing of major instrumentation of the engineering subsystems.

An example of the time sequential operation scheme of engineering components is shown in Fig. XVIII-2.

5. ACCESS TO THE TOKAMAK TORUS

The accessibility of instrumentation to the tokamak machine will be limited at INTOR. It has a thick blanket and shield region and most of its segments will be occupied by large ports for neutral beam injection and vacuum pumping. It is too early to discuss the accessibility in detail. However, since the accessibility of plasma diagnostics and engineering instrumentation to the tokamak torus has a close relation with the configuration of the tokamak machine, joint efforts by physicists, control and instrumentation engineers and machine builders are strongly recommended from an early stage to avoid conflicts with the machine structure.

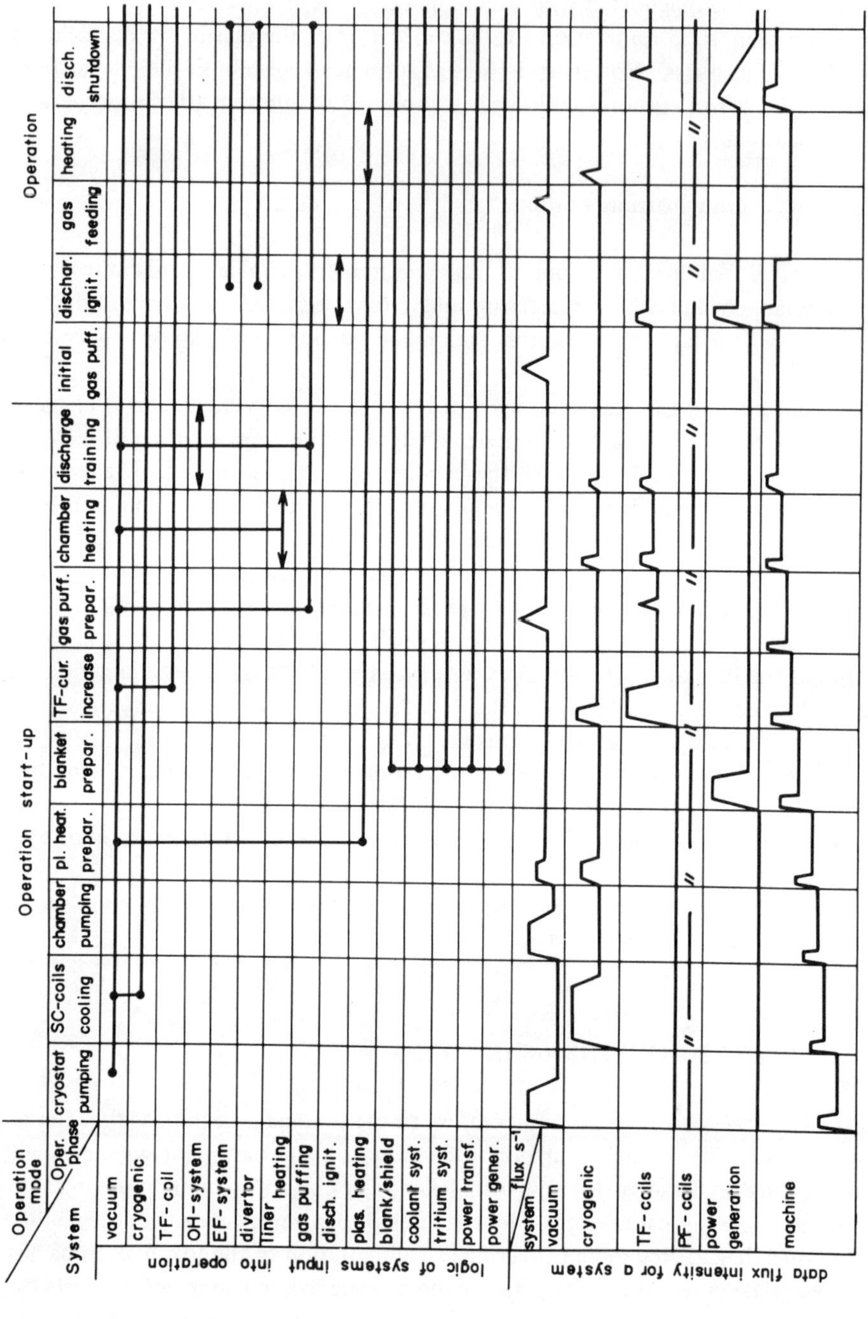
Operation mode
System
Oper. phase
Operation start-up
Operation
cryostat pumping
SC-coils cooling
chamber pumping
pl. heat. prepar.
blanket prepar.
TF-cur. increase
gas puff. prepar.
chamber heating
discharge training
initial gas puff.
dischar. ignit.
gas feeding
heating
disch. shutdown
vacuum
cryogenic
TF-coil
OH-system
EF-system
divertor
liner heating
gas puffing
disch. ignit.
plas. heating
blank./shield
coolant syst.
tritium syst.
power transf.
power gener.
logic of systems input into operation
flux s⁻¹
system
vacuum
cryogenic
TF-coils
PF-coils
power generation
machine
data flux intensity for a system

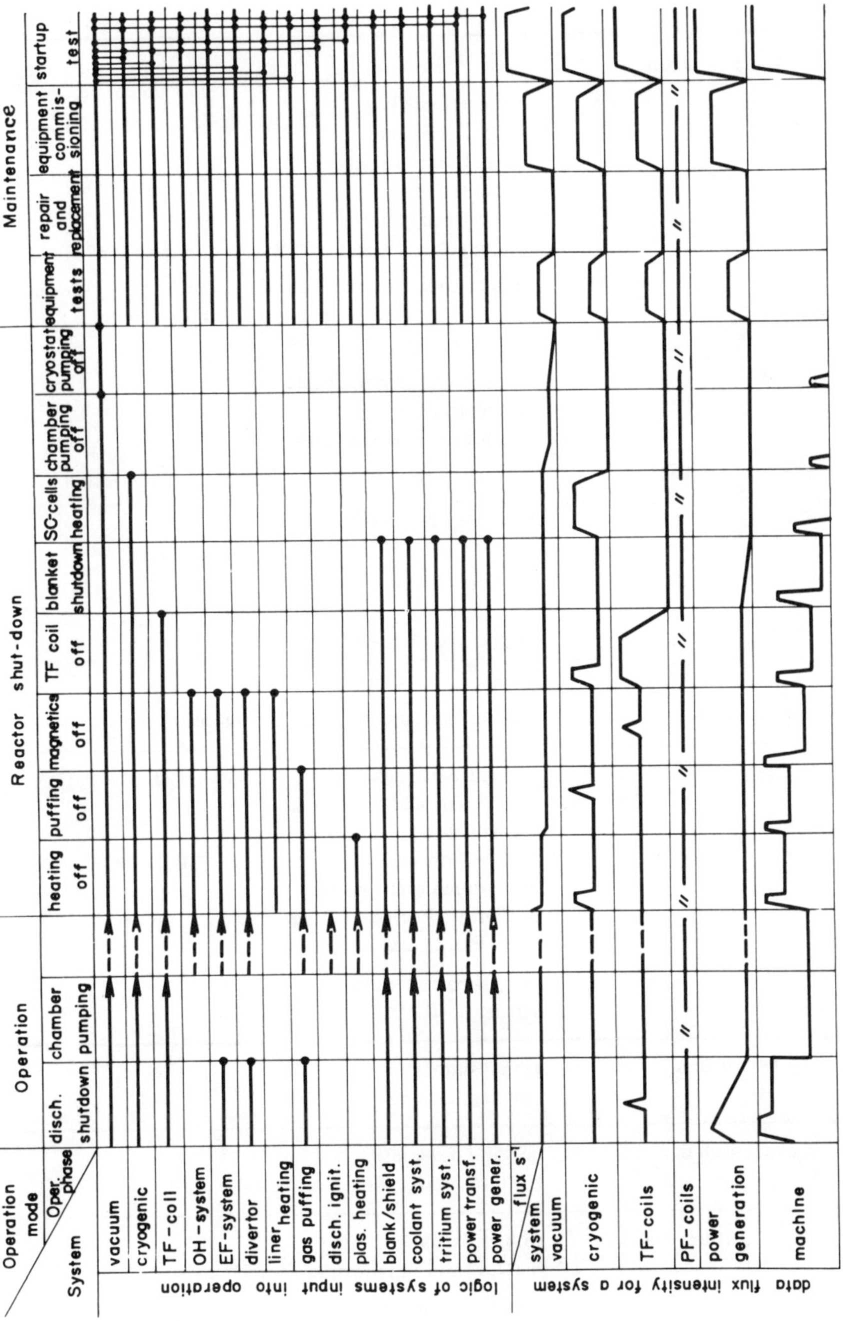

FIG.XVIII-2. Flow of logic and data flux intensity of engineering operation.

TABLE XVIII-5. MAJOR ENGINEERING INSTRUMENTATION

1.	Tokamak machine	
	wall	surface conditions sputtering rate void/crack formation spallation adsorbed material temperature, strain, coolant flow rate, pressure
	blanket	temperature, pressure, strain coolant flow rate, neutron flux, tritium content, activation
	shield	neutron and gamma fluxes temperature coolant flow and pressure
	vacuum vessel	pressure, strain, diffraction, current, leak detection
	port	strain, surface condition leakage
	poloidal field coil	current, voltage, strain, magnetic field, helium coolant parameters, cryostat leakage, flux leakage, volt-second
	toroidal field coil	similar to B_p coils
	coolant system	coolant flow, and pressure temperature
2.	Heating system	grid voltage and currents gas pressure and flow rates cryopanel temperature coolant flow etc. beam dump temperature coolant flow at beam dump beam profile diagnostics outgassing rate shutter position
3.	Fuelling system	gas pressure opening time
4.	Vacuum system	pressure, flow rate mixture composition, leakage valve position
5.	Blanket tritium system	tritium partial pressure gas flow rate and pressure mixture composition

TABLE XVIII-5 (cont.)

6.	Power supply and transfer	current, voltage, volt-second load
7.	Power generation system	coolant flow, pressure, temperature generator parameters heat exchanger parameters
8.	Test module	measurements concerning tests
9.	Power grid	conventional instrumentation
10.	Reactor room	various monitoring including tritium, neutron, γ-ray, X-ray magnetic field
11.	Environment	meteorological measurement tritium, neutron, γ-ray, X-ray
12.	Safety system	interlock system, access monitor

To cover the spatial distribution of essential plasma parameters for control, a few wide-view horizontal mid-plane ports and vertical centre-line ports have to be installed with clever shielding. They will be supplemented by several smaller ports.

Large ports for vacuum pumping or neutral beam injection may be commonly used for diagnostics. The blanket and shield structure will make the distance between the plasma and sensors large; in some cases it will be more than four metres. Clever design is necessary to reach a compromise between easy access and the radiation shielding for openings.

The instrumentation should be at a location allowing for easy maintenance. To keep high availability, convenient configurations for remote maintenance and repair have to be taken into consideration.

6. PHYSICAL ENVIRONMENT OF INSTRUMENTATION

All participants expressed concern about neutron and γ-ray irradiation which may cause radiation damage, induced activity and background noise. Some detectors such as magnetic probes, Rogowski coils, thermocouples or bolometers must be located in the vicinity of the plasma. Some of the diagnostics essentially need direct opening to the plasma or demand close proximity to it. As an example well known from the fission field, thermocouples will change their chemical composition and work function of the materials when irradiated by neutrons, which will change the reading. All the participants' reports mention the vulnerabilit of the optical systems and semi-conductor electronics to irradiation damage effects.

Careful selection of windows and optical fiber materials is necessary to avoid discolouration and excessive stress build-up. Semi-conductors will lose reliability under strong irradiation.

Table XVIII 6 shows the threshold damage levels collected for TFTR. Since INTOR is expected to perform D-T burning on the order of $10^7 - 10^8$ s for its life, these small threshold values imply a serious need for development of new detector techniques and special shielding designs.

Further information available from the fission field must be collected. However, the higher energy spectrum of the neutron, compared to fission reactors, will cause some different irradiation effects.

The following are physical conditions (other than radiation) which the INTOR instrumentation may encounter:

- temperature
- electric fields
- magnetic fields
- electromagnetic condition
- vacuum
- pressure
- space and geometric restrictions

Many of these environmental problems have been met before and will be resolved on the basis of experience at large tokamaks now under construction.

Some experience in deviation of thermocouple function by neutron irradiation are summarized in Table XVIII-7 for the total neutron dose and in Table XVIII-8 for the neutron flux. A strong magnetic field may also lead to additional errors in reading.

Radiation effects on electrical insulator material was assessed and it is found that some materials such as crosslinked polymers will be serviceable after exposure up to 1×10^9 R.

Concerning radiation effects, information from fission reactors, accelerators and the open defense literature are certainly very helpful. One of the participants proposed to form a small working group for the task of assessing the impact of the fusion reactor physical environment on instrumentation systems and identifying all relevant development items. INTOR teams should pay attention to this problem.

7. SAFETY SYSTEM AND RELIABILITY

To achieve a high availability in INTOR, it is necessary to analyse the reliability of each component of the instrumentation and the data processing and control system. Although a fail-safe system must be adopted, too many minor fail-safe

TABLE XVIII-6.* THRESHOLD DAMAGE LEVELS
PERMANENT DAMAGE EFFECTS

Detector or device	Ionization (rads)	Bulk (n/cm^2)	Bulk ($alpha/cm^2$)
Optical isolator (PD)	10^6	5×10^{10}	
Optical isolator (PT)	5×10^4	10^{10}	
Zinc sulfide		10^{14}	3×10^{12}
Indium antimonide (doped 10^{15} atoms/cm^3)	10^8	10^{14}	
Germanium bolometer (doped 10^{15} atoms/cm^3)	10^7	2×10^{13}	
Germanium, PV (gallium doped)	3×10^6	10^{12}	
Germanium, PC (gallium doped)	6×10^7	5×10^{13}	
Diamond		10^{14}	10^{12}
Charge coupled device (SC) (dark current)	10^4	10^{11}	
Charge coupled device (BC) (dark current)	10^4	5×10^{10}	
Silicon Thermistor (boron doped)		10^{14}	
RCA memory CDP1821	5×10^3	10^{15}	
Harris memory HM 16508	10^3	10^{15}	
RCA memory CD4061	10^5	10^{15}	
NMOS memory-4 kilobit	10^3	10^{15}	
Barium titanate	9.6×10^6	7.6×10^{10}	
Lead zirconate titanate	4×10^{10}	2.6×10^{18}	
Quartz		10^{19}	

NOTE: Channeltron temporary fatigue commences after 10^{10} counts, gain can be restored by a clean-up treatment.

* Brucker, G., TFTR Diagnostic Engineering Report No.3 (1978), cited in [3].

TABLE XVIII-7. DEVIATION OF THERMOCOUPLE FUNCTIONS BY NEUTRON IRRADIATION DOSE

Type of thermo-couple	Integral neutron flux (cm^{-2})	Irradiation temperature (°C)	Integral deviations (°C or %)	Type of reactor
Chromel-alumel	2×10^{18}	327	– 0.47	BEPO
Platinum-rhodium platinum 10% Rh	2×10^{19}	100	– 9	BEPO
Chromel-alumel	4×10^{20}	232 327 317	1.7–2.8 – (1–2) – (2.5–5)	ISPRA
Iron-constantan	4×10	232 327 417	2.5 – 1 – 6	
Chromel-alumel	4.5×10^{20}	1000	not observed	MTR
Platinum-rhodium platinum	4.5×10^{20}	1000	3%	MTR
Chromel-alumel	4.2×10^{20}	660	8–12	SM-2
Chromel-alumel	7×10^{21} fast	565	not less than 1%	EBP II

circuits will reduce INTOR's availability. Therefore, a high level of quality control is essential. Safety and availability should be carefully balanced. At the same time, reliability and lifetime of the separate system equipment may be essentially different. Therefore, a need arises to inspect the internals, to put into operation the redundant equipment, and to conduct preventive maintenance and repair of unserviceable equipment while the complex is operating. These functions are the system internal functions and they must be autonomous, i.e. they are not connected with the central control system of the plant. Such an approach allows (from the programme and equipment points of view) an easier solution of the control problems of both the separate systems and the complex as a whole.

The programme of investigations on tokamaks must include the study of the fundamental problems of complex control as well as determination of the computer complex reliability, including programme reliability. Internationally-agreed criteria of quality control and assurance should be established.

TABLE XVIII-8. DEVIATION OF THERMOCOUPLE FUNCTIONS BY NEUTRON FLUX

Type of thermo-couple	Neutron flux density (cm^{-2})	Radiation temperature (°C)	Deviations	Reactor
Chromel-alumel	1×10^{13} (fast neutron)	80	– 5.6°C	Oak Ridge X-10
Chromel-alumel	2.6×10^{13}	100, 232, 327	$\leqslant$ 58 V	AF-NEFR
Iron-constantan Chromel-alumel	1×10^{12}	232, 327	not less than 0.3%	BEPO
Platinum-rhodium platinum	thermal neutrons			
Copper-constantan Chromel-alumel	3.3×10^{14} thermal neutrons	600	+ 26°C	SM-2
Tungsten-rhenium	3.7×10^{14} 9.9×10^{5} P/s	660	– 16°C	SM-2

The safety system must be separated from the control system with hard-wired interlock systems. Each subsystem should have its own hard-wired, fail-safe system whose status is monitored by this system. Essentially, power for the control system and the safety system should be supplied by an uninterruptable power supply. Such kinds of information from large tokamaks are helpful for the safety system of INTOR.

One of the participants also proposed formation of a small working group dealing with systems safety and reliability studies of pertinence to INTOR.

The Safety and Environment Group has made an assessment of personnel safety (see Chapter XIX).

8. R AND D REQUIREMENTS

Required R and D in the field of diagnostics, data acquisition, and control are summarized in Table XVIII-9.

TABLE XVIII-9. R AND D NEEDS ASSESSMENT - DIAGNOSTICS, DATA ACQUISITION AND CONTROL

	Status *	
	Adequate	Inadequate
High-β control, density and α-particle plasma diagnostics	x	
Transient noise and damage measurements of detectors and transducers		x[1]
Radiation damage and thermal studies of optics system, vacuum seals, and insulator materials		x
Blanket nuclear measurement technique	x	
First wall and blanket instrumentation techniques in tokamak environment		x
Feedback control computer algorithms	x	
Reliability evaluation of control system	x	

[1] The plasma and engineering monitors must be assessed in detail to determine precisely what the detailed R and D work will be.

Adequate: Programmes are either planned or could be planned on existing or authorized facilities (with some upgrading if required) and using existing trained personnel.

Inadequate: Either the available and authorized manpower and facilities are inadequate or a substantial increase in effort is needed.

* The ratings are made on the world-wide fusion effort and not separately on the individual national programmes.

9. CONCLUSIONS AND RECOMMENDATIONS

The impact of the general features of INTOR are discussed and the control and data processing systems are qualitatively reviewed. Furthermore, new features provided by long D-T burning, accessibility to the tokamak torus and the reactor environment of diagnostic instruments are briefly discussed. In addition, safety system, reliability, and necessary redundancy are mentioned. Probable issues from INTOR's international features are briefly reviewed.

No essential differences were found in Workshop participants' observations in this field of INTOR. However, this area is particularly dependent upon information and input from all other groups, which are required to make further assessments more quantitative and definite.

The number of diagnostics should be minimized during the latter stages of INTOR to be consistent with proto-typical reactor operation. The control of reactor engineering components such as the blanket, will be similar to that in industrial plant and fission reactor systems.

Time sequential procedures of plasma control and other separate systems were briefly analysed. The diagnostics applicable at each operation step are qualitatively reviewed.

No essential technological problem exists regarding the control and data base hardware because of its current flexibility and the expected rapid advances in this field, although there might be some additional technical requirements such as faster data transition time. More development of software and algorithms for feed-back analysis is required for INTOR.

The philosophy of the total control system should be a compromise between hierarchial centralization and redundant de-centralization to prevent trivial faults from reducing the total system availability.

Investigation of effects of irradiation and the tokamak environments on instrumentation including detectors, sensors, cables, optical system, etc., are strongly recommended. Information from fission reactors, accelerators, and the open defense literature is very helpful for irradiation effects on instrumentation.

Formation of a small working group for the task of assessing the impact of the fusion reactor physical environment on instrumentation systems and to identify all relevant development items is recommended.

The safety system must be separated from the control system with hard-wired interlock systems. Formation of a small working group dealing with system safety and reliability studies of pertinence for INTOR is proposed.

From past experiences, one can expect conflicts between diagnostic requirements and the limitations imposed by the tokamak structure. Joint efforts by physicists, control engineers, and machine builders are required from an early stage in order to provide adequate access to the instrumentation, which must be resolved by further research and development efforts.

Internationally-agreed-upon criteria for quality control and assurance in this field of INTOR are strongly recommended to keep its availability high. Furthermore, this field requires international interchangeability both for hardware and software. It is recommended that INTOR shall impose an international standard for specifications in this field after review of an international group.

Generally speaking, the field of diagnostics, data acquisition and control is feasible for the initiation of INTOR construction in about 1984–5 although there is some remaining ambiguity about irradiation and tokamak environmental effects on instrumentation.

REFERENCES TO CHAPTER XVIII

[1] European Contributions to the 2nd meeting of the INTOR Workshop, Diagnostics, Data Acquisition and Control, Chapter XVI, EUR FU/XII 501/79/EDV 50, June 79 (2 Volumes).

[2] Fusion Research Development Center, INTOR Workshop Report , Group 16, Diagnostics, Data Acquisition and Control, JAERI-M 8514 (1979).

[3] "Diagnostics, Data Acquisition and Control", Chapter inUSA INTOR Report (see Appendix A.5.4.).

[4] USSR Report to INTOR, "INTOR Diagnostics, Control and Data Processing" (see Appendix A.5.3 [44]).

Chapter XIX

SAFETY AND ENVIRONMENT

G.L. KULCINSKI – USA, Chairman
F.N. FLAKUS – IAEA, Expert
D. LEGER – EC
K. SAKO – Japan
G. SHATALOV – USSR

1. INTRODUCTION AND SUMMARY

The purpose of this document is to present information relative to the adequacy of the data base to permit construction and operation of a safe and environmentally acceptable International Tokamak Reactor (INTOR). Both operational safety and safety of the general public are considered.

1.1. General discussion of INTOR safety issues

In any deuterium-tritium burning magnetic fusion reactor there are several safety issues that must be addressed at the outset in that these issues greatly affect the siting, design, operation, maintenance, and ultimately, decommissioning of the reactor. The major issues involve:

(1) Use of tritium in the fuel cycle.
(2) Activation of structural materials, fluid streams, and the reactor hall environment by high energy neutrons produced in the fusion process.
(3) The requirement for use of lithium to breed tritium and the attendant fire potential.
(4) The presence of large magnetic fields and their potential effects on personnel, structures, equipment and the environment.
(5) The potential use of non-radioactive toxic materials, and
(6) Potential hazards associated with an operating plant employing high vacuum technology, large cryogenic systems, and high voltage systems.

The following brief comments are made on the applicability of each of these issues to INTOR.

Tritium. The parameters for INTOR from the various countries specify a tritium flow rate of 700 to 4500 g/d, an 'active' inventory of 0.2–3 kg, 'passive' inventory of 1–2 kg and a maximum annual consumption of 4 to 16 kg. Thus, the plant will have approximately 3×10^7 Ci of tritium on hand at any one time with a maximum annual consumption in excess of 10^8 Ci. This implies that before INTOR can be operated with these quantities of tritium, significant development is required in tritium handling systems, cleanup and decontamination, transportation, storage, accountability, and personnel protection. Much of these needed data will be forthcoming from the Tritium Systems Test Assembly (TSTA), which is to begin operation at Los Alamos Scientific Laboratory (LASL) in 1981. The Japanese Tritium Engineering Test Laboratory, to be constructed in FY1983-84, and the Soviet Vacuum/Tritium Stand will also contribute to this data base. In addition, international agreements on tritium supply and safeguards for INTOR must be negotiated.

Activation products. From the power level (620 MWt), structural materials (steel), neutron wall loading (1.3 MW/m^2), and other operating parameters for INTOR, it is expected that an inventory of activation products of approximately 10^9 Ci will be generated. Also, considerable activated corrosion products may be expected to be circulated in the water coolant streams, and neutron streaming through penetrations may result in activation of exterior systems. To ensure that operating personnel are not subjected to excessive doses from this radiation source, careful design of shielding must be accomplished, neutron streaming reduced to a minimum, and sophisticated remote maintenance procedures and equipment developed. The systems must be designed for a high degree of reliability to avoid frequent corrective maintenance jobs that could increase personnel exposures. During the design of INTOR attention must be given to energy sources that could potentially lead to release of these activation products in accident situations and suitable protective and accident-mitigation systems provided. Experience with operating D-T reactors such as the Tokamak Fusion Test Reactor (TFTR) and the Joint European torus (JET) is vital to the design of INTOR systems for minimizing personnel doses from activation products.

Lithium. It has been indicated that liquid lithium will probably not be used extensively in INTOR. However, it is assumed that lithium will be used in some form in blanket breeding modules. Research will be required with regard to the fire potential for the particular form of lithium to be used. If a fire potential exists, suitable fire prevention techniques such as use of a reactor building inert atmosphere and liners for concrete should be used. Also, additional research may be needed on lithium fire extinguishment techniques. Research underway at the Hanford Engineering Development Laboratory (HEDL) in the US with respect to lithium spill experiments should provide the required data base for this purpose.

Magnetic fields. The reference design of the toroidal field coils specifies a maximum field in the range from 10 to 11 T to be generated by superconducting magnets. In addition, fringing fields from poloidal coils could extend some distance from the device. The current knowledge of the effects on human beings of long exposures to the magnetic fields that will exist at the reactor site is uncertain. Suitable exposure guides and dosimetry methods need to be developed, which may require additional research. The impacts of the magnetic fields beyond the reactor site boundary must be evaluated along with the necessity for magnetic shielding. Consideration must also be given to the effect of the magnetic field on reactor structures, installed equipment, and instrumentation. The large size of the magnets, cryogenic system, stored energy in the field and power required lead to several postulated accident sequences that could do damage to the plant and potentially lead to radioactive releases. Research is required to increase the reliability of the magnets and to develop techniques to minimize the reactor damage that could result from accidents such as loss of magnet cooling or electrical arcing. Also, comprehensive analysis techniques need to be developed and verified for use in magnet safety and accident analysis. Research to be initiated in the Large Coil Programme in 1982 at Oak Ridge National Laboratory (ORNL) in the US and the construction of the T-15 facility in the USSR will provide much of the needed data in the superconducting magnet reliability and safety area.

Toxic materials. The reference parameters did not specifically identify any non-radioactive toxic materials that would be required for INTOR. Such materials will need to be identified during reactor system and blanket module design. For example, beryllium or lead are toxic materials that potentially could be used in blanket modules. The use of mercury diffusion pumps also needs to be examined. It is expected that technology for control of such materials is well within the current state-of-art.

Vacuum, cryogenic and electrical system. The nature of a Tokamak reactor is such that significant occupational hazards may exist owing to the large vacuum, cryogenic, and electrical power distribution systems required. However, careful design employing recognized safety and design codes and standards should bring these occupational safety problems well within an acceptable risk level. Event and/or fault trees should be constructed to identify accident sequences whereby failure of these systems could be initiators that lead to release of tritium or activation products and appropriate design steps taken. Failure detection should be included in the design of the various systems.

1.2. Summary of data base assessment

Regulations appropriate to INTOR. To date, fusion reactors have been experimental devices that have not been subjected to the strict regulatory scrutiny that would be applied to a full-scale commercial power plant. Thus few regulations specific to the fusion field have been developed. Instead, regulations developed for other technologies have been selected on an ad hoc basis, generally by agreement between the government and the contractor for the particular device. The regulations that would be appropriate for INTOR are discussed relative to tritium handling and release, activation product release, radiation exposure, release of toxic materials, magnetic fields, design for natural disasters, and other environmental impacts.

Safety-related design codes and standards. Again, because of the experimental nature of reactors built to date, a generally applicable set of safety-related design guides has not been developed for the fusion field. Because commercialization of fusion is still several decades away, a set of fusion – specific regulatory codes and standards will likely not be available for INTOR design. Nevertheless, INTOR design provides the opportunity to start laying the groundwork for future regulatory guides and standards for the fusion industry. As an example, this document examines the guides and standard presently being used for fission reactors in the US and discusses their applicability to INTOR.

Accident analysis. Considerable investigation has gone into identification of accident sequences for the tokamak fusion industry. Based on conceptual designs of commercial plants, event trees have been constructed to identify accident scenarios as an aid in improving the safety aspects of the next generation of conceptual designs and quantitative assessment of accident consequences. These event trees are not usable at this time to determine defendable values of risk (probability of an accident times the consequences) because of a lack of specific reactor design details and lack of an adequate data base for component failure rates. Such a data base must be developed; however, it will not likely be sufficiently developed in time for quantitative application to INTOR. The state of development of mechanistic computer models for fusion reactor accidents is at an early state. Further development is needed if defendable safety analyses are to be done for INTOR other than "worst case" analyses. The codes will also need to be verified by comparison with experimental data from well characterized and instrumented safety-related experiments, which are just starting to be performed.

Containment requirements. Considerable data, experience and techniques are available from the fission reactor industry to allow design of containment buildings. The designs can be made so that the building can maintain its containment integrity under a variety of adverse natural phenomena (earthquakes, tornados, floods, etc.) and accidents (overpressurization, explosions, missiles). Undoubtedly the most difficult containment task will be containment of tritium. The severity of this task is not known at this time because there are no specific release limits for fusion reactors, and guidance from internationally constituted organizations is urgently needed in this area. A variety of techniques can be employed to control tritium releases including use of glove boxes, double- or triple-walled pipe, ceramic barriers, converters and getters, and separate compartments for tritium handling and storage components. Data from the TSTA and the Japanese tritium Engineering Test Laboratory will provide vital input to establishing the INTOR tritium containment philosophy and approach.

Safeguards requirements. The only material at INTOR that might be potentially subjected to safeguards requirements is tritium. In the United States for example tritium is currently designated as a sensitive rather than a strategic nuclear material. It is considered a limited resource and is quite valuable (4000–8000 $/g); therefore, it is subjected to rigorous materials management programmes. Current control levels in the US are to the nearest 0.01 g. Additional tritium inventory techniques will need to be developed if a 16 kg per year consumption rate in an operating tokamak is to be controlled to this level.

These five areas are discussed more fully in the following sections of this report.

2. SAFETY AND ENVIRONMENTAL PROTECTION REGULATIONS APPROPRIATE TO INTOR

2.1. Licensing requirements

2.1.1. European requirements

Licensing procedures for a fusion device will almost certainly follow the same lines as those for other nuclear installations. A description of the different national systems is given in “Licensing Systems and Inspection of Nuclear Installations in NEA Member Countries” [1].

A general requirement for the licensing of a nuclear installation is that design, construction and operation are carried out in accordance with a Quality Assurance Programme.

Quality is defined as the totality of features and characteristics of a product or service that bears on its ability to satisfy a given need. Quality assurance is defined as all those planned and systematic actions necessary to provide adequate confidence that an item or a facility will perform satisfactorily in service.

The technical standards required for quality, safety and reliability are described in codes [1]. The codes contain specifications for materials which will fulfil the standards under different operational conditions. The codes have a special section concerning materials used in nuclear plants. The first part of a

Quality Assurance Programme is a detailed description of all measures an organisation plans to take (or has taken) to ensure that the quality, safety and reliability of any item of the plant are up to the specified standards. New materials will probably be used in INTOR and it will be necessary to produce a Quality Assurance Programme for these including a research programme.

2.1.2. United States requirements

The legislation abolishing the Atomic Energy Commission (AEC) and establishing the Energy Research and Development Administration (ERDA) and the Nuclear Regulatory Commission (NRC) [4] provided for exemption for ERDA and its prime contractors from reactor licensing requirements. These exemptions were clarified in subsequent regulations governing NRC operation and in legislation setting up the DOE [4].

The codes, standards, requirements and guidelines applicable to INTOR will be those specified in the operating manual of the DOE (formerly the AEC/ERDA manual), most particularly those chapters on environment, safety and health [4]. This manual specifies certain other standards and requirements, e.g., ANSI standards, Code of Federal Regulations (CFR's, Nuclear Regulatory Guides, etc. that are to be met.

The National Environmental Policy Act of 1969 (NEPA) and Executive Order 11752 require an environmental impact assessment for major projects, hence environmental impact statements (EIS, draft and final) would be required for INTOR. Finally, Chapter V of the ES&H Manual (ERDAM 0531) requires a comprehensive safety analysis report (SAR, preliminary and final) that includes operational safety requirements (OSR's) [4].

2.1.3. USSR requirements

The regulations for the design and location of industrial enterprises in the Soviet Union are cited in reference [3]. This document specifies the location of fission installations and other industries which use radioactive substances. The transportation and storage of radioactive materials as well as the air and water purification standards are also included in that report. It is anticipated that fusion reactor will be subject to many (but not all) of the restrictions in reference [3].

2.2. Release of activation products

2.2.1. European requirements

The various European governments all have their own regulations pertaining to the release of activation products. Examples of such regulations in the UK and Federal Republic of Germany are given in the European reports cited in the introduction.

2.2.2. United States requirements

Activation products resulting from fusion reactions may be released to the air and water near the reactor site. The regulations governing acceptable levels of release are found in Chapter XI of the DOE ES&H Manual (formerly ERDAM Chapter 0524). These guidelines are expressed in terms of concentrations of a particular radioisotope (pCi/cm^3) in air or water for both controlled and uncontrolled areas. The guidelines are conservative in that they assume continuous exposure to the concentration for 40 hours per week 50 weeks per year. In the case of exposure to radioactive gases such as nitrogen and argon, the guidelines assume exposure to a semi-infinite cloud, again a conservative requirement for a tokamak environment.

With the exception of plutonium, no DOE standards exist governing the quantity of radioactivity releasable in solid materials (assuming of course that the external radiation standards are satisfied and that the solid is not in an easily dispersable form). Guidance is needed in this area to aid in material selection during design and in disposal of slightly radioactive scrap materials.

2.3. Release of tritium

2.3.1. General considerations of tritium

2.3.1.1. Physical properties of tritium

Tritium (^3H or T) is a radioactive isotope of hydrogen. It decays to helium (^3He) by emission of a beta particle (yield 100%) with a maximum energy of 18.0 ± 0.5 keV and an average energy of 5.69 ± 0.04 keV. The half-life for radioactive decay is 12.36 ± 0.03 years. There are 9620 Ci (356 T Bq) per gram of tritium [5].

2.3.1.2. Toxicity of tritium

Retention of tritium depends on the chemical form in which it enters the human body. The hazard potential of elemental tritium (HT or T_2) of tritium oxide (HTO or T_2O) and of other compounds of tritium (e.g. tritiated methane) must be considered separately.

2.3.1.3. Elemental tritium

More than 98.4% of inhaled elemental tritium is rapidly exhaled by humans [6]. Absorption of tritium gas through the skin is of minor hazard significance. Exposure to elemental tritium in air is limited by consideration of stochastic effects in the lung.

The radiation dose resulting from the brief residence of inhaled elemental tritium in the lung and the slight solubility of elemental tritium in body fluids is considerably smaller than the dose resulting from the inhalation of the same amount of tritium in the form of the oxide.

2.3.1.4. Tritium oxide

Tritium oxide, whether inhaled, ingested or absorbed through the skin is rapidly and completely absorbed and equilibrated with body water and uniformly distributed among all the soft tissues of the body so that the concentration in sweat, sputum, urine, blood is the same.

It is assumed that the intake-rate of tritiated water through skin is equal to half the intake-rate by inhalation in an atmosphere contaminated by tritiated water (Reference Man at work) [7]. Since body water is quite uniformly distributed throughout most tissues a uniform total body radiation dose is delivered by the tritium in body water.

2.3.1.5. Tritium compounds other than the oxide

In an uniformly tritiated environment about half the human tritium intake would be in the form of tritiated organic compounds in food [8]. Most of this organically bound tritium would be converted within the organism to tritium oxide by oxidation or exchange, but some tritiated organic compounds would be incorporated directly into body tissues. Other tritiated organic compounds would be formed within the body by exchange or synthetic processes. This organically bound tritium would add perhaps 50% to the total-body dose from body-water tritium. Two possibilities exist for enhanced effect:

(i) a transmutation effect due to conversion of a hydrogen atom to a helium atom; and,

(ii) an enhanced radiation effect due to origin of the beta ray within the molecule.

Extensive studies of this problem however indicate that the effects of intracellular tritium are mostly due to beta irradiation of the cell nucleus (90%). It appears that tritium incorporation into organic compounds poses no substantially increased hazard beyond that accounted for by its contribution to total-body dose. The chronic radiation exposure from environmentally dispersed tritium in whatever form, is essentially a total body exposure.

2.3.1.6. Conversion of elemental tritium into tritium oxide

Releases of elemental tritium would be of much less concern than releases of tritium oxide, and the question arises how fast elemental tritium is being converted to the oxide by environmental processes? This is clearly an important question since T_2 is probably the most likely release form from fusion reactors. This conversion of elemental tritium to tritium oxide in air either by oxidation or exchange processes is very slow, even in the presence of metal catalysts (years rather than days or hours) [9]. Measurements of HT and HTO ratios following accidental releases offer qualitative information on conversion rate [9]:

Year	Amount	Form	Location
1970	289 kCi	T_2	30 m exhaust stack at Lawrence Livermore Laboratory
1974	479 kCi	T_2	60 m exhaust stack at Savannah River Plant*

The behaviour of tritium in these two cases argues for the low conversion rate. Nevertheless, further research work on tritium oxidation and exchange is needed [10].

2.3.1.7. Relative Biological Effectiveness (RBE) of Tritium Beta Radiation

The linear energy transfer (LET) of weak betas such as those from tritium (18 keV maximum) is higher than that of more energetic beta-, x- or gamma-radiation. A considerable experimental effort has been devoted to determining whether this higher LET is reflected in an increased RBE. ICRP in its 1959 report used a RBE value of 1.7 for tritium. However, this value was reduced to unity in 1969. More recently there has been further evidence presented to justify a value higher than unity but in no case greater than 3 [8, 11].

At the present time the Effective Quality Factor Q = 1 is used for radiation protection purposes [1].

* The highest HTO conversion was found at the centre of the release plume about 5 hours after the release. The total tritium concentration at this point (about 40 km from the release point) was 3.9×10^{-10} Ci/ℓ of which 0.2% was HTO.

2.3.1.8. Biological half-life of tritium

The biological half-life of tritium in man is 10 days (ICRP Reference Man). Under certain medical circumstances, this can be reduced to 1 day.

2.3.2. Other sources of tritium in the environment

2.3.2.1. Naturally produced tritium

The natural production of tritium by interaction of cosmic rays with nitrogen, oxygen and argon has been estimated to be 4–8 MCi/year. This natural production rate results in a total equilibrium inventory in the biosphere of about 17–170 MCi, the most recent estimates being 34 MCi and 28 ± 7 MCi.

About 99% of this tritium inventory is converted into HTO and participates in the normal water cycle.

Measured concentrations of natural tritium in ocean surface water average about 3 pCi/litre. Tritium concentrations in fresh water are usually higher than in sea water (6–24 pCi/litre).

2.3.2.2. Atmospheric testing of nuclear devices

It is estimated that about 3600 MCi and 900 MCi of tritium were released into the northern and southern hemisphere, respectively, from tests of nuclear devices [12].

2.3.2.3. Releases from the operation of Fission Power Reactors

Average discharges of airborne tritium based on operating experience are, for PWRs and GCRs 0.2 Ci. for BWRs 0.05 Ci and for HWRs 15 Ci per MW_e year. The highest airborne tritium release rates reported for a fission reactor are 36 500 and 24 800 Ci/year (HWR, Pickering in 1973 and 1974 respectively) [12]. The rate of tritium production from a fission reactor economy is expected to exceed the rate of production of natural tritium by about the year 2000.

2.3.3. Environmental behaviour of tritium

Approximately 99% of the world's inventory of tritium exists as HTO. Although most of it is in the ocean, a significant portion still resides in the stratosphere. In the troposphere, most of the tritium is in the form of HT, a smaller amount as HTO, and a much smaller but still significant amount as CH_3T. Further, the T/H ratio in the troposphere is higher in hydrogen and in methane than it is in water vapour [9].

For the conversion of elemental tritium, photochemical oxidation may be the predominant conversion mechanism and over larger periods of time may combine with bacterial action to serve as the principle pathways of conversion of HT (or T_2) and CH_3T to HTO or other more reactive forms of tritium.

Tritium released to the environment in the form of tritiated water behaves as ordinary water throughout the hydrosphere and is readily metabolized by both terrestrial and aquatic biota.

Most biological reactions descriminate against the incorporation of tritium in favour of protium. Isotopic effects of tritium in most exchange reactions are negligible. In most ecosystems, tritium concentration tends to become more dilute at levels of greater organization within the food chain [8]. There occurs no significant concentration above that of environmental concentration in water or forage at any level of the food chain.

2.3.4. *European requirements*

If tritium is released to the environment as the only isotope, an amount of up to 10 g (3.6 10^{15}Bq = 97 kCi) can be released, in most places, without giving rise to dose equivalents to the population exceeding any ICRP-limits. Routine releases must be kept far below that value. Accidental releases should be avoided or at least restricted to amounts below 10 g/y.

The doses to the most exposed individuals in the population (i.e. the critical group at 500 m from release point) due to airborne releases of tritium are:

(a) single release: 6.10^{-8} rem/Ci (1.6 10^{-20} Sv/Bq) (only from inhalation pathway);

(b) continuous release: between 3.10^{-7} rem/Ci and 2.10^{-6} rem/Ci (0.8–5.4 10^{-19} Sv/Bq). (obtained from both inhalation and ingestion pathways).

2.3.5. *Japanese comments*

The maximum release of tritium is to be limited to an off-site dose of less than 500 mrem/year. Reducing that exposure level by 100 for an exposure to individuals 0.5 km away from a release 50 m high translates into 10^4 curies per year (1 g/year).

2.3.6. *Comments from the IAEA perspective*

2.3.6.1. Setting tritium release limits for INTOR

Based on the recommendations of the International Commission on Radiological Protection (ICRP) the Agency formulates Safety Standards, operational standards, specialized regulations and codes of practice in the field of radiation protection. Although these mostly represent recommendations for Member States only, they are binding for the Agency's own nuclear activities.

2.3.6.2. The ICRP dose limitation system

In 1977 ICRP has issued its new recommendations of a dose limitation system [17] and has referred to the following three basic principels of protection:

i) justification of a practice; no practice shall be adopted unless its introduction produces a positive net benefit (risk-benefit analysis).
ii) all exposures shall be kept as low as reasonably achievable (ALARA), economic and social factors being taken into account.
iii) limitation of individual risk; the dose equivalent to individuals shall not exceed the limits recommended by the Commission (boundary conditions).

2.3.6.3. Practical implications of the ICRP recommendations

Difficulties are being experienced in implementing the new ICRP recommendations in practice. However, some guidance has already been prepared by IAEA (Safety Series No.45, "Principles for Establishing Limits for the Release of Radioactive Materials into the Environment" [13] and "Topical Seminar on the Practical Implications of the ICRP Recommendations [14, 15] as well as by ICRP (Limits for Intake of Radionuclides by Workers [16]). Further guidance is being prepared [17].

2.3.6.4. Annual dose equivalent limit

The following is an excerpt of the draft Agency's Revised Basic Safety Standards [17].

No individual shall be inequitably exposed, as a result of controlled sources and practices, in excess of the limits laid down in Table XIX-1.

In relation to the dose equivalent limits, the dose equivalent to be controlled shall not be held to include contributions from any medical irradiation of patients or from "normal" natural radiation. In relation to limits authorized for a given practice or installation, only the change in exposure due to the considered practice shall be taken into account and not the pre-existing exposure level.

The limits laid down in Table XIX-1 shall apply to the sum of the dose equivalents resulting from external exposure during one year, and the life-time dose equivalent resulting from that year's intake of radio-nuclides. In the case of occupational exposures, the dose equivalent resulting from one year's intake shall be taken to be the **committed dose equivalent** H_{50} [1].

TABLE XIX-1. DOSE EQUIVALENT LIMITS

	For occupational exposure, mSv	For exposure of members of the public*, mSv
Effective dose equivalent limit	50*** (5 rem)	5** (0.5 rem) [1 (0.1 rem)]

* These limits apply to doses attributable to exposures resulting from justifiable and optimized practices. The exposure from any single source or practice should be less than the annual limits so as to allow for present and future exposures from other sources or practices.
** For stochastic effects, the limit of the annual effective dose equivalent shall be 5 mSv (0.5 rem) and is related to the exposure of critical groups. In any rare cases where the doses to the same individual approach this limit over many years, it shall be prudent to take measures to restrict their life-time dose equivalent to a value corresponding to an average annual effective dose equivalent of 1 mSv (0.1 rem).
*** In case of uniform exposure of the whole body, the distribution of the dose equivalent limit in large occupational groups has been shown very commonly to fit a log normal function with an arithmetic mean of about 5 mSv (0.5 rem) and with very few values approaching the limit. Long continued exposure of considerable proportion of the worker at or near the dose equivalent limits would only be acceptable if a careful cost-benefit analysis had shown that the higher resultant risk would be justifiable.

2.3.6.5. Use of old terms MPC and MPBB

It must be emphasized here that the use of the terms 'Maximum Permissible Concentration (MPC)' of a radionuclide in air or water as well as the term 'Maximum Permissible Body Burden (MPBB)' is discontinued. Instead the terms 'Annual Limit of Intake (ALI)' and 'Derived Air Concentration (DAC)' were introduced by ICRP.

The Commission believes it is sufficient to limit the intake of radioactive materials in any year to the recommended limits and does not recommend any further restrictions on the rate of intake. For the sake of completeness the old values for tritium are recalled here:

MPC_{air}	Elemental tritium in Ci/m^3	Tritiated water
workers, 40 h/week	2×10^{-3}	5×10^{-6}
member of public	4×10^{-5}	2×10^{-7}

MPBB (occupational exposure) = 10^{-3} Ci (corresponding to 0.1 rem/week of $0.028\ Ci/m^3$ in body liquid of standard man)

2.3.6.6. ALI and DAC values for tritium

Guidance on metabolic information, biological models as well as on ALI and DAC values are now in preparation by the ICRP Committee 2 and according to our information are expected to be available soon in published form. The values presented below are preliminary and apply to workers [16].
DAC (Submersion) for elemental tritium:

Exposure to elemental tritium in air during any year is limited by consideration of stochastic effects in the lung.

$$\text{DAC (Bq/m}^3\text{)} = 2 \times 10^{10}$$

ALI and DAC values for tritiated water:

Mode	Limit type	ALI (Bq)	DAC (Bq/m³)
Ingestion	Stochastic	3×10^9	
Inhalation	Stochastic	3×10^9	8×10^5

As far as exposure of the public is concerned, there are at present no ALI or DAC values available. Occupational ALI or DAC values can not be directly translated into values applicable for the public.

2.3.6.7. Doses from tritium already in environment

The chronic radiation exposure from environmentally dispersed tritium, in whatever form, is essentially a total body exposure.

Natural tritium

Assuming that the specific activity of tritium in the body is the same as that in the continental surface waters and using the distribution of hydrogen in the Reference Man of ICRP, the annual doses from natural tritium are found to be of the order of 1 μrad in each of the following four organs and tissues: Gonads, Lung, Bone Lining Cells, Red Bone Marrow [1].

Using the concept of effective dose equivalent as defined in ICRP 26 [17] and using the appropriate weighting factors and effective quality factor, the effective dose equivalent H_e is 0.77 μrem.

Tritium from fall-out

The **dose committment** in the northern and southern hemisphere from fall-out tritium is estimated to be 2 mrad and 0.2 mrad respectively [12].

The **collective dose committment** due to tritium from nuclear explosions is estimated to be about 8×10^6 man rad, corresponding to about 3×10^3 man rad per curie of released activity [12].

Tritium from fission power reactors

The estimated **local collective dose** from **airborne** tritium per unit activity released is 2×10^4 man rad per Ci. The estimated **local collective dose** from **waterborne** tritium (drinking water intakes) is 7.10^{-3} man rad per Ci [12].

Using the above values of collective dose per unit release, the estimated **local collective doses,** for the assumed local situation are as follows:

Reactor type	Local collective dose in man rad/$MW_e \cdot y$ from airborne tritium	from waterborne tritium
PWRs (and GCRs)	4×10^{-5}	7×10^{-3}
BWRs	1×10^{-5}	7×10^{-4}
HWRs	4×10^{-3}	3×10^{-2}

Tritium releases from CANDU stations are of the order of 0.1% of the derived release limits leading to doses of individuals in the vicinity of the station of about 1 mrem/year [18].

2.3.6.8. Doses from hypothetical tritium releases

For a continuous **tritium oxide** release of 1 Ci/day from a hypothetical fusion reactor located at the Oak Ridge National Laboratory (ORNL) site the resultant **whole body dose** is 3.10^{-3} mrem/year to the **closest resident** [25].

If all 3.10^7 Ci of tritium contained in the UWMAK-1 reactor blanket and coolant is released as oxide at a height of 50 m, the resultant **dose committment to the nearest resident** will be 1 rem [19].

A release rate of 1 MCi per year of tritium in the form of HTO to the atmosphere for 40 years leads to an estimated **world wide collective dose** of 2×10^4 man rem in the 40th year and a **total accumulated dose committment** of 1×10^6 man rem [20].

Assuming a release rate of 1 Ci/year of HTO to the atmosphere, the annual **collective dose to the population of the world** was estimated to be 0.024 man rem [20].

2.3.6.9. Approach for setting tritium release limits for INTOR

Environmental models will have to be used to allow establishment of relationships between discharges, environmental levels and resulting doses, making it possible to relate releases of tritium into the environment at the site selected for the location of INTOR to radiation protection requirements, given as dose commitment limits. A derived limit (DL) for release of tritium into the environment, defined as the annual input of radioactivity of specified composition which will result in a dose commitment in the critical group equal to the recommended annual dose limits, will have to be established. The critical group will have to be identified, i.e. the group of people which would likely receive the highest doses (but in which the average dose does not exceed the dose limits).

Furthermore, the establishment of DLs will have to comply with ALARA. This would require optimization of radiation protection by use of techniques such as cost-benefit analysis.

In practice, release limits are usually set at levels which will only correspond to small fractions of the relevant environmental DL. In addition, allowance must be made for releases from other sources, including the foreseen development with time [13].

As far as environmental modelling for the behaviour of radionuclides is concerned, IDRP Publication 29 – recently published – provides substantial theoretical and some practical guidance [21].

2.3.7. United States requirements

The guidelines applicable to the routine exposure of personnel to tritium in air are 5 pCi/cm^3 for controlled areas and 0.2 pCi/cm^3 for uncontrolled areas. Detailed information concerning the meteorological dispersion of releases is necessary in order to convert the above numbers into meaningful release values. The meteorological data must be specific to the tokamak site. Since the above numbers assume that the tritium is in its most biologically hazardous form, i.e. the oxide HTO, they are conservative and guidelines more appropriate to real tokamak releases, i.e. mixtures of gases, HT, and oxides, HTO, need to be formulated.

Continued concern about the exposure of the general public to low levels of radioactivity has resulted in more stringent requirements than the above quoted regulations. While not a DOE requirement, Appendix I of 10 CFR 50 limits routine exposure in uncontrolled areas to 5 mrem/year and the philosophy implied in this regulation is being applied to other areas. For purposes of the INTOR a value of 10 mrem/year at the site boundary from all sources of radioactivity with 5 mrem/year of this reserved for tritium releases, could be adopted.

Exposure of personnel during accidental releases of radioactivity are governed by guidelines specifying 25 rem as a once-in-a-lifetime occurrence. However, DOE guidelines for TFTR specify 5 rem for extremely low probability accidents (10^{-7} per year) and 1 rem for low probability accidents (10^{-5} per year) and these levels would probably be acceptable for INTOR.

It should be assumed for INTOR that the present concern over releases of radioactivity will continue and that, while not required, serious consideration should be given to reducing releases to as near zero as is practicable. Again, no standards exist governing the amount of tritium acceptable in various solids and work is necessary in this area.

2.4. Radiation exposure of operating personnel and of the general public

2.4.1. United States requirements

Chapter XI of the DOE ES&H Manual (formerly ERDA Manual 0524) would be directly applicable to INTOR. This manual chapter limits occupational radiation exposure to the following levels with design objective exposures specified to be below 20 percent of the stated levels (see list on p. 576).

2.4.1.1. Procedural requirements

(1) Restrictions

(a) An individual under age 18 shall neither be employed in, nor allowed to enter, controlled areas in such manner that he will receive doses of radiation in amounts exceeding one-tenth the standards in (1) above.

(b) Students under age 18 exposed to radiation during educational activities shall not exceed 0.1 rem/year. This exposure shall be considered a part of the 0.5 rem/year limit for workers under age 18 and not supplemental to it.

(2) Combining internal and external dose

Current year whole-body internal dose commitment from radionuclides for which the whole body is the critical organ must be combined with the external whole-body dose. Where both the external penetrating dose and internal dose to critical organ are known, they shall be combined for that organ.

(3) Emergency or accidental exposure

Radiation doses received in emergency actions or accidental situations will be chargeable to the radiation exposure status of the individual. However, the

UNITED STATES RADIATION PROTECTION STANDARDS FOR EXTERNAL AND INTERNAL EXPOSURES OF RADIATION WORKERS

Type of exposure	exposure period	Dose equivalent (dose or dose commitment[a] (rem))
Whole body, head and trunk gonads, lens of the eye[b] red bone marrow, active blood forming organs	Year	5[c]
	Calendar Qtr.	3
Unlimited areas of the skin (except hands and forearms)	Year	15
	Calendar Qtr.	5
Other organs, tissues, and organ systems (except bone)	Year	30
	Calendar Qtr.	10
Forearms[d]	Year	30
	Calendar Qtr.	10
Hands[d] and feet	Year	75
	Calendar Qtr.	25

[a] To meet the above dose commitment standards, operations must be conducted in such a manner that it would be unlikely that an individual would assimilate in a critical organ, by inhalation, ingestion, or absorption, a quantity of radionuclide(s) that would commit the individual to an organ dose which exceeds the limits specified in the above table.

[b] A beta exposure below a maximum energy of 700 keV will not penetrate the lens of the eye; therefore, the applicable limit for these energies would be that for the skin (15 rem/year). Note that the tritium beta energy is a maximum of 18 keV.

[c] In special cases with the approval of the Director, Division of Operational and Environmental Safety a worker may exceed 5 rem/year provided his/her average exposure per year since age 18 will not exceed 5 rem per year.

[d] All reasonable effort shall be made to keep exposures of forearms and hands to the general limit for the skin.

decision as to whether the individual will continue to work in a radiation area will be made on a case-by-case basis by plant management in accordance with the advice of the plant health physicist and occupational medical departments and subject to the approval of the head of the DOE field organization.

2.4.1.2. United States radiation protection standards for exposures of members of the general public

Exposures of members of the public shall be limited to the lowest levels technically and economically practicable within the standards prescribed below by the ICRP.

RADIATION PROTECTION STANDARDS
FOR EXTERNAL AND INTERNAL EXPOSURE

Annual dose equivalent or dose commitment (rem)[a]		
Type of exposure	Based on dose to individuals at points of maximum probable exposure	Based on an average dose to a suitable sample of the exposed population[b]
Whole body, gonads, or bone marrow	0.5	0.17
Other organs	1.5	0.5

a In keeping with DOE policy on lowest practicable exposure, exposures of the public shall be limited to as small a fraction of the respective annual dose limits as is practicable.
b See Par. 5.4, FRC Report No.1, for discussion on concept of suitable sample of exposed population.

There are no specific DOE standards for exposures to the general public during accident situations; however, guidance may be obtained from 10 CFR100 and from the radiation dose criteria established for the TFTR project. 10CFR100 defines a minimum acceptable exclusion area to be of such a size that an individual located at any point on its boundary for two hours immediately following onset of a postulated release would not receive a total radiation dose to the whole body in excess of 25 rem. The whole body dose of 25 rem referred to corresponds numerically to the once-in-a lifetime accidental or emergency dose for radiation workers which, according to NCRP recommendations, may be disregarded in the determination of their radiation exposure status. TFTR radiation dose criteria limits the allowable exclusion boundary dose from an extremely low-probability event, (such as a most intense natural phenomena or an event with a probability of occurrence on the order of 10^{-7} per year) to 25 rem with a design objective of 5 rem.

The allowable whole body dose to a member of the general public from a low-probability event (10^{-5} per year) is limited by the TFTR radiation criteria to 5 rem and a design objective of 1 rem.

Operational occurrences that are likely to occur once or more during the TFTR facility life, but not routinely, are limited to a maximum exposure at the exclusion boundary of 100 mrem per year.

2.4.2. European regulations

2.4.2.1. Radiological protection

Basic principles

The International Commission on Radiological Protection has issued a new set of recommendations for dose limitation in 1977, ICRP publication no.26. EEC is this year preparing a directive based on the ICRP-26. Below is given a brief description of the contents of the new recommendations.

The main points of ICRP-26 are:

(a) no practice should be adopted unless its introduction produces a positive net benefit;

(b) all exposures shall be kept as low as reasonably achievable, economical and social factors being taken into account;

(c) the dose equivalent to individuals shall not exceed the limits recommended for the appropriate circumstances by the Commission.

The effective whole body dose equivalent is now to be calculated as the sum of the doses for the exposed organs, multiplied by a weighting factor for each particular organ. The weighting factors are listed in Table XIX-2.

Dose limits

The basic limit for the whole body dose equivalent is the same as previously, namely 50 mSv/v (= 5 rem/y). In Table XIX-3 are listed the dose equivalent limits,

TABLE XIX-2. WEIGHTING FACTORS

Tissue	W_T
Gonads	0.25
Breast	0.15
Red bone marrow	0.12
Lung	0.12
Thyroid	0.03
Bone surfaces	0.03
Other single organs (five in total: 5 X 0.06)	0.30
Sum	1.00

TABLE XIX-3. DOSE EQUIVALENT LIMITS IN mSv/y AND (rem/y).
Ref.: ICRP 26

Tissue or organ	Worker [a, b]	Member of the public
Whole body, uniform radiation or weighted mean	50(5)[c]	5(0.5)[d]
Tissue or organ, irradiated singly:		
Gonads	200(20)	20(2)
Breast	330(33)	33(3)
Red bone marrow	417(42)	42(4)
Lung	417(42)	42(4)
Thyroid	500(50)	50(5)
Bone surfaces	500(50)	50(5)
Lens	300(30)	30(3)
Skin	500(50)	50(5)
Other single organs	500(50)	50(5)

[a] Values derived from the whole body dose equivalent limit and the weighting factor.
[b] Values given in ICRP-26.
[c] Average dose equivalent for a large group should be less than 5 mSv/y (.5 rem/y).
[d] Average dose equivalent for a large group should be less than .5 mSv/y (50 mrem/y).

Note: Footnotes a and b should be regarded as comments rather than modifications of the limits stated in the table.

TABLE XIX-4. DOSE EQUIVALENT LIMITS FOR PLANNED SPECIAL EXPOSURE

In a single event	twice	the relevant dose equivalent limit
In a lifetime	five times	the relevant dose equivalent limit

either recommended by ICRP directly or derived from the 50 mSv/y (5 rem/y) – wholebody limit and the weighting factors, which ever is the most restrictive. For women of reproductive capacity and certain other groups of workers there are specially derived limits. For planned and justified special exposure the limits are given in Table XIX-4.

Organization for dose limitation

Radiation workers should be divided into two categories:

Class A, in which the individual might receive more than 3/10 of the relevant dose equivalent limit. Individual monitoring must be provided for this group;

Class B, in which it is unlikely that an individual will receive more than 1/10 of the relevant dose equivalent limit.

For workers not in these categories it must be made unlikely that any individual will receive more than 1/10 of the relevant dose equivalent limit. The same rule applies to members of the critical group of the public.

Shielding, time and distance should be used as ways of restricting exposure; personal protective equipment should in general be supplementary to these more fundamental provisions. All areas where ionising radiation might give rise to dose equivalents above 1/10 of the limits should be controlled in some way. Access to the controlled areas should be restricted. For areas where the dose limits might be exceeded, time restrictions must be imposed and access should be supervised by a qualified and recognised radiological protection adviser.

ICRP-26 gives no specific radiation and contamination levels for the controlled areas. A radiation level exceeding 2.5 μSv/h (.25 mrem/h) in 40 h/week in 50 weeks/y can give rise to dose equivalents of more than 5 mSv/y (.5 rem/y), and it could therefore be used to specify that an area should at least be controlled.

2.4.3. USSR requirements

The main document in the USSR which specifies the allowable levels of ionizing radiation on personnel and population is the 'Radiation Safety Regulations' HP5-76 [28]. It is expected that these regulations will apply to INTOR.

In terms of dose limits the following category of exposed population are established:

Category A – Personnel (professional workers) responsible for the direct maintenance of the installation

Category B – A limited part of the population which is exposed to radiation because of proximity to nuclear sources

Category C – The rest of the population.

For Category A, the exposure to radiation is calculated for 1700 hours (in the USSR, the working week is 36 hours and a 4 to 6 week vacation is established for the majority of personnel). The irradiation time for groups B and C is 8800 hours.

For Categories A and B, the maximum permissible dose per year; MPD, and the dose limit, DL, are the main criteria. The MPD is the maximum value of the individual equivalent dose per year which does not cause adverse changes in health conditions within 50 years. The DL is the threshold equivalent dose per year for the limited part of the population which is set smaller than the MPD to prevent needless irradiation of the people.

The MPD and the DL values for the whole body, gonads and red bone marrow are 5 and 0.5 rem/year.

Table XIX-5 lists the numerical values of permissible levels of radionuclides that might be particularly important for INTOR.

2.5. Other hazardous materials

In the United States, the Industrial Hygiene Section of Chapter I of the DOE ES&H Manual (formerly ERDAM 0550) Threshold Limit Values for exposure to toxic materials are prescribed. Additional practical guidance is found in **Dangerous Properties of Industrial Materials** [4]. It should be kept in mind that exposures to any hazardous materials should be kept as close to zero as practicable (in keeping with the as-low-as practicable philosophy), since continuing research is resulting in the designation of more and more materials as carcinogens, hence limiting, or even prohibiting, their use in the workplace.

For purposes of this assessment we can assume that the present data base available through the various governmental bodies such as the National Institute for Occupational Safety and Health (NIOSH) is adequate for the safety analysis of hazardous materials associated with INTOR.

2.6. Designing for natural phenomena

There are no prescribed or generic standards for design protection from severe natural phenomena as of this date. In the US, general guidance is provided in Chapter I of the DOE ES&H Manual, formerly ERDAM 0550, and in ERDAM 6301 "Construction Design Criteria".

TABLE XIX-5. SUMMARY OF USSR EXPOSURE LIMITS

	Category A			Category B			
Radionuclide (insoluble state)	Permissible content in critical organ μCi	Max. permissible intake via lungs μCi/y	Permissible concentration in air Ci/l	Limiting intake μCi/y: Via lungs	Limiting intake μCi/y: Via digestive organs	Permissible concentration Ci/l: Air	Permissible concentration Ci/l: Water
Tritium gas	–	4.8×10^6	2.0×10^{-6}	4.8×10^5	–	6.6×10^{-8}	–
HTO or T_2O	1.2×10^3	1.2×10^4	4.8×10^{-9}	1.2×10^3	2.6×10^3	1.6×10^{-10}	3.2×10^{-6}
Sodium-24	–	3.6×10^4	1.4×10^{-10}	36	22	4.9×10^{-10}	2.8×10^{-8}
Chromium-51	60	5.6×10^3	2.2×10^{-9}	5.6×10^2	–	7.7×10^{-11}	1.5×10^{-6}
Manganese-54	36	87	3.6×10^{-11}	8.7	–	1.2×10^{-12}	–
Iron-55	130	2.6×10^3	–	2.6×10^2	–	–	–
Cobalt-60	13	8.7×10^2	–	87	–	–	–

The lower relative toxicity of tritium compounds and the relative immobility of fusion reactor activation products may allow the use of more realistic and site specific design criteria than those applied to commercial fission reactors by the various regulatory agencies, but that remains to be seen.

Where appropriate, the methodology governing the analysis of natural phenomena would utilize the state-of-the-art models employed in design of commercial reactors. However, the use of less conservative parametric information may be permitted.

Natural phenomena could be classified into the two categories as developed for the TFTR, i.e. most intense and most probable.

A most intense event would be defined as the most severe event that would be reasonably predicted for a specific site. A most probable event would be defined as that event which on the average would not be exceeded for a specific site.

The most intense natural phenomena design criteria would require pressure boundary and/or containment integrity for structures, systems, subsystems or components which would limit exposures from radioactive or toxic materials to workers and the general public to levels less than the appropriate guidelines.

The most probable natural phenomena design criteria would require pressure boundary and/or containment integrity for structures, systems, subsystems or components which would:

(1) limit exposures from radioactive or toxic materials to workers and the general public to less than the appropriate guidelines, and

(2) limit property damage to capital equipment or facilities to less than $100 000 as defined in ERDAM 0502 for a Type A accident.

In the United States, the general methodology of the following NRC Regulations and Guides can be employed in determining the most intense and most probable tornado, earthquake, hurricane and flood for a proposed INTOR site. (Site-specific parameters can be substituted for the generalized values used in the NRC models in order to make the analysis more realistic and less conservative):

10CFR50	Licensing of Production & Utilization Facilities,
Appendix A	General Design Criterion 2
10CFR100	Reactor Site Criteria Appendix A
RG1.59	Design Basis Floods for Nuclear Power Plants
RG1.76	Design Basic Tornado for Nuclear Power Plants

2.7. Other environmental impacts

2.7.1. The situation in the United States

There are no DOE requirements directly applicable to INTOR except the very general guidelines dealing with air and water pollution as stated in Chapters XVII and XVIII of the DOE ES&H Manual, (formerly ERDA Manual Chapters 0510 and 0513 respectively), and ERDAM 6301. The INTOR facilities and its associated support systems which directly interface with the environment will be similar to those of an equivalently sized power generation facility. General guidelines for environmental impact assessment and environmental protection related to such facilities can be obtained from the following Regulatory Guides (RG) and Regulations employed by the NRC:

10CFR51	Licensing and Regulatory Policy and Procedures for Environmental Protection
10CFR100	Reactor Siting Criteria
RG4.1	Programs for Monitoring Radioactivity in the Environs of Nuclear Power Plants
RG4.2	Preparation of Environmental Reports for Nuclear Power Plants
RG4.4	Reporting Procedure for Mathematical Models Selected to Predict Heated Effluent Dispersion in Natural Water Bodies
RG4.5	Measurements of Radionuclides in the Environment – Sampling & Analysis of Plutonium in Soil
RG4.6	Measurements of Radionuclides in the Environment – Strontium-89 and Strontium-90 Analyses
RG4.7	General Site Suitability Criteria for Nuclear Power Stations
RG4.8	Environmental Technical Specifications for Nuclear Power Plants (For Comment)
RG4.11	Terrestrial Environmental Studies for Nuclear Power Stations
RG4.13	Performance, Testing, and Procedural Specifications for Thermoluminescence Dosimetry: Environmental Applications
RG4.15	Quality Assurance for Radiological Monitoring Programs (Normal Operations) – Effluent Streams & the Environment (For Comment).

There is an additional regulation recently instituted concerning the situation where INTOR would be built in a country other than the US. Executive Order No. 12114, dated January 4, 1979 (published in the Federal Register, Vol.24 – No.6, dated January 9, 1979) establishes the requirements for review of the environmental effects of a major US action abroad. The thrust of this executive order is that appropriate environmental review documents on INTOR would have to be prepared and approved for continued US participation.

2.8. Magnetic field exposures

The biological effects of exposures to magnetic fields are not well known and standards governing these exposures are non-existent. Certain laboratories, namely the Stanford Linear Acceleration Center and Lawrence Livermore Laboratory, have reviewed the available data and have adopted interim guidelines. These values could be used for INTOR and are (for workers):

(1) Extended-exposure periods (hours): whole body or head, 200 Gauss; arms and hands, 2000 Gauss.

(2) Short-exposure period (minutes): whole body or head, 2000 Gauss; arms and hands, 20 000 Gauss.

Exposure to fields above those specified in (2) above should be avoided and it is recommended that people not linger unnecessarily in magnetic fields regardless of intensity.

For exposure to the general public, again there are no standards, however, approximately the value of the earth's normal field intensity (0.5 Gauss) would probably be acceptable. Although there are several good reviews of the state of the art [22, 23], work is needed in this area.

2.9. Nonionizing radiation effects

Nonionizing radiation is the term generally used to characterize all those forms of electromagnetic radiation wholse primary mode of interaction with matter is other than ionization. In the specific scope of fusion reactor the two most important sources of nonionizing radiation are RF heating and lasers.

The frequencies of the electromagnetic radiations used in the RF heating lies between 150 000 MHz and 0.08 MHz. The measurement of RF field is complex and necessitates specialized knowledge and equipment. It is also uncertain what quantities, units, and measuring methods should be used to establish biologically meaningful RF dosimetry. For practical purposes the most frequent use is the lower power density (W/m^2) sources. The duration of exposure must also be taken into account. The biological effects of the RF are hyperthermia cataracts, disturbances in the nervous system, and some effects on the iron metabolism, hemoglobin synthesis and red blood cell production. Generally accepted rules for RF safety do not exist only the following recommendations have been given [1].

According to the power density in a given area the concept of 4 zones was introduced. For stationary fields these were defined as follows:

(1) safe zone – the mean power density cannot exceed 0.1 W/m^2, human exposure is unrestricted;

(2) intermediate zone – minimal value 0.1 W/m^2, upper limit 2 W/m^2, occupational exposure is allowed during a whole working day (principle 8 hours may be extended to 10 hours for on duty personnel);

(3) hazardous zone – minimal value 2 W/m^2, upper limit 100 W/m^2, occupational exposure time per 24 hours is determined by the formula:

$$t = 32/p^2$$

where t = permissible exposure time
p = mean power density in W/m^2;

(4) dangerous zone – mean power density in excess of 100 W/m^2 (10 mW/cm^2), human exposure is forbidden.

For exposures to non-stationary fields, i.e. intermittent exposure, the following values were adopted:

(1) safe zone – mean power density does not exceed 1 W/m^2 (0.1 mW/cm^2);
(2) intermediate zone – minimal value 1 W/m^2, upper limit 10 W/m^2, occupational exposure is allowed during a whole working day, as above;
(3) hazardous zone – minimal value 10 W/m^2, upper limit 100 W/m^2, the professional exposure time per 24 hours is determined by the formula:

$$T = \frac{800}{p^2}$$

where the symbols t and p designate the same values as above;
(4) dangerous zone – mean power density in excess of 100 W/m^2 (10 mW/cm^2), human exposure is forbidden.

Lasers might be used for diagnostics in INTOR and for lasers extensive classifications and limits have been set by the American National Standards Institute (ANSI) and have received general acceptance. The most important biological effects of lasers are eye injuries, skin burns and possible damage to underlying organs.

2.10. Summary

An adequate data base exists for establishing most of the safety regulatory requirements for the proposed INTOR project. Research is required in order to develop meaningful standards in the following areas: i) the relative toxicity of organic, oxide and elemental forms of tritium; ii) the biological effects of magnetic fields; and iii) concentration guides for disposal of radioactive solids.

The large quantity of information generated during the many years of operational experience at nuclear facilities coupled with the reactor siting and licensing criteria already developed provide the data needed to develop the safety and environmental siting and effluent release criteria in the remaining areas.

3. SAFETY-RELATED DESIGN CRITERIA AND STANDARDS

One of the cornerstones of reactor safety is sound design in accordance with proven engineering principles. For a mature technology this normally implies design in accordance with a well-established and tested set of design criteria and standards. Such a set of criteria and standards provides 1) guidance to the designer, 2) added confidence that the design is based on sound technology and has completeness, and 3) a yardstick for review of the design adequacy. Because of the experimental nature of the fusion field to date, such a set of criteria and standards has not been developed specifically for fusion.

INTOR represents an experimental facility that will precede the first commercial fusion reactor by at least several decades. It, therefore, is not expected to be designed and constructed to the more comprehensive criteria and standards to be used for a commercialized fusion reactor industry; however, INTOR provides the opportunity to start laying the foundation for criteria and standards for fusion reactors early in the time frame of the total fusion development programme. An examination of design criteria and standards presently being used by the fission reactor community can provide guidance for the development of design criteria and standards for commercial fusion reactors. Also, where applicable, these criteria and standards can be employed for fusion reactor design.

3.1. General design criteria for nuclear plants in the United States

The US Code of Federal Regulations, 10 CFR 50, Appendix A provides 64 General Design Criteria for Nuclear Power Plants. Although these criteria were developed primarily for the fission reactor industry, many of them have applicability to the fusion industry. These criteria have been examined for applicability to the fusion field and to INTOR. A synopsis of this preliminary examination is presented in Table XIX-6.

In addition, the US Nuclear Regulatory Commission has developed approximately 150 design guides for nuclear power plants. These guides have also been examined for applicability to the fusion field, with the results that many of them have applicability to fusion and INTOR. In the interest of brevity the detailed results of this examination will not be reproduced here; however, a copy will be in the files of the US Leader of the Team of Experts on Environment and Safety. It is suggested that the INTOR design team carefully examine these design guides as a source of expert design guidance.

3.2. Quality assurance standards for fusion reactors

The quality assurance criteria for nuclear fission power plants and fuel reprocessing facilities in the US is currently contained in 10 CFR 50 Appendix B.

Text continued on page 595

TABLE XIX-6. 10 CFR 50, APPENDIX A, GENERAL DESIGN CRITERIA FOR NUCLEAR POWER PLANTS

Criterion No.	Criterion Description	General statement of applicability to fusion reactors	Applicability for INTOR
1.	Quality Standards and Records	Directly Applicable, (Wording of Criterion is general)	Applicable
2.	Design Basis for Protection Against Natural Phenomena	Directly Applicable	Directly Applicable
3.	Fire Protection	Directly Applicable	Directly Applicable
4.	Environmental and Missile Design Bases	This criterion is applicable to fusion systems. Summary definition of the term "Loss-of-Coolant Accident" may be required for fusion reactors.	Directly Applicable
5.	Sharing of Structures, Systems and Components	This criterion would be applicable if common systems were shared by more than one fusion reactor.	This criterion will probably not be applicable to INTOR; however, this should be reconsidered as the design develops.
10.	Reactor Design	This criterion must be redefined for fusion reactors. Wording pertaining to "core" and "fuel" will require revision.	Same
11.	Reactor Inherent Protection	Not Applicable	Not Applicable
12.	Suppression of Reactor Power Oscillations	This criterion must be redefined to apply to fusion reactors. Plasma oscillations or instabilities are currently considered a fundamental problem and require a great deal of additional study.	Same

Criterion No.	Criterion Description	General statement of applicability to fusion reactors	Applicability for INTOR
13.	Instrumentation and Control	The intent of this criterion is directly applicable to fusion reactors. The actual statement would require revision to eliminate references to the reactor core and the fission process.	Same
14.	Reactor Coolant Pressure Boundary	The concept of reactor coolant pressure boundary must be defined for fusion reactors. Cooling systems are of an equal importance to those of fission reactors, although radically different in design.	Applicable with specific modification.
15.	Reactor Coolant System Design	Directly Applicable	Directly Applicable
16.	Containment Design	Directly Applicable	Directly Applicable
17.	Electric Power Systems	The philosophy of a reliable electrical power system, including dual channel or redundant systems, is applicable to fusion reactors. The wording of the existing criterion would have to be revised to eliminate specific reference to fuel design limits and reactor coolant nomenclature applying to fission reactors.	Same
18.	Inspection and Testing of Electrical Power Systems	Directly Applicable	Directly Applicable

TABLE XIX-6 (cont.)

Criterion No.	Criterion Description	General statement of applicability to fusion reactors	Applicability for INTOR
19.	Control Room	Directly Applicable	Directly Applicable
20.	Protection System Functions	The general requirements of this criterion are applicable to fusion reactors. Specific references to "reactivity control systems" and "fuel" design limits will need revision.	Same
21.	Protection Systems, Reliability and Testability	Directly Applicable	Directly Applicable
22.	Protection System Independence	Directly Applicable	Directly Applicable
23.	Protection System Failure Modes	Directly Applicable	Directly Applicable
24.	Separation of Protection and Control Systems	Directly Applicable	Directly Applicable
25.	Protection System Requirements for Reactivity. Control Malfunctions	Significant Redefinition Required	Significant Redefinition Required
26.	Reactivity Control System Redundancy and Capability	Significant Redefinition Required for Fusion Reactors	Same
27.	Combined Reactivity Control Systems Capability	Significant Redefinition Required for Fusion Reactors	Same
28.	Reactivity	Term reactivity would not apply per se to fusion reactors and this criterion would be replaced by others.	Same

Criterion No.	Criterion Description	General Statement of Applicability to Fusion Reactors	Applicability for INTOR
29.	Protection Against Anticipated Operational Occurrences	This criterion would apply to fusion reactors with the exception of the statement relating to reactivity.	Same
30.	Quality of Reactor Coolant Pressure Boundary	Several different cooling systems on a fusion reactor will probably receive a level of quality equivalent to that applied to a fission reactor primary coolant system. It will be necessary to define specific criteria for the various types of fusion reactors.	This criterion would be applicable to magnet, blanket, first wall, plasma clean-up, and perhaps other cooling systems as well.
31.	Fracture Prevention of Reactor Coolant Pressure Boundary	On a fusion plant, this criterion could reasonably be expected to apply to critical cooling systems, the first wall of the reaction chamber, and critical load-bearing structures. This criterion will be especially important to portions of the fusion reactor system exposed to the extremely high energy neutrons eminating from the plasma.	Same
32.	Inspection of Reactor Coolant Pressure Boundary	This criterion would be expected to apply to critical cooling systems and the plasma containment "first wall" chamber.	Same
33.	Reactor Coolant Make-Up	Systems to supply make-up coolant for the various coolant systems will surely be required for fusion reactors. This criterion statement would require significant rewriting to make it generally applicable to fusion reactors.	The portion of this criterion pertaining to reactor coolant make-up availability will be at least partially applicable to INTOR.

TABLE XIX-6 (cont.)

Criterion No.	Criterion Description	General Statement of Applicability to Fusion Reactors	Applicability for INTOR
34.	Residual Heat Removal	Coolant capacity adequate to remove heat generated during and after the shut-down transient will be required for fusion reactors. Since the heat problem is quite different from a fusion reactor, this criterion statement would require significant revision.	Same
35.	Emergency Core Cooling	The concept of emergency cooling will be vastly redefined for fusion reactors. Emergency provisions must be incorporated on several critical cooling systems and the current criteria pertaining to ECCS must be completely rewritten.	Same
36.	Inspection of Emergency Core Cooling	See Criterion No.35	Same
37.	Testing of Emergency Core Cooling System	See Criterion No.35	Same
38.	Containment Heat Removal	The containment heat removal problem for a water-cooled fission reactor is mainly concerned with adding water in a spray form to condense steam and therefore reduce temperature and pressure in the containment. Other concepts apply when considering an LFMBR. This criterion may be revised for fusion to include such things as fire suppression or tritium recombination after an accident has occurred. Significant rewriting of this criterion will be required.	Same

Criterion No.	Criterion Description	General Statement of Applicability to Fusion Reactors	Applicability for INTOR
39.	Inspection of Containment Heat Removal System	Refer to Criterion 38.	Same
40.	Testing of Containment Heat Removal System	Refer to Criterion 38.	Same
41.	Containment Atmosphere Clean-up	Directly Applicable	Directly Applicable
42.	Inspection of Containment Atmosphere Clean-up Systems	Directly Applicable	Directly Applicable
43.	Testing of Containment Atmosphere Clean-up Systems	Directly Applicable	Directly Applicable
44.	Cooling Water	This criterion is directly applicable to fusion reactors; however, it should probably be revised to address coolants other than water.	Same
45.	Inspection of Cooling Water System	Refer to Criterion 44.	Same
46.	Testing of Cooling Water System	Refer to Criterion 44.	Same
50.	Containment Design Basis	The concept of fusion reactor containment will have some similar characteristics to fission plants but will also differ in some areas. Potential energy sources and radioactive releases certainly must be contained and some type of containment standard is required.	INTO will require a blanket cooling system and will contain Tritium. A definitive containment design basis is necessary.

TABLE XIX-6 (cont.)

Criterion No.	Criterion Description	General Statement of Applicability to Fusion Reactors	Applicability for INTOR
51.	Fracture Prevention of Containment Pressure Boundary	Directly Applicable	Directly Applicable
52.	Capability for Containment Leakage Rate Testing	Directly Applicable	Directly Applicable
53.	Provisions for Containment Testing and Inspection	Directly Applicable	Directly Applicable
54.	Piping Systems Penetrating Containment	Directly Applicable	Directly Applicable
55.	Reactor Coolant Pressure Boundary Penetrating Containment	Directly Applicable	Directly Applicable
56.	Primary Containment Isolation	Directly Applicable	Directly Applicable
57.	Closed System Isolation Valves	Directly Applicable	Directly Applicable
60.	Control of Releases of Radioactive Materials to the Environment	Directly Applicable	Directly Applicable
61.	Fuel Storage and Handling	The concept of fuel handling and fusion reactors. The general provisions for handling the various radioactive materials will apply. Special provisions for Tritium are necessary.	Need redefinition for fusion
62.	Prevention of Criticality in Fuel Storage and Handling	Not Applicable	Not Applicable
63.	Monitoring Fuel and Waste Storage	This criterion would be applicable with some redefinition in the fuel area.	Applicable with redefinition
64.	Monitoring Radioactive Releases	Directly Applicable	Directly Applicable

The criteria contained in this Appendix cover all phases of activity from design to construction, startup, and plant operation. It appears this set of Quality Assurance Criteria is equally applicable to fusion or fission reactors.

3.3. Criteria for protection systems

Standard IEEE 279 establishes minimum requirements in the US for the safety-related functional performance reliability of protection systems for stationary, land-based nuclear reactors producing steam for electric power generation. Two of the more important topics addressed are single failure, and safety channel independence criteria.

It appears that this standard should be directly applicable to fusion reactors. The standard presently does not include requirements for the actuator systems and work is currently underway to expand the standard into this area. In view of the unique activation problems involved with fusion reactors, the inclusion of the actuator systems becomes significantly more important than for fission reactors. Further investigation may show that certain features of fusion reactors are not compatible with the standard; however, that is not apparent at this time.

An early preliminary safety analysis should be performed on INTOR to aid in identifying the reactor protective systems required. For example, for a long-pulse reactor, some means of terminating the burn may be required should a loss-of-flow or loss-of-coolant accident occur during the pulse.

3.4. Fusion-specific guides and standards

There are a number of fusion systems and components that have unique safety-related implications that are not generally found in the fission reactor technology. Examples of some of these systems are:

(1) Superconducting magnet systems
(2) Tritium handling systems
(3) Cryogenic cooling systems
(4) Vacuum systems
(5) Auxiliary plasma heating systems

As the fusion reactor power technology moves toward commercialization, design criteria and standards will be required for these systems. It is recommended that the fusion community should take the lead in this endeavour. As examples, (1) research conducted at the TSTA could form the basis for design criteria and standares for tritium handling systems, and (2) research conducted at the Large Coil Program or on T-15 could form the basis for superconducting magnet system criteria. Thus, research from the base technology programmes can be used to provide interim criteria that can be used in the main line facilities from INTOR through the fusion DEMO. In this manner, valid design criteria and standards will be available as the fusion technology moves into commercialization.

3.5. Safety analyses in Japan

There have been some recent safety studies of the Japanese Experimental Fusion Reactor cooling and fueling systems (JAERI-M7772, 1978 and JAERI-M7964, 1978). Such studies have been extremely useful in the conceptual design phase and should help to devise a more acceptable system.

4. ACCIDENT ANALYSIS

4.1. Accident definition

Accident analysis involves identification of potential accident initiators, accident progression sequences and the ultimate consequences. Safety analysis of an existing system also includes an estimate of the probability of failure of its components. The only fusion reactor which has had such comprehensive analysis is the TFTR [4]. Clark [4] and Okrent et al. [4] discussed generic failure events in Tokamak fusion power reactors. Kazimi et al. [4] provided examples of event trees in the tritium system while Powell et al. discussed various failure events in the magnet system [4]. In addition, failure of various components was considered in several conceptual designs.

The accidents identified in these studies are listed in Table XIX-7. Naturally, the attention is mostly given to accidents involving potential public effects, in the form of radioactivity releases to the atmosphere. On the other hand, accidents involving occupational hazards but no public hazards should not be neglected. Thus, a plant design should take into consideration protection schemes for the workers as well as for the public. A concise review of major accidental events and safety concerns can be found in Ref.[24]. Accident initiators and event trees need to be defined in some detail in order to facilitate assessment of the hazard potential and the appropriate mitigating design features.

The INTOR not being a full-scale power reactor, would not necessarily involve all the accidents cited in the literature (for example, lithium fire would be significant for INTOR only if some lithium-cooled blanket modules were utilized). The TFTR PSAR can serve as a basis for accident identification, but the list needs to be enlarged in view of the several additional systems that will be present (superconducting magnet, blanket and electricity generation components).

TABLE XIX-7. POTENTIAL FUSION REACTOR ACCIDENTS APPLICABLE TO INTOR

1. Tritium System Failures
2. Activation Product Release
   a) Structural Activity
   b) Coolant Activity
   c) Air Activation Products
3. Fires and Explosions
   a) Tritium Fires
   b) Vapour Explosions
   c) Lithium Fires
4. Electrical Systems Failures
5. Magnet System Failures
6. Plasma Disturbances
7. Coolant Disturbances
8. Release of Toxic Material
9. Secondary System Failures
10. Control System Failures
11. External Events (Earthquake, Tornado, Aircraft Impact, etc.)

4.2. Tritium release

In the tritium systems, as in other key systems in a fusion plant, the designer begins with some tentative concepts as a design basis. These then have to be studied, developed into more detailed hardware specifications, and finally tested for potential hazards or accidents both to the worker and to the public and environment. The above statement by Galloway [4] is in an excellent review article that serves as the basis for this section of the data base assessment. Also, the IAEA/NEA International Symposium on the behaviour of tritium in the environment held in October 1978 produced an excellent summary of tritium behaviour.

The tritium process design is subjected to safety/risk analysis. From these examinations the designer learns which and how few chain events will lead to some kind of failure or accident impacting worker, public or environmental safety. The designer, armed with this information goes back to modify the process design to achieve fewer and less likely chain events to a failure. These modifications can include special tritium compatible materials of construction and quality assurance procedures to guarantee compliance with these special materials procurement and handling methods, operating procedures, testing, inspection, etc. These procedures will involve specialized tritium compatible hardware components such as fittings, valves, pressure vessels, dry compressor and pumps, elastomers, catalytic oxidizers, metal getters, adsorption beds, neutral beam injectors, in-line tritium monitors, etc., which have known and hopefully low failure modes and rates. This decision process must also consider consequences of natural phenomena (winds, seismic occurrences, etc.) on the behaviour of the tritium system. Galloway further discusses this approach in terms of a cost/benefit approach as the designer decides which safety systems to include in the tritium facility. A design study and safety analysis for the fuel circulating System of JXFR has also been carried out by a Japanese research group [2].

Once the accident scenarios are developed and appropriate actions taken to mitigate the consequences, it is still necessary to develop the dose commitment related to the accident. This involves knowledge of such matters as atmospheric transport, chemical form (gas or oxide), rate of oxidation in the atmosphere, and the ecological fate [4]. These factors have been addressed in the references cited in [4], but additional work is definitely necessary in the area of atmospheric oxidation rates, transport and in biological consequences of chronic low level radiation exposures.

In 1974 approximately 4.8×10^5 Ci of tritium gas were released from the Savannah River Plant [4]. The release resulted from failure of a process valve at the facility. The release occurred under neutral stability conditions (Pasquill Type D) with a wind speed of approximately 2 m/s. An extensive environmental sampling programme was undertaken to track the cloud and assess the environmental impact of the release. Approximateiy 1000 samples were taken from the atmosphere, vegetation, milk, surface water, soil and human urine. The samples showed that effects of the release were measurable, but insignificant. The calculated maximum dose to a person at the puff centreline on the plant boundary was calculated to be 0.14 mrem and the total population dose was estimated to be 8 man-rem. Measurements showed that less than 1% of the tritium was in the oxide form. Release of approximately 3×10^5 Ci at Lawrence Livermore Laboratory produced similar low consequence results.

Additional data on tritium systems response to accident situations will be obtained at the Tritium Systems Test Assembly (TSTA) [4] under construction at Los Alamos and the Japanese Tritium Engineering Test Laboratory. These facilities will be developing and evaluating components to assure personnel and environmental safety under routine and accident situations. The TSTA will be operational in 1981 and will have established a strong data base by 1985. The Japanese facility will also be producing data by that time. The TSTA and the Princeton TFTR groups have prepared Preliminary Safety Analysis Reports (PSAR's) for their respective systems. In each case the PSAR does contain a chapter devoted to accident analysis. These chapters can provide an initial look at the response of tritium facilities to the various accident scenarios. At TSTA, part of the project charter will be to simulate some of these scenarios and study the system response to these off-normal occurrences.

The areas of tritium storage and transport are worthy of separate mention with regard to accident analysis. The development of adequate shipping requirements (chemical state of tritium, total quantity in one shipment, design of shipping containers, etc.) must be addressed for INTOR. A very good summary of the current status of shipping and storage techniques is given by Sehnert, McKone and Kastenberg [4].

As a summary of the current status of concern over environmental and safety issues, Table XIX-8, cited in Ref. [4], is included. Galloway identified the various on-going construction and conceptual design work in both the magnetic fusion and laser fusion programmes. He then prepared a list of key environmental and safety issues related to tritium and set up the necessary matrix to evaluate where the necessary work is/is not being addressed. This table is a valuable component of the environmental and safety data based assessment. Galloway cites a complete list of references which can be used to obtain additional information. This extensive reference list will not be repeated here.

4.3. Activation product release

Activation Product Inventories. There are three main sources of activation products in any fusion plant. These are:

(1) the activity in structural materials including, first wall, blanket, shield, neutral beam injector walls, limiter walls, etc.

(2) coolant activation products and activated corrosion products which may be circulating with the coolant

(3) gaseous activation products resulting from direct neutron activation of the gases that may be in the reactor building atmosphere.

The methods for activation product inventory estimates are at a reasonable stage of development. Several studies have been performed on the level of radioactivity expected from the first wall, blanket and neutral beam injectors [47–51]. Also, an excellent discussion of activation product inventories and biological hazard potentials for various fusion reactor materials, including 316 SS, is presented by Häfele et al. in Reference [25] and by the EC INTOR team [1]. For stainless steel first walls, Fig.XIX-1 summarizes the results of UWMAK-related studies. The calculated neutral beam injector activity due to neutron streaming is shown in Fig.XIX-2. The total inventory of activity is likelty to be close to 10^3 Ci–10^4 Ci per Kilowatt thermal-year in stainless steel reactors. The exact level can be determined when the design is finalized. For a 600 MW INTOR reactor, the level of activity is therefore expected to be approximately 10^9 Ci

TABLE XIX-8. KEY TRITIUM ENVIRONMENTAL AND SAFETY ISSUES [7]

	J.E.T.	TFTR	Russian T-20	TNS W/ORNL	TNS GA/ANL	EPR ANL	EPR ORNL	DEMO Doublet GA	DEMO Fintor-D	DEMO-ORNL	REF UWMAK-III	REF RTPR LASL	REF MFR LLL	REF TMR LLL	REF MHR LLL	SHIVA–Nova	LFEPR-I LLL	LFEPR-II LLL	REF Saturn W. Ger.	REF LFPP (ORNL)	JXFR
Tritium inventory	10 g	5 g	700 g	3 kg	260 g to 1.8 kg	600 g to 1.3 kg	NA	4.1 kg	10 kg	NA	36 kg	NA	26 kg	10 kg	42 kg	1 mg	10 kg	10 kg	69 kg	NA	0.5
1. Tritium accident analysis																					
a. Fault trees and risk analysis in process design	x	–	–	x	x	x	–	–	x	x	–	x	–	–	x	–	–	–	–	–	–
b. Materials of construction and quality assurance	–	–	–	–	–	x	–	x	x	–	x	–	x	x	x	x	x	x	–	–	
c. Tritium hardware component failure mode and rate (i.e., fittings, valves, pressure vessels, elastometres, catalytic oxidation, getters, adsorbers, cryo-removal systems, neutral beam injectors, etc.)	x	–	–	x	x	x	–	–	–	x	–	–	–	–	x	x	x	x	–	–	x
d. (ALARA) cost/benefit analysis	–	–	–	–	–	–	–	–	–	–	–	x	x	x	x	–	–	–	–	–	
2. Tritium process control																					
a. Low level tritium forms detection (real time)	–	–	–	x	–	–	–	–	–	x	–	–	–	–	–	–	x	x	–	–	
b. Interferences by other beta-emitting radioisotopes	x	–	–	–	x	x	–	x	–	–	–	–	–	–	x	x	x	x	–	–	
c. Dynamic, feed-forward control for key process unit operations	–	–	–	–	–	–	–	–	–	–	–	–	–	–	–	–	–	–	–	–	
d. Fault tolerant, redundant control schemes peculiar to tritium	x	x	–	x	–	–	–	–	x	x	–	–	x	x	x	–	–	–	–	–	
e. Adaptive control and accountability systems capable for real time inventory (MBA) control and its reduction	–	–	–	–	–	–	–	–	–	–	–	–	x	x	x	–	x	x	–	–	
f. Safe-guarding of tritium as a strategic material (classification policy)	–	–	–	–	–	–	–	–	––	–	–	–	–	–	x	x	–	–	–	–	

3. Occupational safety																					
a. Permeation and leak sources	x	–	–	x	–	–	x	–	–	x	x	–	x	x	x	x	x	x	–	–	x
b. Surface contamination	x	–	–	x	–	–	–	–	x	x	–	–	x	x	x	–	x	x	x	–	x
c. Massive releases and spills followed by cleanup and decontamination (air or vacuum containing enclosures)	x	x	x	x	x	x	x	x	x	x	x	–	x	x	x	–	–	–	–	–	
d. Worker emergency egress	–		–	–		–	–	–	–									–	–	–	
e. Tritium protective suits	–	–	–	–	–	–	–	x	x	–	–	–	x	x	x	–	x	x	–	–	
f. Medical and health physics (exposures, quality factors, sensitive individuals, etc.)	x	x	x	x	–	–	–	–	x	x	–	–	x	x	x	–	–	–	–	–	
4. HTO formation rate from the gas phase																					
a. In reactor blanket and coolant streams (oxide, layer control)	–	–	–	–	–	x	–	x	x	–	x	–	–	–	–	–	x	x	–	–	x
b. Practical reactor hall environments	–	–	–	–	x	x	–	–	–	–	–	–	x	x	x	–	–	–	–	–	
c. Avoidance of special catalytic or high conversion reactor hall surfaces	–	–	–	–	–	–	–	–	–	–	–	–	x	x	x	x	–	–	–	–	
d. Accelerated conversion in the environment	–	–	–	–	–	–	–	–	–	–	–	–	–	–	–	–	–	–	–	–	
5. Disposal of tritium contaminated wastes																					
a. Recovery	–	–	–	–	x	–	–	–	–	–	–	x	–	–	–	–	–	–	–	–	
b. Containerization, immobilization, and burial and/or storage	–	–	–	–	x	x	–	x	–	–	–	x	–	x	x	–	x	x	–	–	
c. Deep well injection	–	–	–	–	–	–	–	–	–	–	–	–	–	–	–	–	–	–	–	–	
6. Environmental impact																					
a. Release goals and legislation	x	x	x	x	x	–															
b. Plume dispersion, building down-wash, HTO conversion, and rainout	x	x	x	x	x	x	–	x	–	x	–	x	x	x	x	x	–	–	–	–	
c. Transportation and storage	–	–	–	–	–	–	–	x	–	–	–	–	–	x	x	x	x	x	–	–	
d. Accident release advisory capability (AHAC)																					
e. Global monitoring (short and long term trends)	–	–	–	–	–	–	–	–	–	–	–	–	–	–	–	–	–	–	–	–	
f. Concentration in special metabolisms	–	–	–	–	–	–	–	–	–	–	–	–	–	–	–	–	–	–	–	–	
g. Fate of tritiated organics in the food chain	–	–	–	–	–	–	–	–	–	–	–	–	–	–	–	–	–	–	–	–	
h. Total population commitment	–	–	–	–	–	–	–	x	–	–	–	x	x	x	x	–	–	–	–	–	

x = Subject addressed.
– = Subject not addressed.

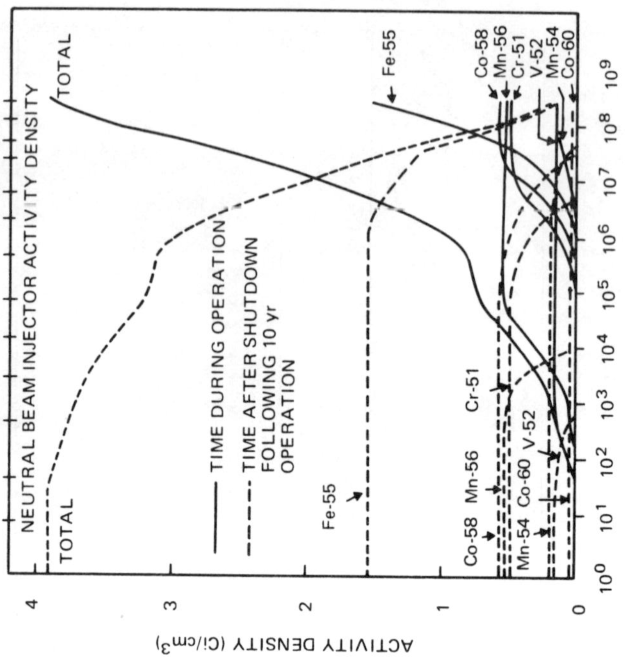

FIG.XIX-2. Neutral beam injector activation.

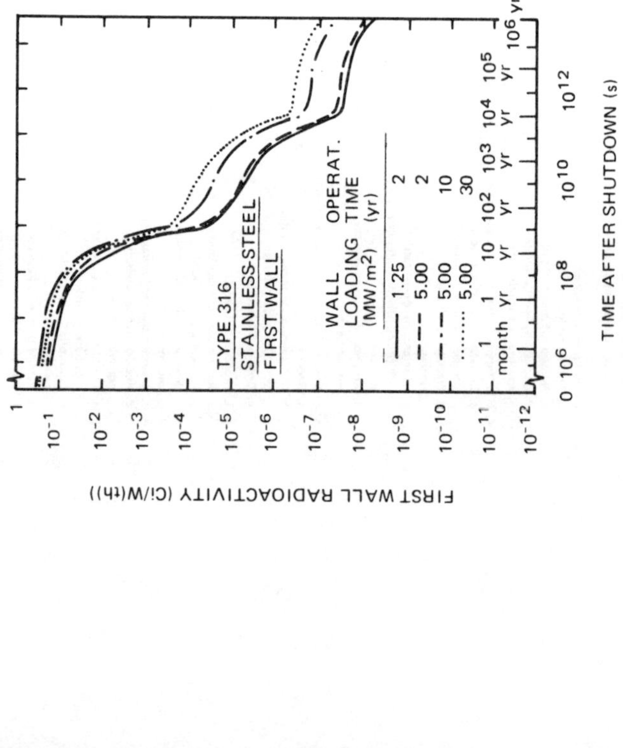

FIG.XIX-1. Comparison of radioactivity levels after shutdown in a type-316 stainless-steel first wall.

which is more than an order of magnitude greater than the level of activity from total tritium inventory (~3×10^7 Ci).

The radioactivity level in structural materials other than the first wall and blanket structure has not been investigated in the same depth and is, to a greater extent, design dependent. Similarly, limited work is available on the expected activity level in the coolant, with the exception of lithium, although tasks are underway to estimate activity levels in helium coolant. Water will be subject to both direct activation and corrosion product accumulation. However, considerable experience is available to draw firm conclusions about the expected level of circulating activity.

If the atmosphere around the reactor is air, considerable radioactivity may be generated due to the presence of argon. Ar-41 equilibrium levels of approximately 1.6×10^{-11} Ci/cm^3 are presented in the report on Shielding and Personnel Access for INTOR. This is about a factor of ten greater than occupational exposure limits. This radioactivity could limit the accessability to the reactor building during operation and after shutdown. A trade-off exists here between cost of shielding required to limit such activation and the dose rate level that will be allowed to exist in the building. Activation of gases does not require data base development beyond the existing information.

Activation product release potential. The potential for release of structural activation products is primarily dependent on the possibility of vapourising the structures or forming volatile compounds such as oxides under accidental conditions [4]. Accidents that result in temperatures above 800°C in the structures and/or the atmospheric gases are potential contributers to mobilization of activation products. The existence of such accidents for some conceptual reactors has been investigated [4]. The accidents appear to involve a chemical fire (such as a lithium fire). However, there is a need to quantify the maximum temperatures reached in the structures under various conditions. Additionally, the rate of oxide formation and volatilization for the structural materials may need to be studied in order to quantify the effects of the presence of alloying materials.

Release of activity associated with the coolant is a relatively more probable accident. It may be expected that pipe leaks and failures will develop. The design should minimize the probability of such failures.

Release of gas activation products is dependent on failures of isolation walls or seals. Such failures could be instigated by external events, internal pressurization or component failure. Since the amount of radioactivity involved is likely to be small, such a release is more of an occupational hazard.

Activation product release consequences. The consequences of release of structural activation products have had only limited attention. Methods for assessment of

TABLE XIX-9. RELATIVE CHEMICAL INHALATION TOXICITY OF SOME MATERIALS IN FUSION REACTOR DESIGNS [28]

Element	Inventory (kg/kW(e))					Industrial threshold limit value (mg/m^3)	Air dilution factor
	UWMAK-I	UWMAK-II	UWMAK-III	LASL-ANL	PPPL		Maximum inventory (km^3 air/kW(e))
Be	None	0.3	None	0.01	0.4	0.002	0.2
B	1.1	2.5	0.47	–	–	15 (B_2O_3)	1.6 10^{-4}
Cr	3.4	2.3	0.81	–	0.82	0.5 (solid compound)	7 10^{-3}
Cu	5.1	6.1	0.65	2.3	1.6	0.1 (fume)	6 10^{-2}
Pb	14	12	18	*	–	0.2	7 10^{-2}
Li	1.2	0.2	0.2	0.4	0.3	2 (as LiOH)	6 10^{-4}
						0.025 (as LiH)	5 10^{-2}
Hg	0.002	None	None	–	–	0.05 (non-alkyl)	4 10^{-5}
						0.01 (alkyl compounds)	8 10^{-6}
Mo	0.39	0.24	0.96	None	None	10 (insoluble compound)	1 10^{-4}
						5 (soluble compound)	2 10^{-4}
Ni	2.6	1.7	0.49	None	0.81	1 (Ni or compound)	3 10^{-3}

* not specified

Note: For comparison purposes, the air dilution factor for 1 kg of HTO is 5×10^4 km^3.

radioactivity dispersion are well established. Less established are the various biological pathways for the source of the isotopes involved and long-term effects due to lack of some of the needed biological behaviour.

Other uncertainties would also need to be resolved, for example, the form of the released radioactivity implies certain deposition behaviour. The implication of such uncertainties on the consequences also need to be addressed.

4.4. Fires and explosions

An ordinary fire is a familiar industrial hazard that any designer will attempt to eliminate and provide protection for sensitive instrumentation. However, certain components of fusion reactors present an inherent potential for fires and explosion which need to be considered in design. Obviously tritium may combine with oxygen explosively under certain conditions. If liquid water is suddenly brought into contact with the cryogenic fluids, vapour explosions may result [4]. If a liquid metal is involved in the blanket as a coolant or breeding material, its contact with air, water or concrete may lead to exothermic reactions that result in high temperatures in the flame area [4].

The above described fires have all been addressed in other industries but are just beginning to be addressed in connection with fusion reactors. Explosive detonation of hydrogen can occur at concentrations in the range from approximately 18 to 59% in air.

Vapour explosion theories have had a history of controversy, but seem to be associated with spontaneous vapourization of superheated liquids. Experiments related to the combination of fluids of interest here have not been reported (e.g. H_2O-He) but may be close to the LNG experiments [4].

LITFIRE, a predictive tool for consequences of lithium fires is being developed at MIT in conjunction with experimental effort at HEDL [4]. Other potential materials for blanket modules involve Li-Pb eutectics and Li_2O, for which experiments are just beginning [4].

4.5. Release of toxic materials

Toxic elements that may be involved in a fusion reactor had been given some consideration in early assessment of the overall risks of fusion power [60]. A list of potentially toxic elements in various conceptual designs is given in Table XIX-9. The relative dilution factor in terms of km^3/kW_e is several orders of magnitude less than that associated with radioisotope dilution factors. Furthermore, in the absence of Hg and Be, the hazard potential is significantly diminished. Perhaps Cu, Li and Pb hazard potentials need to be addressed relative to occupational effects but this should be possible from the existing data base.

TABLE XIX-10. *SUMMARY OF POTENTIAL MAGNET ACCIDENT INITIATORS

Mechanical initiators associated with operation or assembly of magnet
- Fracture of conductor
- Fracture of reinforcement
- Cracking or plastic deformation of stabilizer resulting in high resistivity
- Fracture of superconductor filaments (non-cryogenically-stable conductor)
- Excessive conductor movement (glue fractures)
- Crack in insulation
- Rupture of cryostat
- Failure of magnet support structure

Electrical insulation (non-mechanically caused)
- Metal inclusion in insulator (as fabricated or during assembly)
- Metal particle causes bridging between conductor turns
- Excessive voltage during coil dump or quench
  - Low dielectric strength of heated helium
  - Radiation induced ionization in helium
  - Degradation of insulation by radiation
- Improper installation of insulation by radiation

Coolant initiators
- Break of coolant line(s) from refrigerator
- Coolant channel blockage by broken spacer
- Coolant channel blockage by conductor or coil movement
- Coolant channel blockage by collection of impurities from coolant (from refrigerator or other parts of magnet)
- Liquid He – Hot coolant pipe interaction

Conductor initiators (non-mechanically caused)
- Radiation damage to stabilizer causing high resistivity
- Radiation damage to superconductor (non-cryogenically stable conductor)
- Impurities or inclusions in stabilizer from ingot or conductor fabrication
- Excessive dB/dt (non-cryogenically stable conductor)
  - Sudden loss of plasma current
  - Magnet charge or discharge too rapid
  - Coils pulsed too fast

Lead initiators
- Fracture of lead
- Inadequate cooling of heat resulting in melting

Initiators external to magnet system
- Plasma quench
- Missiles from reactor plant (turbine, coolant line rupture) Flywheel of energy storage system, etc.
- Missiles external to reactor plant (airplanes, etc.)
- Earthquakes
- Sabotage
- Human error

Control system malfunction initiators
- Loss of power to control system
- Failure of sensors
- Failure of interlocks
- Computer malfunction
- Failure of energy dump switching systems

* Combination of Studies, BNL, UCLA & Battelle Studies.

4.6. Electrical system accidents

High voltage system faults are of major concern in the design and operation of a Tokamak reactor. Probable causes of such faults could be due to overheating and overvoltage. The TFTR PSAR [4] lists three possible locations of interest a) a high voltage cable, b) conversion equipment and c) TF field coil fault. It is concluded that "because of the nature and complexity of the equipment involved and from previous experience with similar equipment, it is reasonable to conclude that, despite a great deal of effort expended to avoid them, some high voltage faults may occur and safeguards are necessary".

4.7. Magnet system failures

Substantial amounts of energy (hundreds of GJ) will be stored in the magnetic fields of a Tokamak reactor. Sudden release of such energies due to magnet "quench" has been of concern in the following respects:

a) Sudden energy dump by induction in radioactivity bearing structural materials leading to activation product mobilization
b) Generation of inter-coil forces that lead to structural failures accompanied by missile generation
c) Generation of excessive temperatures internally that lead to local failures accompanied by electric arcs which could melt or fume the coil structure.

Several recent magent studies have attempted to quantify the consequences of magnet accidents [4, 2]. Table XIX-10 presents a summary of potential magnet accident initiators from these studies. The following conclusions can be drawn from these studies:

(1) Magnet coils should be isolated as much as possible from mechanical interaction with the blanket, neutral beam and other radioactivity bearing systems.
(2) Inductive energy transfer to first wall and blanket areas is not likely by itself to induce large temperature uses in the structures. High pressures may result in conducting fluids with long current paths. This can be precluded by design.
(3) Temperature equalization between the conductor and the support structure is needed to prevent local damage (yielding or electrical breakdown) during a quench. Internal energy dump is necessary due to voltage limitation (1000 volts) on external energy dump.
(4) Detection of local hot spots is necessary.

(5) Arcing in INTOR has been examined in the European Report and the results for 16GJ stored energy showed that the liquid helium evaporation rate could approach 24 kg/s while as much as 0.6 kg of steel could be melted per second.

Most of the existing studies, however, do not proceed with the analysis mechanistically to assess the safety consequences of magnet failures. Mechanistic safety-inspired design criteria need to be further developed.

Experience to date is generally based on small magnet systems. Operation of a large magnet safety facility seems to be a necessary step along the way to commercial fusion power. Safety-related tests are planned to be included in the Large Coil Program at ORNL and for the T-15 reactor in the Soviet Union.

4.8. Water coolant system failures

The USSR working group has investigated consequences of a major water coolant system failure in a device like INTOR. The Major Design Accident is initiated by damage to the first wall which releases 300 litres of water into the plasma chamber. It is assumed that approximately 10^5 Ci (10 g) of tritium will be included in the water. The USSR report then examined the procedures which would be necessary to change the damaged sections, remove the contaminated water and wash the surfaces of the machine to remove residual activity. The estimated time for such measures is ~100 days and the probabilities of such an event occurring in INTOR was estimated to be 10^{-6} for a 10 year INTOR lifetime.

4.9. Helium coolant system failures

The Japanese working group has also analysed the effects of a helium coolant failure in a reactor like JXFR; no major problems were identified for a He cooled-Li_2O blanket [2].

5. CONTAINMENT REQUIREMENTS

5.1. Introduction

Containment requirements for INTOR will stem from a number of interrelated factors that must be considered to ensure the health and safety of the general public and reactor operations personnel under a variety of normal, off-normal, and accident conditions. Some of the more significant factors are:

(1) potential for tritium release during normal and accident conditions,
(2) potential for release of activation products,
(3) air activation from neutron streaming,
(4) the quantity of lithium in combustible form that may be used in blanket modules,
(5) potential for buildup of explosive mixtures of hydrogen isotopes,
(6) pressurization potential during failure of components in the cryogenic systems,
(7) site-dependent requirements for protection from natural phenomena such as earthquakes, tornadoes, floods, etc.,
(8) potential for internally- and externally-generated missiles (e.g., aircraft).

Several levels of containment will likely be required for several of these factors, e.g. tritium or lithium which may require double-walled vessels and piping.

The design of containment buildings for light-water fission reactors has developed to a high degree of sophistication to handle a variety of operational and accident conditions coupled with release of radioactivity to the building. These containments have been designed to:

(1) Limit containment leak rate under specified pressurization.
(2) Limit direct-shine radiation by shielding.
(3) Provide positive mechanical and control systems to isolate containment penetrations under accident conditions.
(4) Provide mechanical and structural strength to withstand seismic, wind, missile and flood loadings coupled in certain cases with steam pressurization and other accident-specified loadings.
(5) Provide internal hydrogen recombination systems.
(6) Provide internal water spray system to reduce radioactive species within the containment atmosphere.
(7) Provide internal cooling systems to condense steam and reduce containment atmosphere temperature.

Containment design is also under development for liquid metal fast breeder reactors. Additional conditions for these systems include:

(8) Sodium spill with consequent fire and containment atmosphere temperature rise.
(9) Reactor excursions with consequent energy release leading to missile formation, shock waves and containment pressurization.
(10) Sodium oxide aerosol generation with consequent chemical and physical interaction with materials and components.

Fusion reactor systems can use the results of the above work for much of the containment requirements that are presently foreseen to be needed. There are several differences in the systems that will eliminate many of the fission reactor containment requirements and also a few that may lead to new design conditions. Since this study is an attempt to assess fusion reactor R and D needs, only the new conditions will be addressed.

The large quantities of tritium that will be required for INTOR lead to the possibility of tritium leakage, permeation, adosrption and desorption and large-scale accidental release of tritium to the containment system. Containment design requirements and methods to mitigate the consequences of tritium release need to be developed. The Tritium System Test Assembly (TSTA) under construction at the Los Alamos National Laboratory and the Japanese Tritium Engineering Test Facility are expected to provide specific data related to many of the containment design problems.

The use of lithium in blanket modules represents a larger and more chemically active energy release situation in the event of a spill than sodium systems. Lithium spill experiments at HEDL and lithium fire modelling activities at MIT are expected to provide basic data. This work together with sodium-spill activities in the LMFBR programme should provide the basis for containment requirements and design relative to liquid metal fires inside containments for fusion reactors.

Air activation represents a fusion reactor containment problem because of possible neutron streaming effects through beam lines. It is not known at this time if adequate shielding can eliminate this problem. Special containment ventilation systems may be required to handle this problem. If a vacuum building is used, this will not be a particular problem. Vacuum buildings of the size required for INTOR containment have been built and operated successfully (see NASA publication cited in [4]).

5.2. Containment under normal operating conditions

The INTOR reactor system under normal operating conditions will release small quantities of tritium to the containment by permeation through piping and equipment. Special design of these systems can probably greatly reduce the source of leakage by using data to be developed at Tritium Test Facilities beginning in 1981. The introduction of breeding blanket modules operating at high temperatures may introduce difficulties in preventing the permeation of small amounts of tritium to the containment. It therefore appears desirable to design a tritium-removal system within the containment structure capable of maintaining tritium levels under normal operating conditions well below maximum permissible concentration (MPC) values. The data to design such a system should be available by 1985. The data include: tritium permeation rates through piping and equipment as a function of material, temperature, tritium partial pressure, and coatings; tritium adsorption and desorption from piping equipment and surfaces during maintenance;

tritium cleanup methods and system design data. Lithium breeding modules operating at high temperature may result in large permeation rates of the bred tritium. This problem can be minimized by operating the blanket at very low tritium concentrations. Calculations are required to determine the expected tritium-release rates to the containment from blanket module experiments. Permeation rate data are not available. Upper bound calculations may be required to provide the containment designers the data needed to design the cleanup system.

The activation of containment air by neutron streaming through beam lines may require special ventilation systems within the containment. The reactor shielding design and analysis needs to be completed to determine if air activation is a problem. The major activity is expected to be from Ar-41. Ar-41 levels about a factor of ten above occupational limits may be expected. The short half-life of Ar-41 (1.83 h) may permit continuous venting or hold-up tanks may be used to reduce the vented activity to acceptable levels. The high-energy neutrons escaping the reactor may lead to threshold reactions with components in the air, generating long-lived radio isotopes that will require further investigation.

5.3. Containment under accident conditions

The set of accidents for design of INTOR have not been defined. This is a necessary first step to ensure that the containment is designed adequately to meet established criteria under a wide spectrum of accident conditions. However, certain preliminary assumptions can be made based on present INTOR Reference Concept Parameters. The following represents a preliminary set of accidents for INTOR:

(1) Large tritium release to containment
(2) Lithium spill from blanket module
(3) Earthquake
(4) Tornado
(5) Flood
(6) External Missile
(7) Cryogenic dewar failure with helium liquid spill
(8) Internal Missile

Containment design data for (3) through (6) above are presently available from Fission Reactor Containment design guides [4] and the methodology can be adopted for INTOR. The data for accidents (1), (2), (7) and (8) are specific to Fusion Reactor Systems and will depend on detailed INTOR design information. An examination of the data requirements for these latter accidents follows.

Large tritium spill. The maximum tritium inventory that can be credibly spilled from INTOR needs to be established. The ventilation system for INTOR containment must be interfaced with the cleanup system including containment isolation. Cleanup rate requirements will establish the size of the tritium cleanup system.

Lithium spill from blanket module. Lithium breeding modules may introduce accident sequences that lead to containment pressurization and/or high temperature in the containment atmosphere together with aerosol generation that may contain radioactive species. Analysis of these accidents will require the following data that is presently being developed by HEDL in the Lithium Spill experiments and by MIT in its lithium fire modelling efforts:

(1) A lithium spill analytical model verified by experiments that provides for analysis of containment pressures and temperatures;
(2) Aerosol properties and transport;
(3) Activation material properties and transport.

The results of these analyses can be used to design the containment to ensure acceptable consequences.

Cryogenic dewar failure. Loss of liquid helium to the containment could conceivably cause containment pressurization if coupled with a large tritium release that isolates the containment. Analysis of this type of accident can be performed using a modified version of the CONTEMPT reactor containment code [4].

Internal missile generation. Identification and analysis of the consequences of potential missile sources for fusion reactor systems is not available. Superconducting magnet systems, dewar pressurization from overheating and several coolant accident sequences have been identified as potential missile sources. Detailed INTOR design information is required to establish credible accident sequences. It is expected that methods of analysis are available to analyse this type of accident [4].

5.4. Explosive containment requirements

The possibility of explosions within the containment building is based on the availability of explosive mixtures of material and the presence of ignition sources or conditions leading to self-ignition. Vapour explosion can also occur by interaction of certain liquids at different temperatures.

The INTOR reactor system will employ hydrogen and its isotopes as plasma fuel and the cryo-distillation column which, under accidental release to the atmosphere, could form local concentrations within explosive limits. Ignition sources are plentiful within the reactor system; therefore, calculation of a credible hydrogen explosion will be required. Analysis of the resulting explosion and shock-wave effects on the containment is within present technology [4].

The possibility of vapour explosions between liquids such as liquid helium and the INTOR coolant and/or other available liquids must be evaluated. If the criteria for vapour explosions are met, analysis of the explosion and energy release parameters will be necessary. The data for fusion reactor liquid-liquid interactions are not available and an R and D programme similar to those employed in the LMFBR programme may be required.

6. TRITIUM SAFEGUARDS REQUIREMENTS

6.1. European Community requirements

The main treaty relevant to European Community is the Rome treaty establishing Euratom (25 March 1957). A second treaty in which all the Community countries, except France, are involved is the Non Proliferation Treaty (March 1970) [1]. Among the aims of the Rome Treaty, there are the following:

(Art.2): " .. (d) to ensure a regular and equitable supply of ores and nuclear fuels to all users in the Community (e) guarantee, by appropriate measures of control, that nuclear materials are not diverted for purpose other than those for which they are intended ...".

Point (e) justifies the presence of a safeguard system. However, it must be pointed out that tritium is not covered by any of these treaties. It is a fact that, up to now, a tritium delivery among EEC countries has not required safeguard control.

6.2. United States requirements

The current safeguards regulations in the US for tritium are dictated by the designation of tritium as a "sensitive" rather than a "strategic" nuclear material. As such, tritium is subjected to rigorous materials management requirements because it is considered a limited resource and therefore quite valuable ($4000–5000/g). It therefore must be physically protected in an adequate manner as would any material of this value. Tritium is, however, not treated as a significant proliferation threat at the present.

The shipping and storage of tritium is a question that needs to be addressed. The largest, approved shipping container for tritium gas in the United States is the Savannah River LP-50 which has been approved under Certificate of Compliance Number USA/6678/BL. This is a 50ℓ container approved for shipments of no more than 193 500 Ci ($\cong$ 20 g) of T_2. A 3 kg inventory would therefore require shipment of $\cong$150 such containers. This is a logistics problem, plus the availability of these containers is a matter of concern. It would seem that the most logical direction to proceed would be to develop and obtain a Certificate of Compliance on a new, larger shipping container. The LP-50 is limited to a total pressure of 1250 torr. This is adequate for small shipments and makes for safe, reliable storage of T_2 gas at low pressures; however, perhaps higher pressure containers will have to be developed for the fusion programme. There appear to be no unsurmountable problems here, only the long time delays associated with passing through the bureaucratic licensing procedures. Galloway [4] has written a concise summary of the current shipping and storage techniques.

One area in which new policies are needed is that of shipping large quantities of tritium to an INTOR site in a foreign country. Tritium is currently sold from the US to non US customers. During 1978 for example, the ORNL Isotope Sales Group sold 200–300 g of tritium to non DOE customers, most of it for use in tritium-activated light sources. Of this 200–300 g, approximately half was sold to non US customers. This gas was shipped in quantities not exceeding 30 000 Ci (3 g) per container. So the precedent for tritium shipments outside the US is established, but on a much smaller scale than required for INTOR. Current practices require that the Nuclear Regulatory Commission issue an export license prior to the shipment. This license is issued by NRC only after approval for the license is granted by the US State Department.

Tritium storage has typically been as low pressure (1250 torr) gas in the LP-50 shipping containers. This method of storage has not led to any major problems and should be considered as a prime candidate for the INTOR project. It is possible to store the tritium as a solid on a metal bed. Typically, uranium has been used forming the uranium tritide. This system has been extensively studied with Carlson [1] presenting a good summary of the use of uranium for storage. A major drawback to this technique is that two extra handling steps (the sorption and desorption of the tritium) are required. These are two additional steps where a potential exists for operator error, leading to a tritium release.

Because of its high value and limited availability the materials management requirements are such that very accurate accountability records must be maintained. The Department of Energy requires a biannual physical measurement of tritium inventories. Tritium must be accounted for to the closest 0.01 g. This presents some serious technical problems to a fusion facility with a 3 kg inventory, some of which is tied up in cryogenic vacuum systems and a cryogenic distillation system. The development of new requirements and new techniques for maintaining accountability data is a challenge that must be solved prior to using large quantities of tritium at a fusion facility. The TSTA will start to address these problems in the next few years. Current safeguard regulatiosn for tritium in the US are given in a DOE Manual [4].

The above discussions indicate that an adequate policy for the control and safeguards of tritium must be developed for INTOR. If it is assumed that INTOR is not built in the country providing the tritium, then very extensive policies on safeguarding and accountability will have to be established by the early 1990's.

7. R AND D NEEDS ASSESSMENT

Table XIX-11 shows a summary of the R and D needs assessment for safety and environment.

TABLE XIX-11. R AND D NEEDS ASSESSMENT – SAFETY AND ENVIRONMENT

	Status*	
	Adequate	Inadequate
Develop reliable tritium handling, processing, containment and cleanup systems	x	
Obtain activation product generation, release, transport and shielding date	x	
Develop probabilistic and mechanistic accident analysis techniques		x
Determine tritium conversion rates and biological effects	x	
Obtain lithium and lithium compound fire consequences, prevention, extinguishment data	x	
Obtain magnet accident experimental data		x
Develop definitive magnetic field exposure guidelines for humans		x
Assess the consequences of earthquakes on the safe operation of INTOR		x

Adequate: Programmes are either planned or could be planned on existing or authorized facilities (with some upgrading if required) and using existing trained personnel.

Inadequate: Either the available and authorized manpower and facilities are inadequate or a substantial increase in effort is needed.

* The ratings are made on the world-wide fusion effort and not separately on the individual national programmes.

8. CONCLUSIONS

Regulations pertaining to the control of

(1) radioactivity,
(2) toxic materials, and
(3) industrial procedures associated with:
   (a) high vacuum technology
   (b) large cryogenic systems
   (c) high voltage systems

currently exist in the member nations and they are felt to be adequate for starting the design of INTOR. However, there are few, if any, regulations that are specific to fusion devices, and a concentrated effort to establish these guidelines for INTOR must start as soon as possible. It is likely to take 5–10 years to formulate such regulations in compliance with public review and legislative requirements.

The most critical example of where specific attention must be paid to fusion devices is in the area of tritium containment. The analytical models relating release of tritium to exposure to humans currently exist. In addition, the effects of tritium on the population for a given expore are also fairly well understood at this time. What is required is for a well-established, internationally respected group to put the two together to establish release limits specifically for D-T fusion devices. These guidelines are absolutely necessary before INTOR progresses into the detailed design phase.

There are also two regulations currently in force in some countries which could place undue constraints on INTOR. The first is the 20 g limit on the amount of tritium in approved shipping containers. To supply initial inventories of several kilograms could require an unduly large number of shipments. There is no insurmountable problem here, only the time delays associated with the design of larger containers and regulatory licensing of such units. The second problem is with the current accountability practices. It is currently required (at least in the US) that the entire tritium inventory of a facility be established to the nearest 0.01 g twice a year. Considering that inventories of about 3 kg could exist in INTOR and that 5–10 kg/year might be burned up, such a strict accounting may be impossible. The regulations should be examined to see if such a limit can be raised for INTOR or if such a limit is even appropriate for such a test device.

At the present time, the only material in INTOR which might be subjected to safeguards requirements is tritium. It is currently designated as a sensitive, rather than a strategic, material because of its high value (about 4000 to 8000 $/g). It is anticipated that this status will not change for INTOR but periodic monitoring of IAEA safeguards regulations should be conducted.

Specific regulations for the continued exposure of individuals to high magnetic fields are also urgently needed. At the present, the practice in the field is to use ad hoc values from the accelerator field which may or may not stand up to the close scrutiny of safety review panels.

Guidelines currently exist for the use of non-ionizing microwave radiation, but few, if any, regulations exist for RF heating sources that might be used in INTOR. Such regulations should be established in the next decade.

Considerable investigation has gone into identification of accident sequences for the tokamak fusion industry. Based on conceptual designs of commercial plants, event trees have been constructed to identify accident scenarios as an aid in improving the safety aspects of the next generation of conceptual designs and quantitative assessment of accident consequences. These event trees are not usable at this time to determine defendable values of risk (probability of an accident times the consequences) because of a lack of specific reactor design details and lack of an adequate data base for component failure rates. Such a data base must be developed; however, it will not likely be sufficiently developed in time for quantitative application to INTOR. The state of development of mechanistic computer models for fusion reactor accidents is at an early state. Further development is needed if defendable safety analyses are to be done for INTOR other than "worst case" analyses. The codes will also need to be verified by comparison with experimental data from well-characterized and instrumented safety-related experiments, which are just starting to be performed.

In summary, there are several things that can be done at an early stage to ensure the safety of INTOR. However, neglect of these areas could delay the timely operation of INTOR.

REFERENCES TO CHAPTER XIX

The bulk of this summary was taken from four reports to the INTOR Workshop [1–4]. The reader can consult the original documents cited here and in [1–4] for references and authorship of the analyses.

[1] European Contributions to the 2nd meeting of the INTOR Workshop, Safety and Environment, Chapter XV, EUR FU/XII 501/79/EDV 50, June 79 (2 Volumes).
[2] Fusion Research Development Center, INTOR Workshop Report, Home Task Report of Group 15, JAERI-M 8711 (1979).
[3] USSR Report to INTOR, "Radiation Safety and Protection of Environment" (see Appendix A.5.3 [43]).
[4] "Safety and Environment", Chapter in USA INTOR Report (see Appendix A.5.4).
[5] JACOBS, D.C., "Sources of tritium and its behaviour upon release to the environment", AEC Critical Review Series, US Atomic Energy Commission, Division of Technical Information, Oak Ridge, Tennessee, December 1968.
[6] EASTERLY, C.E., SHANK, K.E., SHOUP, R.L., Health Physics Aspects of Fusion Power, Report ORNL/TM-5461, October 1976.
[7] Draft ICRP publication 10X (in preparation).

[8] HUNGATE, F.P., Balanced Programme Plan. Volume 10. Fusion: Analysis for Biomedical and Environmental Research, Argonne National Laboratory, Report ERDA-116 (Vol.10), June 1976, page 91.

[9] BURGER, L.L., Conversion of Tritiated Hydrogen to Water in the Atmosphere, Report BNWL-2113, September 1976.

[10] PHILLIPS, J.E., EASTERLY, C.E., Tritium Oxidation and Exchange: Preliminary Studies, Report ORNL/TM-5963, May 1978.

[11] DOBSON, R.L., "New radiobiological findings bearing on the 1977 ICRP recommendations", Preprint UCRL-81774, February 1979, Paper presented at the Topical Seminar on the Practical Implications of the ICRP Recommendations and the Revised IAEA Basic Safety Standards for Radiation Protection, IAEA-SR-36/23, IAEA, Vienna 5–9 March 1979.

[12] UNSCEAR, Sources and Effects of Ionizing Radiation, 1977 Report to the General Assembly, United Nations, New York 1977.

[13] IAEA Safety Series No.45, Principles for Establishing Limits for the Release of Radioactive Materials into the Environment, IAEA, Vienna, 1978.

[14] IAEA, ILO, WHO, OECD/NEA, ICRP, Topical Seminar on the Practical Implications of the ICRP Recommendations (1977) and the Revised IAEA Basic Safety Standards, IAEA Vienna, 5–9 March 1979.

[15] LINDELL, B., "Basic concepts and assumptions behind the new ICRP recommendations", paper IAEA-SR-36/51, presented at [14].

[16] Limits for Intakes of Radionuclides by Workers, ICRP Publication No.30 (1979).

[17] IAEA, Revision of the IAEA's Basic Safety Standards, draft. (not yet adopted).

[18] GORMAN, D.J., WONG, K.Y., "Environmental aspects of tritium from CANDU station releases", Paper No. IAEA-SM-232/49, presented at the IAEA International Symposium on the Behaviour of Tritium in the Environment, San Francisco, California, USA, 16–20 October 1978.

[19] SHANK, K.E., OAKES, T.W., EASTERLY, C.E., Assessment of Radiological Releases to the Environment from a Fusion Reactor, Report ORNL/TM-6368, May 1978.

[20] SOLDAT, D.A., BAKER, D.A., Worldwide Population Doses from Tritium Released from Nuclear Facilities, Report PNL-SA-6881, paper IAEA-SM-232/69 presented at the IAEA International Symposium on the Behaviour of Tritium in the Environment, San Francisco, USA, October 16–20, 1978.

[21] ICRP Publication 29, Radionuclide Release into the Environment: Assessment of Doses to Man, Pergamon Press Oxford, New York, Frankfurt (adopted by the Commission in October 1978).

[22] KETCHEN, E.E., PORTER, W.E., BOLTON, N.E., Am. Ind. Hyg. Assoc. J. **39** 1 (1978).

[23] ALPEN, E. L., "Magnetic field exposure guidelines", Proceedings of the Biomagnetic Effects Workshop, April 6–7, 1978, Lawrence Berkeley Laboratory, Berkeley, Calif., USA.

[24] Safety and Environmental Aspects of Deuterium-Tritium Fusion Power Plants: Workshop Summary, CONF-770997 (1978).

[25] HÄFELE, W., HOLDREN, J., KESSLER, G., KULCINSKI, G.L., Fusion and Fast Breeder Reactors, Int. Institute for Applied Systems Analysis, Laxenburg, Austria (1976).

APPENDICES

A.1. INTOR WORKSHOP SESSIONS AND ATTENDEES

1. Workshop Sessions

Steering Committee Meeting:	20–23 November 1978
Workshop Session I:	5–16 February 1979
Workshop Session II:	11 June – 6 July 1979
Workshop Session III:	1–19 October 1979
Workshop Session IV:	10–19 December 1979

2. Workshop Attendees

European Community (EURATOM)

G. Grieger, Steering Committee member
F. Engelmann
D. Leger
P. Reynolds
R. Verbeek, Scientific Secretary
J. Bohdansky, Expert, Session II
G. Casini, Expert, Session III
J. Darvas, Expert, Session III
R. Hancox, Expert, Sessions I–IV

Japan

S. Mori, Chairman, Steering Committee member
T. Hiraoka
K. Sako
T. Tazima
H. Momota, Expert, Session III
Y. Shimomura, Expert, Session III

USSR

B.B. Kadomtsev, Steering Committee member
B.N. Kolbasov
V.I. Pistunovich
G.E. Shatalov
G.F. Churakov, Expert, Sessions II and III

USA

W.M. Stacey, Jr., Vice-Chairman, Steering Committee member
J.R. Gilleland
G.L. Kulcinski
P.H. Rutherford
C.C. Baker, Expert, Session III
V.A. Maroni, Expert, Session III
D.M. Meade, Expert, Session II
J.R. Purcell, Expert, Session II
J.K. Griffith, Administrative Assistant, Session II

IAEA

R.A. Ellis, Head, Physics Section
J.A. Phillips, Consultant, Session III (Head, Physics Section, until July 1979)
H. Seligman, Consultant
V.S. Vlasenkov, Physics Section
F.N. Flakus, Radiological Safety Section, Expert, Session II

A.2. INTOR HOME BASE ORGANIZATIONS

A.2.1. EC (EURATOM) INTOR Workshop Teams

	CATEGORY	COORDINATOR	PARTICIPANT	TEAM MEMBERS*	
1.	Energy and Particle Confinement	F. Engelmann	F. Engelmann	**R. Bickerton**	JET, Culham
				D. Ashby	Culham
				J. Cordey	Culham
				D. Duchs	IPP, Garching
				A. Gibson	JET, Culham
				H. Girard	CEA-FAR
				M. Hughes	Culham
				J. Hugill	Culham
				F. Karger	IPP, Garching
				Ch. Mercier	CEA-FAR
				L. Pieroni	CHEN, Frascati
				J. Taylor	Culham
				C. Thomas	Culham
				H. Watkins	JET, Culham
	Transport Codes	F. Engelmann	F. Engelmann	**E. Minardi**	Culham/ Garching
				D. Duchs	IPP, Garching
				M. Hughes	UKAEA, Culham
				Ch. Mercier	CEA-FAR
				J. Raeder	IPP, Garching
				G. Realini	JRC, Ispra
				A. Taroni	CNEN, Bologna
				M. Watkins	JET, Culham

* Bold-face type denotes team leaders

	CATEGORY	COORDINATOR	PARTICIPANT	TEAM MEMBERS	
2.	First Wall, Blanket Shield	G. Casini	R. Hancox	**R. Hecker**	KFA, Julich
				H. Brockmann	KFA, Julich
				J. Darvas	CEC, Brussels
				D. Filges	KFA, Julich
				S. Forster	KFA, Julich
				J. Mitchell	UKAEA, Culham
				P. Mioduszewski	KFA, Julich
				U. Ohlig	KFA, Julich
				M. Rieger	JRS, Ispra
				R. Verbeek	CEC, Brussels
3.	Impurity Control, Fuelling and Exhaust	F. Engelmann	F. Engelmann	**W. Heiland**	IPP, Garching
				J. Bohdansky	IPP, Garching
				B. Brandt	FOM, Jutphaas
				P. Braun	Sweden
				W. Eckstein	IPP, Garching
				P. Harbour	UKAEA, Culham
				N. Harrison	UKAEA, Culham
				L. Lengyel	IPP, Garching
				P. Mioduszewski	KFA, Julich
				H. Niedermeyer	IPP, Garching
				D. Ohlsson	KTH, Stockholm
				W. Poschenrieder	IPP, Garching
				H. Vernickel	IPP, Garching
				M. Watkins	JET, Culham
4.	Heating	F. Engelmann	F. Engelmann	**F. Prevot**	CEA, Fontenay
				M. Bernard	CEN, Grenoble
				R. Brambilla	CEN, Grenoble
				G. Briffod	CEN, Grenoble
				R. Cano	CEA-FAR
				U. Finzi	CEC, Brussels
				R. Geller	CEN, Grenoble
				J. Girard	CEA-FAR
				J. Jacquinot	CEA-FAR
				A. Messiaen	ERM, Brussels
				F. Santini	CNEN, Frascati
				D. Sweetman	UKAEA, Culham
				G. Tonon	CEN, Grenoble
				F. Valckx	CEA-FAR
5.	Magnetics	G. Casini	R. Hancox	**R. Aymar**	CEA, Fontenay
			P. Reynolds	S. Bobbio	Univ., Naples
				J. Elen	ECN, Petten
				F. Farfaletti-Casali	JRC, Ispra
				A. Knobloch	IPP, Garching
				P. Komarek	KFK, Karlsruhe

CATEGORY	COORDINATOR	PARTICIPANT	TEAM MEMBERS	
5. Magnetics (cont.)			H. Krauth	KFK, Karlsruhe
			C. Leloup	CEA-FAR
			N. Sacchetti	CNEN, Frascati
			W. Spears	UKAEA, Culha
			A. Torossian	CEA-FAR
			G. Vecsey	SIN, Villigen
6. Materials	G. Casini	R. Hancox	**P. Schiller**	JRS, ISPRA
		P. Reynolds	Y. Adda	CEA, Saclay
			F. Brossa	JRC, Ispra
			V. Coen	JRC, Ispra
			J. Delaplace	CEA, Saclay
			K. Ehrlich	KFK, Karlsruhe
			J. Elen	ECN, Petten
			D. Evans	UKAEA, Harwe
			P. Fenici	JRC, Japan
			K. Kaletta	KFK, Karlsruhe
			F. Lanza	JRC, Ispra
			R. Matera	JRC, Ispra
			J. Nihoul	CEN, Mol
			G. Pellegrini	JRC, Ispra
			D. Phillips	UKAEA, Harwe
			H. Uhlmaier	KFA, Julich
			W. Van Witsenburg	ECN, Petten
7. Systems Integration	G. Casini	R. Hancox	**F. Farfaletti-Casali**	JRC, Ispra
		P. Reynolds	M. Bustraan	ECN, Petten
			S. Forster	KFA, Julich
			J. Mitchell	UKAEA, Culham
			T. Raimondi	JET, Culham
			J. Remonville	Luxamton, Luxembourg
			A. Torossian	CEA-FAR
8. Power Supply and Transfer	G. Casini	P. Reynolds	**A. Knobloch**	IPP, Garching
			V. Braunsberger	TU, Braunschwe
			K. Selin	JET, Culham
			A. Van Ingen	FOM, Jutphaas
9. Assembly and Remote Maintenance	G. Casini	R. Hancox	**F. Farfaletti-Casali**	JRC, Ispra
		P. Reynolds	M. Bustraan	ECN, Petten
			S. Forster	KFA, Julich
			J. Mitchell	UKAEA, Culhar
			T. Raimondi	JET, Culham
			J. Remonville	Luxamton, Luxembourg
			A. Torossian	CEA-FAR

CATEGORY		COORDINATOR	PARTICIPANT	TEAM MEMBERS	
10.	Shielding and Personnel Access	G. Casini	R. Hancox P. Reynolds	**C. Ponti** J. Mitchell T. Raimondi K. Verschur	JRC, Ispra UKAEA, Culham JET, Culham ECN, Petten
11.	Stability Control	F. Engelmann	F. Engelmann	**D. Troyon** G. Bosia R. Dei Cas J. Goedbloed G. Lackner G. Maschke Ch. Mercier J. Taylor	CRPP, Lausanne ERM, Brussels CEA-FAR FOM, Jutphaas IPP, Garching CEA-FAR CEA-FAR UKAEA, Culham
12.	Start-up, Burn and Sut-down	F. Engelmann	F. Engelmann	**J. Raeder** R. Bickerton K. Borrass B. Brunelli R. Buende R. Cano A. Cavallo M. Gasparotto E. Minardi A. Sestero	IPP, Garching UKAEA, Culham IPP, Garching CNEN, Frascatti IPP, Garching CEA-FAR CEA-FAR CNEN, Frascati Culham, Garching CNEN, Frascati
	Sensitivity Analysis	F. Engelmann	F. Engelmann	**J. Raeder** K. Borrass R. Buende	IPP, Garching IPP, Garching IPP, Garching
13.	Tritium	G. Casini	D. Leger	**D. Leger** W. Denton H. Dworschak H. Ihle A. Leclair G. Pierini	CEA, Saclay UKAEA, Harwell JRC, Ispra KFA, Julich UKAEA, Harwell JRC, Ispra
14.	Vacuum	G. Casini	D. Leger P. Reynolds	**P. Reynolds** H. Conrads J. Coupland P. Harbour M. Harrison R. Hemsworth H. Henning W. Poschenrieder F. Reiter	UKAEA, Culham KFA, Julich UKAEA, Culham UKAEA, Culham UKAEA, Culham UKAEA, Culham Pfeiffer, Vakuumtechnik IPP, Garching JRC, Ispra

	CATEGORY	COORDINATOR	PARTICIPANT	TEAM MEMBERS	
15.	Safety and Environment	G. Casini	D. Leger	**P. Rocco**	JRC, Ispra
				A. Arendt	KFK, Karlsruhe
				A. Benco	JRC, Ispra
				R. Bergman	NE, Studsvik
				J. Gruber	HMI, Berlin
				P. Komarek	KFK, Karlsruhe
				H. Klippel	ECN, Petten
				P. Mioduszewski	KFA, Julich
				S. Ragusa	JRC, Ispra
				P. Sauermann	KFA, Julich
				J. Schneider	JRC, Ispra
				T. Teague	UKAEA, Warri
				L. Warming	DAE, Risø
16.	Diagnostics, Data Acquisition and Control	G. Casini	P. Reynolds	**H. Persson**	NE, Stockholm
				F. Bombi	JET, Culham
				R. Espefalt	NE, Studsvik
				F. Koechlin	CEA-FAR
				J. Schluter	KFA, Julich
				A. Read	UKAEA, Culha
17.	Cost Schedule, Manpower	G. Casini	G. Grieger	**A. Knobloch**	IPP, Garching
			R. Hancox	R. Aymar	CEA-FAR
				P. Schiller	JRC, Ispra
				A. Van Ingen	FOM, Jutphaas

A.2.2. Japan INTOR Workshop Teams *

	CATEGORY	PARTICIPANT	TEAM MEMBERS**
1.	Energy and Particle Confinement	T. Tazima	**M. Sugihara**
			N. Kasai [f]
			Y. Kinoshita [g]
			H. Momota [h]
			H. Naito [h]
			M. Nagami
			M. Ohnishi [g]
			Y. Shimomura

* Unless otherwise indicated, contributors are staff of the Japan Atomic Energy Research Institute.

** Bold-face type denotes team leaders.

CATEGORY	PARTICIPANT	TEAM MEMBERS
		T. Takizuka M. Tanaka K. Tani T. Tazima T. Tone T. Tuda
2. Blanket/Shield/First Wall	K. Sako	**H. Iida** J. Adachi [c] S. Ando [e] N. Asami [k] T. Aota K. Ebisawa [e] Y. Gomay [e] H. Hamajima [e] K. Hashimoto [e] K. Hida [e] T. Honda [e] Y. Imamura [k] K. Inoue [a] K. Ioki [k] S. Itoh [a] T. Iwamoto [a] A. Kameari [k] T. Kasahara [a] K. Kitamura [k] M. Kitamura [k] T. Kobayashi [a] M. Koizumi [a] H. Kozawa T. Kuroda [c] K. Maki [a] A. Minato [c] S. Mori [c] T. Munakata [e] M. Naganuma [e] I. Nagata [a] T. Naito [k] M. Nakagawa [c] Y. Nakai [c] S. Niikura [k] M. Nishi [a] M. Nishikawa [k] O. Ogino [k] R. Saito [k] H. Sakamoto [d] K. Sako

CATEGORY	PARTICIPANT	TEAM MEMBERS
2. Blanket/Shield/First Wall (cont.)		T. Satow [k]
		Y. Sawada [a]
		Y. Seki
		K. Shinya [e]
		T. Sugawara [e]
		Y. Tanaka [c]
		M. Tokumitsu [c]
		T. Tone
		S. Uchida [a]
		T. Uchida [e]
		N. Ueda [f]
		M. Yamada [k]
		T. Yamada [k]
		M. Yamaguchi [e]
		T. Yamamoto [b]
		H. Yamato [e]
		M. Yamauchi [j]
3. Impurity Control, Fuelling and Exhaust	T. Tazima	**T. Tazima**
		T. Abe
		DIVA Group
		T. Hirayama
		K. Kamada
		M. Kasai [f]
		K. Maki [a]
		Y. Murakami
		H. Makamura
		K. Makamura
		Y. Nakamura
		K. Obara
		M. Saidoh
		K. Sako
		Y. Seki
		S. Sengoku
		Y. Shimomura
		K. Shinya
		K. Sone
		M. Sugihara
		R. Yamada
4. Heating	T. Tazima	**H. Shirakata**
		H. Horiike
		H. Kishimoto
		S. Matsuda
		T. Nagashima
		Y. Ohara
		T. Shibata

CATEGORY	PARTICIPANT	TEAM MEMBERS
5. Magnetics	K. Sako	**S. Shimamoto** H. Tsuji Others are the same as for Group 2.
6. Materials	T. Hiraoka	**K. Shiraishi** T. Kondo Y. Murakami S. Nasu T. Oku S. Takamura
7. Systems Integration and Support Structure	K. Sako	**K. Sako** Others are same as for Group 2.
8. Power Supply and Transfer	T. Hiraoka	**S. Tamura** H. Kishimoto N. Miya [i] R. Shimada K. Shinya
9. Assembly and Remote Maintenance	K. Sako	**K. Sako** Others are same as for Group 2.
10. Radiation Shielding and Personnel Access	K. Sako	**Y. Seki** Others are same as for Group 2.
11. Stability Control	T. Tazima	**T. Takeda** G. Kurita T. Matuura H. Ninomiya R. Shimada Y. Shimomura K. Shinya [e] T. Tsunematsu
12. Startup, Burn and Shutdown	T. Tazima	**T. Tone** M. Kasai [f] H. Kishimoto K. Maki [a] K. Shinya [e] M. Sugihara T. Tazima T. Yamamoto [b] H. Yoshida

CATEGORY	PARTICIPANT	TEAM MEMBERS
13. Tritium	T. Hiraoka	**Y. Obata** H. Katsuta M. Kinoshita H. Kudo Y. Matsuda Y. Naruse
13. Tritium (cont.)	T. Hiraoka	S. Nasu Y. Sanokawa E. Tachikawa K. Tanaka Y. Yoshida
14. Vacuum	T. Hiraoka	**Y. Murakami** T. Abe K. Nakamura K. Ohara
15. Safety and Environment	K. Sako	**T. Tone** Y. Seki
16. Diagnostics, Data Acquisition and Control	T. Hiraoka	**Y. Suzuki** T. Hiraoka A. Ogata
17. Cost, Schedule and Manpower	T. Hiraoka	**T. Hiraoka** M. Kasai[f] T. Tokutake

[a] Hitachi, Ltd., Tokyo
[b] Fuji Electric Co. Ltd., Kawasaki
[c] Kawasaki Heavy Industries, Ltd., Tokyo
[d] Mitsubishi Heavy Industries, Ltd., Kobe
[e] Toshiba Corporation, Kawasaki
[f] Mitsubishi Atomic Power Industries Inc., Omiya
[g] Institute of Atomic Energy, Kyoto University
[h] Institute of Plasma Physics, Nagoya University
[i] Scholarship fellow from Nippon University
[j] Nippon Atomic Industry Group
[k] Mitsubishi Group (Mitsubishi Electric Company, Mitsubishi Heavy Industries and Mitsubishi Atomic Power Industries)

A.2.3. USSR INTOR Workshop Teams

	CATEGORY	PARTICIPANT	TEAM MEMBERS*	
1.	Energy and particle confinement	B.B. Kadomtsev	A.S. Kukushkin	IAE
		V.I. Pistunovich	S.E. Lysenko	IAE
			G.V. Pereverzev	IAE
			V.I. Pistunovich	IAE
			A.P. Smirnov	IAE
			K.N. Tarasjan	IAE
2.	First Wall, Blanket, Shield	B.N. Kolbasov	**V.V. Gurjev**	IAE
		G.E. Shatalov	**G.E. Shatalov**	IAE
			I.Ya. Emel'janov	IAE
			A.M. Epinat'jev	IAE
			B.D. Gornostajev	IAE
			E.M. Kuz'min	IAE
			A.F. Lekomtsev	IAE
			N.A. Osadchij	IAE
			V.P. Smetannikov	IAE
			L.N. Topilskij	IAE
3.	Impurity Control Fuel and Exhaust	V.I. Pistunovich	**V.I. Pistunovich**	IAE
			N.N. Brevnov	IAE
			V.M. Gervids	IAE
			M.I. Guseva	IAE
			Yu.L. Igitkhanov	IAE
			L.N. Khimchenko	IAE
			V.I. Kogan	IAE
			V.A. Krupin	IAE
			V.S. Lisitsa	IAE
			Yu.V. Martynenko	IAE
			S.V. Mirnov	IAE
			V.J. Merezhkin	IAE
			V.S. Mukhovatov	IAE
			K.A. Razumova	IAE
			A.N. Poprjaduhin	IAE
			V.S. Strelkov	IAE
			V.A. Vershkov	IAE
			J.F. Volkov	IAE
4.	Heating	V.I. Pistunovich	**V.V. Alikaev**	IAE
			V.M. Kulygin	IAE
			V.V. Kuznetsov	IAE
			A.A. Panasenkov	IAE
			I.A. Popov	IAE
			N.N. Semashko	IAE

* Bold-face type denotes team leaders

CATEGORY	PARTICIPANT	TEAM MEMBERS	
Heating (cont.)		V.S. Seregin	IAE
		V.A. Vdovin	IAE
		V.N. Vladimirov	IAE
5. Magnetics	G.F. Churakov	**N.A. Monoszon**	ESRIEA
		N.I. Dojnikov	ESRIEA
		D.P. Ivanov	IAE
		G.V. Trokhachev	ESRIEA
		V.E. Kejlin	IAE
		E.Yu Klimenko	IAE
		S.M. Kostenko	ESRIEA
		S.N. Sadakov	ESRIEA
6. Materials	B.N. Kolbasov G.E. Shatalov	**I.V. Altovskij**	IAE
		L.I. Ivanov	BIM
		V.P. Zeljenskij	KIPJ
		L.N. Bystrov	BIM
		E.V. Djomina	BIM
		Z.V. Ershova	
		M.I. Guseva	IAE
		E.S. Ionova	IAE
		L.N. Jopil'skij	IAE
		B.N. Kolbasov	IAE
		I.D. Kulagin	BIM
		Yu.V. Martynenko	IAE
		Yu.M. Platov	BIM
		G.E. Shatalov	IAE
		Ya.I. Shtrombakh	IAE
		N.K. Vinogradova	BIM
		S.N. Votinov	BIM
7. Systems Integration and Support Structure	G.F. Churakov B.M. Kolbasov	**G.F. Churakov**	ESRIEA
		V.V. Gur'jev	IAE
		A.F. Lekomtsev	IAE
		V.A. Loktev	ESRIEA
		N.A. Monoszon	ESRIEA
		V.P. Muratov	ESRIEA
		V.P. Smetannikov	IAE
8. Power Supply	G.F. Churakov	**A.M. Stolov**	ESRIEA
		F.M. Spevakova	ESRIEA
9. Assembly and Remote Maintenance	B.N. Kolbasov G.P. Churakov	**G.F. Churakov**	ESRIEA
		V.P. Smetannikov	IAE
		I.Ya. Emel'janov	IAE
		A.M. Epinat'jev	IAE
		B.D. Gornostaev	IAE

CATEGORY		PARTICIPANT	TEAM MEMBERS	
			V.V. Gur'jev	IAE
			B.N. Kolbasov	IAE
			E.M. Kuz'min	IAE
			N.A. Monoszon	ESRIEA
			G.E. Shatalov	IAE
			E.A. Shiversky	IAE
10.	Radiation Shielding and Personnel Access	G.E. Shatalov	**O.L. Shchipakin**	IAE
			A.V. Kashirskij	IAE
			V.I. Khripunov	IAE
			G.E. Shatalov	IAE
11.	Stability Control	B.B. Kadomtsev	**I.V. Mosin**	ESRIEA
		V.I. Pistunovich	V.D. Shafranov	IAE
			V.V. Arsenin	IAE
			Yu.L. Igitkhanov	IAE
			N.V. Ivanov	IAE
			A.B. Mikhajlovskij	IAE
			S.V. Mirnov	IAE
			V.S. Mukhovatov	IAE
			O.P. Pogutse	IAE
			G.N. Popkov	IAE
			A.M. Stefanovskij	IAE
			E.I. Yurchenko	IAE
			L.E. Zakharov	IAE
12.	Startup, Burn and Shutdown	B.B. Kadomtsev	**V.A. Abramov**	IAE
		V.I. Pistunovich	N.N. Brevnov	IAE
			K.N. Javasjn	IAE
			G.I. Krotova	IAE
			A.S. Kukushkin	IAE
			S.E. Lysenko	IAE
			S.V. Mirnov	IAE
			V.J. Mereshkin	IAE
			V.S. Mukhovatov	IAE
			V.V. Parail	IAE
			G.V. Pereverzev	IAE
			B.A. Trubnikov	IAE
			J.F. Volkov	IAE
			P.N. Yushmanov	IAE
			L.E. Zakharov	IAE
13.	Tritium	B.N. Kolbasov	**Z.V. Ershova**	IIM
		G.E. Shatalov	D.I. Ergrafova	IIM
			V.N. Jelus	IIM
			V.K. Kapyshev	IIM
			S.B. Muhamet-Galeeva	IIM

CATEGORY	PARTICIPANT	TEAM MEMBERS	
13. Tritium (cont.)		B.V. Petrov	IIM
		Rjazantsevann	IIM
		B.V. Shuljatikov	IMM
		V.G. Vasil'jev	IIM
		N.K. Vinogradova	IIM
14. Vacuum	V.I. Pistunovich	**Yu.M. Pustovojt**	IAE
		G.L. Saksaganskij	ESRIEA
		V.K. Kapyshev	IIM
		I.V. Mazul	ESRIEA
		S.B. Muhamet-Galeeva	IIM
		V.N. Odintsov	ESRIEA
		D.V. Serebrennikov	ESRIEA
		V.S. Svishchev	IAE
		V.B. Yuferov	KIPJ
		V.F. Zubarev	IAE
15. Safety and Environment	G.E. Shatalov	**O.L. Shchipakin**	IAE
		A.Y. Kashirskj	IAE
16. Diagnostics, Data Acquisition and Control	B.N. Kolbasov	**L.A. Matalin-Slutskij**	IAE
		J.N. Bolshakova	IAE
		V.I. Donetskij	IAE
		A.N. Fjakhretdinov	IAE
		P.A. Garrilov	IAE
		V.V. Gusjev	IAE
		L.A. Kanunnikov	IAE
		B.V. Lysikov	IAE
		D.V. Orlinskij	IAE
		G.I. Ploskirev	IAE
		V.R. Prozorov	IAE
		I.D. Skorlupkin	IAE
		V.S. Strelkov	IAE
		A.D. Yarochkin	IAE
		A.I. Yefanov	IAE
17. Cost, Schedule, Manpower	B.N. Kolbasov	**E.V. Muravjov**	IAE
		B.N. Kolbasov	IAE
		J.P. Kurkova	IAE
		V.N. Kursakov	IAE
		E.B. Sergeev	IAE

IAE I.V. Kurchatov Institute of Atomic Energy
ESRIEA D.V. Efremov Scientific Research Institute of Electrophysical Apparatus
BIM Bajkov Institute for Metallurgy of Academy of Sciences of the USSR
KIPJ Kharkov Institute of Physics and Technology
IIM All-Union Scientific Research Institute for Inorganic Materials.

A.2.4. US INTOR Workshop Teams

Workshop participants

J.R. Gilleland	GA
G.L. Kulcinski	Univ. Wisconsin
P.H. Rutherford	PPPL
W.M. Stacey, Jr.	Ga. Tech
J.K. Griffith	Ga. Tech (Admin. Asst.)

Consultants

H.P. Furth	PPPL
T. Ohkawa	GA
R. Parker	MIT
P.J. Reardon	PPPL (TFTR)
D.J. Rose	MIT
J. Sheffield	ORNL
D. Steiner	ORNL (ETF)

Concept Advisory Group

C.C. Baker	ANL
D. Cohn	MIT (ETF)
R.W. Conn	Univ. Wisconsin
D.L. Jassby	PPPL
Y-K.M. Peng	ORNL
J.M. Rawls	GA
L. Reid	ORNL
W.M. Stacey, Jr.	Ga. Tech.

	CATEGORY	PARTICIPANT	CONSULTANT	TEAM MEMBERS*	
1.	Energy and Particle Confinement	P. Rutherford	R. Parker	**J. Callen**	ORNL
				R. Goldston	PPPL
				G. Guest	GA
				R. Parker	MIT
				M. Peng	ORNL/ETF
			Modelling Subcommittee	J. Hogan	ORNL
				W. Houlberg	ORNL
				H. Howe	ORNL
				D. Post	PPPL
				C. Singer	PPPL
				R. Waltz	GA

* Bold-face type denotes team leaders

	CATEGORY	PARTICIPANT	CONSULTANT	TEAM MEMBERS	
2.	Blanket/First Wall/ Shield	G. Kulcinski	D. Steiner	**C. Flanagan**	Westinghou
				S. Harkness	ANL
				D. Kummer	MDAC
				R. Micich	Grumman
				T. Shannon	ORNL
3.	Fuelling, Exhaust and Impurity Control	P. Rutherford	J. Sheffield	**R. Conn**	Univ. of Wisconsin
				W. Bauer	Sandia
				K. Burrell	GA
				J. Callen	ORNL
				G. McCracken	PPPL
				D. Meade	PPPL
				A. Mense	ORNL
				R. Miller	GA/ETF
				S. Milora	ORNL
				M. Peng	ORNL/ETF
				D. Post	PPPL
				R. Taylor	UCLA
4.	Heating	P. Rutherford	R. Parker	**D. Jassby**	PPPL
			J. Sheffield	K. Berkner	LBL
				D. Cohn	MIT/ETF
				R. Freeman	GA
				H. Haselton	ORNL
				J. Hosea	PPPL
				J. Rome	ORNL
				J. Scharer	Univ. of Wisconsin
				L. Stewart	PPPL
5.	Magnetics and Cryogenics	J. Gilleland	P. Reardon	**J. Purcell**	GA
				P.Haubenreich	ORNL
				J. Joyce	PPPL
				B. Montgomery	MIT
				J. Rogers	LASL
				B. Wang	ANL
6.	Materials – Radiation Effects	G. Kulcinski	D. Steiner	**E. Bloom**	ORNL
				F. Clinard	LASL
				C. Snead	BNL
	Special Materials			**D. Smith**	ANL
				R. Gold	Westinghouse
				B. Johnson	BNWL
				R. Langley	ORNL

	CATEGORY	PARTICIPANT	CONSULTANT	TEAM MEMBERS	
7.	Systems Integration and Support Structure	J. Gilleland	P. Reardon	K. Wakefield	PPPL
				L. Blumenau	EBASCO
				I. Sviatoslavsky	Univ. of Wisconsin
8.	Power Supply and Transfer	J. Gilleland	P. Reardon	**A. Deitz**	PPPL
				M. Horton	GE
				R. Kustom	ANL
				W. Praeg	ANL
9.	Assembly and Remote Maintenance	J. Gilleland	P. Reardon	**N. Young**	EBASCO
				G. Fuller	MDAC
				F. Puhn	GA
				P. Stompinato	Grumman
10.	Radiation Shielding and Personnel Access	J. Gilleland	D. Steiner	**M. Abdou**	Georgia Tech.
				G. Fuller	MDAC
				E. Hager	GA
				P. Sager	GA
				W. Vogelsang	Univ. of Wisconsin
11.	Stability Control	P. Rutherford	R. Parker	**J. Rawls**	GA
				K. Bol	PPPL
				R. Miller	GA/ETF
				D. Monticello	PPPL
				M. Okabayashi	PPPL
				M. Peng	ORNL/ETF
12.	Startup, Burn and Shutdown	P. Rutherford	J. Sheffield	**J. Rawls**	GA
				R. Hawryluk	PPPL
				J. Hogan	ORNL
				S. Yoshikawa	PPPL/ETF
13.	Tritium	G. Kulcinski	D. Steiner	**V. Maroni**	ANL
				J. Anderson	LASL
				W. Wilkes	Mound
14.	Vacuum	G. Kulcinski	P. Reardon	**J. Watson**	ORNL
				B. Abel	PPPL
				T. Batzer	LLL
15.	Safety and Environment	G. Kulcinski	D. Rose	**J. Crocker**	EG & G
				J. Anderson	LASL
				H. Howe	PPPL
				M. Kazimi	MIT

CATEGORY	PARTICIPANT	CONSULTANT	TEAM MEMBERS	
16. Diagnostics, Data Acquisition and Control	J. Gilleland	P. Reardon J. Sheffield	**K. Young**	PPPL
			F. Bennett	PPPL
			K. Gentle	Texas
			S. Yoshikawa	PPPL/ETF
17. Cost/Schedule/ Manpower	J. Gilleland		**J. File**	PPPL
			P. Reardon	PPPL
SPECIAL STUDIES				
Divertors			**P.H. Rutherford**	PPPL
			S. Attenberger	ORNL
			R.H. Fowler	ORNL
			H.P. Furth	PPPL
			W. Houlberg	ORNL
			C. Ludescher	PPPL
			J.F. Lyon	ORNL
			D.M. Meade	PPPL
			B.E. Nelson	ORNL
			M. Okabayashi	PPPL
			J.A. Rome	ORNL
			J. Sheffield	ORNL
			J.A. O'Toole	ORNL
			T.F. Yang	Westinghous
Size			**P.H. Rutherford**	PPPL
			R.A. Hulse	PPPL
			J. Manickam	PPPL
			C.E. Singer	PPPL
			W.M. Stacey	Ga. Tech
Poloidal Field Coil			**J.A. Schmidt**	PPPL
			S.C. Jardin	PPPL
			M.L. Pelovitz	PPPL
Tritium Breeding			**V.A. Maroni**	ANL
			Y. Gohar	ANL
Electricity Production			**C.A. Flanagan**	Westinghouse
			R.E. Aronstein	Bechtel
			D.A. Sink	Westinghouse
Materials Testing			**G.L. Kulcinski**	Univ. Wiscon
			R.E. Nygren	Westinghouse
Test Blanket Configuration			**G.L. Kulcinski**	Univ. Wiscon

CATEGORY	TEAM MEMBERS	
Cost/Schedule/Manpower	**J. File**	PPPL
	C.C. Baker	ANL
	D. Cohn	MIT
	D. Grumann	GA
	P.J. Reardon	PPPL
	L. Reid	ORNL
	D. Sink	Westinghouse
	I. Sviatoslavsky	Univ. Wisconsin
	J.R. Gilleland	GA - INTOR Participant

ORGANIZATIONAL INDEX

ANL	Argonne National Laboratory
BECHTEL	Bechtel National, Inc.
BNL	Brookhaven National Laboratory
BNWL	Battelle NW Laboratory
EBASCO	Ebasco Services
EG&G	EG&G Idaho, Inc.
GA	General Atomic Company
Ga. Tech	Georgia Institute of Technology
GE	General Electric Corporation
GRUMMAN	Grumman Aerospace Corporation
HEDL	Hanford Engineering Development Laboratory
LASL	Los Alamos Scientific Laboratory
LBL	Lawrence Berkeley Laboratory
LLL	Lawrence Livermore Laboratory
MDAC	McDonnell Douglas Astronautics Company
MIT	Massachusetts Institute of Technology
MOUND	Monsanto - Mound Facility
ORNL	Oak Ridge National Laboratory
PPPL	Princeton University Plasma Physics Laboratory
SLL	Sandia Laboratories Livermore
UCLA	University of California at Los Angeles
U. Texas	University of Texas at Austin
U. Wisconsin	University of Wisconsin
Westinghouse	Westinghouse Electric Corporation

A.3. PREVIOUS TOKAMAK REACTOR DESIGN PARAMETERS

The decision of recently completed conceptual designs of near-term and demonstration/commercial stage tokamak reactor was an important part of the first session of the INTOR Workshop. A summary of certain parameters that characterize these designs was prepared and is included here. The designs summarized in Table A.3-1 had objectives that were similar to the technical objectives adopted for INTOR and thus provided the basis for the choice of the INTOR Guiding Parameters given in A.4. The designs summarized in Table A.3-2 were based on different objectives from those of INTOR and are included only for reference. The designs summarized in Table A.3-3, also for reference, provide some suggestions about the present perception of tokamak power reactor.

TABLE A.3-1. TNS/EPR/ETF

Parameters \ Device	ORNL TNS	GA TNS	PPPL SLPX	ANL EPR	USSR T-20	JAERI PETF	JAERI JXFR
Operating mode							
Power (MWt)	1140	650	530	600	450	260	125
Operating period (s)	560	60	660	75	330	1200	180
Burn time	500	30	34	60	300	30–60	100
No. of pulses/lifetime	0.45×10^6	0.2×10^6	4.0×10^3	1.0×10^6	3.8×10^5	0.08×10^6	9×10^5
Dimensions							
Major radius (m)	5.0	3.6	4.5	4.7	4.5	6.2	6.75
Plasma radius (m)	1.2/1.9	0.95/2.57	1.2/1.92	1.34/2.2	1.0/1.6	1.4/2.1	1.5
Plasma volume (m^3)	225	180	220	337	144	338	300
Reactor radius (m)	11	8.7	8.6	8.7	~ 12	11	13
Reactor height (m)	19	13.3	10.5	13.5	~ 18	15	20
Plasma							
Ion temperature (keV)	12	8.5	13	8	14.5	7	7
Ion density (10^{20} m^{-3})	2	1.9	1.2	1.4	1	1.6	1.1
Effective charge (Z_{eff})	1.5	2	<2	1.7	1	1.2	1.9
Energy confinement time (s)	1.2	1.4	2.0	2.5	0.6	5.6	7.2
B_T on axis (T)	5.3	5.0	6.0	4.5	6.0	4.8	5.5
Safety factor	3.8	2.5	3.0	3.0	3.0	2.5	2.5
Poloidal β	–	1.2	2.8	–	7	3.0	2.2
Toroidal β (%)	7	6–9	3.4	7	3.7	4.0	2.5
Plasma current (mA)	5	11.6	5.4	7.3	3.6	4.9	4.4
TF coils							
Number of coils	12	12	16	16	16	16	16
Conductor	Nb_3Sn	NbTi	Nb_3Sn	NbTi	Nb_3Sn	NbTi	Nb_3Sn
Bore (height/width) (m)	9.8/6.2	9.2/5.8	6.8/5.15	8.7/5.6	7/3	10/6.3	11/7
Maximum field (T)	10.9	10	12	9 (10)	12	8.3	11
PF coils							
Conductor (EFC/OHC)	(Cu/NbTi)/NbTi	Cu/NbTi	Cu/NbTi)/NbTi	NbTi/NbTi	Al (10°K)	Cu/Cu	NbTi
Position	(in/out)/out	inside/outside)	(in/out)/out	outside	inside	outside	outside

TABLE A.3-1 (cont.)

Parameters / Device	ORNL TNS	GA TNS	PPPL SLPX	ANL EPR	USSR T-20	JAERI PETF	JAERI JXFR
Neutral beam							
Beam energy (keV)	150	150	150	180	400	200	200
Injection power (MW)	50	60	40	40	60	50	26
Injection time (s)	6	2–5	6	4–6	10	5–10	120
Vacuum vessel							
Position	containment	first wall	outside shield	outside blanket	outside T_F coils	first wall	inside bulk shield
Material	building	inconel 625	SS	SS	SS	SS	SS
First wall							
Type & material	SS tubes	carbon coating on inconel 625	SS	coolant panel Be coat on SS	SS	SS, C liner	TZM, C liner
Neutron wall loading (MW/m^2)	2.4	1.8	1.3	1.3	1.0	0.5	0.21
Wall lifetime (eff. y)	5	–		2.5	4	10	10
Blanket							
Structure	none	SS**	none	316 SS	SS	concrete, Pb, SS	316 SS
Coolant	–	H_2O	–	H_2O/steam	He, Li	boric water	He
Breeding material	–	none	–	none	U^8, Li, $LiAlO_2$	none	Li_2O
Thickness (m)	–	0.2	–	0.2	0.7	0.6	0.6
Bulk shield							
Material	SS, Pb	W, Pb, SS borated H_2O	SS, Pb, Cu borated H_2O	SS, B_4C, Al, Pb	H_2O, SS	concrete	heavy concrete
Thickness (m)	0.60	0.40/1.1	0.66	0.5/1.0	0.7	1.5	0.45/0.85
Coolant	borated H_2O	H_2O	H_2O	H_2O	H_2O	none	H_2O
Tritium system							
Flow rate (g/h)*	60	8.3	6.5	126	20	0.137 av.	62
Inventory (kg)	5	1.6	0.7	1.3	1–2	0.08	0.5
Consumption (kg/a)	52	18	3.1	16	9.6	0.12	2

TABLE A.3-2. ITR AND MTR

Parameters \ Device	PPPL PITR	MIT HFITR	Wisconsin TETR	Garching/ Frascati ITR
Operating mode				
Power (MWt)	500	80	360	
Operating period (s)	600	25	72.5	
Burn time	9	20	60	
No. of pulses/lifetime	3×10^4	5×10^5	3×10^6	
Dimensions				
Major radius (m)	2.8	1.25	3.25	1.3
Plasma radius (m)	1.35		0.6/1.2	0.47
Plasma volume (m^3)	140		41.6	
Reactor radius (m)	7.5	3.0	7.75	
Reactor height (m)	10.0	2.5	10.5	
Plasma				
Ion temperature (keV)	10	15	12.2	15
Ion density (10^{20} m^{-3})	2.0	3.0	0.77	10
Effective charge (Z_{eff})	<2	<1.5	~ 1	
Energy confinement time (s)	1.0	0.5	0.08	
B_T on axis (T)	4.5	9.2	4.2	9.7
Safety factor	4.0		2.4	
Poloidal β	2.1		2.8	
Toroidal β (%)	8	2.4	6.6	
Plasma current (mA)	6.0	3.7	2.54	3.5
TF coils				
Number of coils	16	bitter	16	
Conductor	Cu	Cu	NbTi	Cu
Bore (height/width) (m)	5.8/4.37		9.2/5.5	
Maximum field (T)	12.5	19.8	8.5	
PF coils				
Conductor (EFC/OHC)	Cu	Cu	(Cu)/(Al)	Cu
Position	outside	outside	inside/outside	
Neutral beam				
Beam energy (keV)	160	150	150	160
Injection power (MW)	55	14	150	
Injection time (s)	2	1	60	
Vacuum vessel				
Position	first wall	first wall	first wall	first wall
Material			316 SS	
First wall				
Type and material	double wall Ti-6% V-4% Al		C curtain + 316 SS tub	bellows/rigid sections
Neutron wall loading (MW/m^2)	1.6		1.4	
Wall lifetime (eff. y)			7	
Blanket				
Structure	none		316 SS	none
Coolant	–		H_2O	–
Breeding material	–		–	–
Thickness (m)	–		0.1	–
Bulk shield				
Material	SS		SS, W, Pb, B_4C	none
Thickness (m)	0.35		1.0	–
Coolant	H_2O		H_2O	
Tritium system				
Flow rate (g/h)*	10		163	
Inventory (kg)	0.23		11.6	
Consumption (kg/a)	0.15		9.9	

* different assumptions about tritium recycling.

TABLE A.3-3. DEMONSTRATION/COMMERCIAL

Parameters \ Device	ORNL DEMO	GA-CPR Doublet	MIT/PPPL HFCTR-DEMO	Wisconsin CPR-NUMAK	FINTOR D	CULHAM MK-2	JÜLICH
Operating mode							
Power (MWt)	1930	2350	2440	2097	1900	5380	5000
Operating period (s)	1260	197	588	245	quasi-steady state	quasi-steady state	quasi-steady state
Burn time	1200	168	500	224			
No. of pulses/lifetime (20 years)	0.5×10^6	3.6×10^6	1×10^6	–			
Dimensions							
Major radius (m)	6.2	6.8	6.0	5.1	8.1	7.4	6.9
Plasma radius (m)/elong.	1.5/1.6	2.1/2.8	1.2/1.5	1.13/1.6	2.2/2.97	2.1/3.7	1.8/3.1
Plasma volume (m^3)	460	2015	317	312	1050		
Reactor radius (m)					17	17	12
Reactor height (m)					26	24	25
Plasma							
Ion temperature (keV)	13	16	12.4	13	11.5	12	10
Ion density (10^{20} m^{-3})	1.7	1.8	5.2	2	1.2	2.9	2
Effective charge (Z_{eff})	1.1	1.7	1.2	1–1.5	3	~1	
Energy confinement time (s)	1.5	0.6	1.0	1.0	1.8	1.5	
B_T on axis (T)	3.4	3.7	7.4	6.1	4.4	4.1	3.7
Safety factor	3.0	2.5	3.0	2.6	2.1	2.5 (wall)	2.1
Poloidal β	–	1.49	–	3.7	3	1.9	3.6
Toroidal β (%)	10.0	10	4.0	6.0	7	9	12
Plasma current (mA)	3.8	22	6.7	7.0	8.9	11.6	8.8

TF coils							
Number of coils	18	16	16	8 (+ 16 Cu coils)	16	24	24
Conductor	NbTi	NbTi	Nb_3Sn	NbTi	NbTi	NbTi	NbTi
Bore (height/width) (m)	11/7	18/11	6.8/5.2	9.5/6.3	17/10		15.9/8.4
Maximum field (T)	8.0	8.6	13.1	11.9	8.0	8.0	8.2
PF coils							
Conductor (EFC/OHC)	(Cu/NbTi)/NbTi	Cu/TbTi	(Cu/NbTi)/NbTi	cryo Al/NbTi	Cu	NbTi	Cu
Position	(in/out)/out	inside/outside	(in/out)/out	(in/out)/out	outside	outside	inside
Neutral beam							
Beam energy (keV)	150	none	(ripple inj.) 120	none			
Injection power (MW)	95	–	100	–	290	113	
Injection time (s)	5	–	6.4	–			
RF heating							
Frequency	none	2 GHz	none	92 MHz			
Injection power (MW)	–	25	–	75–80			
Injection time (s)	–	7.3	–	6.4			
Vacuum vessel							
Position	containment	sec. vacuum enclosure	first wall	sec. vacuum enclosure	reactor vessel	outside of blanket	reactor vessel
Material	building	inside TF coil	Mo alloy-TZM	inside TF coil			
First wall							
Type and material	SS tubes	none	Mo alloy (TZM)	none	modular vessel, SS	mod. cylind. cells, SS	mod. cylind. cells, Mo
Neutron wall loading (MW/m^2)	2.7		3.4		1.34	target 5.6	7.3
Wall lifetime (eff. y)	5		3			~ 2	

TABLE A.3–3 (cont.)

Parameters \ Device	ORNL DEMO	GA-CPR Doublet	MIT/PPPL HFCTR-DEMO	Wisconsin CPR-NUMAK	FINTOR D	CULHAM MK-2	JÜLICH
Blanket							
Structure	SS	inconel 718	Mo alloy (TZM)	Ti-6A1-4V	316 SS	SS	SS
Coolant	He	He	2 Lif·BeF_2	boiling H_2O	He	He	He
Breeding material	Li	Li_7Pb_2/Li_4SiO_4	Li	LiPb	liquid Li	Li compound	liquid Li, Be
Thickness (m)	0.75	0.4	0.6	0.5	1.0	1.0 (min.)	1.0
Bulk shield							
Material	SS, Pb	SS, B_4C	SS, B_4C	B_4C, Pb	SS, B_4C, Pb	borated H_2O SS, B_4C, Pb	SS, B_4C, Pb
Thickness (m)	0.6	0.26	0.6	0.85	1	0.5	1/1.4
Coolant	borated H_2O	H_2O	–	H_2O	He	borated H_2O	H_2O
Tritium system							
Flow rate (g/h)*					77.5		500
Inventory (kg)					4.3		3.4
Consumption (kg/a)					65		177

* different assumptions about tritium recycling.

A.4. INTOR GUIDING PARAMETERS (February 1979)

The data base and R and D needs assessments were done for this set of Guiding Parameters, which was developed early in the Workshop and considered to be sufficiently close to the final INTOR parameters. Note that they are different from the INTOR Suggested Parameters which the Workshop proposes as imput to the definition phase.

Basic parameters			
burn time	30–60 s*		200–500 s**
dwell time	15–30 s*		
availability		25%	
number of pulses during lifetime	10^6*		10^5**
major radius (R)		4.5–5 m	
aspect ratio (A)		4	
plasma radius (r_{plasma}) (z = 0)		1.2 ± 0.1 m	
wall radius (r_{wall}) (z = 0)		1.3 ± 0.1 m	
elongation	1.2–1.6*		1.5**
av. ion temperature (T_i)		10 keV	
av. ion density (n)		1.2×10^{20} m^{-3}	
energy confinement time (t_E)		≈ 2 s	
effective charge (Z_{eff})		1.5	
toroidal magnetic field at axis (B_{T0})		≈ 5 T	
safety factor (q)		3	
β_T		6%	
field ripple		± 1%	
Heating			
power (P)		50–75 MW	
duration		≈ 5 s	
mode		NBI, RF as backup	
neutral beam energy (E)		150–200 keV perhaps some MW needed at higher energy	
Impurity and particle control			
particle fluxes from plasma		–	
fuelling mode		pellet and gas puffing	
fuelling rate		–	
impurity control mode		–	

* Pulsed.

** Quasi-stationary operation.

Basic parameters	
Toroidal field coils	
number	12–16
conductor	SC
bore, height/width	9 m/6 m
pulsed field	0.15 T/s
radiation	10^9 rads
*Poloidal field coils**	
OH current ramp time	> 2 s
OHC conductor	–
OHC location	–
OHC max. field	± 8 T ?
OHC max. field rise	8 T/s
EFC conductor	–
EFC location	–
DFC conductor	–
DFC location	–
divertor field	–
BDC conductor	–
BDC location	–
max. one-turn voltage	1 kV
Vacuum vessel	
location – primary vessel	behind blanket?
material	–
max. temperature	–
coolant	H_2O ?
First wall	
type	–
thickness	–
material	low-Z coated steel?
coolant	H_2O?
max. temperature – structure	400°C?

* OHC: ohmic heating coils, EFC: equil. field coils, DFC: divertor field coils, BDC: breakdown field coils.

Basic parameters	
average neutron wall load (during burn)	1–1.5 MW/m^2
average surface heat load (during burn)	0.3–0.5 MW/m^2
number of pulses/lifetime	(see above under 'basic parameters')
number of disruptions/lifetime	to be determined
toroidal loop resistance	≈ 0.2 mΩ
Blanket (non-breeding)	
number of modules	several
location	outboard only
material	SS
coolant	H_2O ?
max. temperature (structure)	400°C?
thickness	0.25 m ?
poloidal field response time	10 ms
Blanket test modules (T-breeding)	
number of modules	several of different types
type	–
location	outboard only
material, structural	–
material, breeding	–
coolant	–
max. temperature (structure)	–
Bulk shield	
inboard material	–
inboard thickness	0.6–1 m
outboard material	–
outboard thickness	1 m
coolant	H_2O ?
max. temperature (structure)	100°C
Tritium and vacuum	
flow rate	50–100 g/h ?
inventory	2 kg ?
consumption	5 kg/year ?
tritium cleanup time	3 days ?
pre-shot base pressure	≈ 10^{-6} torr
initial base pressure	10^{-7}–10^{-8} torr

A.5. SUPPORTING DOCUMENTATION FOR THE INTOR WORKSHOP

Each of the four delegations published their detailed contributions to the Workshop in home-base reports, as listed below:

A.5.1. European Community – Euratom

(1) European Contributions to the 2nd meeting of the INTOR Workshop, EUR FU/XII 501/79/EDV 50, June 1979 (2 volumes).
(2) European Contributions to the 3rd meeting of the INTOR Workshop, Design Options and Alternatives, EUR FU/XII 501/79/EDV 60, October 1979.
(3) European Contributions to the 3rd meeting of the INTOR Workshop, EUR FU/XII 501/79/EDV 70, October 1979.

A.5.2. Japan

(1) Fusion Research Development Center, INTOR WORKSHOP REPORT – Group 1, Energy and Particle Confinement, JAERI-M 8621 (1979).
(2) Fusion Research Development Center, INTOR WORKSHOP REPORT – Group 3, Impurity Control, Fueling, and Exhaust, JAERI-M 8622 (1979).
(3) Fusion Research Development Center, INTOR WORKSHOP REPORT – Group 4, Heating, JAERI-M 8624 (1979).
(4) SHIRAISHI, K., et al., INTOR WORKSHOP REPORT – Group 6, Materials, JAERI-M 8510 (1979).
(5) Fusion Research Development Center, INTOR WORKSHOP REPORT – Group 8, Power Supply and Transfer, JAERI-M 8511 (1979).
(6) Fusion Research Development Center, INTOR WORKSHOP REPORT – Group 11, Stability Control, JAERI-M 8625 (1979).
(7) Fusion Research Development Center, INTOR WORKSHOP REPORT – Group 12, Start-up, Burn and Shut-down Control, JAERI-M 8625 (1979).
(8) TANAKA, K., et al., INTOR WORKSHOP REPORT – Group 13, Tritium, JAERI-M 8512 (1979).
(9) Fusion Research Development Center, INTOR WORKSHOP REPORT – Group 14, Vacuum, JAERI-M 8513 (1979).
(10) Fusion Research Development Center, INTOR WORKSHOP REPORT – Group 16, Diagnostics, Data Acquisition and Control, JAERI-M 8514 (1979).
(11) Fusion Research Development Center, Engineering Aspects of the JAERI Proposal for INTOR (II), JAERI-M 8518 (1979).
(12) Fusion Research Development Center, Home Task Reports of Groups 2, 5, 7, 9, 10 and 15, JAERI-M 8711 (1980).
(13) HIRAOKA, T., et al., Conceptual Studies of Plasma Engineering Test Facility, JAERI-M 8198 (1979).
(14) SAKO, K., et al., Second Preliminary Design of JAERI Experimental Fusion (JXFR) (Interim Report), JAERI-M 8286 (1979).
(15) SAKO, K., et al., Engineering Aspects of JAERI Proposal for INTOR (I), JAERI-M 8503 (1979).
(16) Fusion Reactor Systems Lab., Evaluation of INTOR Size, JAERI-M 8710 (1980).

A.5.3. USSR

(1) KADOMTSEV, B.B., "On the concept and principal parameters of the INTOR reactor", October 1979.
(2) KADOMTSEV, B.B., "Branch divertor".
(3) LEKOMTSEV, A.F., et al., "Blanket for reproduction of tritium in INTOR" (alternative version).
(4) LEKOMTSEV, A.F., JOPILSKY, L.N., GHATALOV, G.E., "Blanket for electricity generation in INTOR" (alternative version).
(5) ALTOVSKY, I.V., "Studies of materials in INTOR".
(6) KOLBASOV, B.N., et al., "Cost schedule and manpower assessment for INTOR".
(7) DNESTROVSKY, Yu.N., "Energy and particle confinement in INTOR".
(8) EMELYANOV, I.Ya., et al., "Blanket, first wall and shield of INTOR".
(9) PISTUNOVICH, V.I., "Control of impurities, fuel maintenance and removal of reaction products in INTOR".
(10) KUKUSHKIN, A.S., "Plasma physics calculations for INTOR operating conditions".
(11) GEZASIMOV, S.N., KHIMCHENKO, L.N., "Some aspects of control of impurities in INTOR".
(12) IGITKHANOV, Yu.L., "On thermal transfer in boundary layers of the tokamak reactor".
(13) IGITKHANOV, Yu.L., "Some aspects of the near-wall plasma in the tokamak reactor".
(14) SEMASHKO, N.N., "Plasma heating in INTOR. Injection system".
(15) ALIKAYEV, V.V., "Plasma heating in INTOR. RF-system".
(16) MONOSZON, N.H., et al., "INTOR electromagnetic system".
(17) IVANOV, D.P., "Electromagnetic system of J-7".
(18) ALTOVSKY, I.V., IVANOV, L.I., "Materials for INTOR".
(19) ALTOVSKY, I.V., KOLBASOV, B.N., TOPILSKY, L.N., "On the possibility of employing aluminium-base alloys as the first wall material".
(20) GUSEVA, M.I., IONOVA, E.S., MARTYNENKO, Yu.V., "Surface phenomena in the INTOR".
(21) CHURAKOV, G.F., LOKIEV, V.A., MURATOV, V.P., "INTOR system integration".
(22) LIPOV, M.Yu., MURAVIEV, E.V., "On the development of high efficient divertor systems for power tokamak reactors".
(23) MIRNOV, S.V., "Helium removal in the conditions of sorption diaphragm operation in the INTOR".
(24) SPEVAKOVA, F.M., STOLOV, A.M., "INTOR power supply".
(25) GORNOSTAEV, B.D., et al., "INTOR assembly and remote maintenance".
(26) SMETANNIKOV, V.P., "INTOR assembly and remote service".
(27) SHCHIPAKIN, O.L., SHATALOV, G.E., "INTOR radiation shielding and personnel access".
(28) ZAKHAROV, L.E., MEREZHKIN, V.G., IVANOV, N.V., "Stability control in INTOR".
(29) ABRAMOV, V.A., "Start-up, burn and shut-down in INTOR".
(30) MEREZHKIN, V.G., "On current programming in INTOR".
(31) BULYGINSKY, D.G., et al., "On the influence of pre-conization type at the breakdown regime in the tokamak".
(32) VERSHKOV, V.A., KRUPIN, V.A., STRIZHOV, V.F., "Some aspects of start-up, burn and shut-down in INTOR".
(33) KUKUSHKIN, A.S., "Thermonuclear reaction burning control through longitudinal magnetic field crimping".
(34) LEONOV, V.M., "On some conditions for breakdown of discharge in Tokamak-II".
(35) BREVNOV, N.N., KUZNETSOV, N.V., "Conditions for breakdown in T-12".
(36) ABRAMOV, V.A., POGUTSE, O.P., "Some aspects of initial discharge stage in INTOR".

(37) ABRAMOV, V.A., "On the dynamics of admixtures in INTOR".
(38) ERSHOVA, L.V., et al., "Tritium in international reactor tokamak INTOR".
(39) BYCHKOVA, A.D., et al., "Vacuum-tritium test stand facility for tests of fusion reactor tritium cycle gas system".
(40) VASILIYEV, V.G., et al., "Investigations of physical and chemical properties of irradiated nonorganic lithium compounds: oxides, aluminate and silicates".
(41) SAKSAGANSKY, G.L., et al., "INTOR vacuum system".
(42) PUSTOVOIT, Yu.M., "Vacuum system of INTOR".
(43) SHCHIPAKIN, O.L., SHATALOV, G.E., "Radiation safety and protection of environment".
(44) MATALIN-STUTSKY, L.A., "INTOR diagnostics, control and data processing".
(45) GAVRILOV, P.A., et al., "Blanket parameters control".

The reports can be obtained from Kurchatov Institute of Atomic Energy, Moscow, USSR, with the following exceptions:
Reports 16, 21, 24, 41 can be obtained from Efremov Institute for Electrophysical Equipment, Leningrad, USSR;
Reports 38–40 can be obtained from Inorganic Materials Institute, Moscow, USSR.

A.5.4. USA

(1) STACEY, W.M., et al., "The U.S. Contribution to the International Tokamak Reactor Workshop", Georgia Institute of Technology, November 1979.

HOW TO ORDER IAEA PUBLICATIONS

An exclusive sales agent for IAEA publications, to whom all orders and inquiries should be addressed, has been appointed in the following country:

UNITED STATES OF AMERICA	UNIPUB, 345 Park Avenue South, New York, NY 10010

In the following countries IAEA publications may be purchased from the sales agents or booksellers listed or through your major local booksellers. Payment can be made in local currency or with UNESCO coupons.

ARGENTINA	Comisión Nacional de Energía Atomica, Avenida del Libertador 8250, RA-1429 Buenos Aires
AUSTRALIA	Hunter Publications, 58 A Gipps Street, Collingwood, Victoria 3066
BELGIUM	Service Courrier UNESCO, 202, Avenue du Roi, B-1060 Brussels
CZECHOSLOVAKIA	S.N.T.L., Spálená 51, CS-113 02 Prague 1 Alfa, Publishers, Hurbanovo námestie 6, CS-893 31 Bratislava
FRANCE	Office International de Documentation et Librairie, 48, rue Gay-Lussac, F-75240 Paris Cedex 05
HUNGARY	Kultura, Hungarian Foreign Trading Company P.O. Box 149, H-1389 Budapest 62
INDIA	Oxford Book and Stationery Co., 17, Park Street, Calcutta-700 016 Oxford Book and Stationery Co., Scindia House, New Delhi-110 001
ISRAEL	Heiliger and Co., Ltd., Scientific and Medical Books, 3, Nathan Strauss Street, Jerusalem 94227
ITALY	Libreria Scientifica, Dott. Lucio de Biasio "aeiou", Via Meravigli 16, I-20123 Milan
JAPAN	Maruzen Company, Ltd., P.O. Box 5050, 100-31 Tokyo International
NETHERLANDS	Martinus Nijhoff B.V., Booksellers, Lange Voorhout 9-11, P.O. Box 269, NL-2501 The Hague
PAKISTAN	Mirza Book Agency, 65, Shahrah Quaid-e-Azam, P.O. Box 729, Lahore 3
POLAND	Ars Polona-Ruch, Centrala Handlu Zagranicznego, Krakowskie Przedmiescie 7, PL-00-068 Warsaw
ROMANIA	Ilexim, P.O. Box 136-137, Bucarest
SOUTH AFRICA	Van Schaik's Bookstore (Pty) Ltd., Libri Building, Church Street, P.O. Box 724, Pretoria 0001
SPAIN	Diaz de Santos, Lagasca 95, Madrid-6 Diaz de Santos, Balmes 417, Barcelona-6
SWEDEN	AB C.E. Fritzes Kungl. Hovbokhandel, Fredsgatan 2, P.O. Box 16356, S-103 27 Stockholm
UNITED KINGDOM	Her Majesty's Stationery Office, Agency Section PDIB, P.O. Box 569, London SE1 9NH
U.S.S.R.	Mezhdunarodnaya Kniga, Smolenskaya-Sennaya 32-34, Moscow G-200
YUGOSLAVIA	Jugoslovenska Knjiga, Terazije 27, P.O. Box 36, YU-11001 Belgrade

---

Orders from countries where sales agents have not yet been appointed and requests for information should be addressed directly to:

Division of Publications
International Atomic Energy Agency
Wagramerstrasse 5, P.O. Box 100, A-1400 Vienna, Austria